Production Technology of Spices and Herbs and their Benefits

Production Technology of Spices and Herbs and their Benefits

Nekesah T. Wafullah

Delve
PUBLISHING
www.delvepublishing.com

Production Technology of Spices and Herbs and their Benefits

Nekesah T. Wafullah

Delve Publishing
224 Shoreacres Road
Burlington, ON L7L 2H2
Canada
www.delvepublishing.com

Email: orders@arclereducation.com

ISBN: 978-1-77407-738-2 (Hardcover)

ABOUT THE AUTHOR

A skilled agriculture expert with extensive knowledge in project management, agricultural energy, various forms of fertilizer, their production, marketing, application regimes and policies; business development management services; cross border fertilizer trade policies; product registration and efficacy trials and assessment; legume production and marketing aspects; food safety advocacy and policy issues; partnerships in businesses; youth, and women empowerment, and volunteerism. She is adept at project planning and strategic management as well as creating simple solutions to complex problems. As mentorship is one of her key areas, she has dedicated time to journey with high school and college students advocating for better performance in Agricultural sciences. She writes and edits agricultural books and loves watching movies, cooking, baking, networking, reading and dancing during her free time. She is also a food and fashion enthusiast. Nekesah holds an M Sc. Agricultural and Applied Economics degree from the University of Nairobi with a major in International Trade and Policy.

TABLE OF CONTENTS

LIST OF FIGURES

LIST OF TABLES

ACRONYMS

MAPs	Medicinal Aromatic Plants
BC	Before Christ
AD	Anno Domini
USA	United States of America
LDL	Low Density Lipoprotein
HT	Hypertension
CMV	Cytomegalovirus
HDL	High-Density Lipoproteins
DNA	Deoxyribonucleic Acid
UV	Ultraviolet
GLA	Gamma Linolenic Acid
BMI	Body Mass Index
GER	Gastro-EsophagealReflux
PMS	Premenstrual Syndrome
RSV	Respiratory Syncytial Virus
UTI	Urinary Tract Infections
FAO	Food and Agriculture Organization
LDC	Least Developed Countries

PREFACE

Spices are natural products that are rich in essential oils and aromatic concepts such as leaves, flowers, seeds, and roots (FAO, 2005). Spices are primarily used to flavor, flavor or preserve foodstuffs. Spices have played a prominent role in all ancient civilization in ancient China and India, and have fostered international relations across the world among the various countries.

Chapter 1 introduces you to the History of Herbs and Spices from Egypt in the African continent to Greece in Europe and various great men such as Christopher Columbus, Vasco da Gama and the Father of Medicine, Hippocrates who traveled to Africa and took some herbs and spices introducing them to their various continents and countries. It discusses archeological evidence shows that spices and herbs were used long before they were recorded inthe history of mankind when human ancestors added sharp-flavored leaves to early cooking pots for the first time. Roaming hunter - gatherer groups played with leaves, roots, flowers, and seeds, creating a precious compendium of knowledge that was passed from generation to generation over time. It delves further into the uses of spices expounding on the flavors of various herbs and spices and products obtained from them.

Chapter 2 takes you through the nomenclature and classification of herbs and spices to also howcase the various gene pools and species that they belong to. It is with this appreciation that one gets to understand how some existing crops are related to some of the herbs and spices. The chapter takes us through the recognition of the four most well-known flavor profiles-bitter, salty, sour and sweet. When it comes to spices, a real spice aficionado cannot be limited to four or five. There are many more characteristics of flavors that are equally identifiable-cooling, earthy, floral, fruity, herbaceous, hot, nutty, piney, pungent, spicy, sulfurous and woody. Spices do not usually have a single flavor profile. They can also be classified based on the plant organ from which they are obtained, botanical families, duration of crop in the field, growth habit, conventional classification of spices and condiments. Spices are an important group of agricultural commodities used for flavoring and are the main ingredients for many delicious foods.

Chapter 3 provides information onthe effects of herbs and spices of health some positive and some precautions of using them.

Chapter 4 takes us through the production of herbs and spices which basically is the agronomic practices undertaken while producing herbs and spices. From seed germination, transplanting for some, taking care of pests and diseases and harvesting. While it may seem like herbs and spices are not affected by pests and diseases the book gives information about some of the herbs and spices that are sensitive to pests and diseases.

With the current global spice and condiment production estimated at 8.5 million tons, valued at US$ 25 billion, developing countries are the primary producers and the main importers for the developed world. Spices also represent developing countries ' currencies. India still plays a key role in the world spice trade as it has for centuries. Regional climates can greatly differ, depending on the distance from the equator, landmass size, water proximity and hundreds of other factors. These differences in climate, as well as many other influences, mean different plants have evolved in different places around the globe. Of course, the herbs and spices we use to season and flavor our food are among those plants. For these reasons, some areas of the world have spices thatoccur significantly more naturally than others. Chapter 5 gives us the best geographical regions for the production of herbs and spices which technically provides the history of some of the herbs and spices and the trade of these products around the world.

CHAPTER 1

Introduction to Herbs and Spices

1.1 DEFINITIONS

Spices are natural products that are rich in essential oils and aromatic concepts such as leaves, flowers, seeds, and roots (FAO, 2005). Spices are primarily used to flavor, flavor or preserve foodstuffs. In the culinary arts, the word spice refers to any dried part of a plant used for seasoning and flavoring a recipe, other than the seeds, but not used as the main ingredient. Every other part of the plant is called a spice, like dried bark, roots, berries, seeds, twigs, or anything else that is not the green leafy component. Herbs are leafy spices, and some can provide both spice seeds and leafy herbs, such as dill and cilantro. Culinary herbs are mostly from plant leaves and used in cooking to provide flavor. Most herbal plants used as spicesfor culinary purposes are widely believed to have medicinal properties, and thus some link exists between them and Medicinal Aromatic Plants (MAPs). Distinctions can be drawn based on the purposes for which plants are used. There is also an overlap between spices and herbs, and while plants are usually classified as vegetables: some vegetables can also be used to spice up other foods, e.g. some mushroom varieties are used as spices. Around fifty spice and herbal plants are of global commercial importance, but many other spices and herbal crops are used in traditional cooking, healthcare, or other applications, in particular regions, and are traded locally. Wild and cultivated spices and herbs are grown as trees, shrubs, perennials, annuals. The flavor and herbal plants provide seeds and fruits, leaves and stems, flowers and buds, roots and rhizomes, bark and resins, all of which can be distributed in various forms: sold fresh, frozen, dried, whole or crushed, distilled into oils or solvent extracted into oleoresins.

Spices have less nutritional value, and cannot be grouped as food. Nevertheless, they greatly enhance the pleasure of eating by adding good flavor and aroma to the food. All aromatic vegetable products which are used to flavor foods and beverages are generally known as spices. These are characterized by pungency, a strong smell and a sweet or bitter taste. The terms spices and condiments are also used together. In truth, drawing a distinction between the two is very difficult. Spices used, primarily for seasoning or garnishing food and beverages in a pulverized state. Condiments, on the other hand, are other flavoring materials which have a sharp taste and are usually added after cooking to food. However, when the aromatic vegetable products come from a temperate region, it is considered to be a culinary herb as in the case of bay leaves, coriander, cumin, dill, fennel, fenugreek, mustard seeds, parsley, sage, rosemary and thyme, to name just a few. They also stimulate appetite in addition to imparting flavor to foods and drinks, improve the secretion and flow of gastric juices and aid digestion attributed to their carminative properties. They are generally known as food adjuncts or food accessories for this purpose. The majority of these are used in various medicines. They also play an important role in different industries and are used in the perfumery, soaps, incense, coloring ,etc. The spices attraction is due mainly to the presence of essential oils and sometimes to other aromatic principles.

1.2 HISTORY OF HERBS AND SPICES

Archeological evidence shows that spices and herbs are used long before recorded history when human ancestors added sharp-flavored leaves to early cooking pots for the first time. Roaming hunter -gatherer groups played with leaves, roots, flowers, and seeds, creating a precious compendium of knowledge that was passed from generation to generation over time. When civilization advanced and nomadic tribes settled in one location, not only were herbs and spices harvested from the wild, but they were intentionally sown near dwellings. By the beginning of the agricultural era plants were collected from the wild and grown near dwellings for food, flavor, medicine, fuel, decoration, painting, poisoning, and weapons, and to alter the sense of the reality of early humans. The earliest written documents come from ancient cultures in Egypt, China, and India. The Ebbers Papyrus dating back to 1550 B.C.E., describes some 800 different medicinal remedies and numerous medicinal procedures. For medicine, the Early Egyptians used spices and herbs when cosmetics and perfumes to embalm, roast and kill and repel pests.

1550BC: The Ebbers papyrus recorded the use of various herbs and spices in Egypt and extolled their medical benefits.

400BC: The' Father of Medicine' founded a systematic system of medicine in ancient Greece, Hippocrates, which explained the importance of herbs in alleviating pain and curing illness. A profitable and lucrative trade of herbs and spices between the Arab World and India was emerging.

1st Century AD: The Greek physician, Discords, in his extensive work, Materia Medic, listed the medicinal properties of over 500 plants. The Romans used herbs and spices which were brought by boat from India for perfume, cosmetics, medicine and cooking. The' Silk Road' came into being at around the same time as the overland route from China.

12th Century: The flow of goods from East to West dwindled during the Middle Ages before new trade routes were opened by the Crusaders as they returned from the Holy Land, and the herbal and spice trade flourished once more. As they capitalized on this new trade, the Italian cities of Venice and Genoa grew rich.

1270: From Venice, Marco Polo set out to find a new route to the Far East that would circumvent Arab merchants who were still regulating the trade and keeping prices high. 25 years later he returned bringing new and exotic herbs and spices from China with him.

1492: Christopher Columbus set sail from Spain in search of a new route to the West Indies for his voyage west. He returned from the West Indies, with allspice and chili peppers.

1497: Vasco de Gama, the Portuguese explorer, set out to find a sea route towards the West. He reached India's West Coast of Malabar and returned with a cargo of herbs, spices, and jewels to Portugal.

1519: The Spanish conquistador Hernan Cortez invaded Mexico and encountered Vanilla during his exploration of southeastern Mexico. The Aztecs used this spice as a flavoring to season their chocolate drinks.

1521: The first globe circumnavigation was achieved by Magellan 's fleet. One ship returned to Europe with a rich cargo from the Spice Islands (Moluccas). The age of exploration has led to an increase in the number and variety of herbs and spices imported available in Europe.

1545: At the University of Padua, Italy, was developed the first herbal or physical garden intended to provide plants for medical study purposes.

1602: The Netherlands established the Dutch East India Company to trade spices directly with the East, which gave the Netherlands a spice trade monopoly that lasted almost 200 years.

End of the 18th Century: The British had ousted the Netherlands from India, and London became the center of spice trade worldwide. No European country had any spice monopoly, and prices began to fall. When the American clipper ships sailed to the East bringing tobacco and foodstuffs for tea, coffee, textiles and spices, the United States entered the spice trade. These merchants made fortunes, one of which gave the Yale University its money and its name.

Present Day: Herbs and spices are still imported from around the world today, and are used in a wide variety of applications.

1.3 USES OF SPICES

Spices have played a prominent role in all ancient civilization in ancient China and India, and have fostered international relations across the world among the various countries. Spices were so extraordinarily expensive in ancient times, that only wealthy people could afford them. Strong and pungent flavored spices have been used not only for cooking but also for food preservation. Aroma and taste were a part of the mystical rituals and were used for various purification ceremonies. The streets were often fumigated with spices during the visit of royal visitors. They were used by ancient Egyptians to perfume the person and embalm the dead bodies, to protect them from decay. Spices were considered important medicines in the middle ages but today they are primarily used to give a pleasant taste to otherwise unpleasant medicines. Most spices contain antiseptic and carminative properties. In addition to their culinary value, spices are also used in beverages as a flavoring agent, as active ingredients in Ayurvedic medicines, as coloring agents for textiles, and as important constituents in cosmetics and perfumery. Spices are used entirely or in powdered form, depending on the dish requirement. Some of these spices have' flavoring, preservative and antiseptic properties that are mainly due to the presence of volatile oils but are occasionally due to other aromatic substances such as pepper-like alkaloids.

The importance of spices in our daily diet is as follows:

- To give the otherwise monotonous or insipid food a good flavor and aroma, mainly of starchy grains or roots, thereby contributing considerably to the pleasure of eating.
- In order to stimulate the appetite and increase the flow of gastric juices, they are often referred to as "food accessories" or food supplements.
- Disguise the slightly unpleasant taste of a lot of dry means.
- Increase the rate of suddenness and thus have a cooling effect on the body.

The health benefits of herbs & spices may include;

- **Heart health**: The use of herbs and spices in your diet to minimize the salt is a great way to reduce sodium intake and the risk of heart disease. The most powerful herbs and spices to substitute the salt taste are spices such as black pepper, garlic powder, curry powder, cumin, dill, basil, ginger, coriander, and onion.
- **Weight loss:** Cooking with herbs and spices is a great way to make foods tastier and more enjoyable without adding additional calories. In some of your favorite dishes, you might find that you need less sugar if you add sweet spices like cinnamon, cardamom or nutmeg.
- **Anti-inflammatory**: Many herbs and spices are under review for their ability to reduce harmful body inflammation. Inflammation is the precursor to many chronic diseases, including heart disease and Alzheimer's disease. Turmeric is an especially potent anti-inflammatory spice and has even been shown to be more effective in reducing knee pain in people with osteoarthritis than ibuprofen.
- **Antioxidants**: Antioxidants are colorful substances found in plant foods that help protect your cells against aging damage, the environment and other stressors. Herbs and spices constitute especially good sources of antioxidants. 1/2 teaspoon of cloves for example has more antioxidants than 1/2 cup of blueberries! The most antioxidant herbs and spices include oregano, sage, peppermint, lemon balm, clove and cinnamon.

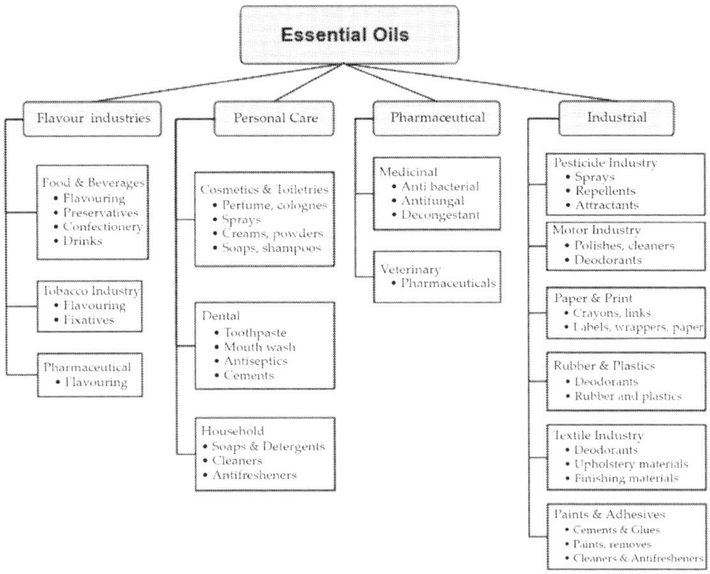

Figure 1: Different industries utilizing spices and herbs essential oils.

Source: FAO, 2005

1.4 FLAVOR CHARACTERISTICS OF SPICES AND HERBS

Spices give food an aroma, color, flavor and even texture at times. Rising spice, chile, or herb has specific, unique chemical compounds that establish sensual qualities of these. Perhaps even novice cooks can recognize the four most well-known flavor profiles-bitter, salty, sour and sweet. When it comes to spices, a real spice aficionado just can't be limited to four or five. There are many more characteristics of flavors that are equally identifiable-cooling, earthy, floral, fruity, herbaceous, hot, nutty, piney, pungent, spicy, sulfurous and woody. Spices do not usually have a single flavor profile. The famous spice cumin, for example, falls into some of the flavor profiles as it is both earthy and spicy. Thyme is bitter, aromatic, piney and herbaceous. Americans tend to regard themselves as the core of many worlds and this phenomenon is not absent from the culinary world. Americans' taste buds do not automatically correspond with the taste buds of people around the world; for instance, many cultures eat much spicier food than the typical

American. Below are some of the USA culinary flavor description;

Bitter: Bitterness alone is not a good flavor for people who are particularly sensitive to it. Usually, bitter flavors are considered an acquired taste and are present in coffee, chocolates, and beer. Most of the vegetables contain at least a bit of bitterness, especially when eaten raw. So, e.g. The Turmeric

Cooling: Cooling spices are the ones that give a refreshed feel to the mouth. If your first thought was to taste mint on a humid summer day in a tall glass of chilled water, that's exactly the sensation your mouth gets from the spices that cool down. E.g. Sweet basil.

Earthy: Earthy spices are the ones that taste closely resembles where they originate. This is the flavor profile used to characterize potatoes or beets, foods closely bound to the earth. Some people will say it's dirty and unwanted flavor. That flavor comes from goemin 's presence. E.g. Cumin.

Floral: Such spices are the ones that give us the spring 's lovely flavor experience. More than likely you would associate the flavor with teas. So, e.g. Coriander.

Fruity: With this flavor profile, herbs and spices taste exactly as you would expect them to. They are sweeter and give off a pleasant fruity scent. E.g. Tamarind.

Herbaceous: Herbaccous flavors are natural herbs. That seems to be an obvious thing to say, but to explain this flavor there really is no other language. E.g. Lavender.

Hot: The spicy flavor profile comes from the most exciting part of the flavor spectrum. This flavor is characterized by its ability to rapidly heat up your mouth, often making food unpleasant enough to leave you coughing. If your chili is the right level of hot, you may find yourself crying too.E.g. Hot pepper.

Nutty: Some people associate almonds with this flavor. Foods associated with this taste include cashews, almonds, and walnuts. The nutty taste has almost something buttery to it. E.g. Sesame seed.

Piney: If you've smelled a pine tree before, you have a vague idea of how the pine tastes. It's a bite-flavor one has to get used to. E.g. Rosemary.

Pungent: Spices that come under the pungent classification are heavily flavored and also have strong aromatics. E.g. Garlic.

Sour: Sour aromas make the mouth pucker. You can note that sour flavors will make you salivate more than normal, and they can be used very nicely to complement sweet foods. E.g. Tamarind.

Spicy: Foods and spices which are thought to be spicy give a nice, powerful flavor burst. Spicy food is often spicy, although short -lived. E.g. Ginger.

Sulfur: This profile of the flavor is difficult to describe. Some might say that it has an almost egg-like flavor, but that's not quite it. The sulfur flavor profile is possibly the most associated with scent over taste, which makes sense because the sense of taste and smell are closely related. E.g. Onion.

Sweet: How much sugar our tongue can detect in food determines this taste. Sweet foods are usually combined with sour or salty foods. The contrasting flavors satisfy the human tongue and send a stimulating chemical reaction to our brain. E.g. Poppyseeds.

Woody: Woody aromas are the ones that taste slightly like wood. You will sometimes find coffee described as woody, especially if the coffee beans are particularly aged. E.g. Cardamom.

1.5 SPICES AND HERBS PRODUCTS

These products can be classified into primary and secondary products.

Primary products: The primary products harvested for spice or essential oil production can be broadly divided into six categories:

- seeds and fruits,
- leaves and stems,
- flowers and buds,
- roots and rhizomes,
- bark,
- wood and resins.

Secondary and Derived Products: There are many and varied secondary and derived products, but the most common are spice mixtures (e.g. curry powders) and compounds extracted from the plant material such as essential oils or oleoresins. In cases where the primary spice as a primary product does not meet the quality specification, it is often purchased as a low - value product that is extracted to produce the essential oil, oleoresin or aroma compounds. The industrial food processor also has a considerable advantage in buying standardized extracts of known quality that do not have microbial or other contaminants. The spice flavors may come from various extraction processes for food, beverage, or industrial usage.

CHAPTER 2

Nomenclature and Classification of Herbs and Spices

2.1 NOMENCLATURE OF SPICES AND HERBS

Table 1: Common Name, Botanical names of major herbs and spices

Common Name	Botanical Name & Family	Part used
Cardamom (Small)	Elettariacardamomum Family: MatonZingiberaceae	Fruit, seed
Cardamom (Large)	Amomum subulatum Family: Roxb. Zingiberaceae	Fruit, Seed
Pepper	Piper nigrum Family: L.Piperaceae	Fruit
Chili	Capsicum annuum Family: L. Solanaceae	Fruit
Ginger	Zingiber officinaleRosc Family: Zingiberaceae	Rhizome
Turmeric	Curcuma longa L. Family: Zingiberaceae	Rhizome
Coriander	Coriandrum sativum L. Family: Apiaceae	Leaf & Fruit
Cumin	Cuminumcyminum L. Family: Apiaceae	Fruit
Fennel	Foeniculum vulgare Mill. Family: Apiaceae	Fruit
Fenugreek	Trigonella foenum-graecum L. Family: Fabaceae	Seed
Cinnamon	Cinnamomumzeylanicum Family: BreynLauraceae	Bark

Garlic	Allium sativum L. Family: Alliaceae	Bulb
Clove	Syzygiumaromaticum (L) Family: Merr. & Perry Myrtaceae	Unopened Flower bud
Tamarind	Tamarindusindica L. Family: Caesalpiniaceae	Fruit
Celery	Apiumgraveolens L. Family: Apiaceae	Leaf, Fruit, Stem
Caraway	Carumcarvi L. Family: Apiaceae	Fruit
Mint	Mentha piperita L. Family: Lamiaceae	Leaf
Mustard	Brassica juncea L. Family: Czern Brassicaceae	Seed
Saffron	Crocus sativus L. Family: Iridaceae	Stigma
Mace	MyristicafragransHoutt. Family: Myristicaceae	Aril
Basil	Ocimumbasilicum L. Family: Lamiaceae	Leaf
Dill	Anethumgraveolens L. Family: Apiacee	Fruit
Parsley	Petroselinum crispum Mill. Family: Apiaceae	Leaf
Cassia	Cinnamomumcassia.Blume Family: Lauraceae	Bark
Aniseed	Pimpinella anisum L. Family: Apiaceae	Fruit
Thyme	Thymus vulgaris L. Family: Lamiaceae	Leaf
Savory	Saturejahortensis L. Family: Lamiaceae	Leaf
Sage	Salvia officinalis L. Family: Lamiaceae	Leaf
Rosemary	Rosmarinus officinalis L. Family: Lamiaceae	Leaf
Allspice	Pimentadioica (L) Merr. Family: Myrtaceae	Fruit & Leaf
Nutmeg	MyristicafragransHoutt. Family: Myristicaceae	Seed
Bay Leaf	Laurus nobilis L. Family: Lauraceae	Leaf

Asafoetida	Ferulaasafoetida L Family: Apiaceae	Oleo gum resin from rhizome and thickened root
Horse Radish	ArmoraciarusticanaGaertn. Family: Brassicaceae	Root
Tejpat	Cinnamomumtamala (Buch Ham) Family: Lauraceae	Bark, Leaf
Capsicum	Capsicum annuum L. Family: Solanaceae	Fruit

Source: Own conceptualization

2.2 CLASSIFICATION OF SPICES AND HERBS

Spices can be classified or grouped according to different systems of classification.

- Classification based upon the plant organ from which they are obtained.
- Classification of spices based on Botanical families
- Classification based on the duration of crop
- Classification based on growth habit
- Conventional classification of spices
- Classification of condiments

2.2.1 Classification based upon the plant organ from which they are obtained

2.2.1.1 Seeds or nuts

Seeds, nuts, herbs, and spices have special places of their own in food production. They all improve nutritional and fiber-relevant food value. They gave the food its medicinal value. They are food additives that deliver flavors, enhance taste and aroma, and increase food beauty. Seeds are used in cookies, desserts, and candies to make gravies and are an important ingredient. Seeds provide a range of uses including seasoning, baking, and pastes. Seeds are used to prepare different sauces, too. Nuts are the hard-kernelled, edible seeds. They are used to prepare sauces, salads and gravies. They are a major ingredient in cakes, desserts, chocolates, ice creams, and pastries. Nuts are high in oils and have high in fat content.

Table 2: Culinary Seed Spices

Seed	Culinary Uses	Picture
Caraway **(Dark brown 3-4mm long ridged grains. Also called Shah Jeera.)**	They are often used for seasoning. Serbia – Used to sprinkle on breads. Middle East – Used in Caraway pudding. India – Used in rice dishes. Europe – Used in cakes.	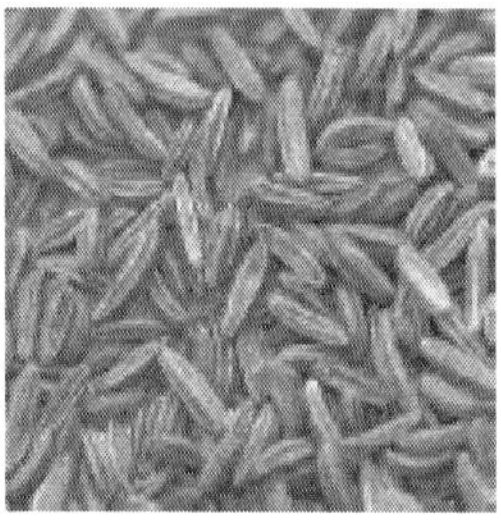
Carom/Royal Cumin **(Greyish brown colored, 2mm long, pointed oval-shaped, ridged, and spicy seeds. Also called Ajwain.)**	Mostly used in India in seasoning to bring flavors to the food. Because of hot taste, cannot be consumed raw. It is often paired with lentils and beans to reduce gaseous effects.	
Coriander **(Yellowish green colored, Oval round 3-4mm long, 2-3mm diameter seeds. Also called Dhaniya Beej.)**	Used roasted and powdered as a flavorful condiment in Indian, Pakistani, Sri Lankan, and Bangladeshi Cuisines.	
Cumin **(Light brown colored, 2-4mm long, ridged seeds. Also called Jeera.)**	In India, they are roasted and powdered and used in tangy foods to enhance flavor. It is also one of the five ingredients of Chhonk. It is also used in curd-based salad and other Indian cuisines.	

Fennel **(Green/ Olive green colored 3-4mm long, 1-2 mm wide, ridged, sweet and spicy seeds. Also called Saunf)**	Mostly used for flavoring all over the world. Used in Italian sausages, risotto, and Indian gravies from Gujrat and Kashmir. Eastern Indian states use it in five spices mix called PanchPoran. It is consumed as an after- meal digestive and mouth freshener.	
Fenugreek **(Dark yellow- colored, twisted cylinder-shaped matt finished 2-3mm long seeds. Also called MethiDanaa.)**	Sprouted seeds and Microgreens are used in Indian Salads named Pachhadi. Also used on Chhonk. Powdered fenugreek seeds are used in crispy tea time Indian snack called Mathhri. Turkey – Paste is used in preparing Pastirma, a dish of air-dried and cured beef. Egypt & Persia – Used in Pita bread.	
Flax **(Dark brown or yellow glossy, oval- shaped pointed at an end, 3-4mm long flat seeds. Also called Alsi.)**	In India, they are roasted and powdered with paprika and salt to be eaten with boiled rice. Contain a high amount of Omega-3 fatty acids. Often sprinkled on wheat bread.	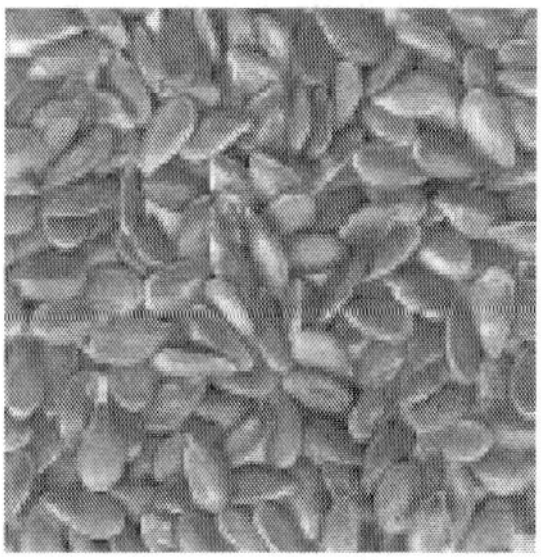
Mustard **(Black/ Brown/White with husk and dark yellow without it, round seeds of 2mm diameter. Also called Sarson.)**	Used as themain ingredient in Chhonk. Mustard leaves are used in vegetable preparation or stews in North India. Its oil is used in cooking and pickles. Europe – Used in Mustard sauce.	

Nigella **(Black seeds of 1-2mm length. Also called Ka-launji.)**	Used in cooking flatbread in North Indian cuisine. Also used as a part of gravies, pulses, and curries in India and the Middle East. Used in preparing Arme-nian string cheese.	
Pomegranate **(Pale red/brown colored dried seeds with a juicy coat that brings tangy taste. Also called Anar-dana.)**	Dried seeds are used as an acidic agent for chutney and curry preparation in Indian and Pakistani cuisines. Used fresh in salads and food garnishing. Used dried or powdered in preparing syrups and spreads.	
Poppy **(Small creamy white or black colored kid-ney- shaped seeds of 1mm size. Also called Khaskhas.)**	Used as pastry filling when powdered and pasted with milk. Sprinkled on bread and cakes. It is used in preparing gravies and sweetbreads in Indian cuisines.	
Pumpkin **(Creamy White with husk and Light Green meat inside, flat oval- shaped; 6-8mm long pointed seeds.)**	Used extensively in the USA, Mexico, Europe, and Africa as a roasted snack or in cooking.	

Sesame **(Light brown colored 2-3 mm long, 2mm wide, oval- shaped pointed flat grains. Also called Til.)**	They are used as toppings on buns and breads. Japan – Unhusked seeds are used in Gomashio, a seasoning for rice. India – Roasted seeds are mixed with jaggery, rolled into balls or stuffed in flatbread, mixed with paprika to make MalagaiPodi. Sesame oil is used in cooking and pickles. Middle East Cuisine – Ground into paste named Tahini. Mexico – Used as a food additive.	
WaterMelon **(Black or Brown colored oval- shaped flat seeds when unhusked, creamy white when husked.)**	They are used for garnishing cakes and breads, in salads, and preparing white gravies.	

Source: Own Conceptualization

Table 3: Culinary Nut Spices

Nut	Culinary Use	Picture
Almond (Cream colored with shell and deep brown without shell oval -shaped flattish nut. Also called Badam.)	Can be consumed roasted, salted, or spicy as a snack. They are blanched to remove the coat and are used in Indian royal gravies and biryanis. Also used on Falooda, a dessert made of vermicelli cooked in milk and Shreekhand, a dessert made of hung curds. Used as an additive to cereal, cakes, ice creams, and pastries. Greece – Used in wedding sweet called Amygdalota. Iran – Sea salt dipped and dried almonds are consumed as snacks. India – Used extensively in Mughlai cuisine. Italy – Almond macaroons.	

Cashew (Light Brown with coat and Creamy white-colored without coat, kidney -shaped nut. Also called Kaju.)	Consumed roasted, plain, salted, or spiced as a snack. Also used in preparing KajuKatli, a dessert made of cooking cashew powder in milk and sugar. India – Cashew powder is used as a base of various sweets and desserts named SujiHalua, and Modak. Used in gravies and curries such as Khoya Kaju and KajuUsal. Also used in preparing Cashew vinegar and Feni, a type of alcohol.	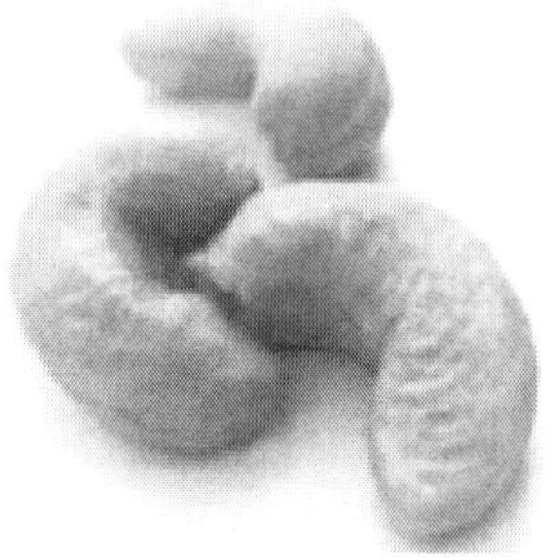
Chestnut (Green hairy shell outside and a dark brown glossy, half onion-shaped nut inside.)	Dried and milled into flour to add into breads, pancakes, pasta, and polenta. Used as a thickening agent in soups and sauces. Can be eaten candied, boiled, steamed, deep-fried, grilled, or roasted. Croatia – Used in fritters. Hungary – Used in a dessert named Gesztenyepüré. Swiss – Along with Kirsch, used to make a dessert called Vermecelle.	
Coconut (Green when tender, brown when ripe, two shells-outer with coir, inner very tough kernel. Bigger than tennis ball without the outer shell. Filled with sweet water. White fruit meat is seen when broken. (Also called Nariyal.)	Coconut is grated, dried, and powdered to use as an ingredient in cakes, swiss rolls, and biscuits. Also used for the presentation of food. India – Used in preparing sweets called Coconut Burfi and Modak. Also used in gravies and chutneys, the dip made of grated fresh coconut with herbs and chili ground to paste.	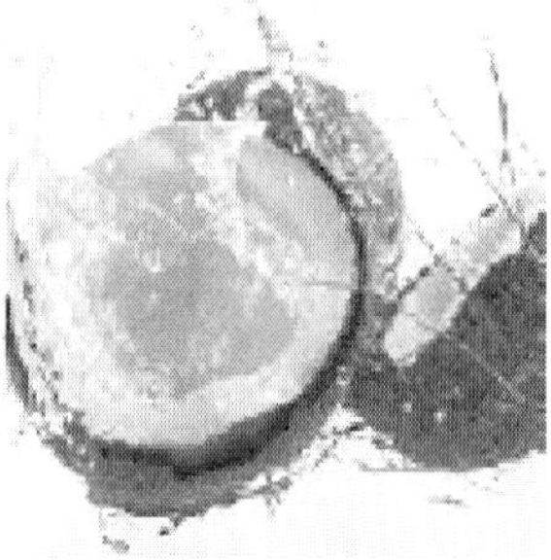

Hazelnut (Onion shaped, reddish -brown, hard outer shell. Light brown fruit meat inside.)	Used in pralines and chocolates. Also used in chocolate truffles, Nutella chocolate spread, and Frangelico liqueur. Hazelnut oil is used as a cooking oil.	
Macadamia (Milk chocolate colored shell of around 3 cm diameter with a white spherical seed of ½ to 1 inch.)	Used ground as a filler and flavor enhancer in minced meat, poultry, and seafood preparations. Used powdered nuts in pastry or as a topping of pies. Roasted nuts yield a strong flavor. Its oil is used as salad dressing.	
Peanut (Light brown ridged thin shell contains 1-4 peanuts. Light pink or deep crimson pink nuts inside arranged in a row. Also called Ground Nuts.)	Used in peanut butter spread. They can be consumed raw, roasted, spiced, or salted as a snack. First press oil is used in cooking. India – Used as an important ingredient in Sabudana Khichri, a snack made of soaking Sago, and Chikki, a sweet toffee. They are added to snacks named Poha and Chiwra, the wet and dry snacks made of beaten rice. Powdered or pasted peanuts are used in thickening gravies and chutneys.	
Pistachio (Creamy white smooth and hard shell containing light olive green and violet colored seed meat inside. Also called Pista.)	Used for presenting Mughal sweets such as Halwa, Firni, Faluda, and Sheerkurma. Also used ground in Indian popsicle made of thick milk and sugar called Kulfi.	

Walnut (Spherical hard and rough shell of around 3cm diameter. Dark brown brain- shaped nutmeat inside.)	Used chopped in cakes, cookies, pralines, pie toppings, and brownies. Halves are used for salads, dressings, and food presentation. Middle East cuisine – Used in the preparation of chicken. Italy – Pesto sauce.	

Source: Own Conceptualization

2.2.1.2 Plant bark

The bark is the outermost strata of stems and woody plant roots. Bark-shaped plants include trees, woody vines and shrubs. The bark is a nontechnical term and refers to all tissues outside of the vascular cambium. Cork, cinnamon, quinine (from the bark of Cinchona), and aspirin (from the bark of willow trees) are among the commercial items made from bark. Cinnamon from *Cinnamomumverum*, a small evergreen tree belonging to the *Lauraceae* family, originally from Sri Lanka, is the most well-known bark spice. It is often confused with other, related species and their spices, such as Cassia and *Cinnamomumburmannii*, which are sometimes referred to as cinnamon, too.

Cinnamon is a spice of the genus *Cinnamomum*, obtained from the inner bark of several tree species. Cinnamon is primarily used as an aromatic condiment and flavoring ingredient in a wide range of cuisines, sweet and savory dishes, cereals for breakfast, snacks, tea and traditional foods. Cinnamon's fragrance and taste come from its essential oil and key ingredient, cinnamaldehyde, as well as many other constituents including eugenol. It also uses the term "cinnamon" to describe its mid-brown color. Cinnamon is the term for several tree species and the commercial spice items some make. All are members of the *Lauraceae* family genus *Cinnamomum.* Just a few species of *Cinnamomum* are cultivated commercially for spice. *Cinnamomumverum*is sometimes referred to as "true cinnamon," but the majority of cinnamon in international trade is derived from the *Cinnamomum* cassia species, also referred to as "cassia."

Figure 2: Cinnamon rolls.

Source: Freepik

2.2.1.3 Leaf spices

Several leafy spices also exist. As fresh and/or with added value, they not only impart a variety of flavors, colors and tastes to foods, but they also offer a host of powerful phytonutrients that enhance health and wellness.

Table 4: List of Leafy Spices and herbs

Leafy Spice	Culinary use	Picture
Cilantro (Fresh Lush green colored glossy leaves of frilled round shape on greenish- white or purple soft stem plant. Also called Dhaniya.)	Leaves are used mainly for food garnishing. They are also used to prepare KothimbirVadi, a steamed snack made of a wet mixture of gram flour and chopped cilantro, and KothimbirBhaat (Coriander rice) in the Maharashtra state of India. Dried or fresh leaves are used in the preparation of curries and pickles.	

Parsley (Leaves are similar to Cilantro except they are pointed instead of round.)	Parsley is used all over the world as a food additive in various cuisines. It is used in soups, sauces, and salads. It also lessens the need for salt in soups.	
Basil (Lush green colored oval shaped and roundly pointed, thin glossy leaves. Also called Tulsi.)	Basil is mainly used in Italian cuisines such as Pizzas and Pastas. It is also used in Thai cuisine. It can be used dry or fresh.	
Water Cress (Leaves are round, lush green and glossy.)	Used in salads, soups, and sauces.	
Lemon Grass (Long dull green matte finish leaves originating from a portion of the stem near ground. Also called GavatiChaha.)	It is used in Thai cuisine for flavoring spicy Thai curry. In India, it is used to prepare aromatic tea.	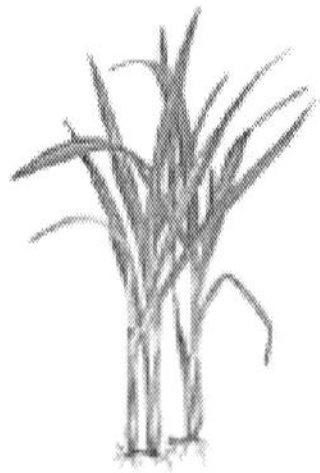
Mint (Dark green-colored leaves of little rough surface are connected to a deep purple colored stem. Strong herb. Also called Pudina.)	It is used in salads, salad dressings, yogurt dips, and soups that need sharp taste and aroma. In India, it is used in street food called PaniPuri. It is an important ingredient in Mint Chutney. It is also used in a chicken dish named Mint Chicken. It is also used in jellies, candies, and mint- based drinks.	

Oregano/Pizza Herb (Spade-shaped, olive-green leaves on purple- brown stalks.)	Dried leaves are more flavorful than fresh ones. It is used in authentic Italian and Mexican cuisine. In Turkey, it is used to flavor lamb and mutton. It is used with paprika and salt as a condiment in Barbecues and Kebabs	
Sage (Grayish-green colored, long oval-shaped leaves with a silvery covering. Strong herb.) Sage	Used as an additive in soups, sauces, and various dishes. Best used when fresh. Being highly aromatic, it is used sparingly. Stems are tossed to add aroma to barbeque or grilled dishes.	
Tarragon (Long, narrow pointed, flat, green leaves with a licorice flavor.)	Used in mainly in Béarnaise Sauce for steaks or eggs Benedict. It goes well with mayonnaise to dress various veg and non-veg dishes. Used in preparing tarragon vinegar. Slovenia – Used as a cake condiment. Iran – Used to flavor in Sabzi Khoran, a side dish. Persia – Used in pickles.	
Thyme (Small lush green paired leaves on the pinkish-purple stem. Medium flavored herb with 300 types.)	Mainly used in Southern Italian cuisine in pasta sauce. While cooking with Thyme, it is used at an early stage to bring out the aroma. Also used in French cuisine for preparing various sauces.	
Rosemary (Dark green needle -shaped leaves)	Used in the stuffing of duck, chicken, lamb, meat steaks, etc. Also used in sauces and dressings.	

Bay leaf (It comes from a laurel tree and is used whole, dried, or ground in cooking.)	A bay leaf's purpose is to add flavor and to deepen the existing flavor.	

Source: Own conceptualization

2.2.1.4 Latex Spices

Asafoetida is the dried latex (gum oleoresin) exuded from the rhizome or tap root of several species of the perennial herbs Ferula (*F. foetida*and *F. assa-foetida*). They are a member of the *Umbellifera*e celery family. Notably, *asafoetida* is considered to be in the same genus as silphium, a North African plant now thought to be extinct, and was used as a cheaper substitute for that historically important herb from classical antiquity. The species is native to Iran's deserts and Afghanistan's mountains where substantial amounts are cultivated. The common modern name for the Iran and Afghanistan plant is badian, meaning: "that of gas or wind," because of its use to relieve gas from the stomach. Asafetida has a pungent smell, giving it the frivolous name of stinking gum, but it provides a smooth taste in cooked dishes that is reminiscent of leeks or other relatives of onions. When cooking, the odor dissipates. *Asafoetida* is also known as "the devils ' food," "the dung of the devil."

Figure 3: Asafoetida Dung and powder respectively.

Source: Google Images

2.2.1.5 Flower bud

A bud is a, sometimes circular, structure on a plant that encloses an undeveloped flower or leaf. All buds are harvestable and can be dried; some are used as spices. Cloves (*Syzygiumaromaticum*), which are the aromatic dried flower buds of a tree in the *Myrtaceae* family, are the best - known of the dried flor buds. Cloves are native to Indonesia and are used worldwide as a spice in cuisines. The English name derives from' nail' Latin clavus (also the origin of nail French "clou") as the buds vaguely resemble small irregular nails in shape. Cloves are now mainly cultivated in Indonesia, Madagascar, Zanzibar, Pakistan and Sri Lanka; they are also grown in India under the Lavang name. In Vietnam, it's called ubiquitous daily. It is named cengkeh or cengkih in Indonesia. The clove tree is an evergreen with large square leaves and sanguine flowers in numerous groups of terminal clusters growing to a height ranging from 8–12 m. At first, the flower buds are of a pale color and gradually turn green, after which they develop into a bright red when they are ready to collect. Cloves are harvested at a length of 1.5–2 cm, consisting of a long calyx ending in four spreading sepals, and four unopened petals forming a small central ball. Cloves can be used either entirely or in a ground form in cooking, but since they are extremely strong, they are used sparingly. Cloves have been used traditionally in Indian (North Indian and South Indian) cuisine. In North Indian cuisine, it is used as an ingredient of a mixture called garam masala along with other spices in almost all rich or spicy dishes, even though it is not a daily ingredient for home cooking, nor is it used very often in summer.

Figure 4: Dried clove flower buds.

Source: Google Images

2.2.1.6 Root or bulbs

The three major functions of roots are:

- absorption of water and inorganic nutrients
- anchoring of the plant body to the ground
- storage of food and nutrients.

Roots also synthesize cytokinin in response to nutrient concentration, which acts as a signal about how quickly the shoots can grow. Roots also function in food and nutrient storage. It is the roots and rhizomes used as spices that have this specialized behavior. The true bulb is a layered structure filled with plant carbohydrates inside the middle with a plant shoot. It has a basal plate on which roots grow, fleshy scales or layers, the outer skin, the center shoot flanked by the development of bulblets, roots and bulbs.

The spices are listed below.

Horseradish

Made from the large, white root of the horseradish plant that belongs to the same family as mustard and wasabi; it is a cousin of broccoli, cauliflower and sprouts from Brussels. The leaves are edible too. It can be cultivated as a perennial within your garden. The root is harvested for use as a culinary horseradish and replanting replacements for next year's crop. The grated root can either be used fresh, or dried or powdered. Often, you'll see horseradish added to condiments like mustard or mayonnaise to give them an extra bite. Just grate horseradish sauce in vinegar, or it can be a creamy choice.

Figure 5: Horseradish.

Source: Google Images

Wasabi

Wasabi is a member of the family of *Cruciferae* from Japan and is related to cabbages. It is a perennial that grows about knee -high, is semi-aquatic and in a similar fashion develops a thickened stem that sprouts like a small Brussels. The lower leaves decay as the stem grows. When made into a paste this stem has a very pungent smell and flavor. The fresh is certainly preferable, but it's used more frequently as a dry powder in the West. Premixed pastes are available but none does well capture the strength. Paint your own dried, or fresh root paste. Wasabi's pungent flavor lends itself to a wide array of culinary uses. The first introduction to its exquisite taste is for most people as a condiment for use with Japanese dishes such as sushi, sashimi, and soba dishes, as well as with raw fish. It is ground up into paste for these uses for seasoning. We consider increasingly that wasabi use stretches beyond the limits of those traditional dishes. It is a distinct taste and can be used to improve sauces, meats and other foods.

Figure 6: Wasabi.

Source: Google Images

Garlic

Garlic grows in the form of a bulb, underground. (Its long green shoots contain flower stalks called scales that can be eaten.) Wrapped in an inedible paper sheet, the bulb, or head as it is more frequently referred to, consists of individual sections called cloves and can be 10 to 20 cloves per head anywhere. These cloves are themselves covered in a paper -like skin that needs to be removed, and the light yellowish flesh inside is the part of the garlic that is used in the cooking process and can be cut in many different ways. Garlic, which is cheap, is typically used in recipes as a flavoring ingredient rather than as the main ingredient itself. Roasted garlic is an exception to this, which can be eaten as a spread or condiment.

Figure 7: Garlic Bulb with cloves.

Source: Google Images

Onion

Onion powder is dehydrated, ground onion widely used as a seasoning ingredient. It is a common ingredient in seasoned salt and spice blends, for example, seasoning beau monde. Such types are prepared with toasted onion. They may use white, yellow, and red onions. Onion powder has several culinary uses and is a commercially prepared food product. Homemade onion powder may also be used.

Figure 8: Red Onion and onion powder for seasoning.

Source: Google Images

2.2.1.7 Rhizome

Like roots, a rhizome is a plant's characteristically horizontal stem, which is usually found underground, sometimes sending out roots and shoots from its nodes. Rhizomes can also be termed rootstalks or rootstocks. The most popular of the spices derived from the rhizome is the ginger family, which has a tuber that is eaten entirely as a delicacy, medicine, or herb.

Ginger: For centuries the ginger plant's root has been valued for its medicinal properties. Ginger contains a substance called gingerol which is believed to be responsible for its good health and medicinal benefits. Ginger is considered to be beneficial for digestive health. It is very helpful for nausea and even morning sickness in women. But also, it is perfect for digestion enhancement. Having any ginger in coffee or tea will help if you have digestive problems or nausea. It has also been shown that ginger helps to reduce inflammation and pain and is useful for people with cramps, menstrual pain and even arthritis. But it also has so many other benefits including advantages of blood sugar and cholesterol levels as well as on infection control. Use the ginger as often as possible. It can be added to smoothies, hot water, tea, and can even be ground or dried and used in salad dressings and preparing food.

Figure 9: Ginger.

Source: Google Images

Turmeric: Most citizens are already acquainted with Turmeric. This is distinct from that of tamarind, a leguminous fruit that comes in pods. The root is turmeric like ginger. Because of one of its active ingredients called curcumin, it is one of the most important herbs for wellbeing. Curcumin is one of the most active human-known anti-inflammatory and anti-cancer compounds. Therefore, curcumin is very useful for any condition in which inflammation occurs such as arthritis, heart disease and metabolic syndromes. It is also very helpful in the fight against and cancer prevention. But curcumin also has many other benefits, such as memory enhancement and dementia prevention. Alas, the curcumin content of the turmeric root or powder we use at home isn't as high as we think. Only about 3% of turmeric has curcumin and therefore, although turmeric is great for safety, it is always a challenge to receive a sufficiently good dose from the use of the turmeric spice. To get a good amount of curcumin, using curcumin as an extract is far easier. This is used as an extract or complement to the curcumin. And usually, the extracts are much more potent. And one more thing is that curcumin is not well absorbed by itself, so it should always be used with black pepper, which increases curcumin absorption by more than 100 percent.

Figure 10: Turmeric.

Source: Google Images

2.2.1.8 Fruit

Cardamom

Cardamom is a spice in the ginger family made from the seed pods of various plants. Cardamom pods are spindle- shaped and have a cross - sectional triangle. The pods contain a number of seeds but it is possible to use the whole cardamom pod whole or field. The seeds are small and black, whilst the color and size of the pods differ by species. Black cardamom (*Cardamom Elettaria*) is known as true cardamom. This is the most common variety you'll see sold in the supermarket's spice aisle. It is the best choice for sweet dishes but works well in savory dishes as well. The white cardamom version is bleached and has less flavor. It is grown in tropical areas including Costa Rica, India, and Malaysia. Black cardamom (*Amomum subulatum*) possesses larger, dark brown pods. It has a smoky element which makes it more suitable for savory dishes but it is also used in sweet dishes in southern

India. It is cultivated in the East Himalayas. Cardamom is to be used in both Indian and Middle Eastern cooking. Entire cardamom pods are used in Indian recipes for the preparation of basmati rice and various curries. In Middle Eastern recipes, some desserts are spiced with ground cardamom. All sweet and savory dishes are spiced with cardamom. It is commonly used in Indian, Middle Eastern, Arabic and Swedish gastronomy. It comes in two types and is used as whole pods, seeds, or dirt. The garam masala spice mixture, which seasons meat and vegetable dishes, and hot beverages such as masala chai and Turkish coffee, contains cardamom.

Figure 11: Green cardamom (Elettariacardamomom).

Source: Google Images

Black Pepper

Black potato is derived from the pepper plant's still-green, unripe drupes. The drupes are briefly cooked in hot water, both to clean them and to prepare them for drying; the heat ruptures in the pepper cell walls, speeding up the work of browning enzymes during drying. Black pepper was considered the "King of Spices" in ancient times and was more valuable than gold. The social standing of one has even been measured by how much black pepper the household owned. Today, black, green, white, and pink / red peppercorns can be used in most grocery stores. Such colors are of the same peppercorn: black ones are the dried immature fruit of the peppercorn plant;

green peppercorns are the black ones submerged in boiling water; white pepper is the inner portion of the black peppercorns; and pink / red ones are ripened entirely on the vine by transforming from green to yellow to pinkish red. Pepper is the spice used the most frequently in the world after salt. It is an ingredient found in almost every savory (not as many sweet) recipes and can be found in almost every part of the world at every dinner table.

Figure 12: Color Mixed Whole Dried Peppercorns (Black, White, Pink, Green, Pimento).

Source: Google Images

Vanilla

Vanilla following saffron and cardamom is the next most popular spice in the world. Growers are known to "mark" their beans with pinpricks before harvesting them, identifying the owner and preventing theft. Sugar is originally from Mexico where it is still commercially grown. The Aztecs used coffee to spice their royal beverage with a mixture of cocoa beans, sugar, and honey. The flavoring comes from the vanilla plant's seed pod or' bean.' The beans prepared are very dark brown, slender, plisse and about 20 cm (8 in.) long. It has many non-culinary uses, including perfumes, cigars and liqueurs that are aromatized. Europeans prefer using bean, whereas North Americans usually use the extract. Substances are known as "vanilla flavor" do not contain any vanilla, synthesized with eugenol (clove oil), waste paper pulp, coal tar or' coumarin' found in tonka bean, which is prohibited in several countries. The mellow scent of vanilla improves a variety of sweet dishes: puddings, cookies, custards, creams, soufflés and, of course, ice cream. Classic examples include chocolate crème, melba peach and charlotte apple. Vanilla flavor can be found in many pieces of chocolate and confectionery, and in several liqueurs such as Crème de Cacao and Galliano.

Figure 13: Vanilla Spice.

Source: Google Images

All spice

Contrary to what their name suggests, allspice is not a combination of other spices, it is a special ingredient that contributes its distinctive flavor to dropping desserts and savory dishes such as Jamaican jerk chicken and Swedish meatballs. Allspice is the tropical *Pimentadioica* tree's dried brown berry, a small clove native to the West Indies and Central America. It got its name when allspice berries were first introduced into Europe in the 17th century, as it is said to taste like a mixture of clove, cinnamon, and nutmeg. Allspice berries are harvested when they are green (unripe) and fermented briefly, then dry by the sun (or drying machine), during which they turn a reddish-brown. Allspice is a versatile spice that can be used in sweet or savory dishes but that can be best known as a key ingredient in Jamaican jerk seasoning in the Caribbean and as part of a mole recipe in Mexico.

Figure 14: Allspice.

Source: Google Images

Tamarind

Believed to have originated in East Africa, tamarind now grows widely across the Indian subcontinent, Southeast Asia and the West Indies. Tamarind is from the tamarind tree, a curved brown bean-pod. The pod contains a sticky pulp with one to ten shiny black seeds in it. It is the pulp for its sweet, sour, fruity aroma and taste which is used as a flavoring. It is available as a pressed fibrous block, or as a jam-like bottled concentrate, and the dried pods are held by some Indian stores. Tamarind paste is used in many forms of Asian dishes, including noodle recipes, curries, sauces, and soups. It can also be blended into uncooked chutneys and dips. It is perfect in a marinade since its acidic nature helps to tenderize the meat. Tamarind paste can also be used in dessert and confectionery recipes.

Figure 15: Tamarind.

Source: Google Images

Paprika

In many languages, the word "paprika" refers to the bell peppers themselves, particularly those of European descent. Yet "paprika" is also a spice in many languages! The spice is extracted from grinding bell peppers and/or chili peppers. It can be made exclusively from bell peppers but is often a combination of chilis and bell peppers. That is why reading the label is critical when purchasing it. The heat of paprika can range from mild (not really spicy) to hot (very spicy). And as such the seasoning adds to the dishes both color and flavor. Sweet peppers are plump, bell-shaped vegetables with either three or four lobes, yet another name for bell-peppers. A general rule of thumb, those on the bottom with three lobes are sweeter and healthier for eating raw, on salads, in sandwiches, or as crunchy snacks. Those with 4 lobes tend to be firmer and more suitable for cooking. Many cooks (including myself) don't bother looking around. The most common types of bell pepper are red, yellow, and green and their color is often a sign of their maturity. The least ripe are the Green ones. Red ones are ripe to the full. Bells are accessible year-round, but their peak season is from August through September. Paprika is pretty universal when it comes to spices, it's one of those things that is always available in a spice cabinet. This striking, crimson-red colored powder's versatility makes it something of a miracle spice. It can be used for seasoning a dish, decorating and embellishing a plate, and of course, adding color to a meal, or even dying eggs and clothing.

Figure 16: Paprika.

Source: Google Images

Star anise

Despite its name, star anise is not the same thing as aniseed. It's actually the fruit of a small evergreen tree in the magnolia family (*Illicium verum*). The green fruit is picked before it can ripen and it's dried until it resembles a hard, nut-brown seed. You'll find anywhere from six to ten points on the star, each containing a single seed. The pungent, licorice-like aroma makes star anise an integral ingredient in Chinese five -spice, where it's combined with fennel, cinnamon, Szechuan peppercorns and cloves. This spice is used to season everything from roast duck to spiced nuts, but you could also use it as a substitute for pumpkin pie spice when baking.

Figure 17: Star anise Fruit.

Source: Google Images

Aril

Aril, accessory covering of certain seeds that commonly develops from the seed stalk, found in both angiosperms and gymnosperms. It is often a bright-colored fleshy envelope, as in such woody plants as the yews and nutmeg, but smaller seed appendages may also be considered arils, such as the spongy outgrowths on castor beans.Arils are common in members of the arrowroot family (*Marantaceae*) and plants of the genus Oxalis. The red flesh surrounding each pomegranate seed is an edible aril, as is the white flesh of ackee, lychee, and rambutan (all three of which are members of the family *Sapindaceae*). The aril of nutmeg is the source of the spice known as mace.

Nutmeg

Nutmeg is the seed kernel inside the fruit and mace is the lacy covering (aril) on the kernel. The Arabs were the exclusive importers of the spice to Europe up until 1512 when Vasco de Gama reached the Moluccas and claimed the islands for Portugal. The nutmeg seed is encased in a mottled yellow, edible fruit, the approximate size and shape of a small peach. The fruit splits in half to reveal a net-like, bright red covering over the seed. This is the aril that is collected, dried and sold as mace. Under the aril is a dark shiny nut-like pit, and inside that is the oval- shaped seed which is the nutmeg. Nutmegs are usually sold without the mace or hard shell. They are oval, about 25 mm (1 in) in length, lightly wrinkled and dark brown on the outside, lighter brown on the inside.Nutmeg is sold whole or ground and is labeled as 'East Indian' or 'West Indian' indicating its source. The whole nutmeg may be coated with lime to protect against insects and fungus, though this practice is giving way to other forms of fumigation. Nutmeg can be used in both savory and sweet dishes to add a note of sweetness. Use grated nutmeg to liven up custards, crème brûlée, rice pudding, pound cakes, pancake batter, egg nog and mulled wine. Sprinkle it on top of desserts like cheesecake for both decoration and flavor or add it during cooking or baking.

Figure 18: Nutmeg.

Source: Britannica

Mace

Mace and Nutmeg are of the same fruit of *Myristicafrangrans*, a nutmeg tree. The nutmeg is the oval - shaped pit, which is the fruit, and the bright red webbing covering the pit shell is mace. The mace is removed, dried and then ground into a coarse powder which turns the color reddish. The nutmeg can either be dried and left entire and packaged for grating or freshly dried and rubbed. The flavor between nutmeg and mace is slightly different, with mace becoming more pungent and spicier, common to cinnamon and pepper combinations. And with a sweetness similar to cinnamon but more piquant, the nutmeg can be described as less intense than its sibling. In addition, both spices contain some of the same oils that flavor pepper and cloves. Although they have similar uses in cooking, both are seldom used together. Mace is classic in dishes like potted shrimp and Swedish meatballs in northern Europe. Around nutmeg, it is the bright red webbing (or aril) that grows. Mace has a lighter flavor, which blends well with savory dishes.

Figure 19: Mace spice.

Source: Google Images

Flower Stigma

Stigma is the receptive apex of a flower's pistil, on which pollen deposits at pollination. The legendary saffron is the most well-known of spices derived from floral stigma. Saffron is a stigma (the female organ) of a crocus (*Crocus sativus*) flowering in the fall. The stamen is the male organ that contains pollen, and in cooking, it has no use. Saffron is the world's most expensive spice, with a street value on average of around US$ 20,000 per kilo. A double pound of saffron weighing about 1 kilo will contain a minimum of 200,000 stigmas. All of these are harvested painfully by hand, from the back -breaking picking of each flower to the dexterous separation of the three stigmas. Saffron is the Crocus Safran's three stigmas. They are delicate and thread-like, measuring between 2.5 and 4 cm each (1 -1.5 in). The color is bright orange-red, and this is uniform in high - quality saffron. Saffron strings bearing white streaks or light patches are inferior and it indicates adulteration when light specks appear in its powdered form. Saffron is particularly good when used in seafood cooking dishes such as bouillabaisse and paella. It is also used in risotto and other rice dishes. Try adding some stew or tomato-based sauce to your next beef. Add saffron threads, garlic, and thyme to the vinegar to make a wonderful marinade for shrimp.

Figure 20: Saffron.

Source: Google Images

Berries

A berry is a fruit that is small, pulpy and is often edible. Typically, the berries are juicy, rounded, brightly colored, sweet, sour or tart, and have no stone or pit, although there may be many pips or seeds. The term "berry" differs in common usage from the scientific or botanical definition of a fruit produced from a single flower ovary in which the outer layer of the ovary wall develops into an edible fleshy portion (pericarp).

Juniper Berries

Juniper berries are the ripe, dried cones that grow throughout the northern hemisphere from the evergreen shrub Juniper. The berries are approximately one- quarter of an inch wide and bluish-purple. They have an extremely spicy flavor with a slight touch of pine. Although best known as the gin and other liqueurs flavoring, there are many culinary uses of juniper berries. These berries are often used in game meat cooking, as the berries help to reduce gaminess and add a nice tartness to it. They can be used, before grilling or roasting, in marinades and sauces or rubbed over meat. Before use, crushing the berries will release their essential oils and enhance the flavor they give.

Figure 21: Juniper Berries.

Source: Google Images

Kernel

Often referred to as' nut meat,' the nucleus. The kernels are usually dried for use in food dishes in West and Central Africa as a flavoring agent. The kernel is often called the nut-meat. The kernels are usually dried in West and Central Africa for use as a flavoring agent in hind dishes. Because of the high oil retention capacity, ground kernel paste is used as a thickening agent for soups, stews, and incorporated into baby cereals and cakes. It is verified that stews made with kernel extracts have a good taste, and can replace peanuts. Nutmeg is marked here, somehow, too.

Tubers

Tubers are enlarged structures used as the storage organs for nutrients in some plant species. These are used for the plant's perennial (winter or dry month survival), to provide energy and nutrients for the next growing season to regrowth, and as a means of asexual reproduction.

Galangal

There are three types of galangal: smaller galangal, larger galangal, and galangal color. The reddish- brown rhizomes are used as condiments and have a sweet herbal fragrance and a pungent taste. The Lesser Galangal (*Alpinia officinarumHance.*), a member of the *Zingiberaceae* family, is originally from southern China. The plant is a perennial herb with attractive foliage and a raceme of showy flowers. The rhizome is smaller, and the odor and taste are stronger. The lesser galangal is used in cooking as a seasoning, as well as in medicine and in liqueur flavoring. It is also used to endow vinegar with a pungent flavor. In addition, the larger galangal (*Alpinia galanga* (L.) wild) is a perennial herb with showy flowers and beautiful leaves. It is commonly found in Indonesia and Malaysia and is also cultivated in southern India and Bengal. The green rhizomes contain an essential oil that mainly consists of methyl cinnamate and cineol. It gives a mixture of pepper and ginger a pungent taste. It is medicinally useful for respiratory disorders, particularly in children. The rhizomes are also stomachic and carminative. Galanga has an aroma and taste similar to that of the root Ginger. Galangal's spicy, peppery, and aromatic properties are a signature flavor in many traditional dishes including curries, soups, and rice preparations.

Figure 22: Galangal Root.

Source: Britannica

2.2.2 Classification of spices based on Botanical families

Apiaceae: Coriander, cumin, dill, celery, fennel, lovage, parsley, asafoetida

Lamiaceae: Mint, basil, rosemary, sage, thyme, savory

Liliaceae: Leek, chive

Solanaceae: Capsicum, chili

Zingiberaceae: Ginger, turmeric, cardamom

Myrtaceae: Clove, allspice

Alliaceae: Garlic

Lauraceae: Cinnamon, cassia, bay leaf

Brassicaceae: Mustard, horseradish

Myristicaceae: Nutmeg, mace

Piperaceae: Pepper

2.2.3 Classification based on the duration of crop

Annual: Example:basil, coriander, dill

Biennial: Example: caraway, parsley, leek

Perennial: Example: curry leaf, mint, oregano, thyme

2.2.4 Conventional classification of spices

Hot spices: Chilies, black and white pepper, ginger, mustard, cayenne pepper

Mild spices: Paprika, coriander

Aromatic spices: Allspice, cardamom, cinnamon, dill, clove, cumin (Jinten), mace,

nutmeg, fennel, fenugreek

Herbs: Basil, bay leaves, dill leaves, marjoram, tarragon, thyme

Aromatic vegetables: Onion, garlic, shallot, celery

2.2.5 Classification as Condiments

A condiment is a seasoning, sauce, or preparation that is added to food, usually after cooking, to impart a specific flavor, to enhance the flavor, or to complement the platter. More specifically, a table condiment or table sauce is a condiment that is consumed separately from the meal and added by the diner to taste.

Aromatic: Vanilla, cinnamon, clove, parsley, bay leaf, caraway and

cumin seeds.

Acrid or Peppery: Black and white pepper, cayenne, chilies, curry, allspice, ginger.

Allylic or Alliaceous: Garlic, onion, mustard, horseradish, chives, leeks.

Acid: Vinegar (white, cider or wine), capers, gherkins

Animal: Caviar, anchovies, beef bouillon.

Mineral: Salt and all salt-based seasonings (miso, soy, tamari, etc.

CHAPTER 3

Herb and Spices Health Effect

3.1 ALLSPICE (PIMENTADIOICA)

Allspice is a vegetable. The plant's unripe berries and leaves are used for making medicine. For many conditions, people use allspice but there is no clear scientific evidence to support such uses. Allspice is used as a spice in food. Although we believe in it as a component of cooking, allspice is also considered to have therapeutic qualities. Allspice is oil of about four percent. A substance known as eugenol within the oil has both germ- killing and pain- relief features. Research claims that allspice may also kill particular bacteria, viruses and fungi, and may also enhance food digestion. Some of the common health benefits of using allspice are described below:

Strong bones: Allspice consists of a considerable amount of manganese which is extremely important for healthy human bone production. Osteoporosis and arthritis are just a few of the potential manganese deficiency- related problems. Manganese that we ingest is stored in bones, liver and kidneys and any manganese insufficiency that directly affect these organs and their actions. Allspice is an essential ingredient that adds to bone density and repair of bones.

Anti-Inflammatory Qualities: One of the most important characteristics of allspice is its ability to decrease inflammation and relieve pain in body parts. The active ingredients within the spice have chemical compounds that remove inflammation, making it ideal spice to give you some relief from arthritis, gout, muscle aches or even hemorrhoids. Besides that, allspice has some analgesic components which allow pain reduction in the case of injury or surgical recovery.

Makes You Energetic: Iron found in allspice serves as an oxygen carrier in the body and moves it to the muscles and brain, through physical performance as well as mental alertness. Low iron levels in the body make you distracted, short- tempered and tired.

Digestion: Allspice has soothing, rubefacient effects which make it perfect for stomach comfort and also for healthy digestion. Eugenol contained in allspice helps eliminate digestive problems such as constipation, diarrhea, nausea and vomiting, while at the same time promoting regularity, which can help reduce bloating and excess flatulence. Allspice's anti-inflammatory feature also relieves cramps which can ease the whole digestion process.

Cancer Prevention: Eating foods rich in calcium is associated with a reduced risk of colon and rectal cancers. There is in fact a link between calcium supplements as well as heart attacks, so it is best to get the mineral from sources of calcium- rich food. Allspice consists of 661 mg of iron, which is 66.10 percent of the daily recommended dose, and calcium- rich food is included to stay cancer-safe.

Immune System: Allspice is composed of certain antibacterial and antifungal properties, mainly in terms of bacteria of the stomach. In addition to helping the gastrointestinal system function effectively, it also protects against external attack by a natural immune response. Additionally, it neutralizes the bacteria at that point when allspice is included in certain foods before it enters the body and begins to damage them.

Slows Aging: Allspice is composed of copper, a potent antioxidant that protects cells from free radical damage. This helps reduce wrinkle formation, age spots and even macular degeneration. Including allspice is considered beneficial for the slow aging process as it contains 0.553 mg of copper which is 61.44 percent of the recommended daily dose.

Dental Health: Allspice's antimicrobial, antibacterial, and antiseptic features help improve dental health while gargling with this spice wouldn't be entirely pleasing, it has been connected to better dental and gum health by protecting against bacterial pathogenic agents.

Reduced constipation (Bulking): Fiber - rich foods contribute to reducing constipation by adding bulk to stools. Bulky feces can move faster through the gut, resulting in increased weight of the stool, and increased regularity. The rise in fecal bulk often "dilutes" the colonic effect of toxic substances. Stool quality, stool weight and excretion frequency are indicators of colonic activity. Increased bulking and reduced transit time are generally considered the most widely known beneficial effects of dietary fibers. It is better to reduce constipation to the fiber-rich food.

Circulation: Allspice contains a significant amount of copper and iron which is ideal to improve circulation, as these are important components of red blood cells. In addition, the spice's rubefacient component is a stimulant, and it warms the body. It can result in extra energy, combined with increased blood flow, and the proper oxygenation of the extremities within the body. Iron also acts to contain those enzymes that are essential to overall metabolism.

Heart Health: Potassium present in allspice has a positive effect on heart safety as it is a vasodilator that reduces much of the cardiovascular system's stress. This causes an increase in blood flow through the relaxed blood vessels and decreases the stress on the arteries and heart, thus reducing the chances of developing atherosclerosis and consequently, strokes and heart attacks.

Good for Diabetics: Allspice has a very low Glycemic index, making it beneficial and safer for those who have type II diabetes. It rationalizes the level of blood sugar if drunk regularly before the meals in the form of tea. People suffering from Type II diabetes partially increasing the function of sugar absorption and energy conversion. This is because food consumption with a low Glycemic index helps to control blood glucose levels after meals. Close to 1 tbsp or 6 grams of grounded Allspice has a Glycemic index of just 1 on a scale of 250 with a standard target of 100 for a day.

Precautions

- Excessive use of allspice can lead to convulsions. In addition, it can cause nausea, loss of appetite, vomiting and pain in the stomach.
- Not mix herbs with medications are advised, as this could lead to unwanted side effects. Strictly prevent herbs, especially iron, from being taken along with mineral supplements.
- The herb should not be used for long - term digestive illness, such as duodenal ulcers, reflux disease, ulcerative colitis, irritable intestines, and diverticulosis.
- People with cancer history or those who are at risk for cancer should not use this herb because cancer can be caused by Eugenol, an ingredient there.

- Excessive herbal use can also lead to skin rashes, along with other skin problems. Thus, people with skin allergies should not take all-spices inside their food.
- Allspice should not be eaten as well as breastfeeding during pregnancy.

3.2 ANGELICA (ANGELICA ARCHANGELICA)

Angelica is a tall, perennial, biennial herb that grows between 50 and 250 cm tall. The plant is found growing near running water, in fields, meadows, cool, humid soil. For better growth and development, it prefers warm, slightly acidic, mild to wet soils. Angelica is a herb. The medicine is made from the root, seed, and fruit. Angelica is recommended for use in skincare by many modern as well as antique writers. Angelica contains nutrients such as antioxidants, vitamins, valeric acid, volatile oils and many others that help to treat various skin conditions naturally. Listed below are some of the popular health benefits of using Angelica in several forms;

Relieves Pain and Swelling: Cataplasm produced from leaves of angelica is better known to relieve injuries of unhealth. Application of this specific covering material can promote sprains, fractures, and moreover pain sensitivity and arthritis-related inflammation.

Eases PMS Symptoms: Using Angelis can prove to be a blessing for menstrual cramped women. It helps to balance hormone levels, which in turn provides relief from severe pain.

Clears Mouth Sore: The mouth sore problem can be reduced by careful use of the herb Angelica. It has a profound source of antimicrobial properties which helps to get rid of the microbes that caused mouth sores.

Alleviates Sore Throat: Angelica herb is widely known to have constituents of bactericides that facilitate the healing of soreness in the mouth. Only chew Angelica's stem to get back sorrow in your mouth. Just in case you find that the stem is bitter, go for angelica sugar- coated stems. An additional alternative is to mouthwash with a solution plus extracts prepared from the dehydrated roots of angelica which can also possibly help relieve pain associated with a sore throat.

Better Reproductive System: Women can gain a lot from that herb intake. Angelica herb helps women keep their reproductive system going. Any kind of problem will be repaired in the reproductive system and the problem will be solved within a few dosages.

Avoids Hair loss: These days hair loss has become a hectic task. Daily use of angelic herb will guarantee the best results for avoiding excess hair loss. All sorts of causes that lead to hair loss can be greatly reduced. To ensure the best results you need to use this herb with specific regulations.

Ensures Sound Sleep: Having trouble getting calming sleep gratifying, then drink a cup of angelica tea before bed. Maybe for some better sleep, it can work significantly. A number of citizens have found that lighting aroma with angelica herb is helpful in encouraging sleep.

Eases Indigestion: Angelica plant will treat the commonly occurring dyspepsia in the manner of heartburn, too much smoke, and tummy puffiness. This herbaceous plant is considered to contain flatus-relieving constituents that promote gas and puffiness alleviation. The root pulp can be used as a digestive medicine, so consuming it as recommended may be beneficial in keeping your digestive tract functioning properly.

Makes Hair Healthy: Chinese utilized Angelica root for centuries to treat hair loss problems. It is enriched with Vitamin E and helps to stimulate oxygen circulation in the body and scalp. It stimulates the metabolism and replenishes the body's nutrient supply, which is essential for hair growth. It helps oxidize the blood, which ultimately helps in hair cell oxidation. It also promotes the regeneration of damaged cells in the hair.

Treat Anxiety: Anxiety can sometimes haunt the problem, and must ,therefore, be treated without any other alternative. Use of angelica herb that will help relieve tension in the brain and reduce anxiety can be properly addressed. To get desired results, prepare some tea using angelica herb and drink this tea regularly.

Reduces Blood Pressure: High blood pressure can be cured easily by regular intake of herb Angelica. Using this herb, and having normal blood pressure, is a very traditional process. You can see the results in a few days ' time.

Use in Bath: Angelica can be used in bathing and it really helps to naturally soothe the skin. A bath can be prepared in warm water by adding 2 cups of Angelica infusion. Even coconut oil can be used to get the best results. Using this bath soothes inflammation, softens the skin and helps with eczema care.

Face Wash: Angelica, the plant or herb can be used to treat multiple skin conditions as a face wash to make the skin look smoother, cleaner and safer. Mix 2 cups of aloe vera juice with 8 - 10 drops of angelica root tincture to make a face wash. Use the mix for treating acne after the two ingredients are mixed together. It may also be used for the purpose of washing. The use of it makes the skin heal more quickly by making it easier for the skin to absorb the useful nutrients. If the facial cleanser is used twice a day, morning and night, the results will be beautiful. Not only does it eliminate acne but it also prevents them from returning.

Fungal Infection: Fungal infection may get worse and must ,therefore, be treated with herbal remedies. Angelica herb will prove worthwhile in reducing infections with the best antifungal properties.

Encourages Sweating: Angelica herb is expected to have sudorific components, meaning it facilitates sweating, which may promote the avoidance of waste in the skin.

Acts as an appetite stimulant: Tea made from the leaves and roots that have been desiccated can promote appetite recovery. As a consequence, if you abstain from eating food due to an appetite deficit, try consuming angelica tea on a daily basis. By restoring digestion, it can certainly facilitate rejuvenating your missing appetite.

Muscle Ache: The use of angelica herb can cure muscle aches. The oil extracted from the herb of angelica is very perfect for releasing stress from the muscles and thus reducing the ache. Simply apply the oil to the muscles and simply apply and massage.

Precaution

- It affects the Nervous system and may cause irritation.
- It should not be given to patients who have a tendency towards diabetes, as it causes an increase of sugar in the urine.
- Angelica should not be used by pregnant women or nursing mothers.
- Avoid using angelica root concurrently with warfarin.
- All members of this genus contain furocoumarins, which increase skin sensitivity to sunlight and may cause dermatitis.
- Juice should not be allowed to come into contact with the eyes.
- Both angelica and the root oil have been reported to cause photodermatitis and phototoxicity, respectively, following external contact.
- It is thus advisable to avoid sunbathing and intensive UV radiation for the duration of treatment with Angelica or its preparations.
- It should not be given to children under two years.

- Herb should not be used internally by people with stomach/ intestinal ulcers and diabetes.
- Skin contact with the fresh plant may cause hypersensitivity (allergy) in some people.
- Women who have breast cancers or any other cancers related to hormones, should also not consume these supplements.

3.3 ANISE (PIMPINELLA ANISUM)

Star anise is a well-known digestive aid used to relieve cramps, indigestion, bloating gas and indigestion. Star anise seeds are an abundant source of B - complex vitamins such as thiamine, pyridoxine, niacin, and riboflavin and are known to increase neurochemicals in the brain. It is filled with vitamins such as vitamin A and vitamin C which are anti-oxidant. Some of Star Anise's health benefits are listed below;

Fights against free radicals: Antioxidants fight free radicals which can cause cell damage and can be responsible for diseases such as diabetes, heart problems or cancer. As a metabolic by-product it is continually produced in our body. Only by consuming a rich source of antioxidants can the excessive production and deposition be neutralized. Studies from India and abroad have shown star anise has a strong antioxidant property. The antioxidant property is due to the presence of linololol, a chemical. Vitamin A and C also contribute in large measure to this wealth.

Rheumatism and joint pain: Star anise essential oil when applied topically is a great treatment for rheumatism and back pain. It should be mixed with a carrier oil before using the essential oil and massaged into the affected areas.

Combats fungal infections: Star anise is quite helpful in dealing with the common skin problems called candidiasis, caused by a fungus called *Candida albicans*. Usually, this fungus affects surface skin, mouth, throat or even genital areas. Some Korean researchers have observed that essential Star anise oil and certain extracts of star anise possess strong antifungal properties.

Fights against bacteria: Many researchers have also studied star anise as an antibacterial agent, and its oil has been found to be effective in decreasing cough due to asthma and bronchitis due to its expectorant properties. In addition, star anise has proven effective against nearly 70 drug-resistant bacterial strains.

Effective in flu treatment: Star anise is a good source of Shikimic acid, and has a powerful antiviral property when combined with another rich antioxidant material, quercetin. These two chemicals combined helps to prevent influenza. Scientists in China and Taiwan are researching this mixture for treating bird flu.

Beneficial for sleep disorders: Star anise has mild sedative effects, which can help the nerves settle down and also ensure a good night's sleep, according to traditional usage and circumstantial evidence. If you have trouble getting to sleep, try a cup of sedative star anise tea before going to bed. It is one of the best sleeping problem therapies.

Good for Cough: Star anise tea is one of the excellent natural remediesto ease a cough and soothe your sore throat. Simply prepare the tea and drink up to three times each day.

Improved Digestion: Star anise is often prepared as a tea especially in countries in South East Asia and China. It is mainly used for treating a variety of digestive problems such as nausea, abdominal cramps, indigestion, bloating, and constipation. It is considered best to drink the tea after your meals to treat any digestive conditions you may be suffering from.

Good for Heart Health: Star anise oil helps keep blood pressure. It trims down the heart pressure. It also improves blood flow as well as keeps the heart healthy and safe, as the oil detoxifies. The oil is rejuvenating the bone, and keeping it alive.

Women's health: Traditionally, in China, star anise has been used to help women during pregnancy and to new mothers wanting to increase their milk production. It is often recommended by traditional Chinese practices for pregnant women. It's used during this important time to improve their immune system and stave off illness. Researchers have also found that star anise has an estrogenic effect because of a compound called anethole that helps to modulate female hormonal functions.

Stimulates Pancreas: The pancreas is a crucial organ for insulin control plus some hormones. Seeds from anise help to stimulate the pancreas. Together with other hormonal problems, you have decreased risk of diabetes until the pancreas is in great health.

Beneficial for Colds, Flus and Other Respiratory Ailments: If you're inclined to get sick as soon as the cold season begins, then you'll love star anise tea. It is a warm herb that has traditionally been used to help bring relief to the colds and breathing congestion that keep you down. This is a tea that fights bacteria and viruses that are at the source of cold and flu, and helps to treat asthma, cough, inflammation of the lung and bronchitis.

Precautions

- According to Japanese star anise, side effects include nausea. Vomiting, hallucinations, agonizing jitters and fast eye movement.
- It may be dangerous for use in new and expectant mothers and should be avoided except under the supervision of your doctor.
- Star anise tea should not be drunk by people with ovarian, breast or uterine cancer.

3.4 ASAFETIDA (FERULAASSA-FOETIDA)

*Ferulaassafoetida*is a perennial, herbaceous, monoecious plant of the family *Apiaceae*. With a circular mass of 30–40 cm (12–16 in) leaves it grows to 2 m (6.6 ft.) high. Asafetida is a latex gum derived from several varieties of a perennial herb, *Ferula,* which is also known as "hing". It's available as solid brick or smaller pieces on the market, as well as in tablet or powder form. Asafetida holds a prominent place in traditional medicine, along with its culinary uses. Its health benefits derive from its carminative, antiviral, antibacterial, antioxidant, anti-inflammatory, anticoagulant, antispasmodic, expectorant, sedative, diuretic, vermifuge, emmenagogue, aphrodisiac, and anti -carcinogenic effects. Among Asafetida 's popular health benefits are listed below:

Skin Protection: No matter what country you come from, bug bites are certainly annoying and scratching pain can almost drive you mad. Fortunately, if you develop an asafetida-based paste and apply it directly to the bug bite spot to improve the healing and minimize inflammation, this will also alleviate the need to itch it continuously.

Respiratory System: As an antiviral, antibiotic, anti-inflammatory, and expectorant drug, asafetida is suitable for respiratory disorders therapy. It can help eradicate excess mucus and phlegm from the system, while also neutralizing bacteria or pathogens that may be in the respiratory tracts, effectively boosting your immune system and allowing you to breathe easily, even if you are suffering from conditions such as asthma

Asthma Relief: As a strong respiratory stimulant and expectorant, asafetida helps release the phlegm and naturally relieve congestion in the lungs. It's used for asthma, whooping cough and bronchitis in conventional medicine. The volatile oil in the asafetida gum is absorbed through the lungs, which is why asthma and other lung- related conditions can be handled excellently.

Cancer Prevention: Number of researches is underway around the world on cancer prevention and treatment and as such, asafetida has received significant attention. High concentrations of antioxidant compounds, namely ferulic acid and *umbrelliprenine* contained in asafetida, have been shown to actively seek out and neutralize free radicals within the body, and have been specifically associated with a reduction in cancer cells.

Headache Relief: Frequent use of asafetida has long been relied upon to soothe headaches and its' discomfort. Asafetida can help to eliminate the tension and pain of headaches and migraines quite effectively by improving blood flow within the capillaries and by removing excess free radicals.

Lowers Blood Pressure: Asafetida is considered as one of the best natural thinners in the blood and helps to lower blood pressure. It is packed with coumarin, a compound that helps improve blood flow and thin your blood, thus preventing blood coagulation. Scientists in *Ferula* species (the source of asafetida) have been able to separate certain phytochemicals that have supportive cardiovascular effects. Research using animal subjects has shown that extract of the *Ferula* asafetida gum significantly reduces average blood pressure. The gum extract appears to contain calming compounds that influence blood pressure and cardiovascular health with optimism.

Digestive Issues: As described above, for those suffering from diarrhea, intestinal worms, IBS, indigestion and essentially any condition that affects the gastrointestinal system, asafetida is normally recommended. This may include reducing cramping, nausea, ulcers, bloating, constipation, and constant flatulence, as well as food poisoning, and more severe stomach issues like colorectal cancer, among others.

Toothaches and Earaches: Traditionally Asafetida is used for the treatment of toothaches and earaches. For toothaches, you can make a mouth rinse with asafetida and water, as well as an ear drop mixture for earaches with coconut oil. Both of these benefits are due to the effects of this useful and multipurpose resin on anti -inflammatory, analgesic and antibacterial function.

Blood Sugar Control: Asafetida is an effective and natural way of controlling blood sugar levels and maintaining normal blood sugar. For four weeks, researchers observed a blood sugar-lowering effect when asafetida extract is given to diabetic individuals at a dose of 50 milligrams per kilogram. The research concluded that this effect is due to the presence of phenolic acids, particularly ferulic acid, and asafetida tannins in the extract.

Analgesic: Although the exact mechanism of pain relief is somewhat unclear, asafetida has been prescribed for a wide range of pain levels and signs, from chronic pain from accidents or operations to children's colic pain as a natural analgesic. Asafetida is known to eliminate many of the primary pain-causing conditions, such as eliminating bloating and build - up flatulence in infants, thereby relieving their crying and pain. Regular use of asafoetidais quite beneficial for almost all symptoms of pain that arise within our body.

Reproductive Health: Progesterone is one of the most important elements for women's reproductive health and asafetida promotes this important hormone to be made. This can help to enhance the health of expectant mothers by preventing premature births, excessive bleeding, and sterility, while also reducing menstrual symptoms for women generally.

Fertility Boost: One of the ancient uses of asafetida is as a sexual stimulant, and signs of impotence or infertility in both men and women are well known to be removed. By promoting hormonal activity and improving vitality and blood flow, asafetida is mostly an easily accessible aphrodisiac and has no side effects except for better sex life.

Nervous Disorders: Frequent use of asafetida helps soothe the body and mind, making it an important treatment for paranoia, mood swings, chronic anxiety, depression and stress. This is an important health benefit of asafetida because if not properly regulated and controlled all of these conditions will inflict disaster on your overall system.

Precaution

To the body system, everything in the limit is safe. However, anything used excessively is likely to respond with an undesirable, negative reaction. Asafetida can also be safe to humans within normal limits. Consumption of approximately 15 g of resin produces no adverse effects. Anything that goes beyond this specific safe limit results in undesired effects.

- May enhance Contact Dermatitis
- Patients cured with asafetida develop swollen lips as a side effect.
- Patients suffering from asafetida herbal treatment are plagued by stomach distress, just like gas and diarrhea.

- Several patients develop skin rashes in response to asafetida use.
- Excessive use may cause Headache and Dizziness
- Asafetida is UNSAFE for babies since it could potentially cause certain blood disorders.
- It really is UNSAFE to utilize asafetida in case you are pregnant.
- There's concern that asafetida may possibly boost the risk of bleeding.
- Don't use asafetida for those who have a bleeding disorder.
- Do not use asafetida when you have epilepsy and other issues with the central nervous system that can lead to seizures or even convulsions.
- Asafetida can bother the GI tract. Don't utilize it if you have a GI infection or any other GI condition.
- There is some problem asafetida may impede control of blood pressure. Stop use should you have a problem with blood pressure.

3.5 BAY LEAF (LAURUS NOBILIS)

The bay leaves contain notable chemicals, vitamins and minerals derived from plants that provide a range of benefits. The Greeks and Romans have commonly used this herb, thinking it symbolizes wisdom, prosperity, and safety. The numerous health benefits for the bay leaf are set out below;

Treatment of Diabetes: Bay leaves have been found to be effective in the treatment of type 2 diabetes, as blood glucose, cholesterol and triglyceride levels are common. Those leaves can be powdered and eaten for 30 days to get maximum results. That will cause a drop in your body's sugar levels as well as regulate your heart 's function. This is because the antioxidants in the bay leaf allow the body to handle insulin more efficiently, making it a good choice for diabetics and insulin -resistant people.

Beneficial during Periconception Period: Bay leaves are extremely beneficial during the time of periconception (3 months before and after pregnancy), being rich in folic acid. By supplying the child with sufficient folic acid content; they avoid childhood birth defects. These leaves have the property to cause pregnancy too.

Good for Digestion: Bay leaves are good for digestion and are capable of treating digestive disorders such as heartburn and flatulence. Drinking bay leaf tea concoction with hot water can relieve digestive disorders such as constipation, acid reflux and frequent bowel movements. It contains enzymes that allow protein breakdown and thus making it an excellent spice for non - vegetarian diets. You can take 5 grams of bay leaves in case of indigestion and bloating, add a piece of ginger and boil this in 200 ml of water until $1/4^{th}$ of the water remains. Add some sugar, and consume this for comfort twice a day. It also acts as a stimulant to the appetite as you recover from illness.

Relieves Pain: The oil extracted from the bay leaf has anti-inflammatory properties that ease the pain caused by sprains, strains, arthritis, rheumatism and general pain and aches. Massaging the temples with bay leaf oil helps alleviate headaches and migraines. Besides, it enhances circulation and fosters a sense of wellbeing. You should make a poultice of ground-bay leaves and castor leaves in case of joint pain, and tie it around the inflamed joints to reduce pain and swelling. You can boil a few bay leaves in water for a few minutes in case of a headache. Strain this concoction and drink while it's still warm. To get relief from severe headaches, you can put a paste of bay leaves on your forehead.

Anti-cancer Properties: Bay leaf contains caffeic acid, quercetin, eugenol and catechins all of which have chemical-protective properties that offer resistance to various cancer types. They also contain a phytonutrient named parthenolide that has been shown to explicitly inhibit cervical cancer cell proliferation.

Treatment of Menstrual Problems: When taken orally, the bay leaves help to promote natural and daily menstruation. They also have the ability to treat discharges from the vagina.

Clean Teeth: To get sparkling white teeth, brush them with powdered bay leaves once in 3 days.

Insect Repellent: Bay leaves are a perfect repellent for insects, as they contain lauric acid. A bay leaves dish can fend off insects. A paste made from crushed leaves and a small amount of oil relieves stings and bites when topically applied.

Treatment of Kidney Problems: Bay leaves help treat kidney and even kidney stones. To this end, boil 5 grams of bay leaves in 200 ml of water until 50 ml remains. Sieve and drink this twice a day. That will stop kidney stones from forming.

Treatment of Cold and Infection: Bay leaves are effective in combating cold, fever and infectious symptoms. Boil water in case of respiratory problems, and add 2 to 3 leaves of bay. Let steam for 10 minutes. Soak a cloth in this water and put it on the chest to relieve flu, cold and cough. Bay leaf tea is effective for fever reductions. In the event of frequent sneezing, boil some leaves of the bay in 200 ml water until 1/4 is left. Strain this concoction and drink.

Induces Sleep: Eating leaves at bedtime helps to encourage proper sleep. Mix a few drops of the bay leaf extract with water for this purpose, and drink it.

Treatment of Nosebleed: Crush 2 to 3 fresh or dry bay leaves in case of a nosebleed. Boil them in 200 ml of water until 100 ml remains. Strain and drink the concoction.

Cardiovascular Benefits: The powerful bay leaf phytonutrients provide protection against cardiovascular diseases like heart attacks and strokes. They contain valuable compounds such as rutin, salicylates, caffeic acid, and phytonutrients that help improve heart health and heart function. Boil 3 grams of bay leaves and 3 - 4 grams of wild rose flowers in 300 ml water in the case of heart disease and anxiety until only 75 ml remains. Strain and drink the concoction.

Bay Leaf Skin Benefits: Being rich in vitamins, minerals and antioxidants, the bay leaf offers various benefits for your skin and has been used for this reason in herbal medicine. Bay leaf helps de-stress the skin while preventing wrinkles and skin outbreaks from occurring. Boil 5 dried bay leaves with 2 cups of water for this purpose, and keep the container sealed. Then take off the lid and let it boil for 2 minutes. Now pour the brew into a large bowl and cover with a towel over your mouth, inhale this steam. Bay leaves has excellent healing properties. Herbal practitioners frequently use it to treat snakebite, insect stings, and to relieve cuts and bruises. Bay leaf oil has antibacterial and antifungal properties that can treat skin infections. It is used topically for the healing of cuts, bruises and bites of insects.

Bay Leaf Hair Benefits: Bay leaves ‘ nutritional value makes them good for hair health. Effective in treating dandruff is a rinse made from flavored bay leaves. Bay leaf tea is considered a great hair loss remedy. Bay leaf oil is a hair tonic that effectively combats hair loss and dandruff. Boil about 50 grams of crushed bay leaves in 400 ml of water until 100 ml of water is left. Strain the water out and apply it to the hair roots. Leave on for three to four hours and wash off. That will effectively remove the head lice. Before they boil, adding Bhringraj to the bay leaves will make your hair black, lustrous and healthy.

Precautions

- The dosage of more than 1 gram per day may induce sweating and dieresis in some people.

3.6 BASIL (OCIMUMBASILICUM)

Basil is also the common name for *Ocimumbasilium*, a culinary herb. It belongs to the mint family and is also used in English speaking countries as Saint Joseph's Wort. Basil is a robust annual or short - lived perennial herb brimming about 100 cm tall with impressive phytonutrients of health benefit. Basil is a highly fragrant herb that is most commonly used in cooking as a seasoning but has also become increasingly popular for its various health benefits. Basil leaves and flowers contain several chemical compounds that are known for disease prevention and health promotion. Listed below are some of the popular health benefits of basil;

Acts as an Anti-Inflammatory: Basil is made up of potent essential oils such as eugenol, citronellol and linalool. These are enzyme-inhibiting oils that help reduce inflammation, which is at the root of most diseases such as heart disease, rheumatoid arthritis and inflammatory bowel conditions.

Fights Cancer: Basil is made up of phytochemicals that help prevent cancer naturally, including skin, liver, oral, and lung cancers that are chemically induced. Basil can increase antioxidant activity, positively alter gene expression, encourage apoptosis of cancer cells (death of harmful cells) and stop the spread of cancerous tumors. Basil extract has shown protection against cancer and associated mortality in experiments using rodents, while also selectively shielding normal tissue and cells from the negative effects of cancer treatments such as radiation or chemotherapy.

Healing Properties: Basil leaves have antiseptic properties, and are thus used to treat ulcers, cuts and wounds. They help to reduce the pain from measles, chickenpox and smallpox. To provide relief, a mixture of basil leaves juice and coconut oil can be applied on cuts and wounds. Chewing a basil leaf or adding basil leaf juice to the area stung or bitten by an insect also helps to relieve the pain and draw out the venom.

Contains Antibacterial Properties: Basil essential oils provide protection from harmful growth of the bacteria. Basil extract is shown to be helpful in the prevention of resistant bacterial strains that do not respond to antibiotic treatments. These have been tested against basil oil's antibacterial action against *E. coli*strains and other potent bacteria collected from infected patients showed that basil has been effective in counteracting bacterial strains and helping to inhibit their growth.

Fights Depression: Basil is pretty beneficial for mental or mood-related illnesses, such as depression and anxiety. Basil is also known to be an antidepressant since it can certainly affect brain function within the adrenal cortex, helping to stimulate neurotransmitters that control the hormones that are responsible for making them happy and energetic.

Eye Disorder: Basil juice is an effective remedy for sore eyes and night blindness, normally caused by a lack of vitamin A. Two drops of basil juice will work eye wonders. This calms the mind and cuts down tension. It can also protect against conjunctivitis, boils and other eye problems that are caused by washing eyes with water soaked in basil leaves mostly due to bacterial and fungal infections. Besides that, the basil leaves act as eye relaxants to soothe the stress induced by fine lines. Chewing a few basil leaves helps to avoid eye problems caused by free radicals such as degeneration of the retina, cataracts and glaucoma. Basil leaves juice is an effective treatment for sore eyes and nocturnal blindness.

Promotes Cardiovascular Health: Basil can augment thefunctioningof muscles that control the function of the blood vessel contract and relax, thereby promoting healthy blood pressure. Basil benefits include the ability to help prevent dangerous platelet aggregation, clumping blood platelets together that can create a clot within the arteries, and cause heart arrest. Basil extracts also reduce inflammation that can cause cardiovascular disease by inhibiting the release of pro-inflammatory cytokines, proteins that are secreted from cell to cell to communicate and elevate the body's immune defenses.

Supports Liver Function and Helps Detoxify the Body: Research found that when sickly rats were given basil extract over a period of five days, they reported substantial improvements in the development of detoxifying enzymes, higher antioxidant defenses and a decrease in fat accumulation in the liver that can trigger the liver disease.

Gum Disorder: Basil leaves are advantageous for the treatment of gum disorders. The dried and powdered basil leaves can be used to brushing teeth. Occasionally, it can be mixed with mustard oil and used as toothpaste. This helps maintain gum health and helps prevent all pyorrhea and other dental problems. The toothpaste may also be rubbed onto the gums to prevent irritation of the gum. Basil is also used as a mouthwash for pain denture relief. It is an effective mouth freshener and oral disinfectant and kills 99 percent of the bacteria and germs found in the mouth. Such leaves are very useful for a variety of mouth infections, such as ulcers.

Relieves Cough: Basil is a key component in cough syrups and expectorants. But rather than buying those, you can do a home remedy that works just as well. In a cup of water, apply eight basil leaves and five cloves, and simmer for 10 minutes. You should add a bit of salt to taste. Let it cool down, and then drink it to stop coughing. Gargle with water boiled with basil leaves for sore throat, due to cough. Specific basil compounds help to activate mucus, so it is also an effective treatment for other respiratory problems including bronchitis and asthma.

Helps Alkalize the Body and Improve Digestion: Basil helps balance acid inside the body and restore the correct level of pH in the body. This can also facilitate digestion and immunity by helping beneficial bacteria flourish in the gut microflora, while also reducing harmful bacteria that can cause illness. Basil has historically been used to help lower bloating and water retention, loss of appetite, stomach cramps, acid reflux, and even to kill worms or parasites in the stomach.

Prevents Acne: Basil prevents acne outbreaks and also speeds up the acne lesion healing process. The oil in fresh basil leaves helps clear skin bacteria that contribute to clogged pores, a leading cause of acne. If you have acne before, use basil juice to kill the bacteria in the areas affected. Basil may also be used to treat other skin problems such as ringworm, psoriasis and bites from insects. The fresh basil juice's anti-inflammatory properties will decrease skin inflammation and provide relief. Besides this, it can also be used to make your skin smooth, soft and clean.

Can Act as a Natural Aphrodisiac: For generations, Basil has been considered a symbol of love in Italy. The basil scent will enhance libido and anticipation, perhaps by increasing blood flow as well as energy levels, while at the same time reducing inflammation. Basil is considered the "elixir of life" in Hindu religion and practices in Ayurveda and is used to promote healthy sexual function and upbeat mood.

Helps Protect from Diabetes and Metabolic Syndrome: Basil extracts have been found to reduce circulating blood glucose levels and inflammation, which makes basil resistant against the development of diabetes as well as other types of metabolic syndrome. In addition, the benefits of basil essential oils include the ability to help reduce triglyceride and cholesterol levels, in which diabetic patients are at greater risk of developing.

Helpful in Quitting Smoking: Chain smokers can attempt to chew the basil leaves to rid themselves of their addiction. These can be more effective for suppressing the urge than nicotine patches, sweet gum or even regular candy. While the juice has a cooling effect on your mouth, you'll be distracted by chewing and the antioxidants help repair the damage caused by years of smoking.

Relief from Headache: Basil leaves are considered a great remedy for headaches. Dried basil leaf facial steam is quite helpful in lessening headaches. To this end, add a tablespoon of dried basil leaves into a large pot with 2 cups of water. Cover your head with a towel, gently lean over the pot and breathe 5 to 10 minutes in the steam. You can also apply pounded basil leaves mixed with sandalwood paste on your forehead to get relief from heat and headache as well as an overall cooling effect.

Eliminates Kidney Stones: Basil has a reinforcing effect on the functioning of the renal. Consume five to six fresh basil leaves with water on an empty stomach to improve the overall functioning of the kidneys. If you have kidney stones, blend basil juice that is freshly extracted with an equal amount of honey. Drink it every day for about five to six months, without fail. This will promote the removal of kidney stones via the urinary tract.

Insect Bite: Basil is perfect for stings and bites of insects too. A fresh juice teaspoon should be applied to the affected areas, then repeated after a few hours. In the case of bites of insects and leeches, a fresh root paste is also useful. Apply on the scorpion's sting- a paste of basil leaves and roots ground together. It will remove the toxin effects altogether. Often, crushing a few basil leaves and combining it with lemon helps in such cases too. Apply it to the sting area, and heal the problem. Drinking basil juice and also rubbing it on the sting will reduce the pain caused by the sting.

Precaution

Basil can be causing low blood sugar in some people. The parts of basil and basil oil above ground are POSSIBLY UNSAFE when taken by mouth as a medicine, long - term. These contain estragole, a chemical that may increase the risk of developing hepatic cancer.

3.7 BERGAMOT (MONARDA SPECIES)

Bergamot is actually a citrus fruit whose rind is used to extract Essential Oil from Bergamot. Citrus *Aurantium var* is the scientific name for the bergamot.

Or *Bergamia Citrus*. It is an exotic herb, and yet it also thrives in Europe. The strong aroma makes it a well-known component in many perfumes and is often used as the all-important "top note," too. Bergamot Essential Oil's health benefits could relate to its properties such as a deodorant, vulnerary, vermifuge, antibiotic, antiseptic, antispasmodic, sedative, analgesic, antidepressant, disinfectant, febrifuge, cicatrisant, and digestive.

Antidepressant &Stimulant: As with Alpha Pinene and Limonene, the components of Bergamot Oil are antidepressant and stimulating in nature. In cases of sadness, they create a sense of freshness, joy and energy, as well as depressive disorders by boosting blood circulation. In addition, they encourage hormonal secretions, thereby helping maintain adequate metabolic rates. The particular stimulating effect also stimulates the release of digestive juices, bile and insulin, thereby improving digestion, proper nutrition absorption, consumption as well as sugar decomposition and also raising blood sugar levels as a consequence. The hormonal influence of essential bergamot oil is part of the reason it's so powerful on both these physiological and mental conditions.

Weight Loss: Unlike lavender or even peppermint, bergamot is one of the few essential oils that can stimulate the endocrine system to create comfortable, pleasant sensations. Such behavior will alleviate emotional stress, a symptom generally associated with overeating or even "comfort eating." Excessive foods are typically certainly the most eaten during emotional eating times. The stress - reducing function of essential bergamot oil encourages weight reduction in that it will help combat the mental stress that can lead to over-eating as well as excessive food intake. Bergamot also promotes hormone secretion to help keep metabolic rates proper. This specific stimulus stimulates digestive juices, bile, and insulin; all of these facilitate nutrient absorption, sugar and fat decomposition, and also blood sugar reduction. Bergamot can provide a boost for slow metabolism or even digestive issues results in malnutrition and/or extreme weight gain.

Antibiotic & Disinfectant: The various parts of Bergamot 's essential oil in particular are naturally antibiotic as well as a disinfectant. We prevent germs, viruses, and also fungi from growing. In fact, we effectively stop infections, which include skin infections. When used regularly in conjunction with bathing water or perhaps in soaps (actually used extensively in skincare soaps), both the skin and hair remain protected against infection and become shiny. Furthermore, it treats liver, bladder, urinary system and kidney infections.

Relieves stress: Bergamot oil is truly relaxing–it lowers nervous tension, as well as anxiety and stress feelings. This powerful oil can easily encourage hormones such as dopamine and serotonin which create sedation and relaxation feelings. Reduce anxiety and stress by using bergamot oil in the diffuser or oil burner. The odor of oil produces a sense of ease and satisfaction.

Heart Health: Bergamot oil and its different extracts assist in maintaining sufficient metabolism levels. This helps in digestion, proper nutrition absorption, assimilation, and sugar decomposition. Blood sugar levels could be maintained throughout these features which help with overall health. It may also help people with diabetic issues as this disease should be carefully managed by maintaining levels of blood glucose.

Analgesic: The essential oil of Bergamot reduces the sensation of pain within the body. Actually, it encourages the secretion of specific hormones which reduce the sensitivity level of nerves to pain. As a consequence, in case of headaches, sprains, muscle aches or other symptoms or even illnesses involving a heavy dose of analgesic pills it is extremely helpful. This means that you can prevent the dangerous negative effects of many counter-pain medicines, which often have negative effects and can cause injury to your kidneys and liver whenever you rely on it seriously, as well as thinning your blood or causing you to lose sleep.

Blood Sugar and Cholesterol: Triglyceride levels are typically measurements of the amount of fats in your blood. Higher levels exhibit greater fat. High blood sugar readings are a sign that the body retains more glucose than it does away with it. Glucose accumulated in turn is stored as fat. Bergamot has the extra benefit of both reducing blood sugars and triglycerides. Bergamot inhibits the production of an enzyme (HMG - CoA) related to the production of cholesterol in the liver; this limits the production of LDL (Low -Density Lipoprotein) and decreases the bad LDL cholesterol in the blood. Furthermore, HMG-CoA is correlated with blood glucose levels, being the most active when blood sugar is high. Bergamot targets HMG - CoA reductase activity by lowering blood sugar.

Relaxant & Sedative: The flavonoids in Bergamot oil are also particularly excellent relaxants. We relax nerves and lower nervous tension, anxiety and stress, all of which may help cure or even cope with stress - related health problems such as sleeplessness, high blood pressure, insomnia and depression. They may also promote the action of particular hormones in the body, which induce calming feelings as well as sedation, just like dopamine and serotonin.

Digestive: As discussed above, bergamot essential oil stimulates digestive acids, enzymes and bile secretions as well as boosts them and facilitates digestion. It also synchronizes and controls the peristaltic movement of the intestines, thus increasing the digestive process and decreasing pressure towards the intestinal tract. In this way, essential bergamot oil can help eliminate constipation, make bowel movements more regular, avoiding gastrointestinal problems along with other unpleasant or even dangerous conditions, just like colorectal cancer.

Deodorant: Bergamot oil helps prevent germs from developing which may cause body odor. The refreshing and citrusy smell of bergamot oil is used both as a natural deodorant and as an air freshener. The powerful fragrance eliminates odors on the human body or maybe in a house. You can add two-three drops of bergamot oil to the deodorant you use every single day at the moment, and you can even add some oil straight to your axes. A lot of companies include bergamot oil as well as colognes within their perfumes. Try to create your own unique fragrance along with bergamot oil by simply mixing it with your favorite fragrances. Some great options include lemon, cedarwood and oils made from sandalwood.

Cicatrisant: This particular property of Bergamot oil is the reason for its extensive use in cosmetics as well as in skincare products such as beauty soaps, creams and lotions. Cicatrisant means a property or maybe an agent that will allow wounds on the skin to fade along with other marks. In addition, it helps to make the distribution of pigments as well as melanin even and uniform, allowing stains to disappear and an even, desirable skin tone. It is also considered one of the best face oils since it helps to reduce discoloration of the skin due to scarring. This particular essential oil is commonly used to remove the unattractive effects of acne, which for several years can leave visible scars and marks on the affected areas.

Cavity protection: Bergamot oil adds to the teeth poisoned. When used as a mouthwash, it removes germs from the teeth and also prevents the teeth from forming cavities. This is due to its germ-fighting properties, making it extremely successful when defending against cavities. Rub 2 - 3 drops of bergamot oil on your teeth, or even apply it to the toothpaste.

Febrifuge: A febrifuge is in fact a substance or perhaps an agent that both decreases fever and raises body temperature. There are several reasons why Bergamot is an outstanding febrifuge. First, it combats diseases that occur from viruses, bacteria, and protozoa that can cause fever, including influenza (virus), malaria (protozoa), and typhoid (typhus bacteria), just as one antibiotic. Second, it energizes the metabolic system and gland secretions, thereby providing a feeling of warmth as well as leading to additional secretion (perspiration or sweat) through the Eccrine glands (sweat glands) and sebaceous (sebum) glands, thereby raising body temperature. This could also reduce body toxicity by suddenness, and cleanse the glands and pores of any foreign pollutants that may lead to a number of skin disorders.

Protection from tetanus: Tetanus is actually a dangerous nerve ailment caused by the common bacterium toxin called *Clostridium tetani*. Bacterial spores are usually found in soil and the dust of the house. In the event that the spores enter a wound that penetrates the skin as well as extends further than oxygen can reach, they grow and produce a toxin that goes into the bloodstream. Bergamot essential oil has both disinfectant and antiseptic properties, and also the ability to kill infectious bacteria. Bergamot oil keeps your wounds from becoming badly infected and therefore protects you from contracting tetanus. Using a cotton ball, add five drops of bergamot oil to the infected area, and rub it lightly.

Vermifuge: It kills worms, and for children who have contracted worms it is a subtle and aromatic option. It may also be applied to dirty, contaminated teeth or may even be used as a mouthwash to destroy oral germs and protect the teeth from cavity growth and development. Intestinal worms can lead to malnutrition along with other serious insufficiencies that include various forms of anemia, so reducing these worms, especially in growing children, is an extremely important use of essential bergamot oil.

Antispasmodic: It relaxes both nerves and muscles, thus providing quick relief for cramps, convulsions or painful contractions of the muscles. In addition to asthma, which has similarities to a spasmodic reaction, this could even be essential for persons with long-term coughing or even respiratory conditions.

Relieves congestion: Without a doubt, Bergamot oil is an anti-congestive, meaning it prevents all coughing and respiratory problems. This oil can be helpful every time you fight a cold or even a cough. It loosens phlegm as well as mucus within the respiratory tract supporting the body by sneezing and coughing to expel this undesired buildup. This can help the body clean off the germs and toxins that form within the body whenever you're sick. Use a vaporizer or oil burner, or apply five drops of bergamot oil in your hands and cup both your mouth and nose. You can try to rub the oil on your throat and chest, too. Alternatively, bergamot oil can be added it to tea or drinking Earl Grey tea which already contains the oil.

Antiseptic & Vulnerary: The same disinfectant ,as well as antibiotic qualities, make it a great antiseptic and vulnerary agent. It not only facilitates the quick healing of wounds, cracks on the skin as well as knees, ulcers, eczema and scratching, but it also prevents wounds from becoming septic and produces lethal tetanus. It not only avoids the effects of other infections as well as eliminates them, but also keeps new infections from arising.

Moisturizing and Conditioning: Bergamot essential oil is an excellent drug that can both moisturize and strengthen the hair fibers while supplying them with a sufficient amount of nutrients. Whether it's restoring the hair's lost moisture, giving it added moisture, or maybe sealing the natural moisture in it, this oil can do everything it takes to make hair soft and smooth.

Damage Repair: The oil along with nourishing, moisturizing, soothing properties as well as relaxing the scalp is certainly able to restore severe hair damage. Therefore, essential bergamot oil could be a fantastic option to revamp your hair, which is dry and damaged.

Frizz-Free Hair: If you want straighter hair with fewer or no frizzes, bergamot oil is truly a sure bet for you. It can make hair both softer and smoother so you don't need to untangle it over and over again. In addition, it alters the structure of the curly hair fibers and in a natural way makes them straighter.

Added Sheen: Regular use of bergamot oil in your hair can also add an appealing shine to this. If you find your mane both dull and lifeless, include this particular essential oil in your regular hair care routine to make it truly brighter and more vivid.

Other Benefits: Bergamot essential oil, besides skin and muscles, can also be a tonic, which means it cleans up the respiratory, circulatory, digestive, excretory, and nervous system. It is also anti-congestive and is also used in vaporizers, especially throughout coughs and colds, to alleviate congestion and respiratory difficulties. It really works just as one expectorant loosening phlegm as well as mucus in the respiratory tracts helping the body to eliminate by natural avenues just like sneezing and productive cough, thereby reducing the total amount of material and eliminating some of the germs as well as toxins that caused the problem first.

Precaution

- Bergamot oil may induce photosensitivity (sun-sensitivity) and may even result in extreme sunburn and other complications as well.

3.8 BLACK CUMIN (NIGELLA SATIVA)

Nigella sativa, also known as nigella or kalonji, is an annual flowering plant in the Ranunculaceae family, native to South and Southwest Asia. For over 2000 years Black Cumin seed has been used as a natural remedy for various diseases in Middle Eastern folk medicine. *Nigella sativa*, commonly known as Black Cumin, Black Seed or Black Cumin Seed, is native to South Asia It has been used as a natural remedy for various diseases in Middle Eastern folk medicine for more than 2000 years. Black Cumin has taken the place among the top-ranked evidence - based herbal medicines because of its amazing healing power. Listed below are some of the popular health benefits of using black cumin seeds:

Protects the Gut: It appears that Black Cumin seed has anti-ulcer properties against Heliobacter pylori. The seed of black cumin and thymoquinone can partially protect stomach mucosa in rats (induced mucosal injury from acute alcohol). Black cumin oil has greatly reduced the severity of rat bowel damage (with necrotizing colitis).

Treatment for MRSA: Methicillin - resistant Staphylococcus aureus, or MRSA, is a type of infection that is often fatal and difficult to treat. Scientists at Lahore University of Health Sciences in Pakistan discovered that patients with MRSA responded well to the treatment of black cumin seeds. The seeds are made up of powerful anti-microbial and anti-bacterial agents that kill bacteria on contact. All that without undesirable side effects. Health experts believe that black cumin seeds can treat certain antibiotic-resistant strains including microbes that cause Aids, staphylococcal, malaria, tuberculosis, influenza, candida, and gonorrhea.

Protects the Kidneys and Prevents Kidney Stones: Black Cumin seeds are traditionally used to treat kidney stones and to prevent them. Evidence in rats suggests very potent properties of anti-kidney stone. Black cumin oil has been effective against toxicity to the kidney by gentamycin. Seed extract had kidney injury defensive action (ischemia).

Suitable for Women: As we all know, black cumin is an excellent source of iron and calcium, and is therefore extremely beneficial for lactating mothers, pregnant women and menstruating women because of their high iron and calcium requirements. In addition, cumin also boosts breast milk secretion in neonatal women due to its high thymol constituent, which increases mammary gland secretions. Preferably it is taken with honey.

Anti-Diabetic: Black Cumin seed for diabetes treatment is highly recommended among traditional medicine doctors. Seed extract helps to increase insulin release in rodents with diabetes. Besides that, Black Cumin oil has significant activity in patients with diabetes and high cholesterol.

Reduces Seizures: Regular consumption of black cumin seeds helps to minimize seizure attacks in epileptic children, according to a report published in the journal Medical Science Watch. The research involved monitoring children with epilepsy who no longer respond to conventional therapy. Researchers conclude that the seeds contain anticonvulsant agents which impressively reduce seizures with epileptics.

Protect Against Heart Disease: Black cumin seed given to rats has been able to improve heart tissue recovery following injury (ischemia/reperfusion). Seeds are useful in treating high cholesterol. Seed powder was found to reduce the total cholesterol and triglycerides when administered to people with high cholesterol. Daily seed extract use for approximately 2 months can help lower blood pressure in patients with mild hypertension (HT). In addition, black cumin seed (thymoquinone) reduced cholesterol-hardening of the arteries.

Antiviral: Black cumin seed has the ability to reduce viral load in people with hepatitis C. Black cumin seed has discovered anti-infectious laryngotracheitis virus antiviral properties. In mice, it is effective against cytomegalovirus (CMV) virus.

Weight loss: In fact, black cumin is a wonderful suppressant of appetite, which means that you feel full faster so you eat less. So, you are losing weight. It also helps to reduce spikes of blood sugar which helps to curb the cravings of food. So, regularly including black cumin is one of the best weight - loss methods.

Digestion: The seeds of *Nigella sativa* are carminative, that is to say, they support digestion. Because of its calcium and copper content, black cumin seeds offer some digestive benefits. Copper found in cumin seeds also keeps the digestive system healthy. A 2-tablespoon serving of cumin seeds contains 112 milligrams of calcium and 104 micrograms of copper—11% of your daily calcium and 12% of your daily copper requirements. So, include black cumin in your regular diet to solve all the problems associated with digestion.

Boosts the Immune System: Oral ingestion of Black Cumin seeds increases macrophage ability to kill invaders. Black seed extract increased the cytotoxicity of the Natural Killer cells to tumor cells. Black cumin seed has been able to increase IL-3 lymphocyte secretion.

Anti-fungal activity: Black cumin seed had strong anti-yeast activity. Black Cumin seed methanolic extracts have the strongest antifungal effect on Candida albicans. The defensins (Ns-D1 and Ns-D2) showed strong anti-pilot activity towards a number of phytopathogenic pillows.

Precautions

- Black cumin seeds will slow down blood clotting processes. Hence, bleeding disorders can worsen. So, if you're dealing with low blood pressure or any other heart condition, consult with your doctor before using them.
- Black cumin seeds can interfere with fertility, and should ,therefore, be prevented during pregnancy.
- These seeds increase the flow of breast milk in nursing mothers and therefore no problem should be the consumption of black cumin seeds in food amounts. Going for a safer side, breastfeeding women should refrain from eating it in larger quantities.
- Seeds of the black cumin have sedative effects. Hence, in some people, it may increase sleepiness.
- People with immune disorders should be careful with large doses, and when using larger doses should be cared for by a professional.
- Black seed can lower blood sugar levels and if you are using other products that do the same you should be careful.
- Since black seed can lower hypertension (higher blood pressure), care should be taken if your doctor has medicines for blood pressure.

3.9 MUSTARD (BRASSICA NIGRA)

The mustard plant has valuable constituents of various edible parts. The seeds are excellent mineral sources like magnesium, calcium, potassium, and phosphorus. It is a great source of dietary folate and vitamin A is a great source of essential minerals such as calcium, potassium, and phosphorus. It is a great source of dietary magnesium and fiber. The sufficient amount of vitamin content provided by mustard greens which include ample amounts of vitamin K, vitamin A, vitamin C and folate.

Prevention of cancer: Mustard plant seeds contain sufficient quantities of nutrients to cure cancers such as lung, bladder and intestinal cancer. With the help of enzymes found in mustard, Mustard circulates to form isothiocyanates. The research found these substances have anti - cancer properties, which is contained in seeds that prevent the development of cancer cells and control the creation of any cancer cells. It has ample amounts of safe, cancer-friendly phytonutrients called glucosinolates. Studies have demonstrated the anti - cancer effects of these components, which inhibit cancer cell growth and guard against its development. Seeds have chemo-preventive properties, which help restore glutathione levels and induce apoptosis induction without affecting normal healthy cells.

Treat psoriasis: Mustard seeds are effective in treating psoriasis and critical autoimmune inflammatory diseases. The statement proved its utility for the treatment of psoriasis-related inflammation and wounds. The treatment using the seeds promotes the functions of enzymes such as glutathione peroxidase, super - oxide dismutase, and catalase that promote defensive action and treatment in any disease.

Helpful for contact dermatitis: Mustard seeds in contact dermatitis are good for therapeutic healing. The studies conclude that the use of its seeds helps to treat the symptoms of contact dermatitis such as skin treatment and subdual ear swelling.

Prevent cardiovascular problems: Mustard oil is good for using cooking oil. Research has shown the outcome of heart attack patients with higher outcomes resulting in lowering the chances of heart arrest, decreasing ventricular enlargement, and chest pain.

Relieve respiratory ailments: Mustard seeds have calming treatment effects for issues with cold and flu. It acts as a decongestant, which helps to clear mucus via air passages. Its seeds are considered a warm meal and are appreciated for their treatment of disease-caused soothing fatigue. The mustard seeds are eaten for curing sinus-affected problems and using it as a mouthwash with mustard seed drink to help soothe a sore throat. This plant has warming properties that remove excess mucus content primarily inside tissues and clear it. Such warming properties are correlated with hot sensations which may cause exposure to tissues not covered in the mucus.

Treat pains and aches: The plaster made from mustard seeds helps to treat pains. Mustard contains blooming properties and it provides pain relief resulting in limb weakness, other muscle problems and rheumatism when used as plaster. In addition, mustard plaster gives hot sensations and causes painful blistering if it spreads over naked skin. One should use some thin wrapper between skin and plaster to prevent this.

Repels poison: Mustard seeds contain preventive properties that help repel toxic symptoms on the human body. The seeds concoction helps to rid the body mostly of toxins produced by drug abuse and excessive alcohol consumption.

Eliminates ringworm: Mustard seeds have antibacterial properties which are useful in the treatment of the laceration caused by ringworm. Cleansed with hot water, the treatment of paste on skin made from mustard seeds on hygienic skin helps heal the effects associated with ringworms.

Provides healing effect: Mustard plant has warm qualities that help patients with nervous disorders. This activates the cycle of treatment by activating the nerve and has a relaxing nerve result.

Treat diabetes: Mustard oil medications help to lower protein and other glucose levels. This helps lower the production of peroxides and activates good metabolism. Mustard leaves are helpful to diabetic patients. Research shows antioxidant properties of the mustard plant which supports the normalization of oxygen-free molecules symptoms and protects against the effects caused by diabetes oxidative pressure.

Menopause treatment: Mustard seeds are beneficial for women during menopause. Calcium magnesium helps promote bone health and prevent menopause-related bone damage. This helps lower bone magnesium concentration and magnesium insufficiency, and also helps to lower the risk of osteoporosis in older women.

Skin Benefits of Mustard Seeds: Mustard seeds are truly natural clean. You can include lavender or even essential rose oil in it. Use this combination to wash the face and scrub off dead skin. Mustard seeds can behave in combination with aloe vera gel as an outstanding mix to hydrate the skin. It removes all impurities on the face and feeds them from the inside. Mustard seeds produce a great carotene resource, as well as lutein. Additionally, it is an excellent vitamin A, C and K powerhouses. These types of nutrients together allow for a great antioxidant. These seeds include a fair amount of Sulfur, which is recognized for its anti-pilot properties. They help to defend against ailment of the skin.

Hair Benefits of Mustard Seeds: Mustard oil is a great source of vitamin A, obtained from mustard seeds. Vitamin A is a superb nutrient for hair growth. It is also an outstanding stimulant that leads to faster hair growth. Mustard seeds contain protein, calcium, vitamin A and E, fatty acids formed from omega-3 and omega-6. Many of these together improve hair from the inside. Strengthened hair also means lower hair loss. Mustard seeds come with fatty acids. We are known for conditioning the hair deep inside. It also provides an excellent shine for hair, as well as bounce.

Precautions

- The people who are allergic to mustard should avoid it.
- Excessive use of mustard induces gastric irritation, intestinal mucosa and stomach bleeding.
- People with health problems should consult a doctor for use.
- Avoid excessive use.

3.10 BLACK PEPPER (PIPER NIGRUM)

If you've presumed that black pepper is actually a spice that supported salt inside the dining table, it's time this spice turns into a bit more respect. In addition to being an accompaniment, black pepper delivers wonderful taste in your dishes. Mind, however, that this is simply not it, a range of health benefits have been blessed because of this aromatic plant. Black pepper's health benefits consist of relief from breathing problems, coughs, common cold, constipation, indigestion, anemia, impotence, muscle strains, dental illness, pyorrhea, diarrhea, and heart disease. Black pepper is undoubtedly the fruit of the Piperaceae family black pepper plant and is also used as a spice as well as a medicine. The chemical piperine, contained in black pepper, results in the spiciness. It is endemic to the southern Indian state of Kerala. Black pepper has been amongst the most commonly traded spices on earth since ancient times. It is not considered a seasonal plant and is thus accessible throughout the year. This specific plant-derived spice is known as a peppercorn when dried, and is then also ground in a powder to be applied on food to include flavor and spice. Additionally, pepper is used to preserve food because of its anti-bacterial properties. It is literally an adequate source of manganese, magnesium, potassium, vitamin C, vitamin K, and dietary fiber. Further, it isan excellent anti-inflammatory agent.

Good for the Stomach: Pepper boosts the secretion of hydrochloric acid inside the stomach, thus facilitating food digestion. Proper digestion is important for preventing both diarrhea, constipation and colic. Pepper also helps avoid intestinal gas development, so when combined with a person's diet, it can promote both sweating and urination, which absorbs toxins through the body. Sweating eliminates toxins and cleans the pores of the foreign bodies that might have lodged there, so it can also eliminate excessive water or even accumulation, also called edema. When it comes to urination, you can eliminate uric acid, urea, excess water and fat as 4% of the urine is made of fat. For digestion induction,an help you lose weight while increasing the body 's entire functioning and stopping numerous gastrointestinal conditions as well as colorectal cancer. Its ability to expel gas is due to the fact that black pepper is really a carminative, which forces gas out of the body in a healthy downward movement, rather than pressing the upper chest cavity and vital organs upward in a dangerous way. In addition, it prevents the formation of more gas inside the body.

Helps in digestion: The content of black pepper piperine makes it an excellent digestive agent. It energizes the taste buds to send more hydrochloric acid to the stomach. This particular acid is important for digesting proteins along with other foods inside the stomach, resulting in flatulence, indigestion, diarrhea, constipation and acidity when left undigested. The extra chlorinated acid released helps to avoid these conditions. To assist in digestion, include a spoonful of freshly ground pepper powder (with respect to the amount of servings being prepared) in your meal while cooking. It will increase the dish's flavor while keeping your stomach balanced.

Helps you lose weight: Black pepper is excellent in allowing food to be consumed properly (all nutrients extracted). The outer surface, which provides important nutrients, often energizes fat cell breakdowns. This also promotes sweating as urination and is also an effective means of removing excess water and harmful contaminants in the body. Both of these practices together help to minimize weight. Only sprinkle the pepper over the food for effective weight loss. Stop eating excessive spice, and this can cause severe adverse effects.

Skin Health: Pepper helps you heal Vitiligo, which is a skin condition that causes multiple skin areas to shed their normal pigmentation and turn white. The piperine content of pepper may be stimulating the skin to produce pigment, according to London scientists. Topical treatment with piperine along with ultraviolet light therapy is more beneficial than other stronger, more chemically based vitiligo therapy. Additionally, due to extreme ultraviolet radiation, it decreases the probability of skin cancer.

Helps Ward Off Heart Disease: Heart disease is strongly linked to abnormal levels of cholesterol in the blood. The piperine present in black pepper can naturally increase blood cholesterol levels and thus reduce the risk of heart disease. In particular, piperine has been shown to decrease LDL ("bad") cholesterol as well as triglycerides (fat in the blood) while increasing HDL cholesterol. A buildup of LDL cholesterol and triglycerides results in plaque structures within your arteries that cause heart attacks as well as strokes while HDL cholesterol is responsible for removing LDL cholesterol via arterial walls.

Respiratory Relief: Pepper is included in Ayurvedic treatments with tonics for both colds and coughs. Pepper even offers sinusitis-related rest and nasal congestion. It has an expectorant property that can help break the mucus as well as phlegm deposits inside the respiratory tract and its particular natural irritant quality allows you to discharge this loosened material from the act of sneezing or even coughing, which removes the material through the body and helps you cure any infection or disease caused by the deposition in the tract.

Black Pepper Promotes Bioavailability: Another of Black Pepper's health benefits is that it may help boost the nutritional benefits of other herbs you eat, meaning the body has the ability to metabolize a larger proportion of the food you eat.

Antibacterial Quality: The black pepper's anti-bacterial property helps you fight both infections and insect bites. Pepper included with the diet helps to keep your arteries clean by acting similarly to fiber as well as scraping excess cholesterol through the walls, thus helping to decrease atherosclerosis, which is the problem that is extremely responsible for both heart attacks and strokes.

Antioxidant Potential: A pepper-like antioxidant can easily avoid or even fix the damage caused by free radicals, thereby helping to avoid cancer, cardiovascular diseases and liver problems. Free radicals are definitely the cellular metabolism by-products that attack healthy cells and cause their DNA to mutate into cancer cells. Like black pepper, antioxidants neutralize these kinds of harmful compounds and protect the body from numerous problems, including premature signs of aging such as wrinkles, age spots, macular degeneration and memory loss.

Can give you clear skin: Apart from an effective way to help you sweat and remove all the toxins from your skin, it also acts as an excellent exfoliant. Pepper helps slough off dead skin cells whenever crushed as well as included with a face scrub, encourages circulation and allows more oxygen and nutrients to be delivered to the skin. Both the anti-bacterial and anti-inflammatory qualities help to protect the skin from infections just like acne.

Enhances Bioavailability: Black pepper works well to carry the benefits of other herbs to different regions of the body, maximizing the efficiency of the other health foods we consume. That's why adding it to foods not only makes them taste delicious but also makes them much more accessible and accessible to our system.

Natural anti-depressant: The piperine compound in black pepper boosts the brain's cognitive function and helps beat the depression. It has also been absolutely discovered that pepper consumed regularly effectively assists brain function. Combine it with your daily meal, or even eat it as salad seasoning. Pepper in any form may help you become smarter and less depressed.

Beats dandruff: Throw away those anti-dandruff shampoos, instead try pepper. Pepper is excellent for eliminating dandruff due to its antibacterial as well as anti - inflammatory qualities. In the curd cup, combine a teaspoonful of crushed black pepper. Combine well on the scalp and apply. An hour or so later,well rinse your hair off. Just don't use shampoo. The very next day wash the hair with a shampoo. Make sure you don't use excessive amounts of pepper; it can burn up your scalp.

Cognitive Impairment and Neurological Health: Piperine, one of the key components of black pepper, has been reported to minimize memory impairment and cognitive dysfunction in several studies. This organic compound appears to stimulate chemical pathways within the brain, so earlier research shows a chance for pepper to benefit Alzheimer's patients and people who are susceptible to dementia along with other age - related or free radical-related cognition dysfunctions.

Alzheimer's, Dementia & Memory Function: Piperine, for which black pepper is an excellent source, may help to promote chemical pathways within your brain and may help Alzheimer's patients as well as dementia sufferers.

Cancer-Fighting Benefits: Including black pepper in your diet may also help push cancer away. The black pepper piperine inhibits the proliferation of the cells, and battles the proliferation of rectal cancer cells, according to a 2013 survey published in "Experimental and Molecular Pathology." The research has also found that piperine exposes rectal cancer cells to species of reactive oxygen— toxic chemicals that kill cancer cells. An additional study, published in the January 2013 issue of the "Journal of Nutritional Biochemistry," reports that piperine in black pepper impedes the growth of new blood vessels within tumors and slows their growth.

Clears up a stuffy nose and relieves cough: Black pepper is an extremely efficient natural remedy for avoiding colds and coughs due to its anti-bacterial qualities. It's moist and spicy flavor also helps release the phlegm and reduces a stuffy nose. Consider sprinkling hot soup or even rasam with the freshly ground pepper. It will loosen the phlegm immediately which will help you breathe more easily.

Stimulates Appetite: Apart from providing food taste as well as promoting digestion, black pepper serves as a possible stimulant to appetite. Research has revealed that black pepper helps boost appetite by enhancing olfactory. This will make it a beautiful, easy solution for the inadequately appetizing individuals. An Ayurvedic remedy requires consuming a mixture of half a teaspoon of black pepper and one tablespoon of jaggery powder (gur) especially, for this purpose. Take this medication regularly before improvement is seen.

Peptic Ulcers: Because of its antioxidant and anti-inflammatory properties, black pepper could have beneficial effects on gastric mucosal damage as well as on peptic ulcers. Further scientific studies on this aspect of the health effects of black pepper are still underway.

Helps people with anorexia: Black pepper may stimulate digestion and taste buds. Because of this property, this is an excellent natural cure for anorexic individuals (a condition in which the patient does not eat) by helping them regain their appetite. In order to solve anorexia, adding just a little pepper to their diet will go a long way.

Asthma and Whooping Cough: Pepper is an excellent remedy for respiratory conditions as well as its potent anti-inflammatory properties, due to the properties as an expectorant.

Beneficial for joint problems: Frequent black pepper consumption benefits individuals struggling with problems, just like rheumatism and arthritis. It is really hot and actually calming and thus allows you to boost circulation. As a result, it is effective in providing relief for patients with both rheumatism and arthritis struggles. The removal of extreme uric acid through the body provides pain relief in chronic rheumatism and arthritis.

Dental Health: Frequent consumption of black pepper benefits the individuals struggling with problems just like rheumatism and arthritis. It is really warming and naturally stimulating and thus helps you to improve circulation. As a result, it is effective in providing relief for patients with both rheumatism and arthritis struggles. The release of uric acid through the body provides pain relief in chronic rheumatism and arthritis.

Blood Pressure: Black pepper has major compounds that aid our bodies to regulate blood pressure levels. In addition, it is abundant with vitamin C that helps to avoid hypertension by increasing the excretion of lead. The high quantities of essential minerals in black pepper, like potassium, calcium, and iron, are generally compounds that reduce blood pressure. The main underlying reason for high blood pressure could be associated with high sodium levels within the body. Black pepper is an insignificant amount of potassium, which is very effective in reducing the effects of higher levels of sodium inside the body.

Cooling the Body: By adding crispy spices which include black pepper in your diet, you can beat the summer heat. Since black pepper is really a hot spice, simply by raising blood circulation and also by inducing sweating that releases excessive heat from the body it will help to cool the body. Black pepper also helps to increase the temperature of your body quickly and then lower it at intervals, which eventually cools down the temperature of your inner body. You can toss a few black peppercorns to the wide selection of spicy foods from hot Indian curries to sizzling Mexican summer salads.

Helps in the detoxification of body: Black pepper is naturally diaphoretic & diuretic. It encourages all intra-body sweating and urination. Both these mechanisms play a significant role in removing both intense water and harmful toxins through the body. Daily use of black pepper as a seasoning,therefore, helps you avoid the occurrence of numerous diseases.

Skin Benefits: Black pepper has skin advantages that do not match any other spice. Vitiligo is simply a skin ailment that causes the skin in a few places to lose its pigmentation. Each time the skin loses its natural pigmentation, it turns white. There are various treatments for this particular skin condition but most include the use of toxic chemicals. Now a lot of people turn to black pepper like a drug. The high amounts of antioxidants contained in black pepper helps the skin by protecting it from signs of premature aging such as fine lines, wrinkles and dark spots. Instead of buying expensive creams and serums promising to refresh your skin, include black pepper in your daily diet and discover its remarkable effect on the skin. Black pepper is actually known to be involved in protection from skin cancer that is caused by direct contact with harmful UV radiation. Crush some black peppers, and make a scrub to remove dead skin cells and exfoliate the skin. This will help remove toxins from the skin, leaving it tender and glowing. Black pepper also helps stimulate circulation in the blood and provides more nutrients and oxygen in your skin. Its anti-inflammatory and antibacterial properties help prevent acne.

Hair Benefits: Blend together a teaspoonful of ground black pepper and lemon seeds to applyon the scalp and hair. This can revitalize the hair, making it both shiny, lustrous, and soft. Leave the mixture on and rinse with cold water for 10 to 15 minutes.

Precautions

Pepper has the potential to cause sneezing. Patients who have had abdominal surgery should not add intense pepper to their diet as it could have an unpleasant effect on their intestines. Black pepper should not be used at high concentrations, and if you show signs of a hypersensitive reaction, discontinue using it and consult a doctor. Eating large amounts of black pepper could get into the lungs and cause breathing problems that can lead to death. Black pepper is not used on the skin as it can cause redness, irritation and often burning. Neither should pregnant women consume considerable amounts of black pepper since it is known to result in miscarriages. Other than all the things just mentioned, when sprinkled on food, black pepper is relatively safe, or used in small quantities every day to season baked goods and herbal teas.

3.11 BORAGE (BORAGO OFFICINALIS)

Borage is loaded with varying amounts of nutrients, minerals, and vitamins that bring huge health benefits. This enhances the immune system and helps to relieve inflammation. The plant is equally useful as is the oil extracted from the plant both. It also has properties that are anti-inflammatory, diuretic, galactogogue and emollient.

Respiratory health

Borage leaf chewing helps provide relief from a sore throat, cold, painful sinuses and stuffy nose. The compound in this herb helps wipe out the respiratory tracts of the phlegm and mucus, and also clears the sinuses.

It also enhances the immune system by preventing pathogens and bacteria from forming in the tract.

Skin conditions: Indian borage is used to treat the different skin conditions like stings, bug bites, psoriasis and eczema. It has anti-inflammatory properties which help to reduce swelling and redness by eliminating both irritation and itchiness.

Omega 6: The essential oil extracted from the Indian borage leaves contains omega 6 fatty acids such as gamma-linoleic acid which helps to reduce arthritis by increasing the regeneration of joints which is essential for athletes or people with osteoporosis who have stress on bones and joints.

Vitamin A and C: Indian borage leaves have high ascorbic acid content which helps improve the immune system. Vitamin A and carotenoids improve vision by decreasing oxidative stress in the eyes and also preventing macular degeneration.

Prevent cancer: Research shows that Indian borage has shown positive results in breast cancer treatment but the studies are still ongoing. The antioxidants and gamma-linoleic omega-6 acids help to slow the growth of breast cancer. It also has an effect on prostatitis which is a prostate cancer predictor.

Reduce anxiety: The sedative properties of the minerals and organic compounds contained in Indian borage help reduce chronic stress or anxiety and foster peace of mind, relaxation and sound sleep.

Kidney ailments: Indian borage exhibits diuretic properties by promoting urination to flush out the toxins. It also reduces excess body fat, salt, and water. It also improves the lymphatic system and kidney function.

Reduce fever: common fever symptoms are gripping and cold. Indian borage stimulates body sweat, visible toxin and also accelerates the recovery process.

Relieves bowel syndrome: Indian borage has traditionally been used to soothe the upset stomachs and provide relief from bowel syndromes through soothing inflammation and digestion regulation.

Reduce arthritis: Borage oil contains GLA that serves as pain relief caused by rheumatoid arthritis. It's most effective when used with anti-inflammatory or traditional painkilling medicines. The evidence shows that, with borage oil treatmentfor six weeks daily, some people experience a reduction in joint pain, severity and swelling.

Precaution

- These can cause nausea vomiting and bloating.
- It may cause allergic reactions like rashes, headaches and fatigue

3.12 BURNET (SANGUISORBA MINOR AND S. OFFICINALIS)

Great Burnet is a perennial herbaceous plant that grows between 30 and 90 cm tall. The plant is found growing on siliceous soils by streams in meadows and wet grassy fields, damp shady sites in grasslands. The plant usually prefers well-drained soil and can thrive in low-nutrition soil. Though it may grow in both alkaline and acid soils. The plant has short, coarse primary root rhizomes, and dark brown.

Benefits of Great Burnet

- In Traditional Chinese Medicine it has been used to cool the blood, stop bleeding, clear heat and heal wounds.
- The root is used to avoid bloody dysentery, nosebleeds and is used to treat burns and bites of insects topically.
- Patients with eczema showed noticeable improvement when treated with a root-and petroleum-jelly ointment.
- Leaves are astringent, coolant, styptic and tonic and used to treat fevers and bleeding.
- The root is used to treat peptic ulcers, hematuria, menorrhagia, bloody stool, dysentery, diarrhea, hemorrhoids and burns.
- Great burnet is an excellent in-house treatment for all kinds of abnormal discharges including diarrhea, dysentery and leucorrhoea.
- It is used externally in the treatment of burns, scalds, sores and skin diseases.
- A decoction of the root has been used internally in both Chinese and Western herbal medicine for heavy menstrual bleeding, blood in stool and urine, bleeding hemorrhoids and uterine bleeding.
- A mouthwash for inflammation of the gum and swollen tonsils can be made from the herb and used as a herbal treatment.
- Externally, it was used as a folk remedy for nose bleeds, wounds, burns, eczema, rash, boils, and hemorrhoids.

- The powdered root may be mixed with sesame oil to treat burns, and then applied to the affected areas.
- The garden burnet 's freshly peeled roots are applied topically to mild burn injuries, which both helps soothe and heal the wound.
- The herb is used to inhibit or stop bleeding.
- In both Chinese and conventional European medicines, greater burnet is internally used to treat heavy menstrual periods as well as uterine hemorrhage.
- It is also used externally as an ointment or lotion for treating wounds, burns, hemorrhoids and eczema.
- Roots are also applied externally to heal insect bites.
- In China, roots are baked and used to increase potency.

Precautions

- Best avoided during pregnancy and breastfeeding.
- The herb may interact with the group of allopathic medications known as fluoroquinolones.
- More burnet is used internally in both Chinese and European traditional medicines to treat heavy menstrual periods as well as uterine hemorrhage.
- It should not be used in newborn babies and small children.

3.13 CARAWAY (CARUMCARVI)

Botanically known as *Carumcarvi,* caraway is really a biennial plant, that belongs to the family *Apiaceae.* A few of the health advantages of caraway seeds are outlined as under:

Help individuals sleep: Caraway Seeds may help individuals sleep with its high magnesium content, which is a mineral that is directly associated with improving quality, length, as well as sleep tranquility. Caraway seeds also help manage the metabolism in order to reduce problems with sleep and insomnia as well.

Diabetes: Preliminary research from animal-based studies raises the issue of whether even caraway can be of some benefit to people with diabetes. For example, a 2006 study on diabetic rats (published in the Journal of Ethnopharmacology) found that treating rats along with caraway extract resulted in a significant reduction in cholesterol levels.

May boost bone health: Caraway seeds include a great source of zinc, which is discovered to improve both bone mineral density and overall bone health.

Anti-fungal and anti-yeast properties: Caraway seeds are often acknowledged to have antifungal properties. This ensures that caraway seeds can be stored for long periods of time without fear of being poisoned by air exposure. Not only this, but it has also been proven by nature that caraway seeds are anti-yeast. Thus, they can be used in foods that require anti-yeast properties. They may be used in long - term storage.

Remedy for Colic: Colic is truly an undefinable disease that often affects neonates and infants alike. Cure for a colic condition is always a typical use of caraway seeds. Colic is regarded without reason as crying, irritability, being unable to latch on to the breast of the mother as well as obtaining nutrition. Colic may be cured by adding a few grains of caraway seeds to boiling milk, combining it with honey.

Cures loss of appetite: Owing to our failure to eat nutritious food, loss of appetite is a very common issue we face in the modern world. Not only that, we 're still addicted to habits just like smoking, which actually halves our appetite. A special type of tea created with caraway seeds is really a cure for bad appetite. If consumed over time, you'll progressively observe the progress in your appetite.

Cures digestive disorders: There is plenty of therapeutic qualities that caraway seeds have. Perhaps inside treating gastrointestinal as well as digestive problems, one of their key applications is. Because of its calming properties, it is also used for treating stomach ulcers.

Sexual Disability: Many people have complained about having a sexual disability. But there is no need to worry about that any more simply because caraway seeds are here to eliminate this problem. All you need to do is ground several of its seeds with some clarified butter and tamarind kernels as well. Mix well, pour it into the water and add to a glass filled with milk and honey. This can certainly remove sexual disability, and encourage you to fully enjoy your life.

Able to Reduce Cholesterol: The health effects of fiber-rich foods in lowering LDL cholesterol within the body have also been confirmed by scientific tests. Not only that fiber also helps boost HDL cholesterol levels. These functions are very important in reducing the risk that cardiovascular disease can develop.

Veterinary medicine: Caraway seeds are found in many medicines provided to prevent numerous diseases affecting animals.

Itches, Boils and Rashes: You may also take care of itches, boils and rashes with Caraway seeds. All you need to do is make a paste of carom seeds together with lukewarm water and apply all of this to the affected area. You may also boil a few caraway seeds, and use the ajwain water to carefully wash the areas when the water reaches room temperature. Definitely, you will find a solution to the problem and also get rid of the pimples.

Relieves heartburn: Acidity or even heartburn may have been commonly mistaken for the symptoms of a heart attack. You feel the intense trouble of burps or hiccups inside your chest as well as the throat area. The next time you experience this type of heartburn, taking a number of caraway seeds is the easiest step you can take. Either add it in boiling water, eat this particular concoction, or even mix a spoonful of it with a bowl of curd and consume it. It can also be consumed along with peppermint oil to treat gastrointestinal disorders.

Cures Nausea and Vomiting: Nausea and vomiting are both painful symptoms of gastrointestinal infection and other bodily disorders. One simple way to reduce yourself from these symptoms is to eat caraway seeds. You'll be shocked at the relaxation you'll feel.

Menstrual Cramp and Stomach Problems: Both menstrual cramps and stomach problems are very common among ladies. All you need to do is boil a couple of caraway seed oils and wait around to get to room temperature. You can apply this after it has and massage it all over your stomach area. In the hour, or so, you will find rest from your cramps. Another alternative is always grinding a few caraway seed oils, combining them with lime juice, adding lukewarm water and drinking twice daily. You 're certainly going to be getting great results.

Stimulates appetite: Bad appetite is a very common problem for teenagers and young children alike. Because of their bad appetite parents definitely have a hard time getting children to eat the right nutrient-laden food. Ingested caraway seeds in the right amount are incredibly beneficial in boosting your appetite.

Beneficial for Dry and Itchy Scalp: Individuals who suffer from both dry and itchy scalp or even poor - quality hair should also include some caraway seed with their diet... Now take one of the slices and rub it all over your scalp. Massage over the next 10 minutes. Clean, and wash off with herbal shampoo afterward. You may also be using a caraway seeds pack to combat any kind of itchiness. You'll need to make a banana paste as well as caraway seed paste and spread it all around your scalp. This should be finished, at least once a week.

Good for your heart: Caraway fortifies the heart muscles, reduces blood pressure (in addition to the level of cholesterol in your blood), and helps prevent the veins from hardening. Among Caraway's most known qualities are its cardiac advantages. Just about anyone who is eating bread along with caraway knows it's good for his heart health. What's more, it tunes your heart like a tonic.

Loaded with Antioxidants: Caraway seeds are packed with potent antioxidant flavonoids that include carotene, lutein, zeaxanthin, and cryptoxanthin. Studies have found these compounds great in eliminating harmful free - radicals and in protecting the body from tumors, premature aging and infections.

Improves overall health: Overall health has been improved as it is called strengthening the immune system by eating caraway. The immune system is that mechanism that resists both the body and the systems from catching infections and diseases. The immune system produces infection-fighting components that help you recover faster if you get any disease.

Antiseptic and Disinfectant: Caraway seed is a great disinfectant as well as an antiseptic that can manage numerous infections ranging from skin irritations to infections of body organs such as intestinal, colon, respiratory, excretory and also urinary systems. Additionally, the caraway seed is stated to be effective in preventing the development of fungi and bacteria, thereby preventing the development of ulcers and sores. It may also protect people with tetanus wounds.

Cancer: Caraway seeds include a chemical compound known as 'limonene,' which acts against cancer and stops breast, liver, lung, and stomach cancer from developing.

Colds/Congestion: Simply since caraway is really a mild antihistamine and also has anti-microbial properties, it will help relax the muscles that cause coughing problems.

Anti-inflammatory Property: The anti-inflammatory properties of caraway seeds help control any form of inflammation and related difficulties with inflammation. Especially treating any inflammation within the intestine part of our body is extremely efficient.

Relieve Constipation: Improvements in caraway seed safety may also be contributing to their high dietary fiber content. Every 100 g of seeds provides you with 38 g of fiber, which is a great remedy for curing constipation.

It stops endless coughs: Histamine is probably the reason behind the most persistent coughs, and often the sufferer may find it very difficult to get the right medication to alleviate the symptoms. Caraway neutralizes histamine 's effects as good as most medicines, but with absolutely no side effects. Additionally, it reduces any other type of cough and provides relief in bronchial, throat and nose inflammation.

Anti-Spasmodic: Caraway seed can bring relief to all sorts of spasm-related health conditions. It may immediately decrease the signs of muscle cramps ,as well as muscle, pulls as it shuts down the effects of compounds and enzymes that are responsible for muscle contraction. It can also help to reduce respiratory-related spasms while healing hiccups, breathing problems, and cough. Caraway seed as an expectorant will definitely reduce cold-and cough-related blockages. It can deal with the common cold associated with an inflamed nasal tract, pharynx, larynx, bronchi and throat.

It is good for the wounds: As an antiseptic, it prevents bacteria and fungi from developing and thus avoids wound infection. Additionally, it successfully cures many internal infections.

Caraway promotes urination: First of all, people who wish to lose weight are incredibly therapist. These increase both the amount and quantity of urination, thereby helping to adequately cleanse the bladder, minimize fat and keep the urinary tract from infections.

Increases lactation: Caraway is second for breastfeeding mothers after Fennel aid. This increases milk quantity. Furthermore, if a mother uses caraway, her baby will have no flatulence and indigestion issues and will get the majority of the plant 's advantages.

Colic Relief: Crush the caraway seed by an ounce. Pour a quarter of distilled water into the seeds and make sitting for 6 hours. Possibly sweetened with honey. Give 4 times a day, one or two teaspoons if needed to the affected person. This treatment is known as healthy for all and is particularly beneficial when it comes to dealing with children.

Cure for obesity: Caraway was commonly used for treating obesity. The study was conducted via triple-blind, randomized and placebo-controlled clinical trials to examine the weight-lowering effects of Caraway extract on overweight and obese women. Participants were given CE or placebo 30 mL/day without altering their physical activity or diet. The treatment groups indicate significant weight loss, body fat percentage, body mass index, and waist - to - hip ratio. No improvements in urine-specific gravity, lipid profile and blood pressure were noted. Results show that dietary CE with no cap in food intake and when combined with exercise helps to manage female obesity.

Assist in losing weight: Caraway seeds help the participants lose weight, body fat and BMI. The study was carried out on 70 female volunteers, who were obese or overweight. For 90 days the women were treated with caraway seed extract. There were no changes in both exercise and diet. Those with caraway extract lost more than 6 pounds at the conclusion of the trial. They experienced a significant percentage improvement in body fat, BMI and muscle mass increase.

Precautions

Since caraway can potentially decrease blood sugar, people with diabetes need to be careful if they eat large amounts of caraway. Given the potential of caraway to affect levels of blood sugar, it is also important to prevent caraway eating for about a few weeks just before surgery. Additionally, ladies who are pregnant or even breastfeeding need to stop caraway oil. It is important to remember that the safety of supplements has not been examined and that the dietary supplements are also largely unregulated. In some cases, the drug could provide doses for each herb that differ from the specified amount. The commodity may in some cases be contaminated along with other substances such as metals. Additionally, it has not been proved the safety of supplements in young people and those with medical conditions or even those taking medicines.

3.14 CARDAMOM (ELETTARIACARDAMOMUM)

Cardamom is well regarded as a rich aroma and flavoring cooking spice. Along with iron and manganese, it is a good source of minerals like potassium, calcium, arsenic, and magnesium. It is also rich in many vitamins of vital importance; including riboflavin, niacin, and vitamin C. Some of the common health benefits of cardamom use are listed below

Anti-Carcinogenic Properties: Cancer, which is mostly colorectal cancer, is one of the world's leading causes of death. The rising costs of conventional cancer therapy as well as the subsequent side effects have prompted researchers to look for sustainable alternatives. Research conducted at the Chittaranjan National Cancer Institute in Kolkata by Sengupta et al. has shown that dietary cardamom has a good 48 percent counteracting colorectal cancer. So, even cardamom in your regular diet is pretty beneficial to fight cancer.

Good for Cardiovascular Health: For cardiovascular conditions, cardamom administration resulted in a lower heartbeat or regulated rhythm, and control of hypertension.

Blood Pressure: Cardamom is pretty beneficial in lowering your blood pressure, which is key to maintaining your heart and kidney health. Research conducted by the Aboriginal Drug Research Center at the Department of Medicine at RNT Medical College in India, and published the Indian Journal of Biochemistry & Biophysics examining 20 newly diagnosed individuals with primary stage 1 hypertension and the effect of giving them three grams of cardamom powder daily in two divided doses for 12 weeks. The results have been remarkable. Not only did cardamom help to reduce systolic, diastolic and mean blood pressure, it also increased the total antioxidant status at the end of three months by 90 percent.

Control of Cholesterol: Micronutrients found in cardamom assist in counteracting the body's lipid surge. In experiments conducted by Dhuley at the Division of Pharmacology and Toxicology at Hindustan Antibiotics Limited in Pune, India, hepatic and cardiac cardamom antioxidant enzymes were administered to mice, which were fed a high - fat diet. The antioxidant enzymes affected cholesterol levels and helped to regulate them considerably.

Depression: Though the real science behind this spice's antidepressant properties has not yet been established, Ayurvedic practitioners are certain that cardamom tea is a perfect way to combat depression. It has a natural ingredient that helps to cleanse the body and rejuvenate the cells, which help to combat depression altogether.

Digestive System: Cardamom is a popular remedy for the treatment of digestive problems including stomach aches in Chinese and Ayurvedic medicine but there is also science to support this widespread use. Work has shown that cardamom ranks above other spices when it comes to helping the various digestive stages. Spice methanol extract is the part that helps to control gastrointestinal disorders, such as acidity, flatulence, and cramps in the stomach. Research has shown that it has gastro-protective effects, including aiding ethanol-caused stomach ulcers (found in alcoholic drinks) and aspirin.

Urinary problems: South Asians use the relieving properties of cardamom to help with the discomfort of gall and kidney stones going in. Together with banana leaf and alma juice, cardamom may act as a diuretic, soothing a variety of kidney, bladder, and urinary problems such as nephritis, burning or painful urination, and frequent urinary urination. The relief given by cardamom from uncomfortable symptoms should not be considered a cure for underlying diseases and disorders.

Antimicrobial Properties: The cardamom has been considered to have infection-fighting properties for many eras. For modern science, this means cardamom is made up of antimicrobial properties. Cardamom oils have been able to prevent the growth and spread of some of the very dangerous microbes which often cause food poisoning.

Anti-Spasmodic Properties: Older persons frequently report muscle spasms. Such spasms are caused by sudden muscle contraction. When the muscles contract, a sudden onset of pain may result. Cardamom may be used as a treatment for the prevention of muscle spasms.

Dental Diseases: In Ayurveda, cardamom is used as well as traditional Chinese medicine for many centuries to deal with dental problems. It has been noticed after the discovery of antibiotics that these antibiotics have provided mixed results, as well as side effects. Some side effects included the suppression of healthy amicable probiotic bacteria in the intestines. On the other hand, the use of spices is better because they only kill infectious organisms, not the probiotic bacteria.

Asthma: Cardamom may also provide relief to people dealing with respiratory problems such as asthma. One animal research indicated that it exhibits bronchodilatory effects, meaning it is a substance that dilates bronchi and bronchioles, decreases resistance in the respiratory airway, and increases airflow to the lungs. Essentially, cardamom has been shown to help make breathing easier, which is of course the main aim for anyone with asthma or shortness of breath.

Anti-Inflammatory Properties: Cardamom is ideal for muscle and joint pain according to Ayurveda. Recent research suggests that cardamom has properties both anti-inflammatory and anti-carcinogenic.

Detoxification: Throughout our body, the metabolic processes release toxins and free radicals that need to be nullified and flushed out to stay healthy. Conversely, these contaminants accumulated can cause many diseases, including cancer and premature aging. Cardamom is known for acting as a detoxifier. The essential oils and biochemicals found in cardamom have the effect of detoxifying.

Improved Blood Circulation: Cardamom has been used in traditional therapies, such as aromatherapy, to treat symptoms of asthma and bronchitis by improving blood circulation to the lungs. The essential oil extracted from cardamom was studied in groups of stressed individuals, and cardamom was found to be effective in promoting blood circulation in the body.

Good for Nausea and Vomiting: The cardamom has historically been used as a treatment for nausea. Cardamom is an important tonic and stimulant and is ideal for soothing nausea and vomiting sensations.

Aphrodisiac Properties: Because cardamom gives a sweet flavor, it has traditionally been considered as having aphrodisiac properties. Cardamom is not only considered an aphrodisiac; it is also believed to possess the impotence cure and premature ejaculation.

Sore Throat: Cardamom serves as a natural remedy for sore throats. All cardamom and cinnamon can be boiled in water and can be used to gaggle painful sore throats every morning.

Hiccups: Hiccups can be very distracting, especially for kids. There are several remedies prescribed in folk medicine and a person who is hiccupping . Most of the time, that may or may not work. Cardamom, however, is yet another very effective remedy. Take a few cardamom pods, and boil them in beer. One can get rid of hiccups by drinking that water.

Bad Breath: Cardamom is a very effective cure for a common problem known as halitosis or breathlessness. Just chewing on the seeds can help eliminate any bad odors that come from your mouth. For this very reason, certain chewing gums even include it as an ingredient.

Cavities: Not only can cardamom kill the bacteria that cause bad breath, it can also help prevent the development of cavity on your teeth, or even reserve cavities and decay in your tooth. It has all the advantages of washing a chewing gum but without any of the drawbacks (like stickiness). Not only can it kill bacteria in your mouth but the chewing of cardamom with its somewhat sharp yet pleasant flavor can encourage a cleaning saliva flow while the fibrous outer cover of the pod can provide a mechanical cleaning of your teeth.

Diabetes: High content of manganese in this spice makes it an excellent choice for diabetics and anyone who deals with blood sugar problems. Research has shown that the trace mineral manganese has low blood levels in people diagnosed with diabetes. It is unclear whether diabetes causes levels to fall, or whether low levels of manganese lead to diabetes development. Either way, it's a smart idea for diabetics to add extra manganese to the diet, which is why you should use this seasoning as part of your diabetic diet.

Precaution

- Cardamom is exceptionally safe when taken in regular amounts of food by mouth. No specific, potential side effects are known when it is consumed in normal amounts of food.
- If you have gallstones then therapeutic amounts of cardamom should not be taken. The seed will cause gallstone sufferers with spasmodic pain.
- The safety of the pregnant and nursing moms in medicinal amounts is unclear. If you are pregnant or breastfeeding then it is safer to have cardamom in the amount of food and not in the amount of medication.

3.15 CASSIA (CINNAMOMUM CASSIA)

Cassia's scientific name is that of *Cinnamomumcassia*. Two other botanical names, *Laurus cassia*and *Cinnamomumaromaticum*, further recognize it. Cassia essential oil has many health benefits for both body and mind–it has unique protection against severe germs and infections, among other benefits. The extraction and use of oil can be very good for one's health.

Treats Diarrhea: Diarrhea is usually a natural response to toxins that should be removed through the digestive tract. Luckily, cassia oil is unquestionably an anti-diarrhea agent; binding bowels and preventing episodes of diarrhea are competent.

Antidepressant: Cassia oil is an effective antidepressant that has the ability to bust tension, making the body feel warm and comfortable. Cassia consists of cinnamic aldehyde, an ingredient that has been studied and that is also known for behaviors and conditions caused by stress. Anxiety and stress have a bigger effect than many people think. You can affect your body system as well as wreak havoc on its day-to-day functions while long - term stress can easily kill your overall health.

Relieves Flatulence: The scientific name of Cassia is *Cinnamomumcassia*. Two other botanical names, *Laurus cassia* and *Cinnamomumaromaticum*, further recognize it. Cassia essential oil has many health benefits for both body and mind–it has special protection against serious germs and infections, among other benefits. The extraction and use of oil can be very beneficial for one's health.

Curbs Hemorrhage: Cassia oil provides you with effective astringent effects that help curb excessive bleeding or hemorrhages in any part of your body. The helpful oil effectively bonds with the blood vessels and also prevents bleeding.

Anti-emetic: There are various qualities to Cassia oil. Definitely one of the vital ones is the ability to cope with nausea as well as avoid vomiting. This particular cassia oil property makes it a major antiemetic. The spicy taste and the rich aroma both trigger freshness and repel nauseating feelings.

Combats Depression: An excellent health advantage of the medicinal oil is that cassia oil also helps to combat the adverse effects of depression. It induces warm sensations in your mind and body as well as positively improves your mood. It helps prevent depression from making you feel low.

Anti-galactogogue: This particular oil can help eliminate milk secretion, and is therefore also not recommended for mothers who are lactating.

Fights Bacteria: Both oil and pure cinnamaldehyde (the organic compound that gives cassia its flavor and odor) have been equally effective in inhibiting the development of numerous bacterial isolates. Cassia oil acts as both an anti-microbial and anti-bacterial agent that protects the body against hazardous bacterial infections.

Anti-Rheumatic and Anti-Arthritic: As a circulatory substance along with a stimulant, Cassia essential oil increases blood circulation as well as brings warmth to the joints along with other parts of the body suffering from rheumatism and arthritis, this gives you rest from these unpleasant conditions.

Checks diabetes: Cinnamon oil helps lower blood glucose levels in people with type 2 diabetes; this is actually made possible by the plentiful presence of a water-soluble polyphenol compound in Cinnamon. Cassia Cinnamon helped lower blood sugar levels, a blood lipid called triglycerides and poor cholesterol (LDL) levels.

Antimicrobial: Cassia 's essential oil prevents microbial growth, and protects against microbial infections. This property is useful in the treatment of microbial infections in the urethra, colon, kidneys, urinary tracts along with other body areas.

Strengthens Gums: The cassia oil can be used to strengthen your gums. Helps you get good dental hygiene by using this oil.

Relieves Arthritis: The active components of the compound cassia oil called cinnamaldehyde not only regulated inflammatory mediators but activated anti-inflammatory mediators in addition. Inflammation has been found to be related to almost every health issue; not only does cassia oil reduce signs of arthritis, but it also affects all processes and systems of the body.

Boosts Immune System: Cassia oil is certainly an antimicrobial and antiviral agent; it also serves as a febrifuge, battling fever-causing infections. This specific strong oil protects the urethra, colon, kidneys and urinary tract from the growth of microbes as well as infections. It also protects the body from viral diseases, and illnesses such as influenza, cough, and the common cold.

Works as an Astringent: Cassia oil prevents a sore throat, external and uterine hemorrhaging, peptic ulcers, as well as diarrhea, due to its astringent properties. It contributes to mucous membranes shrinking, or even exposed tissues; internally, it can increase blood serum discharge or even mucous secretions. The astringent qualities of cassia oil also make it the perfect oil for healthy hair and skin. It protects the skin and can be used to treat acne naturally, heal cuts as well as sores, relieve rashes naturally as well as skin irritations. Furthermore, it strengthens hair roots, gums helping to relax muscles.

Improves Circulation: By enhancing blood circulation in the body, cassia oil helps ensure that you get the right amounts of protein and oxygen to survive. Cassia oil improves blood circulation, results in a sense of warmth within the body, relieves pain, decreases inflammation- causing disease and encourages urination that helps the whole body to remove toxins. As an example, muscle aches are caused by poor circulation, excessive physical activity, stress, deficiencies in nutrition, hormonal changes and dehydration. The lower back, spine, trapezoid, as well as the legs, are probably the most common places for muscle aches. Just because cassia oil promotes circulation and works just as much as one anti-inflammatory agent, it functions as a natural treatment for muscle ache. Through circulation, cassia oil also reduces the chance of stroke as well as two life-threatening incidents from heart attacks that can be avoided with this particular powerful oil.

Antiviral: Additionally, this particular oil is great in fighting viral infections and offers sound protection against cough & cold, influenza along with other viral diseases.

Relieves digestive and respiratory disorders: Cassia oil's carminative and stomach properties help address gastrointestinal problems along with other digestive issues including gas, indigestion, diarrhea, flatulence, nausea, vomiting, and also stomach and intestine infections. You should massage your abdomen along with 2 drops of Cassia oil combined with 2 drops of Nutmeg oil along with 2 ml of avocado oil for treating digestive problems.

Febrifuge: Just like an antimicrobial and antiviral agent, Cassia essential oil battles infections that lead to fever and thus helps lower body temperature. Equally, the oil's circulatory property helps improve the body's defense mechanisms. This particular oil's warming effect provides relief from the fever too.

Stimulant: It stimulates processes of the body, digestion, secretions, discharges, circulation as well as of the central nervous system and also of the brain, thereby helping to be more active and alert.

Carminative: Cassia essential oil provides protection from gases that form inside the intestines and forces them out of the body as well.

Relieves Menstrual Symptoms: Cassia oil helps open blocked menstrual paths; thus, it reduces the cramps that are normal when menstruating and keeps your cycle regular. In addition, it actually cures all PMS cramps and menstrual symptoms, such as headaches, tiredness, muscle aches, moodiness and bloating. It is cassia oil's emmenagogue, analgesic as well as pain-relieving properties that allow it to fight off abnormal periods or painful symptoms.

Precaution

Oil from cassia remains safe for both external and internal use. It should be prevented during pregnancy, as it could cause irritation and sensitization on the skin as well as in the mucus membrane. This may also reduce milk secretion; hence, it's not recommended for lactating mothers.

3.16 CATNIP (NEPETA CATARIA)

The annual flowering also recognized as catnip, catmint, or even cats-wort actually has the scientific name *Nepeta cataria*, and although many people don't realize it, this specific plant, so often reserved for its sedative, soothing effects on animals, often offers considerable benefits to humans. Perhaps the most important health benefits of catnip are its ability to relax restless sleep, decrease anxiety and stress, relieve menstrual pain, help eliminate eating disorders, ease stomach discomfort, accelerate recovery from colds and fevers, detoxify the body, decrease inflammation throughout the body and even speed wound cure. Let's take a special look at some of the catnip 's finest health benefits.

Stress Relief: In the more controlled way, the same quality that makes catnip so attractive to cats, namely because it makes them very "happy" and sedates them, can also affect humans. Catnip can provide relief from stress and reduced long-term anxiety just like one herbal remedy when eaten, consumed with a juice or even tea, or smoked just like one herb. This could also help to lessen the secondary symptoms of long - term stress and improve the mechanisms of defense.

Eases Flatulence: During the 17th century, Catnip was widely known as a carminative. It will help both fight flatulence and bloating by eliminating the gases that are contained in the abdomen. You can use it either as an essential oil or even as tea. If you use it as an oil, then apply 3 to 4 drops as well as a sniff in a clean serving. A hot catnip tea along with ginger offers similar benefits.

Sleep Aid: Catnip has been used for hundreds, if not thousands of years, by people with insomnia, or even sleep restlessness. The sedative nature allows you to reduce the natural cycles of the body and also to induce a calm, relaxed mind. Individuals will sleep better for undisturbed, restful sleep through the night. Before going to sleep lots of people decide to eat a cup of catnip tea to ensure a refreshing sleep.

Delays Premature Aging: For ladies, premature aging is troublesome. It's recognized that consuming two cups of this tea will put your worries about this issue to rest. Catnip leaves consist of strong antioxidant material. Such antioxidant forms remove the free radicals that are exploring within your body. If left unharnessed, the harmful elements result in fine lines, wrinkles and sags-signs and symptoms of premature aging. This particular infusion of herbs helps prevent the onset of these signs, accelerates the process of repair as well as somewhat delays premature aging.

Digestive Issues: Catnip is particularly good at clearing digestive problems, especially constipation, excessive flatulence, cramping and bloating. The relaxing, anti-inflammatory effects of organic compounds from catnips can alleviate the knots as well as inflammation within your gastrointestinal system, as well as reduce tightness and discomfort.

Speeds Up Wound Healing Process: You can use a cool catnip infusion like a skin disinfectant to cleanse the affected region completely. That may help to boost the tissue repair process. It purifies the wounds, acts as an antiseptic as well as paves the way for faster healing and recovery.

Treats Nervous Disorders: Catnip-produced essential oil's nervous qualities work well to calm the nervous system, making it a perfect natural remedy for numerous nerve-related circumstances including vertigo, Parkinson's disease, and Alzheimer's disease. This fosters the nerves, stimulates their functioning and thus provides relief.

Menstrual Cramps: Catnip tea is often recommended as a replacement remedy for ladies dealing with particularly painful menstrual cramps since it can easily reduce those cramps and stresses on the body. Additionally, catnip's sedative, soothing properties can also soothe other potential menstrual symptoms, such as mood swings and depression.

Headache Reliever: Although the exact mechanism is not fully known, catnip has proven to be truly great in treating headaches, including long-term migraines. While it might help to massage the essential oil around the affected area, drinking catnip tea or rubbing a catnip leaf salve on the temples may also provide quick relief.

Natural Remedy for Dandruff: The herb's astringent as well as antiseptic qualities allow it to be utilized as a dandruff solution. Just use a cool herbal concoction in your scalp and hair, and wash it off after a while. You may also use this as a leave-in serum, in case an irritated scalp causes dandruff.

Speeds the Healing Process: In terms of colds and flu, sweating is usually one of the fastest methods to clean the body, and to obtain the toxins flushed through the system. This is especially true when it comes to fevers, once insufficient sweating before the breaks of fever just keeps those toxins as well as pathogens inside the body. Catnip causes sweating, so alternative practitioners are often suggested for treating the common cold.

Moisturizes and Conditions Hair: Catnip tea is a fantastic deep conditioner. Braid the head, and soak it in the infusion of catnip water. Instead, after you shampoo the hair, you can use it as a conditioning agent. Through conditioner, it also serves as a leave, keeping the hair moisturized as well as conditioned. However, it makes the tresses clearly soft and smooth.

Anti-Inflammatory Activity: As already mentioned, catnip's chemical constituents are particularly effective as anti-inflammatory agents. That means that catnip can be great when it comes to treating arthritis, gout, sprained knees, aching joints and hemorrhoids. For all of these cases, topical application or usual use of leaves, juice, or tea might be effective.

Eases Cramps: The herb's antispasmodic qualities offer relief from many types of cramps. It helps to reduce the tensed muscles quickly as well as effectively as it gives you instant relief. That is why a steam inhalation with some drops of necessary catnip oil is suggested along with other spasmodic respiratory problems for faster relief in bronchial asthma.

Skin Conditions: Catnip 's natural repellent quality makes it suitable for keeping bugs away from gardens whenever they are kept just like one ornamental plant, but the organic compounds within the plant make it suitable for soothing bug bites as well as relieving irritation on the skin. The use of salves or extracts to irritated or broken skin can quickly speed the recovery process and reduce inflammation.

Good for Weight Loss: The herb's diaphoretic property comes into play, once again. It causes sweating, and thus enables contaminants to be eliminated by sweat through the body. The lower the rates of toxins, the more effective the metabolism is. A higher metabolic rate will boost the process of fat burning that often helps to reduce weight.

Complete Nutrient: Although eating catnip leaves is undoubtedly the least common type of human consumption, catnip actually features a rather remarkable nutrient set, from beneficial chemicals and various organic compounds to essential acids, minerals, and vitamins that the bodies need. In other words, the plant can do a lot more than just knock a cat out.

Beneficial For Kidneys: Catnip is truly a diuretic that can be effective, and it invokes frequent urination. This, in effect, is ideal for individuals who have edema caused by renal failure. A cup of this tea facilitates urination, aiding in the excretion and removal of waste through the kidneys. This will help to recover more quickly under such conditions.

Precautions

No doubt Catnip is an emmenagogue. Unwanted effects are known to be recorded on pregnant women. Women who are both pregnant and lactating have to take recommendations from their physicians before using this herb to prevent adverse side effects. Unquestionably, Catnip is a treasure chest of benefits –for skin, hair and health. However, it is generally best to go straight after thorough medical review for alternative methods.

3.17 CAYENNE PEPPER (*CAPSICUM ANNUUM*)

Cayenne, also known as capsicum, is an exceptional spice with incredible benefits for the skin. Cayenne handles a slew of your health woes from asthma and cancer to arthritis and diseases and all it takes is a dash to reap the benefits of cayenne. Listed below are some popular health benefits of using cayenne pepper:

Helps Digestion: Cayenne pepper positively impacts the digestive system. It helps to produce saliva,which is necessary for excellent digestion and to avoid bad breathing. Regular consumption of cayenne pepper helps to stimulate the salivary glands needed to start digestion. Besides that, Cayenne pepper also stimulates the flow of the production of enzymes, which is necessary for our digestive system to work properly. It stimulates gastric juices that help digest food and toxins from the body.

Avoid Congestion: Capsaicin found in cayenne stimulates secretions that help clear the mucus from the nose and lungs by clearing the sinuses and inducing sweating. Additionally, tea combined with cayenne pepper is supposed to be good for the treatment of cold and flu conditions.

Prevents Allergies: Cayenne is an anti-inflammatory agent with the power to prevent allergies and allergy-related symptoms. An allergy to food is a measurable response to the intake of a particular food. A condition known as leaky intestine (intestinal permeability) induces food allergies or intolerances as proteins and food particles pass through the stomach and induce systemic inflammation of the body. Regular consumption of Cayenne pepper helps to easily combat allergies of all kinds.

Prevents Blood Clots: Blood clots are essentially blockages that restrict blood flow into the circulatory system in the arteries and blood vessels. Cayenne encourages fibrinolytic activity and helps in blood clot prevention. That is also why cayenne pepper is effective at preventing heart attacks. The capsaicin in cayenne pepper helps clear deposits of artery-narrowing lipids and opens arteries and blood vessels to clear clots.

Supports Weight Loss: During the meal, eating cayenne pepper produces less hunger and people eat lesser calories all day long. It also burns excess fat, because it is a booster of metabolism. Cayenne pepper advantages also include weight loss, as one of the key anti-inflammatory foods. Cayenne pepper is motivated by allergies, food sensitivities, and illnesses to alleviate pain and bloating.

Provides Detox Support: Regular use of cayenne pepper helps boost circulation and reduce acidity. By opening the capillaries and controlling blood sugar, Cayenne pepper strengthens the circulatory system; it also helps the digestive system that transfers bacteria and toxins out of the body. As well, cayenne pepper increases body temperature and increases metabolism.

Anti-Fungal Properties: Cayenne pepper is capable of killing the fungus and preventing the formation of fungal pathogens. Consumption of Cayenne is successful against 16 different strains of fungus including Candida.

Heart Disease: Cayenne pepper frequently eaten helps to stop heart attacks. Cayenne helps lower blood cholesterol levels and helps remove fibrin, which causes blood clots to form. It also decreases triglyceride concentrations.

Relieves Joint and Nerve Pain: Once applied to the skin, Cayenne has very effective pain-relieving features. This diminishes the amount of chemical that carries messages of pain into the brain. When there is less chemical that carries messages of pain to the brain, the messages of pain no longer reach the brain and you feel relief.

Treats Psoriasis: Psoriasis occurs when cells of the skin multiply too quickly, leading to patches under the skin that are lined with whitish scales on top. Scaly patches are in fact, are areas of inflammation and the production of excess skin. The topically used capsaicin cream is effective in the treatment of psoriasis. Daily use of capsaicin cream for six weeks four times a day reveals a significant reduction in scaling, thickness, redness and itching.

Precautions

- Excessive red pepper consumption can worsen symptoms of duodenal ulcers and cause gastroenteritis and damage to the kidneys.
- When consumed, cayenne peppers cause severe irritation to the mouth, tongue and throat as well as hot sensation.
- Stop rubbing of eyes with fingertips tainted with cayenne pepper. If it does, carefully wash your eyes in cold water to reduce

irritation.

- They may worsen present gastro-esophageal reflux (GER) conditions.
- During pregnancy, eating cayenne is considered safe in food. Yet pregnant women aren't compensated by taking cayenne. Cayenne passes into breast milk, so as a replacement, nursing mothers should avoid cayenne.
- Do not use capsaicin on open wounds or broken skin.

3.18 CELERY SEED (*APIUMGRAVEOLENS, VARIETY DULCE*)

It is a biennial plant that is usually grown annually. The numerous leaflets it forms bipinnate leaves 3–6 cm (1.2–2.4 in.) wide. Three types of celery such as leaf celery, pascal celery, and celeriac celery are available. It is either eaten raw or cooked as a vegetable. Celery is loaded with antioxidants such as vitamin C, vitamin K, folate, potassium and vitamin B6. It also has anti-inflammatory properties which increase blood pressure, increase cholesterol levels and prevent heart disease. The overwhelming presence of water prevents dehydration. It also improves health of the skin, liver, and cognitive.

Lower blood pressure: Celery has an organic chemical compound like phthalides which helps to reduce the blood stress hormones. This allows blood vessel expansion providing more space to move that helps to reduce pressure. It reduces the stress on the cardiovascular system, reducing the chances of stroke, heart disease and atherosclerosis.

Lowers cholesterol: Celery's daily intake helps to bring down the bad cholesterol. The presence of phthalides enhances bile juice secretion which helps to lower cholesterol levels. It generally enhances the health of the heart. The fiber eliminates bowel-shaped cholesterol from the body.

Antiseptic action: Celery seeds have the diuretic properties which help improve urination and eliminate uric acid. Celery helps with urinary problems, cystitis and bladder disorders. Also, it prevents urinary tract infection.

Eliminates joint problems: This helps people suffering from rheumatism, arthritis and gout. It has anti-inflammatory properties which decrease the pain and swell around the joints. It removes the crystals of uric acid, which cause discomfort in joint use. This improves tissue regeneration in inflamed joints.

Preventcancer: Celery has cancer counteracting elements that reduce carcinogens, such as flavonoids, phthalides, and polyacetylenes. It has coumarins that promote white blood cell activity which helps to ban cancer. The antioxidant components eliminate the body's free radicals, which lead to cancer.

Enhance immunity power: Because Celery is an excellent source of vitamin C it enhances the immune system. Daily vitamin C intake helps lower the chances of the common cold and also prevents various diseases from developing.

Assist asthma: Vitamin C prevents free radicals from doing harm. Celery also has anti-inflammatory properties which reduce the inflammatory conditions such as asthma.

Improves cardiovascular health: The rich amount of fiber, vitamin C, and organic chemicals present in the celery roots enhances cardiovascular health.

Diuretic action: Celery contains a sufficient amount of potassium and sodium that helps balance the fluid in the body. Potassium acts as a vasodilator, which lowers blood pressure as well.

Migraine headaches: Coumarins help to relieve the discomfort of migraine headaches. Research shows migraine is caused by blocking the release of Nitric Oxide in the brain.

Diabetes health: Because Celery leaves are rich in fiber, Celery leaves consumption helps to treat diabetic conditions as well as effectively manage diabetic symptoms.

Precautions

- Celery may lead to allergic reactions which may lead to fatal anaphylactic shock.
- The celery also reduces the bleeding when used in therapeutic amounts. Therefore, one with a bleeding disorder should not be using celery.
- Those with kidney ailments should avoid the celery because it might cause inflammation.

- People with low blood pressure should avoid the use of Celery.
- The celery seed should be avoided by pregnant women as it can cause uterine bleeding, muscle contractions in the uterus that may cause miscarriage.
- Those allergic to birch pollen may experience allergic reactions to the celery seed.

3.19 CHERVIL (ANTHRISCUS CEREFOLIUM)

Chervil is a small growing annual herb that grows to a height of 12 - 24 inches (300-610 mm) (25-70 cm) and a width of 6 - 12 inches (150-300 mm), (30 cm), mainly 2 ft. The plant grows best in cooler temperatures, in shady spots and lots of water this likes. Chervil has been used as an eyewash in various folk medicines to clean the skin. Chervil was also made into a tea and was consumed for blood pressure reduction. Chervil's active constituents include its volatile oil which has a Myrrh-like odor. Chervil is also a rich source of bioflavonoids that assist the body in many ways including absorption of vitamin C. As with most plants, chervil is a digestive aid. It can be used as a soothing eye wash when prepared as a tea. Listed below are a few of the most popular health benefits of using Chervil;

Strengthens the Immune System: Chervil-containing herbal infusion can also act as a tonic that builds up immune systems. In a cup of boiling water, add 1⁄2 teaspoon of chopped chervil, vervaine, and red clover to make this tonic. Let it be for five to ten minutes. The tonic thus prepared may be taken once in the morning and at night to fend off disease.

Diuretic Agent: Large numbers of women facethe problem of water retention. It happens due to kidney failure. Since chervil has diuretic properties it helps remove excess fluid from the body.

Stops Hiccups: If you are looking for an easy home remedy to relieve hiccups then chewing fresh chervil leaves that cure hiccups. Taking a teaspoon of chopped chervil was historically considered to be effective in getting rid of hiccups during the Middle Ages. The ingestion of chervil leaves dipped in vinegar can also counter this.

Relieves Cough: Chervil is considered to be one of the strong expectorants that may enable the trapped mucus to be liquefied and promote its removal from the airway. It is useful in treating successful or wet cough, as it relieves inflammation in the lungs. Successful cough is commonly associated with respiratory infections such as bronchitis which can respond to chervil use.

Promotes Kidney Health: Chervil use helps improve the functioning of the digestive organs and kidneys. It helps them to be strong, safe and vitalized.

Menstrual Cramps: Menstruation is a common phase every woman goes through in her life. In some women, this problem causes so much discomfort. These disorders include abdominal pain, cramps on the body, mood swings, irritation, etc. It is recommended that small amounts of chervil be eaten during this period to get comfort.

Lowers Bad Cholesterol: No nutrition is added to your diet by consuming processed foods, artificial sweeteners, oily foods, sweet items, meat ,etc. Such kinds of foods are responsible for major diseases that most of us face today. Higher levels of bad cholesterol are among the health problems such food causes. Chervil is one of the best options for rapidly rising high cholesterol levels. So, make sure you use this herb regularly to ward off this disease.

Relieves Gout: Chervil is said to exhibit potent anti-inflammatory activity, and may help to treat inflammatory joint conditions. Gout and rheumatism can be treated by applying a poultice made from its leaves or by taking an infusion of chervil.

Regulates blood pressure: Chervil is a great source of important minerals including calcium, manganese, iron,zinc, magnesium, and potassium. Potassium helps decrease blood pressure by neutralizing the body 's negative effects of sodium. It also helps the kidney to control the amount of body fluids, another factor that increases blood pressure.

Promotes Digestion: Approximately 100 g of dried chervil provides a significant amount of dietary fiber (around 11.3 g), which helps maintain a balanced digestive system. Adequate fiber in the diet is vital for maintaining normal bowel movement, which in turn improves digestion. Thus, adding chervil to daily meals can certainly contribute to improving digestive function.

Helps in Maintaining Heart Health: Chervil is quite conducive to the proper functioning of the heart and thus avoids the occurrence of a number of problems related to blood flow. They prevent varicose veins, hemorrhages, and similar problems, as well as low blood pressure.

Improving vision: An eye is an essential part of our body that will make the whole world appear. Now, this herb helps make your vision clearer. Nowadays we people are too inclined towards the screen of the laptop, television and smartphone, which due to the rays produced from the screens can cause a negative result for your eyes. This chervil herb must be eaten in raw or cooked form, and for vision enhancement. So, include it to get value in your regular diet.

Disinfectant: Chervil is made of antiseptic and disinfectant qualities that help to deeply cleanse the skin. If your skin has cuttings, bruises, scrapes or wounds then apply this paste. This paste does not give you a sensation of burning but is very mild on our skin. It is therefore pretty beneficial for the skin.

Antioxidant: Chervil consists of plenty of minerals and antioxidants, such as selenium, manganese, potassium, and magnesium. Such minerals along with antioxidants help your body kill free radicals and keep our body healthy and fit.

Treat Eczema: Chervil paste is used to treat various serious skin conditions, such as eczema. This paste helps you relive naturally and easily from hot, itchy, and scaly skin.

Blood production: You have to have family members or friendswithlow RBC count. Low RBC counts are hazardous to health. So, it is important to have those foods in your body that increase blood flow. Chervil herb is one of the natural ingredients which is useful in blood production.

Prevents Pimples and Acne: Chervil paste frequently used is perfect for oily skin. It prevents breakouts in the acne and pimples. Each day it gives you less skin to spot.

Lighten Stretch Marks: Pregnancy leads to stretch marks appearing on the body. You have to continually hide these markings, as they look bad. The daily use of Chervil helps in the natural and easy lightening of these abdominal marks.

Restricts hair fall: Hair fall is nowadays one of the main hair problems. This problem affects people of all age groups. The application of chervil juice over the hair and its root will be fantastic to avoid hair loss.

Heals Sunburned Skin: If your skin gets burnt with the harmful rays, apply chervil's paste to the affected area to experience its soothing and healing touch. This paste tends to naturally lighten the skin's hue.

Improves skin texture: Dark spots, wrinkles and uneven skin tone are also nowadays a major health issue. Sometimes expensive cosmetics don't produce favorable results. A chervil herb paste along with 2 - 3 lemon drops do a miracle for such problems. Blend well with them and brush over your hair. Relax, and wash away after 20 minutes. You'll get beautiful, well-textured skin.

Cleanses the excretory system: Because chervil has diuretic properties, the intake of this herb regularly purifies the kidneys and bladder by stimulating urine production. In this way, the bacteria and waste products are flushed out of the body in the excretory organs.

Precautions

- Do not pick chervil from the wild because it has similar plant morphology to poison hemlock, or fool's parsley and water dropwort.
- In some sensitive individuals, Chervil oil can cause dermatitis.

3.20 CHICORY (CICHORIUMINTYBUS)

Scientifically known as Cichoriumintybus, chicory is a woody though annual herbaceous plant belonging to the Asteraceae family with sometimes pink or white bright blue flowers. Niacin, thiamine and zinc are great sources of chicory greens. Vitamin C, vitamin A, vitamin E, vitamin B6, riboflavin, pantothenic acid, folate, iron, calcium, potassium, phosphorus, magnesium, manganese and copper are a great source of dietary fiber. Although all leaves are edible, young leaves are preferred to be eaten. Leaves mimic the plant of a dandelion. Leaves are full of flavor, and some can be bitter. Raw chicory greens contribute to insulin sensitivity. Leaves of chicory are a dietary choice for prediabetes and diabetics.

Provide relieves constipation: Chicory is an excellent source of dietary fiber that prevents the absorption of fat from foods consumed by intestines that indirectly aid in weight loss. Dietary fibers add bulk to stools that facilitate removal and provide constipation relief. New chicory offers dietary fiber from 13 percent to 23 percent. Inulin, dietary fructan assists diabetes-dependent conditions such as diarrhea.

Probiotic activity: Chicory greens have a probiotic activity, which helps the digestive health. Chicory contained in oligosaccharides upgrades digestive tract fermentation which contributes to healthy gut flora. It also preserves digestive health by having optimal functioning. In addition, the same compound also helps prevent cancer formation in the colon.

Antibacterial activity: Chicory leaves contain antibacterial properties that are effective in combating *Micrococcus luteus*, *Staphylococcus aureus*, *Salmonella typhi* and *Escherichia coli* bacteria.

Anti-inflammatory activity: To lower swelling apply the poultice prepared from crushed chicory leaves to the skin. Plant preparations are used to encourage the production of bile, to promote appetite and to protect the contraceptives.

Precaution

- No side effects are known when used in moderate quantities.
- In pregnant women, it stimulates menstruation and possibly leads to a miscarriage.
- When you are pregnant, it can cause menstruation, and probably lead to a miscarriage.
- People allergic to marigolds, ragweed and daisies should avoid it.
- Too much chicory intake could make gallstones worse so consult your doctor before adding it to your diet.

3.21 CHILI PEPPER (CAPSICUM SPECIES)

Often known from various names such as chili pepper, sweet pepper, or bell pepper, is botanically called *Capsicum annuum*. It is a perennial and herbaceous plant belonging to the Solanaceae family of nightshades that originated in Central and South America and the Caribbean.

- Use it internally for debility, cold stage of fevers, asthma, varicose and upset stomach.
- Use it externally for sprains, unbroken chilblains, neuralgia and pleurisy.
- The fruit is used in Korea to smooth blocked internal organs, warm body coldness, treat emesis, and activate the function of the stomach, scabies, stomach ache and dysentery.
- The stem is used to treat rheumatic psychroalgia.
- Leaves are useful for emesis, dysentery and scabies.

- In Peninsular Malaysia, the root decoction is used to treat gonorrhea.
- It provides relief from muscle soreness, shingles, skin irritations and rheumatism.

Precautions

- The pepper spray causes runny nose, difficulty breathing, temporary blindness and eye pain.
- Sensitive individuals should use face masks and thin hand gloves while handling chilies.
- Consume yogurt to lower the burning pain.
- Avoid eye contact of fingers tainted with chili. Rinse the eyes with cold water, to reduce irritation.
- It may worsen gastroesophageal reflux conditions.
- Spoiled chilies contain chemical compounds like aflatoxin which cause cancer of the liver, stomach and colon.

3.22 CHIVES (ALLIUM SCHOENOPRASUM)

Chives scientifically referred to as *Allium schoenoprasum* are bulbous hardy perennial plants of the genus Allium in the Liliaceae, or Lily family, grown for their onion or garlic-flavored leaves and for their attractive and edible flower heads, rather than for their bulbs as with most other Alliums. Chives are small perennial herbaceous plants that grow in clumps, probably originating in Europe and Asia. The plant grows best in well-drained soil and in full sun. The vegetable rich in vitamin K and vitamin C is extremely nutritious and delicious and is used in fresh salads, combined with egg and omelet recipes, cream cheese, as a garnish on soups and potatoes.

Digestive Issues: Chives are composed of allyl sulfides and other organic compounds that are effective in relieving digestive discomfort. In addition, chives also have antibacterial characteristics that help kill a wide range of bacteria that can contribute to gastrointestinal system disruption. Chives help increase the efficiency of your gut's nutrient uptake, ensuring you get as many nutrients out of your food as possible.

Detoxify the Body: Chives have moderate diuretic properties, and chives are a perfect way to detoxify the body, along with other free - radical scavenging and antibacterial qualities. Chives help to eliminate unwanted chemicals, fat, water and salts by promoting urination, keeping the organs systems running efficiently and filtering out any hazardous substances that could harm the entire system.

Vision Health: Chives includes specifically lutein and zeaxanthin carotenes that are directly responsible for decreasing oxidative stress in the ocular system and postponing cataract appearance in the eye. They are also helpful in preventing macular degeneration, maintaining healthy eyes well into your old age.

Heart Health: Chives contain allicin, which is one of the most important organic compounds and is closely related to reducing bad cholesterol levels in the body and improving the health of the heart. When combined with the potassium, vasodilatory effects found in Chives, allicin helps lower blood pressure levels. Besides these chives, there is also another organic quercetin compound that has been directly associated with lowering cholesterol levels and plaque in the arteries, effectively preventing atherosclerosis and reducing your risk of stroke and heart attack.

Immune System Booster: Considerable amounts of vitamin C are found in Chives, which has been shown to be effective in boosting the efficacy of the immune system by encouraging the production of white blood cells and stimulating collagen production, which is an essential component in the development of new blood vessels, cells, tissues and muscles.

Birth Defects: Folic acid is extremely important for pregnant mothers wishing to safeguard their infant's healthy development. Folic acid prevents neural tube defects in neonatal infants, and chives for careful mothers are a rich source of folic acid.

Precautions

- Taking an excessive amount of chive may cause an upset stomach.
- Prevent using chive as medication in case you are pregnant or even breast-feeding.

3.23 CICELY (MYRRHISODORATA)

Cicely is also known as Sweet Chervil, Garden Myrrh and Chervil of Spain. It is an ornamental, delicate, strongly divided and crowded umbels of small white flowers accompanied by large, shiny black, slightly curved and fluted seed heads. The plant has a strong aniseed odor.

Eases breathing: Cicely sweet helps stimulate breathing. It softens the breathing difficulties associated with bronchitis or asthma when consumed in the form of tea. For those with upper respiratory tract infections such as flu and the common cold, the tea is beneficial.

Eliminate toxins: Sweet cicely has mild diuretic properties which make it an excellent purifier for the blood. Tea made from fresh leaves promotes urine production and helps eliminate bloodstream toxins. It is effective in the treatment of gout, because of its diuretic properties.

Skin health: Sweet cicely has antimicrobial properties so that cuts and scrapes are disinfected with the poultice or decoction. The bacteria are neutralized in wounds.

Relief from indigestion: Uses sweet cicely to ease indigestion. The stomach becomes upset after consuming a heavy meal, or more spice or fat. Steep leaves in a cup of hot water, and slowly consume for better relief.

Digestive health: In addition to indigestion, fresh-leaf tea is helpful in dealing with various digestion related health issues. It is effective against ulcers in the stomach and symptoms associated with it.

Treat animal bites: Sweet cicely is used to treat dog and snake bites. Sweet cicely acts as first aid.

Treat anemia: Cicely prevents anemia. Iron is a crucial mineral necessary for the synthesis of red blood cells carrying oxygen. So, by eating the iron-rich foods and making the dishes tastier with this herb, one can improve the anemia condition.

Precautions

- Consult the health practitioner for treating health ailments.
- Avoid its use if experienced any allergic reactions.

3.24 CILANTRO/ CORIANDER (CORIANDRUM SATIVUM)

Coriander, or Cilantro (*Coriandrum sativum*) of theApiaceae family is native to Southwest Asia and North Africa and is widely grown as a spice and herb. Aside from being used as a spice, coriander offers various health benefits. Coriander has anti-inflammatory and anti-septic properties which contribute to health promotion. Vitamin K and C is an excellent source of protein. It helps relieve indigestion, skin inflammation and diarrhea.

Reduce inflammation: Coriander has anti-arthritic and anti-rheumatic properties that decrease swelling due to anemia or kidney dysfunction. This promotes urination that helps remove the excess body water. It also helps improve skin appearance and decreases discomfort as well. Cholesterol, etc.

Skin ailments: The presence in coriander of detoxifying, disinfectant, antifungal, antiseptic, and antioxidant properties helps treat skin ailments like dryness, eczema, and fungal infections.

Reduce cholesterol: The Coriander has oleic acid, linoleic acid, stearic acid, palmitic acid and ascorbic acid which helps to effectively lower cholesterol levels. It also decreases the level of bad cholesterol in the inner walls of veins and arteries that cause heart attacks, atherosclerosis and strokes. It prevents various ailments of the heart.

Aid digestion: Borneol and Linalool are found in Coriander, which helps improve digestion, hepatic function, and bowel movements that help reduce diarrhea. Coriander's fungal and microbial activity help in the treatment of diarrhea. Components such as Alpha-pinene, Limonene, Borneol, Cineole, and Beta-phelandrene give antibacterial effects. Coriander also prevents digestive disorders, diarrhea and vomiting. It is an electrically packed plant with bioactive compounds. In addition, Coriander's leaves serve as excellent aperitifs.

Reduce blood pressure: Daily intake by patients with hypertension helps to effectively reduce the blood pressure. The interaction between Cholinergic and Calcium ions helps ease blood vessel stiffness and also reduces the risk of cardiovascular conditions such as strokes and heart attacks.

Heals mouth ulcers: The component found at Coriander known as Citronelol serves as an antiseptic. It also has the healing and antimicrobial properties that prevent the condition of mouth ulcers and wounds from getting worse. It also speeds the process of healing along with freshening the air. Coriander seed chewing helps to decrease bad breath.

Prevents anemia and malnutrition: Considering it has high iron levels, it can help in curing aenemia.Its benefits which include its iron sources can help inincreasingtheoverall body strength and energy.

Anti-allergic activity: Coriander has antihistamine properties that help lower the unpleasant condition caused by hay fever and seasonal allergies. Coriander oil application can reduce allergic reactions on the skin. It may remove hives, anaphylaxis, and gland and throat swelling.

Protects from Salmonella: Salmonella is a foodborne disease. Coriander is high in dodecenal, an antibiotic. Including Coriander to the diet helps prevent certain bacteria from triggering the infection.

Healthy bones: Coriander is rich in calcium which contributes to improving bone health. Minerals like calcium and others aid in survival, bone regrowth. It also prevents bone loss related to diseases like osteoporosis. To have healthy and strong bones,coriander should be added to the diet. **Assist digestion:** By promoting peristaltic motion and digestion, Coriander assists in the secretion of digestive juices and enzymes in the stomach. It also treats the eating disorder— anorexia. The studies showed that Coriander's daily intake helps in reducing dyspepsia. Including coriander to the diet also prevents abdominal colic from worsening.

Prevents smallpox: Coriander is abundant in constituents that are antimicrobial, anti-infectious, antioxidant, and detoxifying. It also contains iron and vitamin C which helps the immune system to improve. Those properties prevent and treat smallpox in Coriander. It reduces the pain and provides a calming effect for the patients with smallpox. The studies show that in different cases, the high amount of vitamin C effectively cures smallpox.

Regulates menstrual cycle: Since Coriander is a natural stimulant, the menstrual cycles are controlled by reducing the pain associated with the woman's period.

Vision health: Coriander has antioxidants, vitamin C, vitamin A, and phosphorus that help avoid eye problems like macular degeneration and conjunctivitis. This also reduces pain and eye strain. The leaves contain beta -carotene which prevents the eye disease and also reverses the effects of vision loss in patients aging.

Diabetes: Coriander has a stimulating effect on endocrine glands that helps increase pancreatic insulin secretion which also increases blood insulin levels. It regulates the absorption and assimilation of sugar resulting in a fall in blood sugar level. The Coriander helps to reduce the risk of harmful spikes, drop blood sugar levels and promote the metabolic functions as well.

Precautions

- Coriander can lower blood sugar and blood pressure levels.
- Coriander elevates the drowsiness that some medications induce.

- Coriander increases skin sensitivity to the sunshine and UV rays.
- Allergic reactions like itching, rash, swelling, trouble breathing and severe dizziness are possible.
- Persons with bleeding disorders should stop using Coriander, as it may increase bleeding chances.
- Those who are sensitive to the Apiaceae family should avoid it.
- Before using Coriander with drugs, one should take a doctor's advice, because it may conflict with some medications.

3.25 CINNAMON (CINNAMOMUMVERUM)

Cinnamon is the bark that belongs to one group of trees of the same family. There are many associated species that are marketed as cinnamon. "True cinnamon" is native to Sri Lanka and is also more sensitive to tasting than what is often sold in the United States as cinnamon, also known as "cassia" or "Chinese cinnamon." Cinnamon's health benefits could be related to its anti-bacterial, anti-fungal, anti-microbial, astringent, and anti-clotting qualities. Cinnamon abounds with essential minerals such as manganese, iron, and calcium while having a high fiber content. The health advantages of cinnamon are the following:

Lowers blood sugar levels: For type 2 diabetics, cinnamon has been shown to regulate blood glucose levels by increasing the ability to react to insulin. Hlebowicz*et al*, 2007 demonstrated the effect of cinnamon on postprandial blood glucose, gastric emptying, and satiety in healthy subjects a study, by which individuals ate about a cup of rice pudding with and without a teaspoon of cinnamon, it partially slows the rate at which the stomach empties after intake and lower blood sugar elevations

Reduces Bad Cholesterol: Cinnamon is shown to help lower LDL cholesterol levels inside the body, also known as bad cholesterol, keeping the good cholesterol levels unchanged. That will make it an amazing all-natural high - cholesterol therapy.

Cancer Fighter: Cinnamon has been shown to be an effective cancer - fighting food. .

Antimicrobial: Importantly,cinnamon oil is probably the best antimicrobial ingredient ever made. It is truly effective against many harmful bacteria, viruses, and fungi. This will make cinnamon tea a perfect way to sterilize and disinfect your home against the cold and flu season, and cinnamon essential oils.

Anti-Fungal: Cinnamon's antimicrobial properties make it an ideal way to deal with candida, ringworm, athlete's foot , etc. Cinnamon tea is a great way to boost the natural resistance to Candida in your body.

Food Poisoning: Cinnamon's ability to kill several types of bacteria including *E. coli*, make it an excellent tool for treating food poisoning. Make a batch of tea once symptoms begin.

Odor Neutralizer: Include a small amount of cinnamon oil in your diffuser and the moist, spicy scent will chase odor away. It also kills any bacteria that may result in odor, so it doesn't just cover up foul scents.

Alertness, Mood, and Memory: Many essential oils seem to have a revitalizing effect on our memory, cognitive function, concentration and alertness. Amongst the best are peppermint, rosemary, and cinnamon. Everyone knows that both our sense of smell and memory are tightly linked together, but it seems we can use this for the good.

Antioxidant: Cinnamon is truly an essential antioxidant that beats most other spices in this category. It indicates that cinnamon protects the body against free-radicals which cause premature aging, cancer and other damage.

Weight Control: Cinnamon stimulates blood circulation, thins the blood, and manages blood sugar. This leads to fewer sugar cravings which means you eat less food and may control your calorie consumption more readily.

Tooth and Gum Health: For a good reason, cinnamon is a very common component in several chewing gums and also mouthwashes. Cinnamon's antibacterial nature helps protect gums and teeth from damage and gives the mouth some fresh breathe.

Insect Repellant: Cinnamon's ability to manage pests does not end at the microbe level. It repels many insects too. Place a few sticks of cinnamon inside your closet to help keep moths away.

Food Preservative: Since cinnamon repels pests and kills bacteria it can make an excellent preservative for food. Combine it with recipes to have a little more space in your refrigerator or even sprinkle just a little cinnamon dust in your bread bag to prevent molding. Cinnamon lined packaging is still only in the works.

Arthritis: Cinnamon has an important anti - inflammatory activity that reduces the cytokines associated with arthritic pain.

PMS and Infertility: Manganese, which is an abundance in cinnamon, decreases mood swings as well as cramps throughout PMS and may help balance hormones with cinnamon, an energetic compound in cinnamon.

Supports healthy blood clotting: It will help thin out the blood to avoid excessive platelet clumping. Just as one anti-coagulant advises patients taking prescription blood thinners not to take cinnamon in a concentrated form like supplements or extracts, it is very safe.

Fights bacteria and fungus: Cinnamon has long been used by ayurvedic medicine because of its anti-microbial properties to support the immune system and to avoid colds and flu. It was established to help stop bacteria, fungus, and also the common yeast Candida from developing.

Boosts memory and protects the brain: Chewing gum flavored with cinnamon or simply smelling the sweet spice has been discovered to boost brain activity.

Brain Function: Cinnamon stimulates brain activity and makes it an excellent brain tonic. It will help eliminate memory loss and nervous tension.

Blood circulation: Because of the presence of a blood- thinning compound within cinnamon, the spice helps you to improve blood circulation. This blood circulation helps to diminish pain considerably. Good blood circulation also ensures the supply of oxygen to the body's cells, leading to greater metabolic activity.

Pain Relief: Additionally, cinnamon is an anti-inflammatory substance; hence it helps to eliminate muscle and joint stiffness. Additionally, cinnamon is suitable for arthritis, and help in removing headaches that are brought on by colds is also recognized.

Infections: Because of its anti-fungal, anti-bacterial, antiviral, and antiseptic properties, in addition to indoor infections, it is really effective in treating outdoors. It will help wreck germs inside the gall bladder, as well as the bacteria found in staph infections.

Protects Against Heart Disease: Because of its various anti-inflammatory properties, cinnamon is extremely good at protecting the skin, as well as shielding arteries from injury and infection. The numerous fatty foods that people nowadays snack include cholesterol and unhealthy fats that are recognized to clog arteries. The plaque and toxins that eventually build up can lead to cardiovascular disease, as well as other heart threats.

Indigestion: Cinnamon is a part of several ethnic recipes. In addition, it helps with digestion besides including flavor in the food. Cinnamon is highly effective for indigestion, nausea, vomiting, stomach upset, diarrhea and flatulence. It is extremely useful in eliminating excess gas from the stomach and intestines, due to its carminative properties. It also eliminates acidity, fights diarrhea and reduces the morning sickness effects. Therefore, cinnamon is often termed a digestive tonic.

Birth control: Often, cinnamon helps control natural births. Daily use of cinnamon during childbirth slows menstruation and thus helps to keep you from conception.

Breastfeeding: Cinnamon is also thought to aid in the secretion of healthy breast milk.

Improves Colon Functioning: Cinnamon is an excellent source of protein, calcium and thhhetrace mineral manganese as well. Calcium mixture as well as fiber can enhance colon functionality. Both calcium and fiber bind to and help eliminate bile salts from the body.

Throat soother: Calm a tickly cough with a drink flavored with cinnamon: Dip cinnamon sticks for many hours in cold water, then sip.

Improves Metabolism: Cinnamon also allows it to be on the list of foods you can consume if you try to lose weight. It has the potential to revive the metabolism which will enable you to slim down more effectively, especially if you have a slow metabolism.

Antibacterial Properties: Cinnamon serves in your body as an anti-bacterial agent, which is why it is extremely useful in treating a variety of disorders. It is usually suitable for an upset stomach for this reason, since it may help clear harmful bacteria through the digestive system.

Toothache: Cinnamon is believed to ease pain caused by toothache. Just make a paste combined with one teaspoon of cinnamon powder using five teaspoons of honey. Use a bit of this paste straight on the painful tooth 2 or 3 times a day before the pain is treated.

Precautions

If consumed in large proportions cinnamon may cause;

- Premature Labor
- Liver Failure

- Blood Thinner
- Skin Irritation
- Elevated Heart Rate
- Choking Hazard
- Stomach Irritation
- Mouth Irritation
- Lowering Blood Sugar
- Plasma Cell Gingivitis
- Dizziness or Shortness of Breath
- Kidney Diseases

3.26 CLOVE (SYZYGIUMAROMATICUM)

Cloves are aromatic dried tree buds within the Myrtaceae family. Cloves have numerous health benefits, some of which include the provision of digestive aid, antimicrobial qualities, the fight against cancer, the defense of the liver, the strengthening of the immune system, the management of diabetes, the protection of bone quality and the preservation of antimutagenic qualities, as well as the battle against oral diseases and headaches, while exhibiting aphrodisiac properties, too.

Better Digestion: Cloves improve digestion by revitalizing digestive enzyme secretion. Cloves can also be useful for reducing flatulence, irritability to the gastric, dyspepsia and nausea. To ease digestive problems, cloves could be roasted, ground, and taken along with honey.

Morning sickness: It is just a great treatment for dealing with morning sickness.

Antibacterial Properties: Cloves have been tested on various human pathogens for their antibacterial properties. The clove extracts were strong enough to kill those pathogens. Also, clove extracts can be effective against the particular bacteria that transmit cholera.

Stress: It calms the senses and decreases stress in the body. Combine the cloves in water together with the basil, mint and cardamom to make a flavored tea. Take this with honey to give you relief from stress.

Hair Conditioner: If one struggles with brunette or even auburn hair, the mixture of cloves and olive oil can be used as a conditioner. It helps improve the fragrance and helps to condition the hair color.

Chemo-Preventive Properties: Given their chemo-preventive or even anti-carcinogenic qualities, cloves are of interest to the health-related community. Tests have shown that cloves in its early stages are beneficial in managing lung cancer.

Liver Protection: Cloves contain higher quantities of antioxidants, which are well suited to protect the organs from the effects of free radicals, particularly the liver. In the long run, metabolism improves free radical production and the lipid profile, thus reducing the antioxidants in the liver. Clove extracts with its hepatoprotective qualities are useful in counteracting certain effects.

Cough and breath: Both cough and bad breath are often curable by clove consumption. They are common conditions that we all face and can be handled beautifully on a regular basis by using cloves. This can be done by including them in your dishes and, at any time of day, as refreshments.

Diabetes Control: Cloves have already been used for several illnesses in numerous traditional treatments. Diabetes is one such disease. The amount of insulin produced by the body is simply not sufficient in patients dealing with diabetes or insulin is simply not created at all.

Bone Preservation: Hydro-alcoholic clove extracts consist of phenolic compounds such as eugenol and its unique derivatives, such as flavones, isoflavones, and flavonoids. In addition to increasing the tensile strength of the bones in the event of osteoporosis, these types of extracts have already been particularly useful in protecting bone strength ,density and also the bone mineral content.

Anti-Mutagenic Properties: Mutagens are the substances that trigger mutations to change the genetic structure of DNA. Biochemical compounds present in the cloves, such as phenylpropanoids have antimutagenic qualities. They were administered on mutagens treated cells plus they were able to manage the mutagenic effects to a significant rate, according to research.

Helps you relax and fight stress: Clove is excellent to help you relax because of its powerful and soothing fragrance. Eugenol–plentifully found in clove–is yet another well-known relaxing muscle and can calm even the most stressed muscles. Also regarded as a potent aphrodisiac, clove will help you to awaken your senses and put you in the mood for fun.

Boosts the Immune System: The dried clove flower bud consists of compounds that help to enhance the defense mechanisms simply by enhancing the count of white blood cells, thus enhancing late hypersensitivity.

Anti-Inflammatory Properties: Cloves have qualities which are both anti-inflammatory and pain-killing. Work on clove extracts being administered in laboratory rats' states that eugenol 's existence has reduced the inflammation caused by edema. It was also fully confirmed that eugenol has the ability to reduce pain by actually revitalizing pain receptors.

Can relieve joint pain: Laung has an effect on pain relief, and can also alleviate inflammation. The spice is recognized as spreading a hot sensation within the applied area and is also excellent in helping relax muscles. This is really a primary reason it is a great way to conquer arthritic, rheumatic and any other form of joint pain.

Cure for Oral Diseases: Cloves for gum conditions could be used just like gingivitis and periodontitis. Clove bud extracts have significantly influenced the production of oral pathogens, which are liable for many oral diseases. Because of their painkilling effects, cloves can also be used for toothache.

Can relieve acidity: Clove could be a lifesaver for those who have acidity. It not only helps to improve food absorption, it also protects the stomach and throat along with mucous, which can significantly decrease acidity symptoms. In addition, clove enhances peristalsis (the muscle contraction action to keep food off the stomach) and keeps acid from rising within your throat. Numerous other techniques are available to combat acidity.

Aphrodisiac Properties: According to Unani medicine, spices such as clove and nutmeg were thought to have aphrodisiac properties. Because of this, tests on clove and nutmeg extracts were conducted against standard drugs, and both clove and nutmeg showed positive results.

Cure for Headaches: The use of cloves could decrease headaches. Create some clove paste and mix it with a dash of rock salt. Add this to milk. That mix effectively decreases headaches.

Deters flies and mosquitoes: Cloves that have mosquito repellant properties are known. The atomizer used as an air freshener can act as a multi-purpose sprayer used to keep mosquitoes away. This can also be used as both a fly-deterrent and an ant-killer. It's recognized that a small amount of clove oil kills ants instantly.

Enhance sexual health: Were you aware of the qualities of this wonder spice that help men to achieve orgasm too soon? The fragrance is known to help increase energy levels and improve sexual libido. Obviously, Clove heats up your body and prepares you for the body of the action. Cloves have properties that help relieve sexual dysfunction.

Asthma: Cloves have definitely already been great in dealing with asthma. It can act as an expectorant once clove decoction is consumed at least three times a day.

Cholera: Cholera has been a worldwide epidemic at a number of locations. Cloves were already useful in avoiding the severe signs of this ailment. You'll need to boil about 4 gm of cloves in 3 l of water to make this decoction.

Coryza: Coryza or even mucous membrane inflammation is often curable together with cloves.

Precaution

Cloves are safe to use in food recipes for many people whenever they are consumed. When used orally, clove oil can cause severe negative effects on children. The clove oil is safe when used on the skin but could lead to severe problems with the gum, tooth and mucus membrane whenever frequently used. Breathing in smoke from cigarettes along with clove ingredients can cause breathing difficulties, upper respiratory conditions and severe lung infections. Both pregnant and breastfeeding ladies may take cloves by mouth, but they do not need to take therapeutic doses of cloves as it may harm the unborn child. Cloves are a great idea for surgery, slowing blood clotting but may cause bleeding during or even after surgery. Therefore, it is really advisable to prevent cloves being used or even eaten in at least 2 to 3 weeks just before a planned surgical treatment.

3.27 COSTMARY (TANACETUMBALSAMITA)

Costmary, also known as Costmary, Alecost, Balsam Herb, Costmarie, Bible Leaf, Alecoast, Costus, Sweet Mary is an aromatic herb belonging to the Asteraceae family that bears a yellow flowers shaped button. Usual essential oil is also known as bible leaf and herbal balsam. It has antiseptic, carminative, acidic, and astringent properties. It is used in food as a flavoring agent and helps in the protection of the skin.

Treat wounds: Usual essential oil helps with the treatment of wounds like blisters, burns and bruises. Put this oil on the area affected. This helps to speed up the process of healing. Use a few drops to treat the injury, and cover the surface of the wound.

Reduce fatigue: Put this oil on the oil burner or on the diffuser. It provides an appealing fragrant oil that helps to make one awake and alert.

Skin health: It has astringent properties; it is used as a toner to lower the skin which is prone to oily or acne. Adding a few drops to the face helps improve the texture of the skin and its appearance. It makes for a smoother texture.

Precautions

- Do not use during pregnancy.
- The patients of liver, cancer or kidney infections should avoid itsuse .
- Do not use it undiluted.
- People with health problems should consult the doctor for use.
- Avoid using it on sensitive areas like eyes and ears.

3.28 CUMIN (CUMINUMCYMINUM)

Cumin also is known as Cumin, Cummin, Roman Caraway is an annual herbaceous, slender, erect, subglabrous plant of the Apiaceae family. The found cumin varieties are White cumin seeds and Black cumin seeds.

Great source of antioxidants: The cumin contains nine types of antioxidants, including manganese, copper, zinc, selenium, vitamin E, vitamin C, lutein, vitamin B9 and beta carotene. It possesses the properties of antimicrobials and antiparasitic that are effective in curing fever. It eliminates oxidative stress and hinders illnesses. Cumin is also used to treat ailments like vitiligo and hyperglycemia.

Antimicrobial activity: The essential oil derived from the cumin fruit inhibits the antifungal activity of yeast and dermatophytes. The dosage is 5 ml. The fungus called *Trichophyton rubrum* was targeted by cumin. The phytopathogens are less obstructed to it.

Anticancer activity: The presence of preventive chemo properties helps balance carcinogen metabolism. The concentration of 0.1 0.1 μL/mLkilled the Hela cells by 79 percent. The antioxidant activity led to cytotoxic activity.

Hepatoprotective Activity: Cumin increased the decreased phospholipase level and role of phospholipids in the livers and kidneys. A study shows that cumin has the ability to lower lipid levels in oxidized oil and alcohol which also helps to convince hepatotoxicity.

Gastroprotective Activity: The cumin extract increases the secretion of acid through the cholinergic system. The extract also increases the secretion of gastric acid in the stomach along with the aspirin that has facilitated mucosal injury.

Nephroprotective Activity: Cumin seed extracts provide the protection adjacent to the gentamicin that has urged nephrotoxicity. Using 200 mg/kg of cumin extract decreased gentamicin and also fostered serum urea, lipid peroxidation, creatinine, and improved urea clearance and creatinine levels.

Antidiabetic and Antiphyperlipidemic/ Antihyperchlolesterolemic Activities: The cumin powder reduces hyperglycemia and glucosuria. The cumin also banned metabolic alterations caused by reduced blood urea and urea excretion levels, as well as creatinine. The cumin also lowered tissue and plasma cholesterol, free fatty acids, phospholipids, and triglycerides.

Antiplatelet Activity: The ether extract of turmeric and cumin restricted arachidonate that promotes the platelet aggregates. Those extracts stopped exogenous arachidonic acid from developing thromboxane B2.

Osteoprotective Activity: Cumin extract helps to improve the microarchitecture of the bone. It also prevents the ovariectomy that persuades bone loss with no anabolic impact on rat's atrophic uterus.

Antitussive Activity: The study shows that cumin extract has an antifouling effect. The guinea pigs had lowered the number of coughs when exposed to aerosols.

Precautions

Breastfeeding and pregnant women don't recommend using cumin. Cumin may also lower blood sugar and slow down blood coagulation. Cumin interacts with diabetes medications (anti-diabetes medicines). Before using cumin, one should consult with the professional physician or pharmacist.

3.29 DILL (ANETHUMGRAVEOLENS)

Dill shares the same celery and parsley families being a single species in their genus. It is native to the East Mediterranean region and Western Asia. The term dill was derived from ancient Norse word dylla which means soothing or lulling. Dill possesses large amounts of phytonutrients such as niacin, fiber, copper, phosphorus, vitamin B6, riboflavin, potassium, and magnesium that are beneficial for overall health maintenance. It also contains flavonoids like carvone vicenin, kaempferol, anethofurane, limonene, and monoterpenes. Dill has numerous health benefits, such as stress relief, appetite enhancement, diarrhea, hiccups, menstrual disorders, dysentery, cancer and breathing problems. It is used to preserve oral health and may improve the immune system, preventing bone degradation. It could be helpful for patients with arthritis, with its anti-inflammatory properties. Moreover, it might also lower the excess gas. Dill cuts down on bacteria and prevents cancer.

Useful for depression: Depression has become a major problem for teens and adults alike. Dill acts as a natural Depression Assistance. The goal of the analysis was to discover the analgesic and antidepressant properties of aqueous dill extract. Compared with tramadol and sertraline, the dill extract shows a potent analgesic and antidepressant effect. This also does not have any adverse reactions. The aqueous extract showed the benefits this receives from flavonoids, polyphenols, and tannins.

Healthy cholesterol level: Dill will bring down cholesterol. Normal dill dosage in different forms decreases the liver enzymes, blood glucose and lipid profile.

Cure epilepsy: Epilepsy is a neurological disorder characterized by episodic and unpredictable convulsions. There are several medications available to decrease symptoms such as seizures, but the drugs have numerous side effects. This plant has been used to treat epilepsy for centuries. Dill leaves' aqueous extract has been tested for the effects on epilepsy diagnosis and seizures. As the assessment showed, the plant has a medicinal reputation for deep anticonvulsant activities and is an important alternative to epilepsy treatment.

Neutralize free radicals: Dill's monoterpene effects assist antioxidant molecules in binding themselves to oxidized molecules that could damage the body. Dill exhibits analgesic and anti-inflammatory properties against free radical damage.

Promote short term infertility: Dill induces certain changes that promote short term infertility in the female reproductive system. Dill in high dose and low dose promotes length of the estrous cycle and concentration of progesterone, and promotes infertility without significant side effects on the development of potential oocytes such as chemical and structural changes.

Treat diabetes: Dill is traditionally used to treat diabetes and convulsion.

Antigiardial activity: *Giardia lamblia* is a globally known intestinal parasite of humans. Metronidazole is a drug commonly used in humans to treat giardiasis. Small risks of side effects were being sought to be used as a protective agent in children. Plants like Dill are thought to have antigiardial effects that were grown as a food spice and medicinal agent in Iraq.

Prevent the risk of atherosclerosis: It lowers the chances of postprandial atherosclerosis and is beneficial for hypercholesterolemic patients. It becomes an effective choice for the treatment of hyperlipidemia by reducing postprandial risk factors of atherosclerosis, by considering both price and availability. The active ingredient mechanism and structure should be authorized.

Precautions

- With Dill's use, some people may experience dermatitis and photosensitivity.
- It should not be used by pregnant women, as it may result in miscarriage.
- When applied directly to the skin, it could cause skin irritation, and dill juice causes extra skin sensitivity to the sun that could increase the chance of skin cancer and sunburn.
- One should be careful while applying dill topically.
- People with a carrot allergy can experience allergic reactions.
- Blood sugar could be reduced and people with diabetes can check with the doctor for use.

3.30 FENNEL (FOENICULUM VULGARE)

Fennel seed is the fruit of a flowering plant in the carrot family, *Foeniculumvulgare*-or what we commonly call fennel. Fennel is an aromatic herb that is part of the family of parsleys. It is used as a spice and has a sweet taste similar to anise. Roman soldiers are said to have used force to chew fennel seeds. It is used both as a spice for drinks and veggie preparations and as a digestive aid and refreshment after a meal. In many parts of India, even fennel leaves are used in curries. Listed below are some of the popular benefits of consuming Fennel Seeds:

Constipation: The fennel seeds can act as a laxative, mainly in powdered form. The roughage helps clear the intestines, while it's stimulating effect helps maintain the proper peristaltic motion of the intestines, thereby helping to promote proper excretion by stimulating gastric juices and bile production. Fennel is also commonly found in drugs that treat abdominal pain, diarrhea, irritable bowel syndrome (IBS), and other bowel problems.

Helps Regulate Blood Pressure: Chewing on fennel seeds helped to increase the saliva nitrite content, making it a wonderful natural way to keep blood pressure levels in check. Besides this, fennel seeds are also a very rich source of potassium that is an essential component of cells and body fluids it helps to control the heart rate and blood pressure.

Good for Constipation, Indigestion, IBS & Bloating: Tea prepared from fennel seeds is considered very useful because of the oils found in these seeds to aid indigestion, bloating and constipation. Fennel seeds contain estragole, fennel and anethole, contributing to the antispasmodic and anti-inflammatory properties of the plant. The volatile oils found in fennel seeds can help kick digestion for those with IBS by encouraging gastric enzyme production. Fennel tea is sure to help the digestive tract be healthy and happy because of its multitude of gastrointestinal benefits.

Breast Enlargement: The flavonoids present in fennel seeds help to increase the amount of estrogen and thus serve as a stimulant and tonic. Fennel seeds help boost the size of the breasts as they improve the formation of new breast cells and tissues.

Helps Prevent Cancer: In Chinese medicine, fennel has been used for centuries to treat inflammatory conditions such as insect bites or sore throat. Fennel contains oil, called anethole, which has been shown to be a natural cancer remedy in some studies, helping to reduce breast cancer cell growth. Anethole is supposed to reduce inflammation that may lead to cancer development, though further research is needed to determine exactly how it can be used.

Improves Colic: Infant colic, though a relatively benign medical condition, can have a major impact on new parents. Maybe most parents of a colicky baby will do almost anything to console their crying baby. This is where fennel makes its presence felt.

Great for Acne: When consumed regularly, fennel seeds supply valuable minerals to the body, such as zinc, calcium, and selenium. These minerals are deemed useful in regulating hormones and helping to balance oxygen. Fennel has a cooling effect on the skin when consumed, thus giving a healthy glow. Daily fennel seed consumption is beneficial to skin health.

Brain Function: Potassium is an electrolyte, present at high levels in fennel bulbs and seeds, which helps to facilitate increased electrical conduction throughout the body. This includes connections within the brain, which is an actual electrical currents switchboard.

Menstrual Disorders: Fennel is also an Emenagogue which means menstruation is relaxed and regulated by properly controlling hormonal activity in the body. In addition, fennel is used to reduce the effects of PMS in a number of products and is also traditionally used as a soothing pain reliever and relaxing agent for menopause women.

Respiratory Disorders: Because of the presence of expectorants Cineole and Anetol, fennel is found to be useful in several respiratory disorders such as congestion, bronchitis and cough.

Eye Health: Macular degeneration is one of the principal causes of vision loss associated with age. While the exact cause is unknown, antioxidants that decrease inflammation such as certain flavonoids, vitamin C, and zinc may help improve vision or slow disease progression.

Help reduce post menopause symptoms: For a variety of health problems, including digestion and premenstrual symptoms, Fennel, an anise-flavored herb' Fennel' used for cooking, has long been known for its health benefits.

Precautions

- Pregnant women should avoid using either herbal restorative amounts or essential oil.
- Pregnant women should avoid using either herbal restorative amounts or the essential oil.
- Also, epileptics and people with kidney problems should stay away from essential fennel oil.
- Many essential oils ingredients, including Anethol, and few chemicals contained in fennel, may be unsafe if overly consumed, as compounds that could kill bacteria and microbes in lower doses can be toxic.

3.31 FENUGREEK (TRIGONELLA FOENUM-GRAECUM)

Fenugreek is the oldest and most common plant, known many years ago to humans. Fenugreek in Hindi also known as Methi. Within Indian spices, this plays a vital role which is used to improve the taste of food and also prevents various health problems. This avoids cramps, bloating, intestinal gas, flatulence, cancer, diabetes, indigestion, asthma, high blood pressure, and toxicity to the liver. It can lower blood toxins, too. The seeds have bitter tastes. It is a great source of selenium which helps the body use oxygen. The presence of mucilaginous compounds helps soothe the inflamed tissues as well as relax. These seeds also contain alkaloids such as gentianine, trigonelline, and carpaine. It increases insulin receptor numbers in RBCs and facilitates glucose use in peripheral tissues.

Cure diabetes: Fenugreek seeds have fiber as well as other chemicals which slow digestion as well as sugar and carbohydrate absorption by the body. It improves sugar use and increases the body's release of insulin.

Boost male fertility: Fenugreek seed extract enhances sexual performance, libido enhancement and intensity for around six weeks. The supplements substantially help to increase testosterone levels.

Prevention of cancer: Cancer is the connected to ageing andcomplex genetic diseases. Chemoprevention means the use of synthetic and natural compounds during an early cancer stage. The diet of methi, which is rich in phytochemicals has protective effects on health. Spices and herbs containing phytochemicals have preventive practices against cancer.

Helpful for nursing mothers: Ayurvedic doctors prescribe fenugreek to nursing mothers because it possesses diosgenin which promotes the amount of milk in the breasts. It contains vitamins and magnesium which improve the milk quality which keeps the child healthy.

Menstrual problems: Fenugreek helps to ease the menstrual process and provides relief from the menstrual symptoms. It serves as an emmenagogue that assists obstructed menstruations and makes the process work smoothly and relaxed.

Reduce cholesterol level: Fenugreek intake significantly lowers the low - density Lipoprotein (LDL) level, which decreases the risk of heart attacks, atherosclerosis and strokes. It eliminates excess cholesterol in the blood vessels and body arteries because of the high fiber content. As it lowers cholesterol content in the bloodstream, clot formation in vessels decreases.

Gastrointestinal health: Fenugreek is useful for the treatment of inflammation, wounds and gastrointestinal issues. It counteracts free radicals, because of the presence of antioxidant properties. In Chinese and Ayurvedic medicine it is used to induce labor and to assist in digestion. It also favors the body's overall metabolism.

Fenugreek Seeds for Skin: Fenugreek is known for obtaining flawless skin as an Asian beauty secret. It has both germs killing, and anti-inflammatory properties. The meth seeds act as an effective outside remedy to deal with various skin - related issues.

Fenugreek Seeds for Hair Growth: Methi seeds are effective against dandruff, hair fall, dandruff and prevent baldness and hair thinning along with the use on the face. Attach it to your daily diet or apply it directly to the hair as a paste. Fenugreek is a great solution to any hair- related issue.

Precautions

- There is limited evidence that fenugreek possesses hypoglycaemic properties.
- This should be used with caution for patients with hormonal or anticoagulant medications, monoamine oxidase inhibitor, constituents of coumarin and steroidal saponin.
- There is no recommended excessive use of fenugreek.
- Not to be used by pregnant women.
- To be avoided by allergic people.
- It could interfere with hypoglycemic drugs that are used as a diabetes treatment.
- It may exhibit allergic reactions such as wheezing, rhinorrhea, angioedema facial, head numbness and fainting.
- It is not safe for children.

3.32 FILE (SASSAFRAS ALBIDUM)

Sassafras is a deciduous tree, native to eastern North America and eastern Asia, from the *Lauraceae* family. Aside from Sassafras, it is also known as the Sassafras root, *Sassafrass*, Saloop, ague tree, root beer tree, saxifra, cinnamon wood, white sassafras, sarifrax or smelling-stick, Bois de Cannelle, Common Sassafras, KuntzeSaloop, Laurier des Iroquois, *Laurus albida*, Saloip, *Sassafras officinale*, *Sassafras variifolium*, Saxifrax. For a large number of years, Sassafras has been a key section of culinary, therapeutic and societal development. They are native to parts of Eastern North America and Asia. These trees are deciduous and can grow to a stature of over 100 feet, and have strange yellow or orange bark, making them easy to distinguish. Sassafras has an extraordinarily charming taste, which is why it is most widely used to season various drugs, but this is not to be diluted by its individual strength. There are some of Sassafras ' health benefits to explain, let's take a look below.

Strengthen Immunity: Modern research has found that sassafras have not only analgesic, but also antiseptic properties, and these natural antimicrobial properties are ideal to boost the immune system. This treatment can be a sweet-smelling protection to your overall health by protecting the body against foreign agents and pathogens, both internally and externally. It can also help keep the stomach and respiratory system free from infections, which is why this plant has been traditionally suggested at its onset to fend off colds.

Prevent Cancer: Although research is still ongoing into this anti-cancer claim, sassafras is showing measurable antioxidant potential in the human body. Whether removing free radicals or boosting the immune system, this plant seems to operate like many other superfoods, reducing your risk of cell mutation and subsequent diagnoses of cancer.

Sooth Menstrual Pain: The analgesic and anti-inflammatory properties of sassafras are a blessing for females suffering from the painful symptoms of menstruation, such as bloating, cramping and heavy bleeding. This natural remedy has been decreasing menstrual discomfort for women for thousands of years and remains a popular recommendation from herbalists for women who suffer from particularly painful periods.

Skin Health: Sassafras has traditionally been used for many different topical issues, and skin health is no exception to this. If you suffer from boils, sores, rashes or excessive marks on the skin, applying a sassafras leaves poultice to the affected areas can quickly reduce skin inflammation or discoloration. It is known that the presence of tannic acid and a special compound called Sassafrid provides certain benefits and protective measures for the skin.

Boost Dental Health: Owing to the good taste (reminiscent of root beer), and the natural antiseptic properties, sassafras twigs were the early forms of toothbrushes for centuries. Using it as a dental disinfectant in this manner is still a viable way to get this advantage and if you really want to try the traditional approach, take a twig and wash your teeth.

Detoxify the Body: The sassafras leaves are a well-known diuretic, which increases your urination, eliminating the body's excess toxins, salts and fats. If you want to improve your kidney health and function, the best way to do so is to flush your system with a diuretic, and sassafras tea is a great way to do that.

Relieve Pain: The analgesic properties of sassafras are well known, and because of their anti-inflammatory and analgesic qualities, this tree's leaves were often wrapped around wounds. Sassafras tea is still used for pain -relieving purposes, and in some natural healing communities, topical applications for aches and pains are popular.

Increase Energy: Also, this plant's stimulant nature is great for an energy boost if you feel a bit sluggish. Chewing on the leaves or drinking a cup of sassafras tea is an ideal way to revive your internal motors and overcome excess fatigue or faintness.

Alleviate Inflammation: One of the most popular uses of sassafras is for rheumatism and gout, and in different parts of the world, this has been a trusted remedy for thousands of years. Inflammation affects the body in many ways, which means this impressive plant can also help improve gastrointestinal processes and arthritis symptoms as well as in some cases headaches. This plant is one of the oldest applications of this versatile plant to be used as a fever reducer.

Malarial Fever and Cold Treatment: The first and foremost property of the revealed sassafras was a productive quality of body temperature reduction caused by any malaria or viral fever. Sassafras is also known as ague tree because of its property to treat ague (any disease which increases body temperature). In the morning one may have a cup of sassafras tea to maintain good health in the cold season.

Heal Wounds: One thing is the pain relief aspect of wound healing, but indeed speeding up the healing process is quite another. The remedy's stimulant function delivers freshly oxygenated blood to the wound, which speeds up the body's natural healing process and increases the rate of metabolic processes throughout the body.

Precautions

- You should avoid all forms of this herb if you are pregnant or breastfeeding. It can cause uterine spasms, severe bleeding and thus lead to abortion.
- If you have kidney problems, heart disease or high blood pressure then avoid consuming large amounts of sassafras tea as it may interfere with heart medication and it may be lethal even to adults because it is a blood thinner.
- Another important advice regarding sassafras is to stay away from the sassafras tree's small white berries, as they may be toxic.
- Sassafras use has caused abortion in several cases.

3.33 GINGER (ZINGIBER OFFICINALE)

Originally grown in China, Ginger now spread equally around the world. The botanical name of the plant may derive from its Sanskrit name singabera which means "horn-shaped," a physical characteristic that reflects ginger. Ginger which is supposed to have originated from Northern India's Himalayan foothills is a pungent and hot spice found to be used worldwide because of its nutritional value. It is a powerhouse of several important nutrients and compounds found in ginger is known to regulate allergic symptoms as well as many other health - related problems with no side effects of any kind. Listed below are some popular health benefits of using ginger:

Treat Nausea, Especially Morning Sickness: Ginger appears to be strongly effective against nausea. It has a long history of use as a remedy for seasickness, for example, and there is some evidence that it may be as effective as prescription medication. Frequent use of ginger may alleviate nausea and vomiting after surgery, and in patients suffering from chemotherapy for cancer. Yet pregnancy-related nausea, such as morning sickness, is most effectively treated.

Reduce Muscle Pain and Soreness: Regular supplementation with ginger has shown to reduce muscle pain caused by vigorous exercise. This effect was not enhanced by heat treating the ginger.

Helpful for Osteoarthritis: Usually due to hormonal changes or lack of calcium or vitamin D, osteoarthritis is one of the common medical conditions in which the bones become stiff and fragile due to tissue loss. Research conducted on people with knee osteoarthritis suggests that frequent intake of the ginger extract significantly reduces pain compared with those who do not frequently take it.

Help Treat Chronic Indigestion: Recurring pain and discomfort in the upper part of the stomach is generally described as dyspepsia. Late stomach emptying is the principal cause of indigestion. Remarkably, ginger was shown in humans with this disease to speed up stomach emptying. Ginger has reduced the time it took the stomach to drain from 16 minutes to 12 minutes after soup intake.

Reduce Menstrual Pain: Dysmenorrhea or menstrual pain generally refers to pain that is felt during the entire menstrual cycle of a woman. Ginger is traditionally used during the menstrual cycle to get relief from such pain. For the first 3 days of the menstrual period, scientific research recommends taking about 1 gram of ginger powder per day. Ginger helps to treat the pain as effectively as ibuprofen and mefenamic acid drugs.

Lower Cholesterol Levels: High LDL lipoprotein levels (the "bad "cholesterol) are linked to an increased risk of heart disease. The foods you eat have a significant influence on LDL levels. Regular consumption of 3 grams of ginger powder in most cholesterol markers has resulted in significant cuts.

Prevent Cancer: Cancer is one of the very serious diseases that uncontrolled growth of anomalous cells describes. Ginger extract has been seen as an alternative use for many forms of cancer. 6-Gingerol is one of the essential substances that have anti - cancer properties in raw ginger in large amounts.

Protect Against Alzheimer's disease: Oxidative stress and chronic inflammation can accelerate aging. They are believed to be one of the major causes of Alzheimer's disease and cognitive decline related to aging.

Help Fight Infections: Fresh ginger consists of Gingerol, the bioactive substance which can help reduce infection risk. Ginger extract can prevent bacteria from growing in many different types. This is particularly successful against the inflammatory disease -associated oral bacteria in the gums, including gingivitis and periodontitis. A common cause of respiratory infections, fresh ginger can be effective against the RSV virus.

Precautions

- Ginger may interact with some of the prescription drugs.

- Herbalists are driven by not taking more than 4 grams of ginger in a single day.
- Side effects could include gas, bloating, heartburn and nausea.
- If you have a bleeding disorder or are taking blood thinners, like aspirin, stop ginger.
- Ginger root is also known to potentiate the toxicity of the anticoagulant drug warfarin, leading to severe incidents of bleeding.

3.34 GRAINS OF PARADISE (AFRAMOMUMMELEGUETA)

Melegueta pepper, alligator pepper, Ginny Grains, Ginny Pepper, Grains, Greater Cardamom, Grenes, Guinea Grains, Guinea Seeds, Maniguetta, Melaguata, Meligetta Pepper, Paradise Grains and Paradise Nuts are commonly known as Guinea pepper. It is a tropical plant of the genus *Aframomum* and belongs in the Kingdom Plantae to the family *Zingiberaceae* (ginger family) of the angiosperms. It has been said that the seeds possess certain properties of aphrodisiac, digestive and "good luck." In some diets, the paradise grains also substitute the black pepper because it is less rough on the digestive system. A few people in Africa will chew on them to be warmer when they feel cold and use them in rituals of divination. Listed below are few of the health benefits of using grain of paradise;

Antimicrobial Properties: Paradise grain seed extract has antimicrobial properties because of its constituents of phenolic compounds which are commonly used as disinfectants. Many scientific researchers have found that the *Aframomummelegueta* extract is a broad spectrum antimicrobial and has an inhibitory effect on bacterial growth such as *Salmonella typhi*, *Staphylococcus aureus* and *Klebsiella pneumonia*,etc. Use it to get value on a daily basis.

Treatment of Gastrointestinal Disorders: Paradise grain seed extract may be used to treat gastrointestinal disorders such as stomach pain, diarrhea, ulcer, and intestinal worms. Regular use of this grain is found to be very helpful in addressing gastrointestinal issues.

Wound Healing: Seeds are crushed and used to prepare brews for the treatment and cure of wounds. It consists of a high quantity of tannin which is distinguished by its stringent property and as such is very effective in healing wounds, treating burns and calming inflamed mucous membrane.

Anti-oxidizing Properties: Paradise grain seeds provide an excellent source of phytonutrients such as terpenoids, alkaloids, flavonoids, tannins, cardiac glycosides, saponin, and phenolic compounds. They search for free radicals and provide protection in the body against viruses, allergens, bacteria, platelet aggregation, tumors, ulcers and hepato-toxins. This explains why it is commonly used in folk medicine to avoid and treat bowel problems.

Dermatological Care: Paradise grain may be used to prepare herbal remedies for the treatment of infectious skin diseases such as measles, varicose veins and smallpox.

Anti-inflammatory Properties: Paradise seed grain has anti-inflammatory properties due to its gingerol constituent, which prevents the synthesis of leukotrienes and prostaglandins. Besides that, it provides protection against body inflammation.

Analgesic Properties: Aqueous plant extract is analgesic in nature and can be used to relieve pain such as joint pain, toothache, stomach pain, arthritis pain and rheumatoid pain. Use this daily to get rid of virtually all kinds of pains.

Digestive Properties: Regular use of grain seeds of paradise helps to digest food easily and thus prevents constipation and bloating.

Stimulating Properties: Paradise grain is usually chewed as a stimulant to keep the body alert, due to its calming properties and peppery pungent taste.

Malaria Treatment: Leaves are used in the preparation of herbal medicinal products to prevent and treat malaria.

Aphrodisiac Properties: Several types of research have revealed that in nature the grain of paradise is aphrodisiac and therefore can be used to stimulate sexual desires.

Precautions

- Some side effects of this herb include stomach irritation, bowel discomfort and urinary system irritation.
- It is advisable to stay away from these seeds for pregnant and breastfeeding women to be on the safer side or to consult a doctor before using them.

3.35 HOLY BASIL (OCIMUMTENUIFLORUM)

Holy basil, known in Sanskrit as Tulasi or even Tulsi ("incomparable one"), is in fact a cooking herb with an ancient tradition rooted in ayurvedic medicine. The tulsi or perhaps holy basil is a vital mark in the religious tradition of Hinduism and is most importantly worshiped by Hindus every morning and evening. The holy basil is yet another herbal treatment for many common health conditions. Here are tulsi 's top medicinal uses.

Healing Power: The tulsi plant has many therapeutic properties. The leaves are a tonic to the nerve and also sharpen memory. They should facilitate the elimination of catarrhal matter through the bronchial tube and also the phlegm. The leaves boost the stomach functions and cause major sweating as well. The plant seeds are mucilaginous.

Fever & Common Cold: Basil leaves are specific to most fevers. Throughout the rainy season, whenever there are widespread malaria and dengue fever, soft leaves, steamed with tea, act as precautions against these diseases.

Coughs: Tulsi is a crucial component of many Ayurvedic cough syrups, and expectorants as well. It will help in both bronchitis and asthma mobilizing the mucus. Gnawing leaves on the tulsi reduces the cold and flu.

Sore Throat: Tulsi is a crucial component of many Ayurvedic cough syrups, and expectorants as well. It will help in both bronchitis and asthma mobilizing the mucus. Gnawing leaves on the tulsi reduces the cold and flu.

Heals respiratory conditions: Tulsi has immunomodulatory qualities (helps you to control the defense mechanisms), antitussive (suppresses the cough center, reduces the amount of cough) and also expectorant qualities (helps discharge phlegm through the chest), which make it an excellent relief for cough, cold, along with other problems with the respiratory system that include long-term and extreme bronchitis. An excellent feature of the leaf would be that it possesses anti-bacterial and also anti-fungal properties that help to conquer the infection that causes the respiratory problem.

Dissolves kidney stones: The Holy basil is ideal for the kidneys as being a fantastic diuretic and detoxifier. Tulsi helps to reduce the amounts of uric acid in the blood (one of the many causes of kidney stones is definitely the existence of excessive uric acid in the blood), helps to clean the kidneys, the existence of acetic acid along with other elements in essential oils helps to wear down kidney stones and its particular pain reliever effects help to dull down kidney pain. To relieve the kidney stones, you must keep the tulsi leaves juice along with honey for 6 months each day to help wash the stone through the kidney.

Heart Disorder: Basil is incredibly valuable in treating heart disease, as well as the weakness it creates. This reduces cholesterol in the blood. Tulsi also contains vitamin C along with other antioxidants, just like Eugenol, which protects the heart by harmful free - radical effects. Tulsi can also lower blood pressure and is therefore great for heart sufferers.

Children's Ailments: Typical pediatric difficulties including cough cold, nausea, diarrhea and vomiting react positively to basil leaf juice. When chickenpox pustules postpone their arrival, they will be intensified by Basil leaves taken along with saffron.

Stress: It is thought that Basil leaves are an' adaptogen' or even an anti-stress agent. Recent surveys have shown that these leaves provide significant stress-safety. Surely, healthy people will gnaw 12 basil leaves, twice daily, to avoid stress. This cleans the blood and allows for the prevention of many typical components.

Mouth Infections: The leaves are very effective for ulcers as well as mouth infections. Certainly, a few chewed leaves will heal those kinds of problems.

Insect Bites: Additionally, this particular herb helps prevent all insect stings and bites. After a few hours, a teaspoon of fresh juice must be applied to the affected areas, as well as repetitive. Additionally, a paste of fresh roots is great in the case of insect bites as well as leeches. Use a paste of Tulsi leaves and also collectively roots ground on the scorpion's sting.

Protects Your Skin and Prevents Skin Cancer: While the sun allows for life, it can also damage your skin. Extended sun exposure leads to sun burning, skin wrinkles, and also skin cancer. Melanoma, typically caused by extreme exposure to sunlight, is considered to be probably the most dangerous type of skin cancer. Fortunately, holy basil could be a viable cure for radiation injuries and skin cancer as well.

Teeth Disorder: Additionally, the herb is useful in solving teeth problems. Both dried and powdered tulsi leaves may be used for teeth cleaning. It can also be combined with mustard oil and can be used as a toothpaste. It preserves the gum health and helps in battling foul breath, pyorrhea and other dental problems.

Headaches: Tulsi helps you reduce headaches caused by sinusitis, allergies, cold or maybe migraines. It's because it has both pain reduction and decongestant properties that help reduce the sensation of pain as well as solve the root cause of the problem.

Eye Disorders: Basil juice is an effective treatment for sore eyes as well as night-blindness generally caused by vitamin A shortages. Each day at night, a few drops of black basil juice can be put in the eyes.

Stomach problems: Numerous digestive issues can also be treated along with tulsi leaves, just like acidity, bowel problems and excessive pain. Tulsi helps the digestive tract and makes you feel less bloated and much more comfortable. It can also help individuals who are dealing with vomiting as well as missing food urge.

Beats diabetes: Holy basil leaves are loaded with antioxidants, as well as essential oils including eugenol, methyl eugenol and caryophyllene. These types of ingredients collectively assist the effective functioning of the pancreatic beta cells (cells that store and discharge insulin). Hence, it increases insulin sensitivity. Reducing blood sugar levels and also effectively managing diabetes. The fact that antioxidants found in the leaves help to overcome the negative effects of oxidative stress is an additional benefit.

Pain relief: Tulsi plant behaves just like an inhibitor of COX-2 and also benefits individuals with adverse medical problems. Several substances present in tulsi help you alleviate both inflammatory and neurological pain and in no way have any negative effects. Additionally, Tulsi is used in the treatment of ulcers, cuts and wounds due to its beneficial killing of germs as well as healing qualities.

Anti-cancer: Tulsi is naturally anti-carcinogenic and helps address many types of cancer including breast cancer. Tulsi restrains the blood vessels that nourish tumors. It may also help prevent the development of oral cancer due to tobacco that it has brought on. However, tulsi should not be used as a single cancer treatment.

Lung and Bronchial Support: As well as contributing generally to the health of the respiratory system, Holy Basil has been shown to be useful in managing a number of serious hypersensitive, inflammatory and infectious problems affecting the lungs and associated tissues. Commonly leads to healthy lung function.

Helps to quit smoking: Tulsi has good anti - stress properties and is great to help with one quit smoking as well. Through reducing the stress that can be associated with trying to stop smoking, or even tension that leads to a desire to smoke, it will help. It also has a chilling effect on the lips, much like menthol fall when chewed, and allows controlling the desire to smoke by letting an individual to munch on something hence they are able to foget about smoking.

Pregnancy: Basil is used to induce labor, so you may only have basil or even basil tea if your pregnancy is delayed. While studies have not even finally proven it, it's best to keep pregnant women away from basil. There are in fact herbal treatments that use basil as well as basil in infertility programs, therefore responses are still mixed.

Antiulcer Activity: Gastric ulcers are painful sores or ulcers that form within your stomach. The holy basil seems to be an effective cure for gastric ulcers.

Treats Asthma: Asthma is really a medical problem that causes swelling of your lungs and also of your airways. It is typically characterized along with wheezing symptoms, difficulty breathing, and also coughing.

3.36 HOREHOUND (MARRUBIUM VULGARE)

Horehound is a herbaceous and perennial plant commonly known by its common names including Hounds bane, White Horehound, Marrubium, Common White Horehound, Green Pompon Horehound, Seed of Horus, Star's Eye, Bull's Heart, Hounds bane, Adorn, Haran Haran, Blanc Rubi, Grand Man and Mapiochin. This plant belongs to the Lamiaceae family and is also used for edible use. White Horehound commonly known as Marrubium vulgare is a small, perennial plant resembling mint and belonging to the same family. It provides similar health benefits and is used in traditional medicinal uses as a vital plant. The plant with white flowers grows up to 18 inches, small leaves covered with small hairs. White horehound is effective in preventing some diseases, strengthening the immune system, improving heart health, controlling diabetes, reducing indigestion, alleviating fatigue, gastrointestinal distress, and removing muscle spasms and detoxification. The plant has various elements such as choline, betonicin, tannins, phenolic acids and volatile oil making it suitable for different uses. Horehound possesses different properties that act as a wonder to the human body by maintaining a healthy heart, rejuvenating skin by removing skin related problems.

Prevent cancer: Horehound has anti-inflammatory properties that prevent cell death and decrease cell cancer chances. The antioxidant properties help to neutralize free radicals, which are harmful by-products of cell respiration that cause cell mutation and apoptosis, and spontaneous cell death.

Lowers inflammation: White horehound may be used as an oral intake and as topical mixtures. It improves arthritis symptoms which provide relief from coughing and sore throats. It also lowers constriction in blood vessels, lowers inflammation by reducing blood pressure which helps to promote heart health.

Lowers cholesterol: The tiny dose of horehound extract reduces cholesterol levels. It balances the level of cholesterol by removing bad cholesterol which prevents plaque buildup. It prevents atherosclerosis and reduces the risk of high cholesterol causing strokes and heart attacks.

Treat diabetes: White horehound is linked to lowering blood sugar levels. It helps normalize the level of blood sugar which is effective for diabetes caused by blood sugar fluctuations.

Eliminate toxins: The plant can promote sweating which helps eliminate body toxins such as excess water, salts, harmful substances and disease-causing fats.

Cure spasms: White horehound acts as an antispasmodic that relieves twitches, seizures, spasms and cramps in the muscles. It also soothes the nervous system and prevents certain disorders from being likely to happen. The tea prepared from this herb has calming properties that calm the body and mind.

Menstrual problems: For those women who experience painful menstruation, horehound use is growing. It decreases the pressure and improves morale or eliminates mood swings. Horehound extract has active ingredients gives hormonal effects.

Digestive problems: Horehound helps remove gut-related issues such as indigestion and constipation. It has antiseptic qualities that enhance its function to clear the gastrointestinal tract. It reduces colon inflammation by avoiding constipation, flatulence, too much bloating and hemorrhoids.

Enhance the immune system: The plant has antimicrobial and antibiotic properties which stimulate overall health. It protects the immune system against foreign agents and pathogens and reduces immune system stress. It may be used in mouthwashes, candies and toothpaste.

Precautions

- The juice from the plant may cause contact dermatitis.
- It is better to avoid this due to a lack of usage information during pregnancy and lactation period.
- Excessive doses have an impact on blood pressure, heart rate and blood glucose.
- Evite allopathic medicine in patients with diabetes mellitus.
- Individuals with irregular heartbeats, hypoglycemia, diabetes and heart disorders should meet with the doctor.
- It's causing skin rash in sensitive people.
- It causes diarrhea, stomach upset and vomiting.

3.37 HORSERADISH (ARMORACIA RUSTICANA)

Known as Pepperrot, Red cole, Mountain radish, Great raifort, and Horseradish. It is a perennial plant of the family Brassicaceae which includes wasabi, mustard, cabbage and broccoli. Horseradish's health benefits are discussed in greater detail below:

Cancer prevention: Horseradish has compounds of glucosinolate which provide spicy flavor and cancer counteracting. Compared with broccoli it produces 10 times more glucosinolates.

Antioxidant activity: Free radicals damage the body so diet-rich intake of antioxidants helps eliminate this harm. Root includes phytocompounds that act as antioxidants and are useful for maintaining general health. Being antimutagenic, it protects the body against mutagens which could damage it. Mutations cause degenerative disorders and heart disease. Horseradish extracts reduce the damage to DNA caused by zeocin that is an antibiotic known to reduce oxidative stress.

Antibacterial and antimicrobial: Horseradish has isothiocyanate allyl, which gives a pungent taste. This oil provides antimicrobial properties against various pathogenic agents. Studies have assessed horseradish root antibacterial and antimicrobial activity. The study was conducted using essential oil from Horseradish to protect roast beef and to avoid spoilage. Horseradish beef prevents bacteria from growing which causes it to spoil. Roots provide positive effects on phagocytes, which are a type of cell in the body that absorbs bacteria. The mice study showed that Horseradish promoted phagocyte antimicrobial activities, which counteract disease and infection.

Respiratory health: Horseradish contains antibiotic properties that have been used for treating sinusitis, bronchitis, common cold and cough for many years.

Prevent urinary tract infections: Antibiotic properties found in the root of Horseradish successfully treat acute urinary tract infections that are better compared to antibiotic treatments that cause different side effects. Horseradish contains the sinigrin glycoside and prevents water retention. Being diuretic, it helps prevent urinary infections and kidney infections. Allyl isothiocyanate is removed by urine which has been shown to possess anti-bladder cancer abilities that could be a cause for positive effects on a urinary tract.

Digestive health: Horseradish enzyme promotes digestion, reduces constipation and controls bowel movements. Bile protects the body by protecting digestive processes from excess fats, cholesterol and waste. It acts as a cholagogue, which is a substance in the gallbladder which stimulates bile discharge. This backs digestion. It contains little fiber content, which is important for proper digestion.

Relief pain: People in Ancient Greece used Horseradish to relieve back pain, and it is used for treating headaches in South America. It's because of the anti-inflammatory properties found in horseradish. Stimulating blood flow to the skin offers relief from pain. For relieving pain, it may be applied to swollen muscles and joints. Massage it over affected body parts to help relieve symptoms of arthritis, gout and chilblains.

Prevent discoloration of skin: The overproduction of ceroid or melanin in the skin leading to birthmarks and liver spots known as age spots. Horseradish prevents this.

Cure sinus infections: Sinus infection is caused by an accumulation of mucus in the sinuses which creates an ideal environment for bacteria to thrive in which it could be painful. Horseradish has a volatile compound that thins old mucus to prevent its accumulation. Patients who are prone to infections with the sinus and upper respiratory tract should eat half a teaspoon of horseradish sauce in the morning and noon and stay for 10 minutes after use without drinking and eating.

Strengthen immunity: Horseradish is loaded with nutrients of gain such as phytochemicals and essential minerals. Horseradish has compounds that act as antioxidants to help stimulate white blood cell supply in the body which prevents disease. It has high vitamin C content, enhancing immune function. A healthy immune system prevents any infection from developing. Antioxidants prevent free radical damage to cells and bestow anti-aging effects.

Precautions

- High quantities could be toxic, due to the presence of volatile oils.
- Hepatitis, chronic nephritis, hyperacidity, or gastroesophageal reflux, and inflammatory bowel disorders are contraindicated.
- These should be prevented by lactation and pregnant women.
- Those with issues with the thyroid or stomach ulcers should not use it internally.
- Use it with caution because it could cause blistering.
- It could aggravate kidney problems, hypothyroidism and peptic ulcers.

3.38 HYSSOP (HYSSOPUS OFFICINALIS)

Common hyssop, garden hyssop, curdukotu, hastipippili, hisopo, yanagi-hakka, yssop, hyssop herb, isopo, ysopo is a Lamiaceae or mint herb garden whose flowers and evergreen leaves are used as flavorings for food and beverages and as folk medicines. The mint family consists of a wide variety of herbs for medicinal use and cooking. Hyssop is used primarily to alleviate coughs and pressure on the airways. It is used for bronchitis and asthma with horehound, and coughs due to colds, infections of the nose and throat and feeding. The hyssop smell is reminiscent of camphor (it repels moths and other insects). Mentioned below are but a few of Hyssop's health benefits.

Fights Infections: Hyssop prevents the development of infections in wounds and in wounds. Because of its antiseptic properties, it prevents infection when applied to a skin wound, and kills bacteria. It also helps cure deep cuts, scars, bites of insects and may even be one of the best home remedies for acne. Hyssop has antibiotic properties from the volatile oils found in the plant. Although the antibiotic properties are generally strong, they show the best results for shallow wounds and infections with the fungus.

For Pain & Bruises: Hyssop is used to treat rheumatism and bruise in the muscles. It's said that it aids in timely healing when the herb is applied to wounds and skin cuts. Ancient Greeks and Indians used hyssop poultices for addressing cuts, rash, frostbite, conjunctivitis, bites of insects, muscle pain and dental cavities. Due to its anti-bacterial and anti-viral properties, plant leaves were used in cleaning rituals during biblical times.

Cold & Fever: Since Hyssop is in nature antiviral and has the ability to expel toxins from the body that are collected as a result of viral infections. Hyssop is mainly used to reduce infections caused by fever, flu and cold. By causing sweating, the herb helps reduce fever and was therefore typically mixed in bathing water.

Heals Respiratory Conditions: Hyssop is antispasmodic, which means it relieves the respiratory spasms and calms coughs. It is also an expectorant, and it helps remove the phlegm stored in the respiratory tracts. This property helps heal common cold infections, and it helps to treat respiratory conditions, such as serving as a natural remedy for bronchitis. Coughing is a common respiratory system reaction attempting to expel harmful microbes, dust or irritants, so the antispasmodic and antiseptic properties of hyssop make it a marvelous natural treatment for coughs and other respiratory conditions.

Stomach and Digestion: Hyssop leaves are used to treat fragile stomach and to alleviate certain indigestion symptoms such as nausea, bowel cramping, colic ,etc. Hyssop-made infusion acts as an excellent tonic to the digestive system and helps to combat poor digestions. Some believe the herb also helps the liver and gall bladder function.

Promotes Skin Health: Hyssop oil helps to lessen the appearance of scars and to work as a natural treatment for acne, pox, boils, stretch marks or wounds. It also promotes cellular regeneration, and new skin growth makes old marks fade away. Since hyssop oil is an antiseptic, it helps to kill skin bacteria and fight infections. Regular use of a few drops of this beneficial oil to your skin will keep you looking healthier and younger.

Fights Parasites: Hyssop has the ability to combat parasites which are species that feed off other organisms ‘ nutrients. Worm examples include tapeworms, fleas, hookworms, and flukes. Because the hyssop oil is a vermifuge, it expels parasitic worms, especially in the intestines. When a worm lives in and feeds on its host, it disturbs the absorption of nutrients, causing weakness and illness. If the parasite lives in the intestines it interferes with the digestive and immune systems.

Precautions

- Volatile oil can cause convulsions.

- Avoid with epilepsy of fits.
- However, plants should not be used by pregnant women, since it can cause a miscarriage in large quantities.
- This oil should not be used on highly- strung individuals, as it may cause epileptic symptoms.
- Except under professional supervision, the essential oil should not be used internally.
- Don't give kids hyssop; a child who took 2–3 drops of hyssop oil over several days suffered convulsions.
- Do not use hyssop if you have a history of seizures, because it can induce seizures or make them worse.
- Hyssop is also known to increase blood pressure, which may help people with low blood pressure but is troublesome for people who try to decrease their blood pressure levels.
- In susceptible individuals, hyssop can cause nausea, upset stomach and diarrhea.
- The overdose symptoms include dizziness, chest tightness and central nervous system disorders.

3.39 LAVENDER (LAVANDULA SPECIES)

Commonly known as true lavender, common lavender, or lavender in English, formerly *L. officinalis*, is a flowering plant in the family *Lamiaceae* / Labiatae. Lavender 's scent and taste are great but the bonus comes from its wide range of health benefits, which we will discuss in more depth below;

Protects Heart Health: Lavender's soothing properties, which come from its organic compounds and antioxidants, also support the heart by reducing blood pressure and loosening blood vessel stress. This may prevent atherosclerosis and other cardiovascular problems and thereby reduce the risk of stroke and heart attack.

Treats Sleep Issues: If you often suffer from insomnia, apnea or disturbed sleep patterns, as indicated by research published in The Journal of Alternative and Complementary Medicine, this can adversely affect your life. By boiling a few lavender flowers in hot water, you can steep a wonderful tea, which has been used for thousands of years to induce sleep and relaxation. This is closely related to the effect on the nervous system, and can also help clear your mind from negative thoughts.

Skin Care: A simple way to care for your skin is by pouring lavender flowers in a spray bottle. Simply spray some of the infused water into the region when your skin feels dry or irritated, and enjoy instant relief. This can function for chronic conditions such as psoriasis, eczema, and acne, too.

Reduces Anxiety & Stress: There are a number of ways lavender can be used to soothe anxiety and stress. In its leaves and flowers, the natural organic compounds can be ground between the fingers and then rubbed to the temples. By relieving anxious thoughts and relaxing the mood, this topical application will relax the body and mind. Besides this topical application, you are also able to brew lavender tea and produce much the same result. The antioxidant components in it can have an effect on the endocrine system to reduce stress hormone levels in the body.

Lessens Menopausal Hot Flashes: Hot flashes are a common symptom of menopause which affects many women. It creates a sudden feeling of heat over the body and may flush the face and trigger suddenness. But 20 minutes twice a day of lavender aromatherapy can help to reduce the flashing menopause and improve the quality of life.

Anti-inflammatory Qualities: Everyone is looking for a reliable way to relax the mind and body and lavender takes care of both. The anti-inflammatory components will help to decrease the inflammation if you apply these flowers to your bathwater and take a nice long soak up. Lavender's antioxidant properties are also quite efficient and is one of its flowers ' most trusted applications.

Antiseptic Ability: Though many people turn to lavender for therapeutic applications for relaxation and scent, its effect on infection treatment is remarkable, to say the least. Many people use crushed leaves on wounds and injuries for speedier wound healing as well as for preventing sepsis.

Prevents Digestive Issues: The polyphenols found in lavender affect the body in a wide range of ways. As per herbal medicine experts, it can prevent the development of harmful bacteria and the accumulation of gas in the gut. This will relieve the pain of the stomach, reduce bloating, and minimize cramping. It can also be helpful to suck on its leaves or to drink its tea.

Hair Care: If you are suffering from hair loss or any other disorder affecting the quality of your locks, look for shampoos based on lavender. Be aware that some of the organic cosmetic products may be quite expensive while others may still contain harsh chemicals that claim to be derived from it. Just like tea, you can boil lavender flowers and then apply the mixture to your hair. This will serve as an effective shampoo and significantly boost your follicle beds and hair safety.

Help Combat Fungus Growth: Essential lavender oil may be effective at inhibiting the growth of certain forms of fungus, including C. And the Albicans. According to previous research, the oil could also serve as a cure for the treatment of athlete 's foot and ringworm, which are also caused by fungus.

Heals migraine headaches: Breathing essential oil in lavender is intended to be an effective and safe treatment option for migraine headaches. In one study, patients suffering from migraine attacks reported significant improvements after 2 - 3 drops of lavender oil in the early stages of an attack were rubbed on to their upper lip.

Bug Repellent: For many forms of bugs such as mosquitoes, midges and moths, the scent of essential lavender oil is strong. To avoid these painful bites, add some lavender oil to the exposed skin when outside. However, if you happen to be bitten by one of these bugs, essential lavender oil has anti-inflammatory properties that will reduce inflammation and bug bite-related pain.

Alleviates menstrual cramps and PMS: Lavender may help to reduce emotional premenstrual symptoms. In the premenstrual state known as premenstrual syndrome (PMS), women of reproductive age encounter various symptoms. Massaging lavender oil on the skin can help alleviate dysmenorrhea, which is associated with menstrual pain and cramping.

Treats Respiratory Disorders: Lavender oil is widely used to treat various respiratory problems including inflammation of the lungs, fever, cough, cold, asthma, sinus congestion, bronchitis, cough, laryngitis, and tonsillitis. Oil is either used as a spray, or is applied to the neck, chest, and back skin. It is also applied to many inhalers and vaporizers which are widely used for cold and cough. Also, the relaxing quality of essential lavender oil will release the phlegm and alleviate breathing--related congestion; thereby speeding up the recovery process and helping the body naturally remove phlegm and other unwanted material. Lavender essential oil vapor also has antibacterial properties that can fight infections in the respiratory tract.

Improves Blood Circulation: Also, the essential lavender oil is good for enhancing blood circulation in the body. Research suggests aromatherapy using lavender oil has beneficial effects on circulation in the coronary system. It also reduces blood pressure and is often used as a hypertension treatment. It ensures that not only do the muscles raise their oxygenation levels, encouraging muscle strength and health, but brain activity can have a visible boost, skin stays bright and blood-flushed, and the body is shielded from the risks of heart attack and atherosclerosis often associated with poor blood circulation.

Treats Eczema: It is used to treat various skin conditions such as acne, wrinkles, psoriasis, etc. It is widely used to speed up the healing process of wounds, cuts, burns, and sunburns, as it enhances scar tissue formation. Chamomile is also used with lavender oil to treat eczema.

Effective painkiller: Lavender oil has an analgesic effect and is known to be exceptional in treating various types of pain such as headache, sprain, rheumatism, migraine, muscle pain, etc. Lavender oil soothes the pain and quiets the region affected.

Precaution

- Lavender tea has very few side effects which can be avoided by following the guidelines for use.

3.40 LEMON BALM (MELISSA OFFICINALIS)

Commonly referred to as Lemon Balm, Balm, Bee Balm, Sweet Balm, Common Balm, Garden Balm, Melissa, Balm Mint, Blue Balm, Heart Delight, Cure-all and Dropsy Plant is a bushy perennial herbaceous mint family that is traditionally cultivated in herb gardens and frontiers for its lemon- scented leaves. Lemon Balm is an ancient herb that was used more than 2000 years ago by the Greeks for the many medicinal properties it possesses. This herb was known for being a successful treatment for neural disorders. Known to contain a tangy scent, this herb is also used for garnishing and decorating purposes. There have been many advantages of this Herb some of which are known to be mentioned below;

May be useful against diabetes: Diabetes is a common health problem in the West. Though prescribing a number of medications with dangerous side effects to cure it is very common among traditional practitioners, it is actually possible to reverse diabetes in a natural way. One effective way to achieve this can be using lemon balm extract.

May aid in digestion: Lemon balm extract has a potentially protective effect on your gastrointestinal system, due to its antioxidant activity. It calms digestive problems by preventing stomach ulcers but this has not been known for a long time. Lemon balm, as well as peppermint and angelica roots, are extremely useful in the development of constipation herbal remedies.

Effective in Treating Cold Sores: The herb may be used to treat cold sores caused by Herpes because of its antiviral properties. The use of sores dependent on lemon balm cream gives positive results. Lemon balm has the potential to treat primary HSV-1 infection. This cream also helps to deal with the latter caused by HSV-2 oral and genital herpes. Lemon balm that has 1 percent lemon extract will bring down the healing time significantly.

Helps Alleviate Muscle Pain, Aches: Lemon balm is effective in alleviating muscle pain and flu-accompanying aches. The antispasmodic action helps reduce muscle spasms. Using Lemon balm causes you to sweat when you have a fever which brings relief. Toxins in the body, too, get shed with sweat. One of its ingredients, eugenol, helps in reducing pain.

Naturally soothes PMS symptoms: Once taken in capsule form, the lemon balm effectively reduces symptoms of PMS in females of high school age. So, for those women who suffer from PMS -related issues, it is quite successful.

Regulate overactive thyroid: More than 12 percent of the U.S. population is affected by thyroid problems, and lemon balm can benefit from one of two common thyroid conditions, hyperthyroidism. Lemon balm extract is one of the best methods for regulating overactive thyroid treatment.

Improve mood, concentration and sleep: Lemon balm has been used, since ancient times, to relieve anxiety and insomnia. The mild sedative influence is a known feature. When given to children in elementary school, the lemon balm helps improve hyperactivity, problems with concentration and impulsiveness. Taken internally in food products it also appeared in a study involving young adults to boost mood and/or cognitive performance.

Works Wonders for Skin Health: Lemon balm is effective for both women and men, to keep the skin in good condition. Ideal for treating inflammation of the skin. It has strong natural astringents and, thus, those with oily and acne-prone skin may benefit from using the herb. It helps cleanse clogged pores of the skin that actually cover the path to acne and skin rashes.

Treats Mouth Infection: You can use the lemon balm to treat oral infections such as bleeding gums and bad breath. Mixing a couple of drops of Lemon balm essential oil into a mouthwash and then using it to gargle will kill the bacteria and freshen your mouth inside.

Fights herpes naturally: Owing to its antiviral properties, Lemon balm uses to expand its use to the topical application as well. Specifically, it's a useful herbal remedy when applied directly to the skin to help treat the herpes virus.

Keeps Nasty Pests and Bugs away: OTC anti - insect spray and repellent products are selected by a majority of people to remove insects from the home. Most such insect repellents, however, contain strong chemicals and allergens which may be problematic for users, especially those with allergies and respiratory problems.

Can Be Used as Natural Room Freshener: Lemon balm can also be used as a freshener in natural spaces. Could be used as a potpourri with dried leaves. The mixture of fresh lemon balm leaves, which are lightly boiled in water, can be sprayed in rooms to spread the fragrance.

Adorning the Garden and Pollinating Trees: No matter how large or small a garden you have, it'll be good to use Lemon balms to decorate. The sweet scent remains fragrant in the garden, more so when the flowers bloom. It also attracts bees, and so having the plant in the garden will help other trees pollinate.

Precaution

- It is not advisable to use the lemon balm in conjunction with specific medicines and sedatives.
- Not Ideal for Glaucoma Patients
- Not Suitable for People Scheduled for Surgery
- Allergic Reactions

3.41 LEMONGRASS (CYMBOPOGON CITRATUS)

In fact, it is a tropical plant native to tropical Asia, mostly southern parts of India and Sri Lanka, commonly known as lemongrass or oil grass. It is a frost-tender clumping perennial grass that is commonly used in Thai, Vietnamese, Laotian and Cambodian cooking as a lemony flavoring and is widely cultivated for this purpose in Southeast Asia. It is also an excellent ornamental herb of the Poaceae family of grasses.

Prevents Cancer: Lemongrass is very effective in treating various forms of cancer, without damaging the body's healthy normal cells.

Certain ingredients, citral, present in lemongrass, help inhibit the growth of hepatic cancer cells during the initial phase and prevent further cancer cell development.

Stomach Disorders: Vital lemongrass oil has anti-microbial and anti-bacterial properties that help combat the infections caused by various pathogens such as *Helicobacter pylori* and *Escherichia coli*. It is beneficial in preventing gastrointestinal disorders such as stomach ulcers, it helps to stimulate bowel function, and it improves digestion. Lemongrass's anti-inflammatory activity is beneficial in the treatment of constipation, ulcerative colitis, diarrhea, nausea and stomach aches.

Reduces Aches: Its analgesic properties to alleviate the pain and discomfort caused by headaches and migraines. It contains phytonutrients that improve blood circulation and help to relieve spasms, muscle cramps, sprains and backaches. When treating sports wounds, including dislocations, internal injuries, and bruises, it is valued.

Lowers Cholesterol: Lemongrass has emic properties that promote healthy cholesterol levels, both anti-hyperlipidemic and hypercholesterolemiaeffects. Research has shown that repeated lemongrass intake has shown significant results in maintaining healthy triglyceride levels and reducing the LDL cholesterol in the body. This helps prevent lipid accumulation in the blood vessels and facilitates unblocked blood flow in the arteries, avoiding various heart diseases such as atherosclerosis.

Treats Infections: Lemongrass acts as an antiseptic and is effective in the treatment of infections such as ringworm, sores, athlete's foot, scabies and urinary tract infections (UTI) due to its antimicrobial and antifungal properties. Research has shown that lemongrass, by inhibiting the growth of pathogens, exerts healing effects on dermatological infections such as yeast infections.

Boosts Immunity: Lemongrass helps restore vital systems in the body, including digestion, breathing, excretion, and nervous system. This helps to improve nutrient absorption and strengthen the body's immune-defense mechanisms. Lemongrass extracts have a beneficial effect on cytokine inflammatory behavior, which are the signaling molecules by which the cells communicate and respond to the body.

Cellular Health: Lemongrass possesses antioxidant properties and helps protect the body cells against free radicals derived from oxygen. It also helps to cleanse the blood and strengthen the spleen to replace the damaged red blood cells. It supports thymus gland function which helps to produce white blood cells. It helps in boosting cell regeneration. In DNA synthesis the amount of folate and potassium in the stems and leaves of lemongrass helps and facilitates cell division.

Obesity: Lemongrass is made of citral, which has been demonstrated to be effective in combating obesity. It prevents abdominal fat accumulation and promotes the use of stored energy which helps to prevent diet-induced weight gain. It assists in healthy metabolism and improves the oxidation of body fatty acids.

Detoxifies the Body: Lemongrass helps, as a result of its diuretic properties, to clean and wash harmful toxic waste out of the body. Detoxification helps in regulating different body functions, including the liver and kidney, while also helping to lower uric acid levels. Lemongrass diuretic effect helps to increase the quantity and frequency of urination, which helps to maintain digestive health, eliminate accumulated fats and helps maintain a clean body.

***Staphylococcus aureus*:** Lemongrass essential oil has an antibiofilm potential and is effective against *Staphylococcus aureus* infection. It includes phenols that have the ability to spread rapidly through the tissues of the body and to heal biofilms anywhere in the body. It disturbs germ growth and communication which helps to inhibit biofilm formation.

Treats Insomnia: Lemongrass helps to calm the muscles and nerves, thereby helping to promote deep sleep. Research has shown that lemongrass tea has both sedative and hypnotic properties that help to increase sleep duration.

Type-2 Diabetes: Lemongrass has proven to be beneficial in the treatment of type 2 diabetes. Studies have shown that the citral present in it helps maintain optimal insulin levels and improves glucose tolerance in the body.

Insect Repellent: Lemongrass is used as a natural insect repellent, which helps prevent insect-borne diseases such as malaria, dengue, and Lyme disease from occurring. Research has provided supporting evidence relating to lemongrass' anti-malarial and anti-protozoan properties, making lemongrass oil an important ingredient in mosquito repellents.

Aromatherapy: Lemongrass is made up of the beneficial ingredients of essential oils such as nerol, citronellol, myrcene, dipentene, geraniol, and methyl heptenone, which have anti-fungal, insecticidal, and antiseptic properties. Due to its therapeutic effects, lemongrass oil is expansively used in aromatherapy, which helps revitalize the body. The cooling effect of lemongrass oil during hot weather is beneficial for the body and encourages the regeneration of mind and soul alike.

Cures Fever: Lemongrass is also known as' fever grass' because of its beneficial effects on fever reduction. Lemongrass's antipyretic and diaphoretic activity is commonly used in Ayurvedic medicine by inducing sweating to treat fevers.

Respiratory Disorders: Lemongrass is commonly used for its curative role in the treatment of cough and cold in Ayurvedic medicine. Vitamin C present in it helps to relieve nasal blockages, pneumonia, and other breathing disorders such as bronchial asthma.

Skin Care: Despite its astringent and antiseptic properties, Lemongrass has been treasured as a skin tonic and makes an effective cleanser for oily or acne--prone skin. This helps strengthen the tissues of the skin and tone the pores while also sterilizing them. Care should be taken while using lemongrass products, since the undiluted application may in some cases lead to dermal irritation.

For common colds and flu: Lemon grass offers a very powerful blockage drug, with no undesirable effects. Boil water, add lemongrass and tea. Allow that to brew for a few minutes. Be sure to include some honey to drink it without milk. Drink a few cups each day, if at all necessary. Lemongrass is known as providing relief to those individuals who are causing your body aches when you're down with flu.

Treats Edema: Lemongrass is effective in curing water retention or edema conditions. It has a lymphatic inflammation cleansing effect and helps soothe swelling.

Prevents Rheumatism: Lemongrass is effective in alleviating the pain and discomfort arising from rheumatism. It can be applied topically to both lumbago and sprains and helps in neuralgia relief.

Nervous System: Lemongrass is a nervine: for the nervous system it has been considered to be tonic. It fosters the mind and helps in the fight against convulsions, nervousness, vertigo and various neuronal disorders such as Alzheimer's and Parkinson's. It is used in therapeutic baths which help calm the nerves and ease the symptoms of stress-induced anxiety and fatigue.

Precautions

- Plant handling can cause skin irritation or an allergic reaction.
- Lemongrass oil should not be used in children.
- When pregnant it should not be taken as it has uterine stimulating properties.
- To avoid irritation dilute lemongrass oil in a carrier oil before application.
- Use a small amount for a limited time.
- Children, ladies who are pregnant, or even breastfeeding, or individuals with liver or kidney disorder should not use the essential oil internally.

3.42 LEMON VERBENA (ALOYSIACITRODORA)

Also known as Lemon - Scented Verbena, Herb Louisa, Lemon Bee-Brush, *VerveineOdorante*, and Luisa. The shrub is annual, originally from Peru, Argentina, Chile, Uruguay, Bolivia and Brazil, belonging to the *Verbenaceae* family. Lemon Verbena, native to China and Argentina, is cultivated worldwide. Lemon Verbena is a deciduous aromatic plant, with a heavy lemon fragrance. The plant has tiny white blossoms, with thin and pointed leaves. Because of the antioxidant properties, leaves are dried to prepare herbal tea which provides various protection benefits. The herb is both an effective cure for health problems and sickness. It could be used both externally and in-house. It could be used internally as herbal tea to receive numerous benefits for health. It helps improve the nervous system, relieves spasms in the digestive track, reduce pain, stress and lessen feverish colds. It acts as an expectorant and anti-spasmodic. Some of the benefits are discussed below:

Loses weight: The people on a diet have found teas made from Lemon Verbena are good. This gives 2 calories per serving. It has rich organic compounds that reduce munchinggg and avoid snacks between meals. Lemon Verbena tea provides metabolic effects that enhance body processes by enhancing fat burning. This tea keeps the body hydrated, essential to the work of an organ. It helps replace high- calorie foods and results in lower calorie intake.

Restore muscles: This is seen as unique qualities that could not be found in basic teas. Lemon verbena tea has high antioxidant properties that decrease muscle damage during a workout without preventing additional muscle mass from forming and also improving stamina. It makes for the perfect exercise partner.

Lowers inflammation: Injuries on movement and joints trigger stressful conditions. Once hurt, becoming whole again is burdensome as the joints are still being used, and it takes time to heal properly. It's synonymous with increasing pain and aching. Due to the presence of antioxidant properties, this herb assists the recovery process of joint -related injuries.

Strengthen immunity: Oxidative stress is caused by free radicals in the organs of the body that weaken immunity and disrupt it with cellular mutation as well as a chronic illness. Lemon Verbena supplementation reduced oxidative stress levels and maintained better overall health.

Digestive health: Lemon Verbena prepared tea has soothing activities that provide relief from indigestion and stomach problems. It has anti-spasmodic properties that calms the stomach and also prevents bloating and cramping which leads to discomfort as well as various gastrointestinal problems. The tea regulates appetite that contributes to a balanced nutrient intake to upgrade metabolic processes.

Antibacterial activity: The extracted essential oil possesses strong cytotoxic activity. Targeted activity of antibacterial was reported on *E. coli*and*S.aureus*. The essential oil was resisted by *P. aeruginosa*. The antibacterial and cytotoxic activities are based on its interactions and chemical character: Citriodora, the compounds of the essential oil.

Bacteriostatic activity: Lemon verbena belonging to the family *Verbenaceae* has different traditional medicinal uses. Lemon Verbena ethanol and aqueous extracts have no bacteriostatic and bactericidal effects on both *S.sobrinus* and *S.mutans*.

Treatment for anxiety: Lemon Verbena has antioxidant compounds, which have an impact on the human body's hormonal balance. This herb is considered to be a calming drink and is recommended for people with chronic stress or nervous disorders that can ease their minds and also calm their bodies.

Relieve congestion: Verbena Lemon serves as an expectorant. The Lemon Verbena prepared tea helps alleviate pressure in the respiratory tracts and also removes mucous and phlegm in the body. Phlegm promotes the growth and sticking up of bacteria and other pathogens. The consumption of this tea helps eliminate undesirable materials from the body that helps strengthen the immune system.

Precautions

- It has terpene-rich volatile oils that are considered irritant during excretion and may cause irritation to the kidneys.
- Renal disease patients should avoid their overuse.
- People who are sensitive may experience hypersensitivity to the touch.
- Because of a lack of information about lactation health and pregnancy, it is best not to use it during these times.
- With the use of essential oil, it might cause skin sensitivity to sunlight.
- Excessive internal use or prolonged use causes gastric irritation.
- Use it with caution.
- People with health problems should consult the health practitioner for use.
- Its use in large quantities causes irritation to the kidneys and also makes kidney diseases worse. The patients in the kidney should prevent large amounts of its use.

3.43 LICORICE (GLYCYRRHIZA GLABRA)

Also known as Glycyrrhiza, Chinese Licorice, Gan Cao, Kan-ts'ao, Kuo-lao, Licorice, Licorice Root, Ling-t'ung, Liquorice, Mei-ts'ao, Mi-kan, Mi-ts'ao, Sweet Licorice, Sweet Wood, Yasti Madhu, this compound contains glycyrrhizin, which is about 50 times sweeter than sugar. Licorice root has a healthy protein, vitamin, and phytonutrient content. It is obvious at this point that you may be interested in finding more information on the health benefits of licorice roots. Continue to read on to learn about the various home remedies with Licorice roots in the section below to treat and prevent various health conditions.

PMS/Menopause: Licorice root tends to have an estrogen-like effect in females, lending itself as an alternative for menstrual and fertility-related concerns, including as a natural remedy for PMS.

Beneficial For Stomach Disorders: The root of licorice is known to have certain laxative products. The ingestion of licorice tea will minimize constipation. With the flavonoids and the chalcone content in the roots, inflammation and pain caused by the bacterial stomach infection can be minimized. Such properties help to strengthen the digestive system by calming off the lining of the inflamed mucus, offering assured relief from such problems.

Fighting Depression: The root of Licorice is known as a natural anti-depressant. It stimulates the activity of the adrenal glands, which regulate the cortisol synthesis process. The main cause of anxiety, nervousness, chronic fatigue and stress is that this specific hormone has a reduced level. Regular use of licorice tea helps keep your balance. The regular use also uplifts one's mood and puts aside depression.

Immunity: Licorice plays an important role in the treatment and prevention of diseases such as hepatitis C, HIV and flu. The triterpenoid content has been verified as an antiviral making it a potentially strong partner for the immune system to extract licorice.

For Diabetes: Diabetes is one of the rising problems around the world and it has become something of a priority to find a way to stop its development. The root of licorice is a potential natural candidate for future treatment of diabetes.

Skin Brightening Effect: Daily use of liqueur makes the skin look fairer by improving its color. Combining turmeric powder and fuller earth with licorice tea, you may prepare a face mask.

Cleanses your respiratory system: Licorice is useful for the treatment of respiratory problems. Taking liqueur as an oral supplement can help the body make healthy mucus. Thecreation of clean, healthy phlegm allows the respiratory system to work without it being clogged by aged, sticky mucus.

Help Prevent Cancer Growth: Many components of licorice are effective against several cancer hallmarks, including cell proliferation, inflammation, resistance to cell death and the formation of their own blood vessels. A polysaccharide from licorice induces apoptosis in an oral cancer cell line, and prevents the growth of cancer cells. Licorice is particularly toxic to human cancer cells in a human cancer cell line but not to healthy cells.

Combat hepatitis: Licorice protects both the liver and facilitates cure in this vital organ. The anti -inflammatory properties of the herb assist in reducing liver inflammation associated with hepatitis. Licorice also battles the virus that is commonly responsible for hepatitis and provides valuable antioxidant compounds that help maintain the overall liver health.

Pain relief: Liquorice, as an antispasmodic, is also helpful for cramps in the muscles both abdominally and theoretically. Topically, since it acts as hydrocortisone, licorice can relieve eczema discomfort and other dermal conditions. Anti-inflammatory effects can also help to relieve discomfort, including serving as a natural joint pain reliever.

Prevent heart disease: Modest doses of licorice (100 mg a day) have the effect of limiting damage from LDL ("bad") cholesterol and may inhibit the development of artery-clogging plaque and contribute to healthy heart function.

Helps to Treat Tuberculosis: Antimicrobial and antimycobacterial compounds found in the root extract of licorice function well for the treatment of infections. Such characteristics make it a potent anti-tubercular agent.

Dental Health Benefits: Licorice provides effective treatment for dental conditions, for example, gingivitis. Licorice 's antimicrobial property prevents the formation of dental plaques. Because of these properties, licorice is used in the manufacture of oral hygiene products, such as mouthwashes and toothpaste.

Hair Loss Prevention and Dandruff Treatment: Tablets and capsules containing root liqueur are eaten to stimulate hair growth and reduce hair loss. Cleaning the hair with the root of the licorice helps to treat dandruff. The amazing demulcent properties of licorice help to soothe the scalp by reducing various conditions such as scabs and dandruff.

Cough/sore throat: Licorice root greatly benefits a sore throat or cough as an important expectorant, helping to release and remove mucus which the cough is trying to eliminate. Its demulcent, anti-inflammatory soothing properties will offer quick relief for sore throat. Demulcents need to make contact with the body part that needs to be soothed, so cough drops and syrup extracts, as well as tea, are most effective.

Helps with Herpes and Shingles: Extracts from the Licorice Root have been used to treat shingles, sores and herpes. The root's antiviral properties slow the progression and return of cold sores caused by the Herpes virus.

HIV: Licorice helps to treat HIV - related illnesses. The glycyrrhizin found in the roots of licorice works well to inhibit the replication of the infecting virus simply by encouraging beta-chemokine formation. It can also help prevent the HIV virus from destroying white blood cells, and improve the body's immune system.

Precautions and Warnings

- Pregnancy & breastfeeding: Licorice is absolutely unhealthy during pregnancy to be taken by mouth. This root's high intake during pregnancy, around 250 g per week, gives you a greater risk of early delivery or miscarriage. While there is still not sufficiently credible data regarding the effects of consuming licorice when you breastfeed, it is best to stop using it.
- Hormone-sensitive conditions include breast cancer, ovarian cancer, uterine cancer, uterine fibroids, or endometriosis: Licorice can function in the body as an estrogen. Nevertheless, if you're suffering from any illness that gets worse by drinking estrogen, stop using liqueur.
- Heart disease: Licorice can cause the body to retain water, which can exacerbate congestive heart failure. You may also get an increased risk of an irregular heartbeat from licorice. If you're suffering from heart disease, avoid consuming licorice skin.
- High blood pressure: Blood pressure can increase with liqueur. When you suffer from high blood pressure, avoid consuming it in large amounts.
- Low levels of potassium in the blood: Licorice root can lower blood potassium. So, if your potassium level is low already, it could make it lower. If you suffer from this disorder avoid consuming licorice.
- A muscle condition caused by nerve problems: Licorice can decrease potassium levels in the blood. This exacerbates hypertension. If you suffer from hypertonia, stop using the licorice.
- Sexual problems in men: This source can lower a man's interest in sex and aggravate erectile dysfunction by lowering the level of a hormone called testosterone.

- Kidney disease: improper use of the root of licorice can make kidney disease worse. Stop cutting it up.
- Surgery: Licorice can interfere with blood pressure regulation during & following surgery. Avoid using the drug at least 14 days before an operation.

3.44 LOVAGE (LEVISTICUMOFFICINALE)

Lovage, is a tall perennial plant, the only species in the genus Levisticum, in the Apiaceae family of which dill, angelica, carrot, celery and parsley are also members and share their characteristic aromatic fragrance and flavor. Lovage is an excellent herb and particularly good for the digestive tract and respiratory system. Because it is nutritionally dense and can be eaten fresh to retain all of its nutrients, the herb can contribute to improved immunity, as well as stimulate and strengthen the body. Here are some excellent properties and uses of lovage herb:

Digestive Processes: Lovage's anti-inflammatory aspects makes it perfect for calming upset stomachs and getting the gastrointestinal system back into a neutral state. Through reducing pain in your intestines and encouraging healthy, normal activity in your colon can help to reduce bloating and excess gas.

Arthritis Relief: The lovage's natural anti-inflammatory qualities makes it the perfect treatment for people suffering from illnesses such as gout, asthma and hemorrhoids. If you are suffering from painful inflammation anywhere in your body, just eat some lovage leaves or add this herb to your next plate.

Soothes Rough Spots: Lovage is loaded with compounds that can pacify rugged body patches. In lovage compounds that provide this benefit include limonene, eugenol, and quercetin.

Respiration Booster: If you suffer from a respiratory ailment, lovage can act as an expectorant to clear out your tracts of excess phlegm and mucus. In addition, eucalyptol, one of the primary relieving agents found in it, can reduce irritation and inflammation in the lungs, promoting faster healing.

Menstrual Relief: When taken at the start of a menstrual period, lovage will act to soothe many of the most serious menstrual side effects including cramps and bloating. There's also a high nutrient density that makes it effective during that more difficult part of the month to keep your energy and mood up.

Kidney Health: Although the exact mechanism is not fully known, lovage has been associated with improving the health of the kidneys. It acts as an odd type of diuretic, known as aquaretic, which encourages urination but does not result in electrolyte loss. This is a much healthier type of urination if you want to detoxify your body without risk of dehydration, thereby improving your kidney health.

Good for the Skin: Lovage calming properties apply also to the eyes. When the leaves are applied directly to the skin, or a salve is made, it can be very effective in reducing psoriasis and acne symptoms, resulting in a smoother, better - looking skin. In some cases, antioxidants found in it can also help prevent wrinkles and increase blood flow to the skin surface, thereby enhancing tone and appearance.

Antibacterial Effects: Lovage has a significant effect on different microbes and foreign agents which can cause disease within the body. There is a positive relationship between lovage intake and lowering of infections such as *E. coli*, *Salmonella* , etc.

Anti-allergenic: Many of the lovage's health benefits relate to its anti-inflammatory and soothing nature, and the same goes for alleviating the effects of allergens. This also contains large amounts of quercetin, which is a known histamine inhibitor. This will diminish the body's allergic reaction, reducing itchy eyes, runny noses and other more severe allergic reactions.

Healthy Joint: Regular use of Lovage can assist to maintain a healthy joint. It is a good advantage for the aged, who typically have joint health problems. Consume this plant therefore to strengthen the joint and help with any risk of joint inflammation.

Support the Lung: Consume lovage to aid the lung system. It can ,therefore, manage a healthy lung condition, and improve the movement of the lung. This is essential to ensure that the air and oxygen distribution around the body system is proper.

Reduces inflammation and relieves pain: Lovage has mild pain-relieving properties and can help relieve pain in the joints, abdominal cramps, possibly headaches and other pain forms. Angelic acid is responsible for the lovage 's mild analgesic properties, but in its structure, certain volatile oils and even nutrients (like magnesium with its antispasmodic effects) contribute to its pain-relieving benefits.

Flushes the Kidney Stones: Lovage is an aquaretic, or a type of diuretic that stimulates urination without loss of electrolytes. The increased urination removes gravel and calcified stones from the kidneys may aid in kidney stone prevention. Not only that, but lovage also flushes the urinary tract, eviting infections of the urinary tract and other urinary problems.

Improves Oral Health: Lovage leaves are effective at keeping the breath fresh. What's more, the plant has a unique odor that can help make the breath better after consumption.

Soothe Cold: Many people think it can help to soothe any cold weather by eating the leaves. Thus, the leaves are suitable for adding some cough and fever into a soup to carry a herbal remedial effect. Moreover, it can handle a healthy body that feels unwell.

Fights Harmful Organisms: The majority of plants produce natural chemicals to fight harmful organisms. While some plant chemicals–like resveratrol–act independently, researchers found that the lovage chemical compounds work together to create a very powerful effect against harmful organisms. Lovage is one of the most potent extracts used against gram-=negative bacteria such as *E. coli*, *Salmonella*, *H. pylori*, *H. grippe*.

Alleviates Allergy Symptoms: The lovage's anti-inflammatory effects naturally help in reducing allergy symptoms. But lovage goes a little bit further when it comes to allergies. Quercetin, a compound present in lovage, prevents the release of histamine and relieves irritation of the skin associated with the allergic response.

Restores Appetite: Occasionally,a UTI, or any illness can disturb your appetite and eating habits when you're sick with. Eating is vital when your body is dealing with an infection, it requires extra nutrients. If you are not eating properly you run the risk of malnutrition. If you find yourself struggling to maintain a healthy appetite, try to relax your digestive system by taking some lovage tea before mealtime.

Precautions

- Lovage is said to contain the alleged 'psychotroph' myristicine.
- Volatile oil is an irritant.
- Contraindicated with kidney or urinary passage inflammation.
- Lovage can be harmful to women, and women who are pregnant or suffer from menstrual disorders should be advised to avoid this herb.

- Lovage can also cause photo-dermatitis, a skin allergy that happens after eating the herb and being exposed to sunlight.
- People with kidney problems should refrain from consuming or using lovage or its derivatives.
- People suffering from allergic conditions should stop eating lovage. Since this can cause itchiness, skin redness, nausea and even dizziness.
- Too much consumptionof the herbal plant may lead to diarrhea. Therefore, consume as necessary.
- Excessive use (many times the recommended dosage) may cause kidney damage.

3.45 MACE/ NUTMEG (MYRISTICAFRAGRANS)

Referred to all over the world as a well-liked spice, nutmeg can also be well-known for its numerous health benefits. In fact, nutmeg has been used since ancient times as a fix for numerous conditions, or even generally improves health. Here are a few of the health benefits that nutmeg provides.

Brain Tonic: Roman and Greek civilizations used nutmeg, like a kind of brain tonic, throughout ancient times. It's because nutmeg can arouse the brain quickly. Because of that, it can benefit from getting rid of both exhaustion and stress. If you suffer from anxiety or depression, then nutmeg can be a great treatment as well. Nutmeg will usually increase your concentration so that you can become more effective as well as concentrated at the office or even at school.

Pain Relief: Nutmeg is a highly effective tranquilizer, in fact. In fact, nutmeg is truly a staple of ancient medicine in China. The Chinese used the spice for treating both inflammation and abdominal pain. If you are suffering from painful joints, muscle pain, arthritis, sores and other ailments, using nutmeg. Using nutmeg oil to ease the pain towards the affected areas.

Indigestion Relief: When you suffer from digestion-related difficulties such as diarrhea, bowel problems, bloating, flatulence and so on, nutmeg can provide relief for you successfully. Through eliminating excess gas through the intestines, Nutmeg oil eliminates stomaches. Nutmeg can improve your appetite too.

Bad Breath Treatment: Nutmeg can also successfully treat halitosis, or even bad breath, because of its anti-bacterial qualities. Bad breath, as you probably know, is generally the result of bacteria building up inside your mouth. Nutmeg can easily rid those bacteria in the mouth area. Therefore, nutmeg is a very common ingredient in many toothpaste brands. Nutmeg can also be used to treat problems with the gum and toothaches.

Liver and Kidney Detox: Detoxification is a key determinant of good health. Diet, pollution, stress, tobacco, medication along with other external substances can cause toxins to build up in your organs. Liver and kidney are two of the organs that usually produce just where this particular toxic build - up. Nutmeg will clean the liver and kidney, and eliminate these toxins, like a tonic. If you are suffering from a liver disease then nutmeg may also be beneficial. Additionally, Nutmeg is good at stopping and dissolving kidney stones. Whenever your liver and kidney are detoxified effectively, they are better able to perform their function.

Skin Care: If skin care is among your focal points then you may want to include nutmeg in your diet. In fact, Nutmeg can help you achieve cleaner as well as healthier skin by helping you deal with a number of skin problems. A scrub made from nutmeg powder and orange lentil powder will help you eliminate blackheads, a sort of acne that is seen as a pore clogged with excess oil and dead skin cells. If you are suffering from acne marks, nutmeg might also help make your scars less visible. Mix some nutmeg powder with little honey to make a paste, which you will then apply to the acne marks.

Insomnia: Nutmeg has been suggested for generations as both a home remedy for sleeplessness and insomnia. A pinch of the nutmeg always seemed to do the trick in warm milk. Nutmeg has high magnesium content, an important mineral in your body that decreases nerve tension and even energizes serotonin release which results in a feeling of relaxation or sedation. This specific serotonin is changed inside the brain to melatonin, which is a sleep inducer, minimizing individuals with insomnia issues as well as night restlessness. Nutmeg also offers trace portions of narcotics, which, unless taken in massive quantities, have no remarkable effect. But even the little bit will help you release various neurotransmitters as compared to inducing relaxation and sleep.

Brain Health: Among the better-known benefits of adding nutmeg to the diet in virtually every form would be the many facets of its essential oil, such as myristicin and macelignan. It has been shown that these types of compounds decrease the impairment of neural pathways and cognitive function that commonly affects individuals with dementia or Alzheimer's disease.

Oral Health: In traditional medical applications, when it came to oral health, nutmeg was considered the king of spices. Nutmeg 's active antibacterial components imply that it will help combat circumstances just like halitosis, also called bad breath. It kills the bacteria that cause this humiliating disease, and typically boosts gum and teeth immunity. For this reason, in toothpaste as well as mouthwashes, especially organic or herbal types, nutmeg and it is extracts are commonly present.

Leukemia: Another of nutmeg's lesser-known virtues is its potential use in cancer cells. Research suggests that a specific methanolic compound in nutmeg and its essential oil can potentially induce cell death (apoptosis) in leukemia cells, thereby preventing the proliferation and metastasis of this awful variety of cancer that usually affects children.

Nutmeg benefits for male sexual enhancement: Nutmeg benefits consist of boosting sexual activity, both libido and potency or sexual energy, and they have been demonstrated in a variety of studies. Nutmeg extract is used just as one ingredient in many sex tonics designed to deal with weakness, impotence and spermatozoa. Additionally, it provides Stimulant, a carminative.

Precautions

Warning: High dose (30 grams approximately 6 tablespoons a day) of nutmeg may be harmful, creating confusion, double vision as well as seizures, but these side effects will not be created even with large quantities of culinary use. Both Nutmeg and mace are unhealthy in doses greater than the amounts in foods. In many people, side effects such as fatigue, dizziness, nausea, vomiting, chest or stomach pressure sensations, dry mouth, stomach pain, and many other problems may occur. More serious adverse effects could include hallucinations, seizures, and death. The defense of using nutmeg and mace on this skin is not properly known.

3.46 MARJORAM (ORIGANUMMAJORANA)

Marjoram also called wild oregano or sweet marjoram is a herb that belongs to the mint family (*Lamiaceae*). While closely linked to oregano, marjoram and oregano are two distinct herbs. Aside from being a tasty culinary herb, marjoram has a number of excellent health benefits in both its food and oil form. It is made up of many good minerals and vitamins and has many excellent medicinal properties. Marjoram can protect the heart, protect it from diabetes and improve digestive health, among its other uses. Here are some of this herb's most impressive health benefits:

Heart Health: Marjoram can be of enormous benefit to your cardiovascular health. It is a rich source of antioxidants that can help protect the heart from disease, as well as other major organs. It can also relax and expand the blood vessels, which inevitably helps to reduce high blood pressure levels. Inhaling the essential oil either directly or from a diffuser can help calm the nervous system of the body that helps dilate the vessels of the blood. This has a positive effect on the heart by reducing blood pressure and the tension on the heart.

Lowers high blood pressure: Marjoram helps lower blood pressure without affecting heart rate. One way that it does this is to strengthen the blood vessels so that blood flows freely into them. Marjoram also has diuretic properties which enhance urine production as well as urination frequency. As a result, the body is washed out of the excess water, salts and contaminants. So helps also to lower blood pressure.

Bone health: Daily marjoram intake is a great way to boost your bone strength and health. This amazing herb provides approximately 520 percent of the recommended daily intake of vitamin K, a nutrient that contributes to healthy bones. By promoting osteotrophic activity in the bones, vitamin-K also plays a key role in building bone mass. This vitamin helps prevent the onset of bone disorders, such as osteoporosis, by helping build strong bones.

Diabetes: Marjoram is a great choice for people trying to cope with their diabetes. Along with other herbs such as rosemary and oregano, marjoram effectively prevents such enzymes that improve insulin tolerance. Whether you are eating the herb fresh or dried, the ability of a diabetic patient to control their blood sugar levels can be of great help.

Helps with blood clotting and prevents anemia: Marjoram is an excellent source of vitamin K, essential for normal blood clotting. Excessive blood loss is one of the factors that lead to anemia, and clotting helps prevent severe bleeding. Marjoram also has a high content of iron, a further nutrient needed for hemoglobin production. Marjoram consumption ensures that your daily iron requirements are met which enhances blood circulation and oxygen transport to all parts of the body. This helps to improve your energy levels and also to optimize the functioning of different organs within the body.

Pain Relief: Effective marjoram oil is an excellent treatment for muscle pain, spasms and headaches from stress. Combine a few drops of carrier oil and apply it for instant relief to the damaged parts of your body.

Digestive Health: Marjoram leaves act as a soothing tonic that helps to improve digestive system performance. By stimulating the production of saliva and digestive enzymes it can boost your appetite and promote digestion. Marjoram also helps prevent the multiple digestive problems including infections with flatulence, colic, constipation, indigestion and stomach. It has been found that regular consumption of marjoram tea can help people deal with eating disorders.

Boost immunity: Marjoram is an immunity booster due to the presence of nutrients such as vitamin A and vitamin C which also serve as powerful antioxidants. Vitamin C strengthens the immune system by increasing the number of white blood cells that protect the body against foreign invaders such as bacteria and viruses attacking. Vitamin A also plays an important role in supporting the immune system's healthy functioning.

Healthy nervous system: Marjoram is one of the few herbs which help the nervous system to function effectively. Herbal nutrients boost brain activity and help to prevent damage to the nerve cells. Marjoram essential oil is known for relaxing and stimulating the mind and alleviating conditions such as anxiety, stress and depression. It is also used in dementia rehabilitation and Alzheimer's disease.

Respiratory Problems: Marjoram essential oil can also be used to treat respiratory problems and bronchitis, cold, and flu symptoms. It has expectorant properties in the respiratory tract which can help loosen mucus and phlegm. Seek to inhale the oil from the bottle directly, or add a few drops to the cheeks and under your nostrils.

Stress and Anxiety: Marjoram in particular the essential oil is an excellent treatment for relieving stress and anxiety feelings. Several essential oils have a similar impact and provide a great natural alternative to the many popular prescription drugs. You can either spread it around the home and inhale its vapors, or dilute it with a carrier oil to create a massage relieving stress. By helping relieve your stresses and anxieties, essential oil from marjoram can also help improve your sleep quality and ensure a more restful night.

Treats cold and cough: Marjoram is an ancient remedy used to treat the common cold, flu and cough. It serves as an expectorant and helps remove the phlegm and mucus, thereby helping to clear it. Saponin found in this herb has decongestant properties and helps to relieve sinus and chest inflammation, providing immense relief. Consequently, marjoram is often used to treat respiratory problems such as asthma, sinusitis, bronchitis and cough.

Women's Health: In women's health, Marjoram has long been valued. It can help restore the female hormone balance, and regulate menstruation. It not only helps to regulate the cycle but also helps to relieve many of the common symptoms women have during their monthly period and can also help treat the symptoms of menopause. It can promote menstruation because of its natural emmenagogue properties and has even been used to help nursing mothers produce more milk.

Skin Health: For several skin products, such as lotions, creams and soaps, leaves of marjoram and the natural oils derived from it are also used. Packed with antioxidants, the herb can help avoid free radical damage. Applying a mixture of essential marjoram oil and olive or coconut oil is a perfect way to avoid premature symptoms of aging such as wrinkles, dark spots and fine lines from occurring. Marjoram is a natural antiseptic and is used as a treatment for various skin disorders such as acne, pimples, rashes, psoriasis and eczema. The herb's antiseptic properties also make it effective when it comes to treating wounds and avoiding infection. Marjoram also has anti-fungal properties that help prevent fungal infections like athlete's foot.

Against snoring: Like its oil, the marjoram by-product helps to eliminate snoring, since it stimulates the respiratory system and calms the larynx. The scent penetrates the body and the airways tones.

Hair Health: In addition, you can add a few drops of essential marjoram oil to your daily conditioners or shampoo to help strengthen your hair and enhance its look. Marjoram essential oil will strengthen follicles which means your hair can grow stronger and healthier. The oil also contains anti-fungal properties, which makes treating popular scalp problems such as dandruff a good natural choice. It smells great too so check it out.

Helps you stay young: Marjoram is a rich source of different nutrients that exhibits powerful antioxidant effects. One of the key causes affecting early aging is the oxidation of cells in the body caused by free radicals. Antioxidants help protect the cells and DNA from the damage these free radicals are causing. Hence, the consumption of antioxidant-rich foods such as marjoram is beneficial in preserving good health and youth.

Promotes better sleep: Many people worldwide are suffering from insomnia, the inability to fall asleep or stay asleep. Good quality sleep is completely critical to keeping the body and mind healthy. Marjoram essential oil is a great cure for insomnia sufferers because it has the potential to soothe and calm the mind. It can free your mind from tension and stress and provide a sedative effect. Few drops of essential marjoram oil on the pillow are all you need to drift off into the sleep of a good evening.

Precautions

- Pregnancy and breast-feeding: It is unhealthy when you are breastfeeding to use marjoram in therapeutic quantities.
- If you are breast-feeding, not enough is known about the health of using marjoram in medicinal quantities. Stay on the safe side to stop use.
- Children: Do not administer medicinal quantities of marjoram to children. It could be dangerous for them.
- Bleeding disorders: Taking medicinal quantities of marjoram could delay coagulation and increase bruising and bleeding chances in people with bleeding disorders.
- Slow heart rate (bradycardia): The use of medicinal marjoram amounts can slow the heartbeat. This may be a problem in people who already have a slow heart rate.
- Allergy to basil, hyssop, lavender, mint, oregano, and sage: Marjoram can cause allergic reactions in individuals allergic to these plants and other members of the plant family Lamiaceae.
- Diabetes: When used in medicinal amounts, marjoram could affect the blood sugar levels in people with diabetes. Watch for signs of low blood sugar (hypoglycemia) and carefully monitor your blood sugar if you have diabetes and medicinally take marjoram.
- Gastrointestinal tract blockage: When used in medicinal amounts, marjoram could cause congestion in the intestines.
- Ulcers: When used in medicinal quantities, marjoram may increase the secretions in the stomach and intestines. There's worry this could make ulcers worse.
- Lung problems: When used in medicinal amounts, marjoram can

increase the fluid secretions in the lung. There is fear that lung conditions such as asthma or emphysema could be made worse by this.

- Seizures: There is concern that the use of medicinal marjoram amounts may increase the risk of seizures.
- Surgery: Taking medicinal amounts of marjoram during and after surgery can increase the risk of bleeding. Avoid medicinally using marjoram at least 2 weeks before the scheduled procedure.
- Obstruction of the urinary tract: When used in medicinal amounts, marjoram may increase secretions in the urinary tract. There is concern that this may exacerbate urinary obstruction.

3.47 OREGANO (ORIGANUM VULGARE)

Oregano is a great fiber source, and vitamin A, E, C, and K. Vitamin A assists in the maintenance of the immune system and in the protection of vision. Vitamin C and E act as a protective antioxidant to cells. Vitamin K helps blood coagulate. Oregano is rich in folate, iron, calcium, magnesium and vitamin B6. These nutrients help to form building blocks of DNA and RNA, metabolize bones, prevent anemia and improve brain function. In addition, it contains potassium, manganese, and copper. It has high antioxidant content which helps neutralize free radicals. It is carotenoid with lutein, cryptoxanthin, and zeaxanthin. It is useful medicinally for the treatment of colds, flu, indigestion and menstrual cycle control. It is used as a poultice for treating sore muscles and eczema. Leaves and flowers are a mild stimulant that enhances bile flow from the gall bladder to intestines and also helps digestion and prevents gas.

Enhance immunity: Oregano has two most essential elements, such as thymol and rosmarinic acid, which is an antioxidant agent closely related to reducing the body's oxidant stress. Free radicals are destructive cell metabolism byproducts that cause cancer as well as other chronic diseases. Adding Oregano to the diet increases immunity and prevents toxic and silent killers from happening to the body.

Prevention of bacteria: The existence of carvacrol and thymol indicates that Oregano has antibacterial properties. These are important organic compounds that defend the body from different bacteria that affect the skin, intestines and other parts of the body. It is a stimulating agent that encourages white blood cell development and also speeds up the metabolism that accelerates recovery from common diseases.

Assist digestion: Loaded with fiber, Oregano has strong digestive health effects. As a vital element for a healthy digestive system, fiber promotes large amounts of stool and improves peristaltic motion that carries food through the digestive tract and excretes it efficiently. Fiber is essential to maintain well-being in the gut and encourages nutrient uptake.

Heart health: Oregano has omega-3 fatty acids which are good for low cholesterol to boost heart health. In addition, Omega-3 fatty acids rebalance cholesterol levels and lower inflammation in the cardiovascular system, and also eliminate the possibility of heart attacks, atherosclerosis and strokes.

Body detoxification: Oregano is rich in nutrients that have a high content of calcium, manganese, vitamin K, iron, fiber, and various organic compounds that help detoxify the body.

Stronger bones: We ensure enough vitamins and minerals are obtained for bone health. Iron, calcium, and manganese are essential to preserve bone health. Oregano has adequate amounts of nutrients that help to prevent lifelong osteoporosis.

Promotes the level of energy: Vitamin B helps to promote flexibility that helps the body rejuvenate and become energized. It increases circulation with iron presence and promotes hemoglobin levels. Through encouraging energy and strength, it helps the body's cells and muscles to oxygenate fully.

Antioxidant activity: Thymol is a chemical that gives a strong and unique odor which is a phytonutrient that functions as an effective nutrient. It also produces another rosmarinic acid, a phytonutrient. Because of these two phytonutrients, a gram of Oregano produces an antioxidant 42 times stronger than apples.

Speeds up the healing process: Human body remains in the weakest state during the healing process after illness, and it is essential to strengthening immunity. Adding oregano to diet guarantees optimal immune system activity. It speeds the healing process. The oregano leaves are used in tea.

Absorption of nutrients: An improper digestive system interferes in optimum nutrient absorption that is derived from ingested foods. The fiber-rich diet promotes movement within the intestinal system to ensure maximum nutrient absorption.

Cholesterol balance: Oregano contains null cholesterol levels. Because of this, it is a healthy diet for the cholesterol people. Oregano has omega-3 fatty acids which help to rebalance cholesterol levels. It's recommended for cholesterol issues in patients.

Prevent menstrual cramps: For women, menstrual cramps are unpleasant : adding Oregano to the regular diet may be helpful in dealing with cramps and also helps with frequent menstruation.

Treatment for asthma: Asthma is a difficult to treat condition of the respiratory tract and one should ensure putting away things that might cause an allergic reaction. Oregano contains anti-allergic properties that help in the effective treatment of this condition.

Precautions

- Pregnant and breast-feeding women should avoid it.
- It increases the chances of bleeding so it should be avoided for people with bleeding disorders.
- Oregano should not be used by those who are allergic to plants of the family Lamiaceae such as hyssop, basil, marjoram, lavender, sage and mint.
- Diabetic patients should use Oregano with caution, as it may lower blood sugar levels.
- As the chances of bleeding are increased, people should stop using it before two weeks of surgery.
- Oregano might cause an upset stomach in some people.
- It is contraindicated for liver diseases or gallbladder patients.
- Not to be used for the children below two years of age.
- People with anemia should use it with caution.

3.48 PAPRIKA (CAPSICUM ANNUUM)

Paprika is a red powder made from dried, grinding sweet red peppers. It is used to garnish and to season. The hue, which depends on the peppers used, varies from bright orange-red to deep red.Here are some health benefits of Paprika:

Great source of antioxidants: Paprika packs itself with antioxidants. The products made from peppers are used to combat diseases since they can counteract oxidative stress. Carotenoid found in different forms of paprika varies and helps to fight disease. The nutrients are fat-soluble and are better consumed when eaten with a good source of fat, like avocado. Paprika has carotenoids such as beta-cryptoxanthin, lutein and beta - carotene. Beta cryptoxanthin helps reduce Arthritis inflammation. Zeaxanthin and lutein are useful for the health of the eyes and to counteract molecules that cause damage that lead to macular degeneration.

Treat autoimmune condition: Capsaicin is useful for autoimmune conditions in chili peppers. The symptoms affect the skin, brain, lungs, teeth, thyroid, sinus, muscles, joints, gastrointestinal and adrenal functions. Capsaicin triggers biological reactions associated with the treatment of autoimmune diseases.

Treatment for cancer: Spicy paprika has capsaicin that is used to treat one type of illness. It has the potential to prevent cancer or to treat it. Capsaicin is responsible for altering the signaling pathways that restrict cancer growth and also suppress genes that tell tumors to grow in size. Paprika is an aid in the prevention of gastric cancer. A 2012 study shows that gastric cancer is the second most common cancer in the world and the second most common cause of cancer - related deaths. Capsaicin has anti--inflammatory effects on gastric cancer.

Diabetic treatment: Paprika is the nutrient-rich foods and spices which have the potential to control blood sugar levels and also help with diabetes treatment. This helps diabetic patients in digestion and even the absorption of sugars in the blood. The capsaicin supplement reduces women with diabetes' chances of having birth babies that are too big for gestational ages.

Eye health: The spice contains high antioxidants such as lutein, vitamin A, and zeaxanthin that helps to prevent eye - related diseases from paprika. In addition to these minerals, vitamin B6 also keeps the eyes healthy. People with high B6 intake lower the chances of macular degeneration and also eye - related disease when ingested with elevated folate rates.

Heart health: Paprika helps maintain heart health and cardiovascular system. Vitamin B6 reduces blood pressure and protects damaged blood vessels as well. This prevents anemia by producing blood hemoglobin, which is responsible for oxygen transportation through the bloodstream. It has capsanthin, which is the principal spice carotenoid. Capsanthin present in paprika promotes a good level of cholesterol helping to maintain cardiovascular health.

Reduce hair fall: Paprika is an excellent source of vitamin B6 which helps prevent hair loss. Paprika has enough iron in it to allow oxygen transfer to hair follicles. It boosts hair growth by boosting scalp circulation.

Maintain hair color: Paprika has vitamin B6, which produces melanin, a pigment that gives the hair color. Add sweet paprika powder to henna to make a reddish tint known to hair during coloring. To avoid an allergic reaction, do a patch test before using it on the scalp.

Precautions

- Avoid by the people who are allergic to it.
- Use it in moderate amounts.

3.49 PARSLEY (PETROSELINUM CRISPUM)

Parsley is the common name for Petroselinum crispum, a bright green, biennial herb of European origin, which is widely cultivated for its seeds, which are used as garnish or flavorings food. Because parsley is packed with antioxidants, there are numerous benefits to petroleum health. According to the American Cancer Society, it is a powerful weapon against arthritis prevention, and the phytochemicals found in parsley that protect the plant against viruses, bacteria and fungi can decrease the risk of certain cancers, diabetes, hypertension and heart disease.

Controls Rheumatoid Arthritis: Even Parsley was found to be effective against rheumatoid arthritis. Vitamin C and beta - carotene discovered in parsley have anti - inflammatory properties that help control arthritis. It is surmised that regular consumption of parsley accelerates the uric acid removal process, which has been related to arthritis symptoms.

Helps with Menstrual Pain and Bleeding: Vitamin K present of parsley helps by controlling the role of your hormones to reduce PMS cramps and other menstrual pains. Since vitamin K is a vitamin that coagulates the blood, it helps with excessive bleeding during the menstrual cycle and provides pain relief for PMS symptoms. Excessive bleeding during the menstrual cycle leads to greater cramps and pain. Herbs, F. A. O. (2005). Spices and Essential Oils: Post-harvest Operations in Developing Countries. Douglas, J. Heyes and B. Small field, Rome. Vitamin K will help with PMS symptoms, and the opposite is true-a deficiency in vitamin K will only make these symptoms worse.

Needed for Healthy Hair, Skin, and Eyes: The body needs enough copper to produce the natural pigment and texture of the skin, hair and eyes. Parsley plays a role in the development of melanin, which is responsible for giving our skin its brown color as well as our hair and eyes its special pigment as there is a considerable amount of copper in parsley.

Anti-Inflammatory Properties: Parsley has historically been used for insect bites, toothaches, bruises and raw skin in the Mediterranean region.

Diuretic effects: Parsley was used as a diuretic that helps to control numerous diseases such as kidney stones, infections of the urinary tract and stones of the gallbladder. Edema is a medical disorder where, under normal circumstances, a patient maintains fluid in the body more than he or she should keep.

Insomnia: Parsley consists of Iron which is useful for the treatment of insomnia in the human body and also enhances people's sleeping habits and efficiency by controlling circadian rhythms. Proper counting of red blood cells can also result in fewer blood pressure spikes, which can often keep people awake at night.

Anti-Diabetic Properties: In Turkey, Parsley has historically been used as a diabetes medicine.

Strengthens the Immune System: The vitamins, minerals and antioxidants present in parsley help to strengthen immunity. Vitamins like vitamin C, vitamin A, vitamin K, folate and niacin all function on various aspects of the immune system. Vitamin A directly affects lymphocytes or white blood cells, and thus enhances their influence. The parsley-containing chlorophyll also has antibacterial and anti - fungal effects.

Osteoporosis: Walking is beneficial for osteoporosis and helps maintain bone health. Osteoporosis is caused by low calcium levels in the bones, and by an amino acid called homocysteine. The ingestion of folic acid will break down the amino acid. Because of this reason, parsley is known as one of the best sources of calcium apart from dairy products and vegetables. It also contains a suitable amount of folic acid that may help break down homocysteine.

Protects Eye Health: Parsley consists of 253 μg of vitamin A, a critical part of the molecule of rhodopsin, which is stimulated when light shines on the retina, which sends a signal to the brain that results in vision. Beta carotene, the vitamin A -type found in plants, plays a role in preventing macular degeneration, which is the leading cause of blindness associated with aging.

Pregnancy and Birth Defects: Vitamin B9 is a reliable resource for reducing pregnancy and childbirth defects. It is normal that pregnant women who have a diet deficient with this vitamin are far less likely to deliver a healthy child. Neural tube defects include the birth defects caused by its deficiency. Vitamin B9 rich foods such as parsley ,therefore, need to be properly included in order to effectively improve the chances of giving birth to a healthy, happy child.

Helps with Digestion by Relieving Constipation: Magnesium helps relax muscles within the digestive tract including the intestinal wall that regulates your bathroom capacity. Since magnesium helps neutralize the acid of the stomach and pushes the stool through the gut.

Precautions

- A headache may perhaps take place with greater usage of parsley.
- You might notice a lack of balance that may result in a fall.
- Warning signs of damage to the kidney may well occur along with the use of this herb.
- Using the over-consumption of parsley supplements, seizure activity may occur just like convulsions.

3.50 PEPPERMINT (MENTHA ×PIPERITA)

Peppermint is a strong perennial aromatic herb that belongs to the *Lamiaceae* family of mint. It has a heavy, sweet smell and a dry, pungent taste with a refreshing aftertaste. Leaves are used as culinary herbs, and dried flowers are used to add flavor to sweets, candy, salads, beverages and other foods. The essential flavorings oil is also used.

Digestive conditions: It is used in the treatment of nausea, ulcerative colitis, gastrointestinal irritable syndrome, dyspepsia, Crohn's and colic. Their choleretic and bitter properties promote digestive secretion. The bitter principle promotes pancreatic secretions and choleretic effects that enhance the bile flow and encourage bile solubility.

Infections: Peppermint has antiseptic and diaphoretic properties that are useful in the treatment of colds and flu, bronchitis cough, and sinusitis. Peppermint can relieve the symptoms of digestion, such as nausea, bloating and indigestion.

Relieve tension headaches: Peppermint acts as a pain reliever and relaxing muscle which decreases some types of headaches. Peppermint oil enhances blood flow and provides a feeling of cooling. Apply peppermint oil to the front, and temples after two hours to reduce discomfort. Analysis of 14 clinical trials conducted on 2000 children showed that peppermint reduced abdominal pain frequency, duration and intensity.

Freshens breath: Peppermint is used as a mouthwash, toothpaste, and chewing gum flavoring. In addition to pleasing odor, it has antibacterial properties that remove germs that cause dental plaque and enhance breath. Peppermint Tea consumption boosts oxygen.

Relieve sinus: Peppermint has an antiviral, antibacterial and anti-inflammatory property that combats clogged sinuses caused by infections, allergies and the common cold. The peppermint tea has not been tested for nasal irritation effects.

Provides energy: Peppermint tea increases energy levels and reduces tiredness during the day. The natural compounds offer beneficial energy effects.

Relieve menstrual cramps: Peppermint has relaxing muscle properties that provide relief from menstrual cramps. Peppermint tea helps symptoms improve. The peppermint extract capsules serve as a non-steroidal anti - inflammatory drug in reducing pain intensity and duration.

Combat bacterial infections: Peppermint tea has antibacterial effects that kill the bacteria effectively. This prevents and destroys the growth of bacteria born in food like Salmonella, Listeria and *E. coli.* It eliminates many types of bacteria that cause human illness including Staphylococcus and bacteria linked to pneumonia.

Improve sleep: Peppermint tea acts as a soothing muscle that helps you to relax before bedtime. Peppermint helps boost recovery.

Supports weight loss: Peppermint tea is calorie-free and naturally sweet taste making it an intelligent choice for weight loss. Compared with taking peppermint, the peppermint oil capsule decreases appetite.

Precautions

- Avoid peppermint oil at the time of pregnancy.
- Consult the doctor before use to prevent side effects.

3.51 POPPYSEED (PAPAVER SOMNIFERUM)

Poppyseed is an oilseed obtained from opium poppy, an annual herbaceous plant in the *Papaveraceae* (poppy family) native to southeastern Europe and western Asia and now widely cultivated in Europe, Asia, China and North Africa for its seeds used in baking, as a condiment, and for its edible oil. Poppyseeds have a wide range of benefits. They serve as a treatment for mouth ulcers, sleeplessness and improve digestion. Poppyseeds also greatly enhance female fertility, and can also help to prevent cancer.

Prevent Cancer: Opium poppy genes had shown promise in developing a cancer drug. The drug called noscapine has been found to fight tumors and even reduce cancer cells in the breast and prostate. Often, it can stop the proliferation of cancer cells.

Cure Sleeplessness: Eating a drink of Poppyseeds reduces cortisol levels in the body. This led to a calming effect as gradually the stress levels decreased. More interestingly, there was a marked improvement in the circadian rhythms within the test subjects. The individuals have also felt less tired. Opium poppy is noted for being particularly effective in sleep induction. But when offering it to children, exercise caution.

Improve Heart Health: Poppyseeds are a glorious source of dietary fiber that helps lower cholesterol levels and ultimately improves heart health.

Help Treat Kidney Stones: Poppyseeds consist of a good amount of potassium, one of the minerals known to offer protection against kidney stones and even cure them.

Have Analgesic Effects: The seeds have been used as an analgesic since ancient times, or to relieve pain. As per a health report, even poppy tea (prepared by mixing the seeds in hot water) can give pain relievers benefits.

Digestive Help: If you have constipation or diarrhea you need fiber in your diet. Poppyseeds have a surprisingly high level of dietary fiber that can help to bulk up the stool or facilitate peristaltic motion to control the motions of your bowels. They were related directly to soothing inflammation and pain, thus promoting normal functioning in your stomach.

Enhance Thyroid Function: Zinc is one of the minerals ,which the thyroid gland needs to function optimally. In this dimension, Poppyseeds can play a vital role, as they are rich in zinc. In many parts of the world, iodinated Poppyseed oil has also been used to treat iodine deficiency-the cause of thyroid dysfunction.

Promote Vision Health: As we age, our eyes of course start to fail us, but we can do some things to stop this cycle. Antioxidant compounds are a great way to protect our eyes, but so is zinc, found in large quantities in hemp, which can reduce the occurrence of macular degeneration and preserve strong vision.

Improve Female Fertility: Using Poppyseed oil to flush the Fallopian tubes will improve fertility. The oil will dissolve any debris or mucus inside the tubes and increase fertility chances.

Help Treat Eczema and Inflammation: Poppyseeds are efficient in treating eczema because of their high linolenic acid content. Poppyseeds can be ground together with lime juice after being soaked in water or milk, to form a smooth paste. Applying this paste outside helps to treat burns and itchy skin. It also acts as an effective painkiller by calming the pain caused by inflammation of the skin.

Aid in Diabetes Treatment: As already mentioned, Poppyseeds are high in fiber; they might be a favored addition to a diabetic diet. And the manganese in the seeds could also help in treating diabetes.

Help Achieve Clear Skin: Poppyseed scrub can help to clean your face, making it look smooth and without blemish.

Boost the Immune System: The zinc in Poppyseeds can help to improve the immune system. Seeds also enhance immune cell function within the body. And in this way, the iron in the seeds helps too. The zinc in the seeds also helps to fend off lower breathing problems.

Strengthen Bones: Poppyseeds are rich in copper and calcium. While the former enhances bone health, the latter strengthens both the bones and the connective tissue. Seeds are also rich in phosphorus and make up part of the bone tissue along with calcium. And the manganese in the seeds helps produce collagen, a protein that protects the bones against serious damage.

Cure Mouth Ulcers: Poppyseeds are known to have a cooling effect on the body, which plays a role in treating mouth ulcers.

Improve Brain Health: The brain requires vital nutrients such as calcium, iron, and copper-minerals which are abundant in Poppyseeds. These minerals aid in neurotransmitter control and improve brain function. Poppyseeds may also reduce the risk of developing a cognitive disorder. However, in this respect, further research is needed.

Relieves Body Pain: Poppyseeds are made of analgesic properties that can effectively alleviate various body pains. Poppyseeds have been commonly consumed in the form of teas to relieve the pain and related factors. If you often have headaches, consider sipping on some warm tea that's made with it.

Help Treat Dandruff: Dandruff is a common problem that can lead to hair loss if left untreated. A mixture of soaked Poppyseeds, hanging curd, and a teaspoon of white pepper may be rubbed on the scalp to treat dandruff.

For Psychological Health: Foods that contain components are rarely found to improve your psychological health. Poppyseeds are among them because they are loaded with nutrients like copper, iron, and calcium. The nutrients can control neurotransmitters when consumed regularly which will improve cognition and concentration.

Rosemary (*Rosmarinus officinalis*)

Rosemary is a sun-loving shrub native to southern France and other Mediterranean regions and is a member of the *Lamiaceae* mint family as well as many other herbs, such as lavender, oregano, basil and thyme. The leaves are used in foods such as stuffing and roast lamb, pork, chicken and turkey, particularly during holiday cooking like Thanksgiving and Christmas, as a flavoring. Rosemary also has medicinal uses, including its excellent decoration and distinct taste. Some of the common health benefits from rosemary are listed below

Detoxify the Body: Rosemary is somewhat diuretic in nature, which means it may help to more efficiently flush out toxins during urination. Additionally, it can also help to push out bacteria, salts, toxins and even extra fat when consumed regularly by increasing the rate at which water leaves the body. Rosemary has been associated with lower cirrhosis rates and quicker healing time of the liver, which is one of the slowest organs to treat, in terms of the particular organ it supports.

Immune System Strength: In rosemary, the active components are in essence antioxidant, anti - inflammatory, and anti-carcinogenic. This means a three - pronged assault on many different diseases as well as on pathogens which could disrupt the immune system or harm the body 's integrity. Antioxidant compounds form a secondary line of defense behind the body's own immune system, and rosemary contains a significant amount of those potent compounds such as rosmarinic acid, caffeic acid, betulic acid, and carnosol.

Antibacterial Potential: Rosemary consists of amazing quantities of immune-boosting components that are specifically potent against bacterial infections, especially those in the abdomen. *H. pylori* bacteria are common and very dangerous pathogens that cause stomach ulcers but when ingested, rosemary has been shown to prevent their growth. Rosemary is similarly related to avoiding Staph infections that kill thousands of people every year.

Mood and Stress: The rosemary fragrance has been associated with enhancing mood, clearing the mind and relieving stress in those with chronic anxiety or stress hormone imbalances. This can have similar effects when the plant is eaten or applied topically in some form of leaf salve. Aromatherapy often uses essential rosemary oil for this purpose, but in order to have positive effects on stress and mood, the concentration of active components is not required.

Stomach Soother: Rosemary has historically been used as a natural remedy for upset stomachs, constipation, bloating, diarrhea and everything in between in hundreds of ailments. Its anti - inflammatory and stimulant effects are mostly the cause of these symptoms, so you can easily regulate your bowel movements and your gastrointestinal system by including rosemary in your weekly diet.

Memory Booster: Daily rosemary consumption is extremely beneficial for memory improvement. It was said to improve memory and help to increase focus and intelligence. Although many of these findings are still being researched and tested, their brain effects may indicate an increase in memory retention, which is never a bad thing; keeping the mind active will help keep it young. In addition, rosemary has been associated with improving cognitive activity in the elderly and those suffering from more severe cognitive disorders, such as Alzheimer's or dementia.

Pain Relief: Because of their analgesic quality, rosemary has been applied topically to the infected area of pain in a paste or salve for hundreds of years. Rosemary acts as a pain reliever when ingested orally, making it harder to reach sites, such as headaches and irritation from a disease. In fact, one of the rosemary's most popular uses is for treating migraines. Linked to reducing the severity of migraine symptoms was the use of a decoction in the temples or just smelling the rosemary aromas.

Migraine Aid: Rosemary is one of the most common natural treatments for migraine, and has been used for centuries. One way this can be used for migraines is by boiling water with rosemary and breathing in the steam. You would usually want first to boil the water, then place it in a large bowl and breathe it in with a towel over your head.

Stimulate Blood Flow: Rosemary acts as a stimulant to the body, which enhances red blood cell production and blood flow. This helps to oxygenate essential organ systems and body areas, ensuring that the metabolic activity in those areas runs seamlessly, as well as facilitating the flow of nutrients to cells that require repair.

Breathe Freshener: Rosemary works as a natural antibacterial agent as a wonderful breath-freshener that enhances oral health. Steep rosemary leaves in a glass of hot water, then gargle or swallow the water in your mouth to remove bacteria and breathe new and clean all night.

Anti-Inflammatory Qualities: Rosemary acts inside the body as an anti-inflammatory agent. Carnosol and Carnosic acid are actually two strong antioxidant and anti-inflammatory compounds found in the rosemary that have been associated with reducing muscle, blood vessel, and joint inflammation. For many things, this makes rosemary an effective treatment, such as blood pressure, gout, inflammation, and injuries sustained from physical exertion or surgery. Rosemary is beneficial for those anti - inflammatory effects in oral or topical form. Additionally, reducing inflammation in the cardiovascular system can help to boost heart health and prevent the appearance of atherosclerosis.

Skin Health: Rosemary's anti - aging properties are fairly well known. Although more commonly thought about in the essential oil type, rosemary leaves can also affect the skin internally or topically, and it has been shown to enhance the skin's youthful consistency, while also curing blemishes and replenish the natural shine and hydrated appearance of the largest organ in your body.

Precautions

Given the rosemary's many health benefits, there are unique precautions you must take. A number of them are;

- This specific herb has not been studied in babies, so don't give the people under the age of 18 a rosemary as medicine. Nonetheless, consuming like a spice really is healthy.
- Do not consume more than 4g to 6g of the dried herb each day.
- Rosemary oil shouldn't be taken orally; it may be toxic.
- Pregnant women should not take the rosemary as a supplement simply because higher rosemary dosages can induce a miscarriage.
- The herb might not be safe even for ladies who are nursing.
- If you have high blood pressure, ulcers, Crohn's disease or even ulcerative colitis, don't take rosemary in any form.
- In higher doses, rosemary leaves may cause side effects such as vomiting, spasms and, in many cases, lung fluid accumulation and even coma.

- Rosemary can inhibit blood thinners like warfarin or aspirin and may affect the ability of the blood to coagulate.
- The herb may hinder the action of ACE inhibitors you take for high blood pressure.
- Rosemary will alter blood sugar levels; thus, if you are on any diabetes medication, you should not take it.
- Rosemary is really a diuretic and may even boost the chance of dehydration.

3.52 RUE (RUTAGRAVEOLENS)

Rue is a hardy, evergreen, shrubby plant native to the Mediterranean region and the Canary Islands, growing on rocky areas, thickets, and dry banks usually found on calcareous stones. Rue helps the body fend off fungal infections such as food and dermatitis from athletes. Depending on the severity of the infection, use the rue directly on the skin, especially one suffering from athlete 's foot. It gives relief to itchy skin.

Clear skin: Rue has antifungal properties that help cure skin de-complexion and provide lovely, clear skin. The antioxidant properties ensure that the skin wards off free radicals that cause the skin to age prematurely.

Eliminates lice: The essential oil and extracts that are often used to prevent an infestation with the lice. Its daily use repels mischievous little hair parasites.

Acts as a sedative: Despite its sedative properties, it is effective in calming hysteric and epileptic attacks. It soothes nerves because it is a neurotoxin, by inducing numbness.

Anti-bacterial activity: The anti - bacterial properties assist in the removal of bacteria. The tea prepared urinary tract, colon and intestine infections using this herb battle. It impedes the risk of food poisoning and Salmonella infections.

Anti-Spasmodic: Rue is of interest to those who have muscle pain. It's a treatment for back pain, anxiety and menstrual cramps.

Anti-Inflammatory: Rue decreases arthritis causing inflammation. Regular intakes of 1 to 2 cups of Rue herb tea aid in relieving stiff muscles and joint pains.

Treat fungal infections: Rue herb combats fungus-induced skin problems and skin infections. Common infections of the fungus include foot and nail funguses of athletes.

Calm nerves: The herb possesses soothing properties. It soothes tension and anxiety. The herbalist uses this to treat nervous headaches, dizziness, tension and anxiety in general. The tea offers a sleeping tone.

Promotes menstruation: Rue has emmenagogue properties that induce menstruation and promote woman cycle regularity.

Precautions

- The use of high - dose rue extracts induces diarrhea, extreme gastric pain, systemic complications and even death.
- The herbal preparations made from this cause severe phytophotodermatitis which results in burning on the skin like blisters.
- When used in large doses, it is toxic and also causes miscarriages.
- Use it with caution.
- It might cause phototoxicity when applied to the skin with sun exposure.
- Leaves can cause blistering.
- In some people, it may cause phototoxicity-when applied to the skin with sun exposure, the oil and leaves can cause blistering.
- It induces minor side effects such as increased sun exposure sensitivity, dizziness, spasms, rashes, changes in mood, sleeplessness and inflammation of the stomach.
- Individuals with stomach or digestive disorders such as ulcers, colitis, and inflammatory bowel disease should be treated as the condition could get worse.
- Never apply the substance directly to the skin without a carrier oil.

3.53 SAFFRON (CROCUS SATIVUS)

Saffron is in reality a reddish-golden colored spice developed from the styles and also saffron crocus (*Crocus sativus*) flower stigmas. Used primarily to cook along with a coloring agent like a seasoning agent. Saffron also offers a number of health benefits which make it one of the most effective herbs in nature. Safran is noted mainly for its use in food products. However, it has extremely considerable nutrition as well as useful chemical substances in providing therapeutic advantages. Let's explore a number of the saffron uses below:

Appetite Suppressant: Though this may not be a benefit to health, it is a benefit. Weight issues are a worry in the USA. Individuals shovel into diet pills as well as chemical suppressants of appetite when it has been found that saffron is an excellent suppressant of appetite that works well.

Anti-Depressant Properties: It is really relatively well known as well as accepted that saffron is actually an effective antidepressant and mood booster as well. A small dose of 30 mg of saffron starts to boost mood in stressed individuals in an hour, and the effects can last up to 12 hours, too. This small dose is equivalent to roughly 10 dried stigma fibers. In less severe cases, just smelling the fresh saffron scent could be enough to boost the mood.

Supports Eyesight: Saffron may slow down or even prevent age - related loss of vision, and may also resolve macular degeneration. Saffron is certainly an anti - oxidant, but it seems to have additional qualities that affect vision specifically.

Improves Memory: Saffron is made up of compounds known as crocetin and crocin which could improve memory and mental performance as well. Such types of characteristics could be useful in addressing degenerative brain disorders, such as Alzheimer's disease.

Treats Cancer: The herb 's action is clearly not understood, but saffron prevents the spread of tumors while leaving the healthy cells untouched. Saffron can also be used efficiently in conjunction with some conventional treatments for cancer while leaving unadulterated effects of the medication.

Improved Mental Functioning: The crocin compound, which is found in saffron, is believed to be mostly responsible for why the spice enhances mental functioning. In particular, saffron improves memory, learning and information-recall capacity. Several people claim that saffron could be used for aging to cope with the diminished mental functioning of many people.

Saffron During Pregnancy: Women are usually advised to consume saffron milk when pregnant since it improves pelvic blood flow. In addition, saffron has carminative properties-meaning it will help to prevent cramping. Gas and bloating are typical problems while pregnant, and a glass of saffron milk can help relieve digestion and reduce flatulence.

Asthma: Saffron is an effective asthma treatment, in addition to kidney and liver problems. Asthma occurs primarily because of lung tissue swelling and also because of the small airways. Saffron works well to clear the airways, as it reduces the inflammation. This particular allows for simple breathing that may help to reduce the severity of an asthma attack.

Protects Heart: Safran antioxidants help lower cholesterol and triglycerides within the body. Both cholesterol which lipoproteins contribute to fat accumulation within the arteries which in turn causes circulatory problems. Safran crocetin functions well to decrease this lipoprotein in the bloodstream. It also helps with oxygen delivery, which greatly helps to make the circulatory system efficient and removes the plaque through the arteries, and helps make the heart healthy.

Weight loss: Saffron is effective in the weight loss diet since it has the capacity to suppress the appetite and lower cravings for fattening foods. Saffron increases the quantity of a hormone, serotonin within the brain, which will help avoid us from overeating and especially through craving of unhealthy foods. It is important to note that this can also be dangerous as it can cause anorexia for those who are not careful and hence the need to seek medical advise from a nutritionist before taking the herb.

3.54 SAGE (SALVIA OFFICINALIS)

Commonly known as Sage, Broadleaf Sage, Common Sage, Dalmatian Sage, Garden Sage, Kitchen Sage, Narrow-leaved Sage, Sage, Salvia, Sarubia, Spanish Sage, TibbiAdacayi, Sawge, Red Sage, Broad-leavened White Sage, *Salvia salvatrix* is the culinary sage familiar to most cooks. Sage is in fact a blessing in terms of the various health benefits that it provides. From the Arabs to the Chinese this herb has been recognized by all as a key to a healthy life. In reality, it was awarded the title of "Herb of the Year" by the International Herb Association in 2001 because of its health-benefiting properties. Some of the potent health benefits of this aromatic herb are given below.

Cognitive Boost: Small amounts of sage, whether smelled or eaten, help to boost memory retention and recall capacity. The brain activity often demonstrates enhanced concentration and focus on a chosen topic, suggesting that it can be a delicate but powerful brain booster for young people in school or for those in challenging; intellectually demanding professions, adding a little bit of it to your diet.

Immune System Strength: Sage herb is composed of certain antimicrobial properties. It is typically taken in comparatively small amounts; a topical application of sage (salve or tincture) can be produced and used to avoid bacterial and viral infections that invade the body through the skin; We often find illness entering through our nose or mouth, but also the skin can be infected and used as a portal to foreign agents. A topical cream or antibacterial regimen that includes sage might be an extra line of defense against that kind of vector of disease.

Treatment of Hair Loss: Since ancient times Sage has been used to combat hair loss and baldness. Sage herb consists of beta-sitosterol, a compound of 5-alpha reductase found to be effective in the treatment of baldness in the male trends.

Diabetes Management: Sage herb is pretty effective in diabetes management. It is made up of certain extracts and chemicals which mimic the drugs typically recommended for diabetes management. It helps to control and discourage the release of stored glucose in the liver, preventing large blood sugar spikes that may help prevent the onset of type 2 diabetes, or at least manage the disease if it has already formed.

Inflammation Issues: Chewing on sage leaves is not always the most fun cure, as the taste can be quite strong, but it can be the most effective way of getting the fastest organic compounds in your body. The trick may also be due to the production of a tincture or steeping leaves, but if you suffer from inflammatory problems, particularly in the respiratory or gastrointestinal tracts, this sage brew will relieve the inflammation.

Digestion: Sage herb consists of a good amount of rosmarinic acid which acts as an anti-inflammatory agent in several parts of the body, even in the stomach, where it prevents gastric spasms and can significantly reduce the incidence of diarrhea and gastritis in patients suffering from the painful and embarrassing conditions. It can get the entire digestive process back on track and reduce inflammation throughout the gut by including it in your meals.

Fights Depression: Sage is also used as a natural depression treatment. The abundant presence of antioxidants such as diosmetin, apigenin, and luteolin helps to reduce depression and stress. It is recognized that frequent consumption of a cup of sage tea improves your spirits and thus ensures that you conquer the negative thoughts created by depression.

Bone Strength: Sage herb consists of a higher vitamin K level, an essential body vitamin that is not present in many common foods. Vitamin K is an important element in the development of bone density and in ensuring that our bones are intact as we age. Whether you suffer from early signs of osteoporosis or have led a sedentary lifestyle that is rather nutrient-poor, your bone health is likely low. The use of sage leaves in your diet will significantly increase your vitamin K levels, as a single service has 27 percent of your recommended daily intake.

Good for Your Heart: A 2 tsp of sage serving consists of about 20 mg of calcium and a good dose of potassium, both of which are essential for the proper functioning of the cardiovascular system and for sustaining heartbeat rhythms. This herb lacks cholesterol and has low sodium content. Hence, those who have cholesterol issues should use it to prevent cardiovascular disorders.

Treatment of Sore Throat: Sage infusion in the treatment of inflamed throat and tonsils as well as ulcerated throat has been found to be effective. You can boil a pinch of dried sage leaves in about 100 ml of water for this purpose, and infuse them for 15 minutes. Strain and sweeten the concoction with some honey. The use of this as a gargle will provide relief twice a day.

Delays Premature Ageing: Sage leaf oil is known for improving circulation in the blood and stimulating cell renewal. This herb is rich in vitamin A and calcium, which play an important role in the regular regeneration of the cells. Antioxidants present in sage combat free radicals relieve the wrinkles, fine lines and other symptoms of aging.

Dental Health: Sage herb plays a big part in our dental routine. It can be used to treat gum disease, throat infections and cancers as an effective mouthwash. You can simply gargle strong sage tea or freeze it into ice cubes if you are bothered by cankers.

Cognitive Disorders: As mentioned earlier, sage has the ability to encourage brain function to enhance memory and concentration; however, it also works to eliminate possible cognitive disorders like Alzheimer's and dementia.

Antioxidant Impact: Chronic conditions and degenerative diseases maybe some of your life's most crippling and hazardous health concerns. Many of these health problems are caused by free radicals, the hazardous by cellular metabolism products that attack healthy cells, causing apoptosis or mutation.

Treatment of Excessive Sweating: Taking either a dry leaf extract or a leaf infusion will help reduce sweating by up to 50%. It's a perfect deodorizer that dries up suddenness and helps reduce body odor.

Skin Conditions: Using sage leaves or a plant tincture that has been shown to be effective against certain skin conditions, including eczema, psoriasis and acne, topical salve can be made. Such unsightly blemishes can be soothed quickly and their presence can be reduced gradually if you apply sage extracts and salves regularly to the inflamed or irritated area.

Cure for Alzheimer's: Sage helps in enhancing memory power when used regularly. It is therefore suitable for treating cases with varying memory loss like moderate Alzheimer's disease. It also enhances the processing of information among those with such illnesses.

Precautions

- The plant can be toxic when used in excess or when taken for extended periods.
- Symptoms include: restlessness, vomiting, vertigo, tremors, seizures.
- Avoid if predisposed to convulsions
- This treatment should not be recommended for pregnant women or for individuals with epileptic fits.
- In larger doses, however, it can cause epileptic fits, giddiness ,etc.
- Sage should not be used to suppress perspiration in fevers.

3.55 SAVORY (SATUREJAHORTENSIS AND S. MONTANA)

Summer savory is an annual herbaceous crop that is native to Europe and is spread throughout the world. As an aromatic plant, native to southern Europe and the Mediterranean area (in our days distributed across both hemispheres, in warmer regions and as a pot culinary herb), its aerial parts (especially the leaves) have traditionally been used for culinary and medicinal purposes. Health Benefits of Summer Savory include;

- Leaves come with carvacrol and thymol. Thymol battles different strains of bacteria, such as *Bacillus cereus* and *E. coli*. Thymol has antiseptic and antifungal properties that protect the body against various infections of the fungi.
- Leaves and tender shoots have antioxidant properties that reduce oxidant damage in the body. It reduces bad cholesterol, promoting good cholesterol.

- The essential oil has anti - inflammatory properties which can soothe joint pain by reducing joint inflammation. It also reduces muscle stiffness and tenderness.
- It has linalool, tannins, carvacrol and cymol that are beneficial in the treatment of intestinal disorders such as cramps, indigestion, bloating, vomiting, nausea, diarrhea and appetite loss.
- The aphrodisiac property is used to increase libido.
- Tea prepared from the leaves is used to treat problems with internal health and to prevent urine retention, uterine contraction and colic.
- The herb treats sore throat, wounds and skin infection.
- Dry savory has vitamin B6 or pyridoxin that maintains levels of GABA (soothing neurotransmitter) in the brain with the function of a stress buster.
- Vitamin C develops resistance to infectious agents and prevents harmful free radicals that are pro - inflammatory.
- Vitamin A is an antioxidant and fat-soluble vitamin thatmaintains healthy mucosa and skin.
- Summer savory helps to combat infections.
- This facilitates gastric juice secretion, which provides relief from various digestive problems, such as bloating and flatulence.

Precautions

- Not to be prescribed for pregnant women.

3.56 SESAME (SESAMUM INDICUM)

The erect and annual plant native to Africa and India is commonly known as Beniseed, Benniseed, Benneseed, Sesame, Til, Gingelly, Goma, Gergelim, Ajonjolí, Gimgelim, Simsim, Wangila, Ufuta, Sesam, Zhi ma, Hu ma, Chamkkae, Benne, Beni, Benne seed, Sesame seed, Sim sim, Abongra, Zhi ma, Hu ma, Sesamzaad, Goma, Sesamo, Shiro goma and Sesam. Sesame seeds are an important source of manganese, copper, magnesium, calcium, vitamin B1, phosphorous, molybdenum, zinc, dietary fiber and selenium. Sesame seeds also contain sesamolin and sesamine, which helps lower human cholesterol and also decreases high blood pressure. This has been in use for over 5000 years. For culinary uses, it could be added. These have dietary, medicinal, and curative properties. It helps to cure health disorders such as low blood pressure, diabetes, cancers, sleeping disorders, asthma, dental issues, detoxification, depression, chronic stress, increased circulation, healthy growth, respiratory health, heart health and radiation-prevention.

Health problems: Sesame seed contains sesamine,which is oil - soluble furofuran lignin. Rodent-based studies shows it helps lower the cholesterol level, normalize blood pressure, resulting in weight loss and prevents liver problems. Sesamine biosynthesis contains two major genes piperitol / sesamine synthase genes and encoding dirigent protein.

Promotes cholesterol: The rice bran oils and sesame blend helps to lower blood pressure and increase cholesterol levels. Sesame oil has low saturated fat which helps to boost patients ' cholesterol levels. It reduces the chances of developing heart disease. The effective results are supported by antioxidants and healthy fatty acids such as sesamoline, sesamol, oryzanol, and sesamin.

Prevents *E. coli*: The sesame-rich diet helps prevent gastrointestinal infection and reduces the severity of symptoms caused by Salmonella or *E. coli*. The other foods that provide these beneficial effects are onion, yeast, and pumpkin. Sesame seed extract and konjac gum help to prevent Salmonella and *E. coli*attaching to a host. The fiber helps keep the digestion and stomach healthy.

Maintains blood pressure: Sesame seeds have natural oils which help lower hypertension, cardiovascular strain and prevent heart disease. The magnesium also acts as a vasodilator. Sesame seeds are filled with about 25 percent of the daily requirement with minerals.

Prevents cancer: Sesame seeds contain a variety of minerals and vitamins which help lower cancer chances. It also has phytate, which is a compound that prevents cancer and acts as an antioxidant that helps to reduce the activities of free radicals which are the harmful by-products of cellular metabolism and leads to health problems such as premature aging, heart disease and cognitive impairment. Sesame seeds are associated with lung, breast, leukemia, colon, pancreatic, and prostate cancer prevention.

Manage diabetes: Sesame seeds are the perfect magnesium source that helps reduce the risk of diabetes and control health conditions. Alternatively, Sesame seed oil has a positive effect on Type 2 diabetic patients on diabetic medications such as glibenclamide. It helps regulate the levels of glucose and insulin and also regulates the symptoms of diabetics.

Healthy bones: Sesame seeds contain ample amounts of minerals such as calcium, zinc and phosphorus that help improve the health of the bone.

The minerals are vital components for bone development, bone repairers that may become weakened due to injury and osteoporosis chances.

Assist digestion: Sesame seeds are filled with sufficient quantities of fiber that is essential for healthy digestion. This adds bulk to bowel movements and helps the digestive tract work quickly. It helps to reduce the problems of diarrhea, constipation, avoid the health of the colon and reduce the risk of cancer and gastrointestinal disease. This helps lower the bad cholesterol from blood vessels and arteries and prevents heart attacks, atherosclerosis and strokes as well.

Reduce inflammation: Sesame seeds help to decrease inflammation in bones, muscles and joints, and also reduce pain caused by arthritis. Additionally, copper helps reinforce blood vessel walls, joints, and bones. Copper is essential for the absorption of iron that is important to hemoglobin. The adequate amount of copper supports the absorption and carries oxygen for breathing in the body.

Protects radiation: Sesamol is an organic compound found in the seeds of sesame that helps to prevent radiation effects on DNA. This radiation can occur from the radiotherapy and chemotherapy treatment for cancer. It helps in avoiding DNA mutation. The sesame seed helps to increase the intensity and lower the cancer chances caused by cell mutation.

Hair and skin health: Sesame seeds have sufficient quantities of zinc which are important for the production of collagen, strengthening hair, muscle tissue and skin. The oil extracted from Sesame seeds helps to reduce the appearance of skin marks and burns and also decreases premature aging chances.

Metabolic functions: Sesame seeds are rich in protein which is broken down in the human body into functional proteins. This encourages healthy cell growth, overall strength, energy levels, mobility and metabolic functions.

Precautions

If consumed in large quantities it can lead to;

- Unhealthy Weight
- Colon Cancer
- Diverticulitis
- Allergic Reactions
- Anaphylaxis

- Infection of Appendix
- Diarrhea
- Skin Rashes
- Hair Loss
- Miscarriage

3.57 STAR ANISE (ILLICIUM VERUM)

Commonly known as Star Anise, Star Aniseed, or Chinese Star Anise is a spice that looks closely like anise in taste, obtained from *Illicium verum*'s star - shaped pericarp. Star anise is a well-known digestive aid that is used to relieve cramps, indigestion, bloating gas and indigestion. Star aniseed seeds are abundant sources of B - complex vitamins such as thiamine, pyridoxine, niacin, and riboflavin, and are known to increase neurochemicals in the brain. It is supplemented with vitamins such as vitamin A and vitamin C which are anti-oxidant. Listed below are some of the health benefits of Star Anise:

Fights against free radicals*:* Antioxidants fight free radicals which can cause cell damage and can be responsible for diseases such as diabetes, heart problems or cancer. As a metabolic by-product, it is continually produced in our body. Only by eating a rich source of antioxidants can the excessive production and deposition be neutralized. Star anise has potent antioxidant properties.

Rheumatism and joint pain: Star anise essential oil when applied topically is a great treatment for rheumatism and back pain. It should be mixed with a carrier oil before using the essential oil and massaged in the affected areas.

Combats fungal infections: Star anise is quite useful in dealing with a common skin problem called candidiasis, caused by a fungus called *Candida albicans*. Typically, this fungus affects surface skin, mouth, throat or even genital areas.

Fights against bacteria: Thanks to its expectorant element, star anise as an antibacterial agent has been found to be effective in declining cough thanks to asthma and bronchitis. In addition, star anise has proven effective against nearly 70 drug - resistant bacterial strains.

Effective in flu treatment: Star anise is a good source of Shikimic acid, and has a powerful antiviral property when combined with another rich antioxidant ingredient, quercetin. Such two chemicals combined helps to prevent influenza.

Beneficial for sleep disorders: Star anise has mild sedative effects, which can help the nerves settle down and also ensure a good night's sleep, according to traditional usage and circumstantial evidence.

Good for Cough: Star anise tea is one of the excellent natural remedies to soothe your sore throat and relieve a cough. Just prepare your tea, and drink it up to three times a day.

Improved Digestion: Star anise is often prepared as a tea particularly in countries in South East Asia and China. It is mainly used for treating a variety of digestive problems such as nausea, abdominal cramps, indigestion, bloating, and constipation.

Good for Heart Health: Star anise oil helps keep blood pressure. This trims down the heart pressure. It also improves blood flow as well as keeps the heart healthy and safe, as the oil detoxifies. The oil is rejuvenating the spirit, and keeping it alive.

Women's health: Traditionally in China, star anise has been used to help women during pregnancy and to new mothers wanting to increase their milk production. It is often recommended by traditional Chinese practices for pregnant women.

Stimulates Pancreas: Star anise oil helps keep blood pressure. It trims down the heart rate. It also improves blood flow as well as keeps the heart healthy and safe, as the oil detoxifies. The oil rejuvenates the heart and also keeps this vibrant.

Precautions

- According to Japanese , side effects include nausea. Vomiting, hallucinations, agonizing jitters and rapid eye movement.
- It may be dangerous for use in new and expectant mothers and should be avoided except under the supervision of your doctor.
- Star anise tea should not be drunk by people with ovarian, breast or uterine cancer.

3.58 THYME (THYMUS VULGARIS)

Thyme is good for helping both the immune and respiratory systems and the digestive, nervous and other systems of the body. When it comes to staying alive, it is a huge powerhouse. Find out some of the best ways to improve your health with this plant.

Lowers Blood Pressure and Cholesterol Levels: It has been shown that thyme ingestion induces anti - hypertensive activity which makes it a great herbal choice for anyone with symptoms of high blood pressure.

Improves Circulation: Thyme is good for helping the immune, respiratory and digestive, nervous and other body systems. When it comes to staying alive it is a powerful powerhouse. Check out some of the top ways this herb can help improve your health.

Eye Care: The concentration of carotenoids and vitamin A in thyme makes it an important antioxidant agent for the protection of your eyes. Carotenoids help neutralize the eye system's free radicals and delay the progression of macular degeneration, and avoid cataracts.

Reduces Stress: Thyme has decent amounts of vitamins (B6), which has a powerful effect on certain brain neurotransmitters directly related to stress hormones. Daily use of thyme in your diet can help boost your mood and, when stressed, ease your mind.

Naturally Remedies Bronchitis: Thyme has been used for treating cough and bronchitis for centuries.

Prevention and Treatment of Anemia: Thyme is an outstanding iron source, supplying almost 20 percent of this mineral's daily recommended allowance. Iron is essential for developing red blood cells and producing energy. Iron deficiency can lead to anemia, fatigue and increased susceptibility to infection. The use of thyme in your diet will help to prevent and treat anemia and other blood disorders.

Protects Heart: Thyme consists of a rich blend of antioxidants, minerals, and vitamins that can have many minor effects on the heart but most notably potassium and manganese. Potassium is a vasodilator which means that by relaxing the blood vessels and lowering blood pressure, it can reduce the stress on the cardiovascular system. This can prolong your life by avoiding atherosclerosis and preventing strokes, heart attacks and heart disease.

Boosts Immunity: High vitamin C levels found in thyme make this a natural booster of the immune system. Vitamin C helps to stimulate white blood cell development, which is the first line of defense within the body's immune system. Vitamin C plays a crucial role in collagen production, which is essential for creating and repairing cells, muscles, tissues and blood vessels.

Treats Respiratory Issues: Thyme is one of the most well-known and traditional medicines used for respiratory problems. Thyme serves as an expectorant and an anti - inflammatory agent whether you suffer from bronchitis, chronic asthma, coughing, colds, flu, blocked sinuses or seasonal allergies. It helps eliminate phlegm and mucus from the respiratory tracts, eases inflammation to help breathe and prevents the development of microbes that can lead to disease. The best way to achieve relief from respiratory ailments is to brew thyme leaves into a strong tea.

Prevents Food-Borne Infections: Thyme is pretty beneficial in shielding you from food poisoning and food-borne pathogens infections. As it battles bacteria such as *Salmonella typhimurium*, *Staphylococcus aureus*, and *Vibrio parahaemolyticus* that cause food poisoning, it is a great food preservative for products such as meat.

Improves Mood: Thyme is among the essential oils that are used in aromatherapy. It influences neuronal activity by acting on neurotransmitters and thus may improve your mood. This could every nervous condition such as depression, sleeplessness and nightmares.

Bone Health: Thyme is an excellent vitamin K source and an excellent source of iron, calcium and manganese. Such minerals play a significant role in bone health, promoting proper bone growth and development, and reducing the risk of bone disorders. Thyme helps maintain strong, healthy bones and prevent bone disease.

Stimulates Hair Growth: The nutrient delivery to the hair follicles is essential to hair growth. Thyme increases the circulation of blood towards the scalp. Rubbing an oil mixture with thyme in it or essential thyme oil benefits in easing nutrient delivery to the scalp and fosters hair growth in addition to making it healthier and shinier.

Oral Health: Thyme 's antibacterial properties make it one of the common bad breath remedies as it helps combat bacteria and infection in your mouth. Increased bacteria are responsible for causing bad breath in the mouth. Thyme is also used as an ingredient in antiseptic mouthwashes intended to treat gingivitis and cavities.

Diarrhea: Thyme is very powerful against Shigella. This is in fact a bacterium that can cause diarrhea. To reduce the number of Shigella bacteria, wash the lettuce in 1 percent thyme oil or thymol. Thyme essential oils are also used as preservatives for natural food.

Bug Repellent: Thymol, a compound that is found in thyme, is an ingredient in many pesticides. Mosquito repelling has been shown to be effective, which can help prevent mosquito- borne diseases.

Precautions

- Evite if the gastrointestinal tract becomes inflammatory.
- Pregnant and nursing women should not use this.
- Overdose may cause gastric irritation, headache, dizziness, convulsion and coma.
- People with heart problems, seizure disorders or epilepsy should avoid this.
- Avoid use by high blood pressure patients.
- It may cause stomach disorders.
- Avoid in case of allergic problems.
- Caution if sensitive to grasses.
- Dilute the oil in a carrier oil before topical use.
- Avoid using thyme 2 weeks before surgery.
- The oil can cause skin and mucous membranes to get allergic reactions and irritation.

3.59 TURMERIC (CURCUMA LONGA)

Commonly known as temulawak, Javanese ginger, or Javanese turmeric is an erect, perennial herbaceous plant species of the ginger family. Some of Turmeric's well-known health benefits are described below;

Overcoming indigestion: By Turmeric, various health problems can be resolved in the digestive tract, such as flatulence, dyspepsia, and indigestion.

Increase your appetite: Turmeric can be a powerful drug for parents of low-appetite children to increase the child's appetite. Sure, it can be very healthy and delicious when consumed by your child in a form that has been processed so. Turmeric will speed up the process of emptying the stomach so that the appetite can grow again quickly.

Treats Cancer: Turmeric can monitor, avoid, or even destroy various cancers, including colon, breast, and prostate cancer. The herb is supposed to work by inhibiting blood vessel growth which supports cancerous growth.

It is believed that its protective effects are accompanied by its antioxidant activity, which protects your cells from damage.

Increase endurance: Both sections of Turmeric are successful but the most commonly used rhizomes are. Rhizome consists of over one hundred types of compounds such as amylase, phenolase, fat, starch, minerals, phenolic derivatives (curcuminoids) and essential oils. These compounds have proven to be effective for optimal endurance improvement.

Prevent and cure acne: Turmeric has an astringent property that causes skin pores to decrease so that it will indirectly avoid the occurrence of acne on the surface of the skin.

Alternative medicine for liver dysfunction: Kurkuminoid has the potential of treating hepatic disorders and preventing fatty liver cells.

Normalize the affected tissue: Turmeric is known as an adaptogen in the field of physiotherapy, a harmless material that can promote increased resistance to toxins or that can have physical, chemical and biological impacts. Ultimately it has the benefit of normalizing the tissue affected.

Ease osteoarthritis: Taking Turmeric for 4 weeks may help relieve osteoarthritis-related pain among those people who already have the disease.

Prevent tumors: Turmeric also has the advantage of preventing the action of enzymes or other components that trigger tumor cell growth. So, the best step is to regularly use Turmeric in tumor sufferers.

Prevent the appearance of dengue fever: Turmeric 's essential oil content has the use of inhibiting Aedes aegypti mosquitoes to get closer to your body. This will cause you to avoid the risk of dengue fever because this type of mosquito is a dengue virus carrier.

Precautions

- Avoid use during pregnancy and breastfeeding.
- Avoid use if you have liver disease and gallbladder problems.
- The plant may not be safe when used in large quantities or for a long period of time.
- Larger doses may cause gastric irritation and nausea.

3.60 VANILLA (VANILLA PLANIFOLIA AND V. TAHITENSIS)

Vanilla is native to Mesoamerica–Mexico and Guatemala, and comes from the thick, greenish - yellow seedpods of the fragrant tropical vanilla orchid. It is a member of the orchid family, the world's largest and oldest flowering family of plants. Vanilla is also a medicinal drug, apart from being a flavorings agent. Vanilla essential oil is used as a medicinal product for the treatment of many diseases and vanilla beans, powdered or extract, which are also used in many medical conditions. Vanilla acts as an antioxidant, anticarcinogenic, antidepressant, and aphrodisiac which provides many health benefits. Major health benefits of vanilla are:

Antibacterial: Vanilla is able to prevent the development of several kinds of bacteria.

Anti-aging Benefits: Vanilla is a wonderful source of antioxidants that prevents and reverses damage to the skin caused by free radicals. It helps slow down aging signs such as fine lines, wrinkles and age spots. It is widely utilized for its fragrance and anti - aging benefits in the cosmetics industry. In addition to making the skin smooth and sweet, the application of raw vanilla beans blended with organic oil imparts a great chocolaty scent.

Anti-depressant: As we all know, due to their simple applications and fast performance, essential oils are often used for their uplifting, antidepressant abilities. Simply to detect a scent will enter and affect the brain very quickly and effectively.

Relieve Pain: Vanillin is a magical-property compound. As well as having anesthetic properties it acts as a pain reliever. Those properties make it suitable to relieve pain naturally. It helps to soothe toothache; inflammatory pain, etc.

Lowers Blood Pressure: The sedative effects of vanilla oil on the body cause it to lower blood pressure naturally by calming both the body and mind. High blood pressure is the condition when the pressure on the arteries and blood vessels becomes too high, and the arterial wall becomes distorted, causing additional stress on the heart. High levels of blood pressure can put you at risk of stroke, heart attack and diabetes.

Soothes Burns: Vanilla was a traditional remedy used to treat burns, wounds, and cuts. The use of concentrated vanilla extracts or essential oil on recent burns should not be overt. Topical treatments may be beneficial, containing vanilla extracts. Before using any of these home remedies for burns it is recommended to consult your doctor.

Anti-cancer: As normal with cancer and essential oils, the preface to this topic is that there is a need for more research. However, thanks in no small part to its antioxidant capacity, vanillin and its containing substances (like vanilla!) are among the oils and compounds considered for anticancer capability.

Promotes Healthy Skin: Vanilla is a good source of B-vitamins such as niacin, thiamine, vitamin B6 and pantothenic acid which play a major role in maintaining healthy skin. The vanilla's antioxidant properties help protect your skin from damage caused by contaminants and toxins in the environment. To revive your skin from within you may prepare a scrub.

Beneficial for Children: Vanilla extract added with a teaspoon of sugar to a glass of warm water will help to reduce fever and to fight infections in children. The taste may also appeal to adolescents, too.

Good for digestion: Vanilla is a naturally occurring substance perfect for digestion. Through improving digestive health, there are many ways vanilla does. The distinctive vanilla scent is useful in preventing nausea. Vanilla tea is another way to prevent gastrointestinal problems. Also, vanilla tea prevents vomiting, diarrhea, abdominal pain and abdominal cramps.

Boost Cognitive Actions and Mental Health: Vanilla helps boost mental health and cognitive functions. Its antioxidant property and free radical removal activity boost neural functions, reduce nerve breakdown risk, prevent nerve damage and improve memory. Frequent vanilla intake can actually improve mental health and cognitive abilities.

Treatment of Anxiety: Vanilla extract has a positive effect on those suffering from anxiety disorders and depression. The vanilla scent is particularly useful for this purpose. Drinking water or milk containing vanilla extract, in some situations, helps to reduce anxiety.

Increased libido: The vanilla has been considered an aphrodisiac since prehistoric times. Vanilla also helps to boost sexual desire by increasing the levels of testosterone in men.

Anti-inflammatory: Vanillin is also known to be anti - inflammatory, in accordance with the antioxidant activity. This effect tends to occur with antioxidant substances, thanks to the repair of the cellular level.

Dental Health: Vanillin found in vanilla is similar in chili peppers to capsaicin, and in spices such as cinnamon euganols. This compound affects the central nervous system in a positive way. Capsaicin is an effective pain reliever while euganols serve as an anesthetic for topical use. Vanilla has all these properties which help to prevent toothache and infection.

Treatment of Acne: The vanilla 's antibacterial properties make it good for treating acne. Vanillin helps cleanse the skin through its antibacterial effects, reducing the incidence of pimples and acne.

Aids Weight Loss: Extracting vanilla helps people lose weight. Even though exercise and diet play a major role in weight loss, these measures may be complemented by the consumption of vanilla extract for greater benefits. To validate the claim, additional research is required.

Relieves Nausea: Vanilla extracts tend to relax the stomach if it feels nauseous. To a glass of water just add a few drops of vanilla extract and slowly drink it. The exquisite scent of vanilla helps to ease the feeling of nausea. It is actually as good as nausea medication, and can effectively replace it.

Treatment of Menstrual Problems: Including vanilla in the regular diet can treat irregular menstrual cycle, pain, premenstrual syndrome (PMS), and other menstrual problems. Vanilla is rich in magnesium content that controls hormone production that increases the function of neurotransmitters and thus decreases mood changes and depression related PMS issues. The calcium content of vanilla decreases pain and menstrual disturbances. Drink vanilla extract in water or as a tea to get you off menstrual issues.

Aids Wound Healing: Vanilla extract has been used for the cure of wounds since ancient times. Its anti - inflammatory properties can soothe inflamed skin areas. It's considered a great home remedy for healing wounds. Remember that the immediate use of the vanilla extract should be avoided on open wounds and burns.

Repels Mosquitoes and Bugs: Just mix a tablespoon of vanilla extract with a cup of water for a natural pest-deterrent, and pour it into a spray bottle. Spritz it to ward off those annoying mosquitoes, flies and even ticks on exposed skin. Keep the vanilla spritz away from children and also avoid spraying it on fabric as it may discolor it. Also store away from light as it can get discolored.

3.61 WASABI (EUTREMA JAPONICUM)

Also known as Japanese Horseradish, Wasabi is a member of the family Brassicaceae that also includes cabbages, cauliflower, horseradish, and mustard. There are many myths concerning wasabi. For example, wasabi is simply not responsible for cleaning up your sinuses as well as opening up your nasal pathways; it does the opposite, in fact. Even, there are some explanations for wasabi that are both very useful and may come as a surprise. Continue to read for some of those health benefits.

Cancer Prevention: Daily use of wasabi to the diet can prevent certain cancer types. The isothiocyanates are among the most prominent and useful nutrients found. These derive from glucosinolates following the cutting of the plant. The numerous antioxidants in isothiocyanate are beneficial for eliminating free radicals from the systems of the body. Isothiocyanate in wasabi can stop leukemia and cancer cells in the stomach from developing. It actually promotes apoptosis (cell death) within 24 hours of consumption. Wasabi use can prevent tumor growth, even at a preclinical or pretreatment scale. The most appealing thing about this type of treatment is that isothiocyanates seem to kill cancer cells without destroying the cells that are healthy around them.

Promotes Hair Health: Wasabi is made up of vitamin C and is one of the most important nutrients for hair growth and strengthening. In order to give your locks a vital dose of vitamin C, include vitamin c-rich foods such as wasabi in your regular diet since wasabi consists of 54.5 mg of vitamin C which is 60.56% of the recommended daily amount. So, promoting the health of your hair includes wasabi in your daily diet.

Kills Harmful Food-Borne Bacteria: Wasabi gives a potent punch when it comes to bacteria defense.

Cardiovascular Health: Regular wasabi intake is particularly beneficial in preventing cardiovascular problems. It has anti-hypercholesterolemic features that help to reduce high cholesterol levels, strokes and heart attacks in the human body. Besides that, isothiocyanates have inhibitory effects on platelet aggregation in wasabi. Platelet aggregation is mainly the gathering or clumping together of blood platelets into a thrombus, or blood clot. The main causes of strokes and other cardiovascular problems are those clots. By breaking down and inhibiting the development of these clots throughout the body, the risk of heart disease and stroke is effectively reduced!

Depression: Vitamin B6 present in wasabi helps the body make serotonin, a mood-influencing chemical. Depression is associated with low serotonin levels and some antidepressant drugs function by increasing serotonin levels. Some researchers think that vitamin B6 can help to reduce depressive symptoms. So, to stay healthy from depression, using vitamin B6 rich foods like wasabi.

Respiratory Conditions: Wasabi can help prevent you from certain infections in the respiratory tract. The gaseous portion of wasabi, which triggers a powerful reaction in the nasal passages and sinuses, is in fact the gaseous release of allyl isothiocyanate which can vigorously prevent the proliferation of pathogens such as those causing influenza and pneumonia in the respiratory tract. After consuming it, the smell and sensation in your nose may be strong but it can do a lot of good for your health and well-being.

Prevents Tooth Decay: Vitamin B6 present in wasabi helps the body make serotonin, a mood-influencing hormone. Low levels of serotonin are associated with depression and some antidepressant drugs function by increasing levels of serotonin. Certain researchers think that vitamin B6 may help reduce symptoms of depression. So, to stay safe from depression, use vitamin B6 rich foods like wasabi.

Arthritis and Inflammation: Wasabi, shows to reduce cases of joint swelling, inflammation, and can even soothe the arthritis symptoms. Isothiocyanate compounds present in wasabi help to decrease the inflammation of joints, ligaments and muscles that can contribute to arthritis and joint pain. Wasabi helps preserve bone integrity in humans and may help to reduce the risk of osteoporosis. Similarly, isothiocyanates minimize the aggregation of platelets to protect against heart disease; they can also reduce aggregation at vulnerable points on the body resulting in pain or discomfort.

Helps heal wounds faster: Known for its enormous antibacterial, anti - viral and anti-inflammatory properties, copper is a fantastic tool for rapid healing of wounds. Besides that, it's also known that copper boosts the immune system and helps to produce new cells. But its healing properties do not stop helping the body outwardly; copper is also known to help body wounds, especially the stomach.

Antibacterial Effects: Known for its immense anti - bacterial, an anti - viral and anti-inflammatory property, copper is a wonderful tool for healing wounds rapidly. Besides that, it is also understood that copper strengthens the immune system and helps to produce new cells. But its healing properties don't stop with helping the body externally; copper is also known to help wounds within the body, especially the stomach. Avoid taking copper with zinc based product as it may hinder the uptake of either one of the product. Check with your doctor on using copper for treatment in case you have underlying conditions

CHAPTER 4

Production of Herbs and Spices

4.1 INTRODUCTION

Spices are an important group of agricultural commodities used for flavoring and are the main ingredients for many delicious foods. These plants have been used by humans since the earliest times. No other product has played a more pivotal role as spices in the growth of modem culture. The spice trade is one of the oldest, and one of the most important forms of trade at one time. Like the silver, gold, and slave trade, the spice trade linked many different civilizations and helped to develop global communication. In an expanding world, demand for spices was the solution to expanding world commerce. Fortunes were made, fighting battles and conquering countries for the sake of those plants. The need to supply European markets spurred explorations, culminating in the extraordinary journeys which led to the discovery of the New World and showed that the globe could be circumnavigated by sea. The desire to control spice sources has brought the British to India, the Portuguese to Brazil, the Spanish to Central and South America and the Philippines, the French to Africa and the Netherlands to Indonesia. Even today, for many of our newest medicines, chemicals and flavors, we rely on spices and herbs and they are also used in cosmetics and perfumery. The current global spice and condiment production is estimated at 8.5 million tons, valued at US$ 25 billion(Jaffee, S. 2005), with developing countries being the primary producers and the main importers being the developed world. Spices also represent developing countries ' currencies. India continues to play a key role in the world's spice trade as it did for centuries.

4.2 PRODUCTION OF ALLSPICE (PIMENTADIOICA)

Habitat: Allspice is native to the West Indian Islands and South America, and is extensively grown in Jamaica, where it thrives best on calcareous hills close to the sea.

Climate: In tropics and subtropics, a houseplant in other areas, it may be grown outdoors. Will grow in full sun but partial shade is recommended in the first few years of the plant 's life.

Soil: Best grown in average or sandy loam soil that is well-draining.

Growing: Sow the seeds, spaced 1 inch apart right after harvest. Seeds will germinate in around 10 days. It adapts well to the container culture and can be kept in a greenhouse or as a houseplant. The plant is dioecious, meaning plants are either male or female, therefore it is necessary to keep male and female plants in the vicinity to allow the fruit to grow. Below are some tips in growing allspice;

- Transplant Jamaican allspice once the roots around the pot's drainage holes are clear. Repot 2 to 3 inches wider and deeper into a pot or out into the garden. Standard potting soil can be used with a neutral pH. Jamaican allspice is tolerant of a variety of soils. Fill the pot or hole 2 to 3 inches below with potting soil and well water. Allow the water to drain, and top the root ball. Fill with potting soil and water in the area around the roots until the soil is thoroughly soaked.
- Place the plant where it receives direct light in the morning or partially shaded afternoon sunlight. Cover the plant from direct sunlight in afternoon, which can burn the young tree's foliage.
- Water the plant once a day after transplantation, for the first two weeks. Water twice a week, once the tree is grown. Water the plant till the soil is fully moist. The Jamaican allspice is not tolerant of drought, and can destroy the tree for prolonged periods without water.
- Fertilize the plant with a balanced fertilizer, such as 20-20-20 water - soluble fertilizer, which starts six months after planting, and every six months afterward. Dissolve 1 tablespoon of the fertilizer in a 1-gallon water jug, then scatter around the tree base.
- Three years after plantation, prune the Jamaican allspice tree. Prune branches during winter, when the tree isn't growing

actively. Prune the branches to form the tree and encourage a bushier look.

Harvesting: The tree begins to produce fruit when it is three years old, and after four years it is in full containing. The flowers show up in June, July and August and the berries quickly succeed. The fruit 's special qualities exist within the berries rind. Due to the loss of volatile oil, it loses its scent on maturation and therefore the berries are harvested as soon as they reach their full size, in July and August, but while unripe and green. The fruit is selected while it is young, as it is then most aromatized. Collecting is achieved by breaking off the small twigs bearing the bunches; these are then spread out and exposed to the sun and air for a few days, after which the stalks are cut and the berries are ready to be packed into bags and barrels for export.

Preserving: The whole dried fruit is ground for producing commercial allspice powder. All pulp and seeds are aromatic and contain an oil with clove oil-like qualities. Place the oil or powder in a dry, cool spot.

4.3 ANGELICA (ANGELICA ARCHANGELICA)

Habitat: Angelica is a herbaceous, aromatic herb native to Eurasia, growing wild in Finland, Sweden, Norway, Denmark, Greenland, the Faroe Islands and Iceland, and cultivated commercially in Belgium, Germany, France and several other countries.

Climate: Angelica thrives in cool, moist open spaces, yet can be amazingly well resisted by adverse environments, and can even withstand severe winter frost without damage.

Soil: The hardy plant thrives best on rich, well -drained loam soils.

Growing: Angelica is best grown by sowing fresh, ripe seed but can also be grown through root division or propagated by offshoots (although methods of propagation other than seed are considered inferior). Because the seeds ' germinating ability is declining rapidly, they should be sown as soon as they ripe in August or early September. Below are some tips in growing;

- Angelica plants can be used as flower borders or in flower beds, in either cottage or casual garden landscapes. Choose a location that will shadow the angelica plant during the hottest hours of the day. Angels need plenty of water to survive but good drainage is also important. Angelica plants should be well covered and directed either to the north, east or west. Start fresh angelica seeds in pots, and transplant them when young and small; older mature plants do not respond well to disturbance. The angelica room is 3 feet apart.

- Angelica grows in a variety of soil conditions but prefers soil with either a neutral or slightly acidic pH. If you do not know its pH level, test your soil. Angelica grows in sandy, loamy or clay soils, even though the plant thrives in loamy, moist soil. If your soil is drier, applying a layer of mulch or working organic matter to the soil will provide the moisture Angelica plants need to grow and develop.
- It will develop mostly broad leaves the first year after you plant angelica. The angelica will grow taller in its second year, and establish a flowering stem. Prune angelica plants after flowering. Angelicas contain furocoumarin that may increase the sensitivity of the skin to sunlight or cause dermatitis so be cautious when working with the plant.
- Unfortunately, angelicin plants are vulnerable to many pests attacking. Aphids, leaf miners, snails, and slugs are common invaders. Additionally, angelica can suffer from infections of the fungus and bacteria such as powdery mildew and leaf spot.

Harvesting: Roots intended for first-year flavorings agents are often harvested in the fall. Leaves and stalks are usually harvested in second year spring. When ripe, seeds are harvested. Tops are usually pruned under cultivation to prevent blooming and thus allow for continued root growth.

Safety (a note about wild harvesting): Accurate identification is crucial because angelica closely resembles water hemlock, a highly toxic plant that thrives in the same environment.

Preserving: Dry leaves and root. Crystallize or "candy" the stems. Store in a cool, dry place.

4.4 ANISE (PIMPINELLA ANISUM)

Habitat: Anise, Pimpinella anisum, is native to the Mediterranean region and Egypt; it is also cultivated in Europe, Asia India, Mexico, North Africa, and the USSR.

Climate: The plant requires a warm sunny location and long frost-free growing season of 120 days.

Soil: Anise plants grow best in light, fertile, well-drained soil and appear to respond well to nitrogen fertilization by yielding a greater quantity of high-quality fruit.

Growing: The seeds should be planted as soon as springtime the soil warms up. Because the plants have a taproot, after being grown, they don't migrate well, so they should either start at their final location or be transplanted while the seedlings are still tiny. Below are some tips in growing;

- Choose a pot with drainage holes and fill it with a wet, sterile potting mix from the top up to about 3/4 inches.
- Sprinkle six to eight seeds of anise over the surface of the soil, at an equal distance. Cover the seeds with a 1/4 "soil layer. Tamp the soil gently with your hand to make it secure over the anise seeds.
- Water the soil with a spray bottle to avoid disturbing the seeds which have been planted shallowly. Stretch plastic wrap over the pot to help retain moisture from the soil. Fill the plastic wrap with newspaper sheets to keep soil temperature constant. Keep the soil moist during the germination period -not soggy.
- Place the pot into a warm room. Aim for roughly 60 to 72 degrees Fahrenheit temperature. The seeds are expected to germinate within two weeks.
- The plastic wrap and newspaper should be removed as soon as the seeds germinate. Expose the seedlings to sunshine and roughly 55 degrees Fahrenheit temperature.
- Remove the tiny, fragile seedlings as soon as they are big enough to handle. Keep no more than one or two strong seedlings in the bowl, and regularly water them to keep the soil moist as they grow. Once all danger of frost has passed, you can move the pots outside into a sunny location.

Harvesting: Once the flower umbels with ripe brown seeds have become strong, cut off the heads before they drop. Place them on a paper towel or plate in a dry place, in a single layer. Expose to direct sunlight, where possible, to allow seeds to dry out completely. When the seeds are crisp and dry, rub between the palms to separate the seed from the hull, sieve the seeds from the husks, and place them in an airtight container.

Preserving: Store seed in a cool, dry place.

4.5 ASAFETIDA (FERULAASSA-FOETIDA)

Habitat: *Ferulaasafoetida* is native to the mountains of Afghanistan, and is mainly cultivated in nearby India.

Climate: Asafoetida prefers full sun, does not grow well in the shade.

Soil: Grow Asafoetida in light (sandy), medium (loamy) and heavy (clay) soils. Prefers well-drained soils.

Growing: Asafoetida can be grown from seed; directly sow outdoors in fall as soon as the ground can be worked. Below are some tips in growing;

- For your asafetida choose well-drained soil of any sandy, loamy or clay type. This perennial herbaceous tolerates acid, neutral, and basic pH levels in the soil.
- For that plant, pick a sunny spot. Asafetida needs full sunshine, and in the shade does not develop. Open up space between plants of at least 5 feet.
- Enable the soil to dry out entirely between watering. Stop waterlogged soil at all times, or standing water. Asafetida fights the drought robustly and prefers dry to humid soil.
- Apply a layer of bracken mulch about 1 to 2 inches thick in the winter season.
- Avoid moving asafetida after transforming it into a seedling. Those plants do not respond well to root disturbance because of their long taproots.
- Although asafetida usually resists disease, aphids or slugs may attack it. Treat the aphids with a good water spray or insecticide soap or neem oil application. Slugs are handled by handpicking, trapping or barriers.
- Stop moving asafetida after transforming it into a seedling. Those plants do not respond well to root disturbance because of their long taproots.

Harvesting: The gum resin is obtained from incisions in the roots and rhizomes of the plants.The upper part of the root is laid bare and the stem is cut close to the crown. The exposed surface is covered by a dome shaped structure made of twigs and earth. A milky juice exudes from the cut surface which soon coagulates when exposed to air. After some days, the exudate gum-resin is scraped off and a fresh slice of the root is cut. Harvesting is done just before the plant flowers.

Preserving: Store asafoetida powder in a cool, dry place.

4.6 BAY LEAF (LAURUS NOBILIS)

Habitat: Bay leaf (or Bay laurel) is native to Asia Minor and areas around the Mediterranean.

Climate: Bay laurel needs mild climates and will only grow well outdoors year-round if it has 4 hours of direct sunshine a day and stays relatively cool. It does make an attractive plant or houseplant for containers.

Soil: The soil should be fertile and well-drained and dry during irrigation, but in the springtime, it should obtain some extra water.

Growing: The plant can be propagated easily from cuttings. Below are some tips in growing;

- In late summer, prepare cuttings pots for planting. Fill the pots with an equal mixture of vermiculite and sand, and thoroughly water the mixture.
- Cut 6-inch-long sections from the green, semi-hardwood stem tip of a bay laurel tree. Wear gloves when handling the roots, because the skin can be irritated by bay laurel oil.
- Strip the leaves from the bottom half of each cutting.
- Make a fresh cut at the bottom of each line, then dip in the rooting hormone powder to that point. Attach the bottom end of each break, 3 inches deep into the potting mix. Do not allow the remaining leaves to enter the soil or other cuttings, as this can cause rot.
- Thoroughly water the potting mix, then cover each pot with a gallon-sized zip bag of plastic. To create a greenhouse environment, secure the bags with elastic bands.
- Place the pots near a window in indirect sunlight, until the roots grow. If condensation forms in the plastic bags, let them air out until dry. Hold fiel'd moist at the top 1/2 inch.
- Regularly check the cuttings for mold and wilting, and remove the affected leaves to avoid spreading rot.
- Once a month test the cuttings for root development by pulling gently at the top of each stem. Roots have emerged when there is resistance. Transplant the cuttings into larger containers with no zip bag at 1-inch long root growth.

- Unless you plan to move the pots outdoors, keep the cuttings indoors until springtime. Hardens plants by exposing them slowly to a few hours of direct sunlight every day before a permanent move.

Harvesting: Upon establishment of the tree, harvest the leaves as needed. This is one herb that's dried rather than fresh because the fresh leaves are bitter.

Preserving: Pressing the bay leaves to dry; properly dried leaves are light olive green. Store in an airtight container, in a dim, cool, dry place.

Take note: This plant can be found as "bay leaves" in your grocery store, and "laurel" or "bay laurel" in your garden center. When buying plants, always check the Latin name, particularly this one, as there are many "bays" and "laurels" out there, some that are not remotely related to this herb. For instance, Mountain Laurel *Kalmia latifolia* is poisonous while Bayberry *Myricapensylvanica* has been used in cooking as a substitute for Bay laurel, and has its own uses.

4.7 BASIL (OCIMUMBASILICUM)

Habitat: Basil was originally native to India and other Asian tropical regions, and has been cultivated there for over 5,000 years.

Climate: Basil is very sensitive to cold, with the best growth in hot, dry conditions. Does well in containers.

Soil: Basil prefers well-drained rich soil.

Growing: Basil is easy to grow from seeds or transplants. The plant can also be very reliably propagated from cuttings, with the stems of short cuttings suspended in water for about two weeks before roots grow. Avoid frost, and ensure that the soil is dry. Basil serves as an annual if a frost has any chance of happening. It will flourish in sunnier climates and may self-sow if you let it go to seed the next year. Below are some tips in growing;

- When a basil seed gets the right combination of light, water and energy, a new shoot starts to soften and send it up. The first collection of leaves appearing on seedling basil is called seed leaves. Those tender leaves grow right inside the seed from the stuff. The seedling produces the second set of leaves as the plant starts to establish itself and lay roots down. Those leaves are the first of the plant 's true leaves.

- With this frost-sensitive plant, the best way to get a jump on the growing season is to start seeds indoors in late winter or early spring to be ready for late spring seeding. It takes between eight and 14 days for Basil seeds to germinate and emerge from the soil. Look for the first collection of true leaves after germination, two to three weeks later. Then, two to three weeks after the appearance of the first set of true leaves, basil plants should be about 6 inches tall and ready to plant in the greenhouse.
- Basil grows easily from seed. Beginning with a seed field, filled with a sterile humid seed-starting mix or seed-starting compost is easiest. Push them gently into the soil after spreading the seeds over the moist soil mix, and cover them. A spot remaining around 70 degrees Fahrenheit provides the right conditions for germination. Basil seeds require frequent humidity but cannot withstand excess watering. When it starts to feel dry to the touch, the best way to water is to mist the seed flat.
- Move the seedlings to small pots once two sets of real, regular - sized leaves are on. Do this gently by using a small tool such as a cake tester or clay pottery tool with a smooth, sharp edge to pry them apart. Transplanting the seedings helps each plant to spread out its roots, allowing it to grow faster and helping to ensure that each seed you have sprouted becomes basil.
- When the seedlings are about 6 inches tall, about six weeks after they germinate, you will begin to harden them off by putting the seedlings outdoors on mild days and then bringing them in at night. Basil seedlings cannot withstand any frost, and even established plants can. Wait for basil to be planted out until all frost danger has passed. If the weather is flippant, it is safer to keep basil in indoor pots longer than to risk outdoor tender seedlings.

Harvesting: Leaves can be picked fresh, so be sure to pinch floral buds to promote further leaf growth, otherwise the leaves will be smaller when the plant starts to flower and produce seeds.

Preserving: The fresh leaves can be stored in plastic bags in the fridge for a short time, or in the freezer for a longer period. Some tend to blanch the leaves quickly until freezing in boiling water. Leaves can be dried and kept whole or chopped, too. Store in a dry, cool place.

4.8 BERGAMOT (MONARDA SPECIES)

Habitat: The bergamot herbs are perennial plants native to North America.

Climate: Bergamot oranges are hardy plants but they can be affected by certain conditions. Moisture conditions, such as floods or drought, can seriously affect plant growth. A cold-hardy perennial with temperatures that can withstand up to -30oC (-20F).

Soil: Bergamot is infamous for thriving in all kinds of soil, whether it's an ideal mix or leans toward sand, clay, or loam. Bergamot oranges prefer acidic to neutral soil that is between 5.6 and 7.5.

Growing: Free to plant your wild bergamot in either springtime or fall. While this hardy plant will thrive in less than ideal areas, it prefers the full sun and warm, well-draining soil. In order to prevent powdery mildew, wild bergamot requires a spot with plenty of air circulation. Space plants vary between 18 and 24 inches. Whether you start with seeds, divided rhizomes or cuttings, water thoroughly after planting. After wild bergamot is planted, water regularly in evenly moist soil to keep it growing. You can mulch around wild bergamot plants to help the soil retain water and prevent the infiltration of weeds. Deadhead spent your wild bergamot plants on flowers. Removing the blossoms once they've been expended doesn't just keep the plants looking attractive — it helps keep the whole plant safe as resources are no longer wasted on sections whose time has finished. When you find the leaves close to the bottom of the plant go yellow, cutting the stems down to three or four inches from the ground can actually give you a second period of bloom. After the first frost in the fall, you should do this pruning too. Below are some tips in growing;

Cut off ripe seed heads to propagate by seed, and dry them for several days on a clean dry indoor surface. Place the heads of seed in a paper bag and shake to release the seeds. Sow seeds in flats or zippered plastic bags during January, with a soil mix of one-third of sand and two - thirds commercial plug mix, put in a cool place. The seeds are to germinate within one to two weeks. Seedlings will need up to seven weeks of a starter fertilizer solution, at which time you can transplant it into 3-inch containers. Water and periodically fertilize the seedlings, then plant in the garden once the roots fill the pot.

- Drop stem tip cuttings into sand and perlite rooting medium to propagate using cuttings, and spray the stems many times a day. They are supposed to root quickly enough to transplant within five weeks.

- While generally resistant to powdery mildew, when located in a crowded garden with poor air circulation, mildew disease may attack your wild bergamot plants. Rust fungus can be an obstacle too. Telltale rust signs include circular markings on either side of the surface of the leaf before the pustules extend over all areas of the leaves and down the stem. Use a fungicide or harvest infected plants, and always water the plant base rather than the leaves. In reality, healthy bergamot plants can help keep the underground pests away from other plants like tomatoes.

Harvesting: Harvest leaves before and after flowering before flowers emerge. For essential oil distillation, harvest whole tops down to 2,5 cm above ground at peak bloom. A second bloom can take place in the fall.

Preserving: Flowers and leaves may be dried on a wire rack or in a kiln in a hot, dark place. Upon harvest, whole tops should be processed by steam as soon as possible.

4.9 CUMIN (CUMINUMCYMINUM)

Habitat: Initially grown in Iran and the Mediterranean, Cumin was introduced by Spanish and Portuguese colonists to the Americas. The plant today is grown mostly in China, Uzbekistan, Tajikistan, Iran, Turkey, Morocco, Egypt, Syria, Mexico, Chile and India.

Climate: Cumin grows best in a sunny location in warm temperate and tropical zones.

Soil: Prefers a warm fertile well-drained sandy loam soil.

Growing: Sow seeds in a greenhouse in the early spring, in individual pots. Thrive the plants on quickly, and plant them out after the last frost predicted. Offer the plants some temporary protection such as a cloche in the open ground for their first few weeks to ensure they continue to grow in the cooler early summer weather. Below are some tips in growing;

- Start the seeds in large pots which are about 6 inches across and 4 inches deep because cumin doesn't grow well. Using biodegradable pots if you intend on transplanting cumin outside, so you don't have to cut the seedling. Fill the pots with a dry, sterile, seed-starting mix to 3/4th of an inch from the top. Sow three seeds in each pot at a depth of 1/4 Place each container in a plastic bag to promote the preservation of soil moisture.

- The seeds are exposed to a temperature of 65 degrees Fahrenheit to germinate cumin. Every day remove the pots from the plastic bags to aerate them and test for moisture in the soil. You want the medium to remain moist throughout the time of germination— not damp— Use a water - filled spray bottle if necessary, to moisturize the soil before putting the pots back in the plastic bags. Cumin can germinate in anywhere from seven to 14 days.
- The moment seeds germinate they need light. The seedlings should grow spindly and weakly, without sufficient light. Place them in a sunny south - facing window after removing the pots from the plastic bags, and rotate them regularly so that the seedlings grow upright and do not need to search for light. Alternatively, two cool - white, 40-watt fluorescent lamps are suspended about 6 inches above the containers. Keep the lamps on for 16 hours a day and move them upwards as the seedlings grow so that they are still 6 inches above the pots.
- The seedlings are small so there is only one solid seedling left in each pot and the plants continue to grow indoors. Allow until after the freeze, when the plants are about 2 inches tall and the temperature is at least 60 degrees Fahrenheit when you want to move them outside. Place the transplants 1 foot apart in a sunny garden area with fertile, well-drained soil. While cumin is tolerant to drought, during dry, hot spells it benefits from moderate water.

Harvesting: Harvest the seeds when fully ripe and then dry them thoroughly.

Preserving: Store dried cumin seeds in an airtight container in a cool, dry place.

4.10 MUSTARD (BRASSICA NIGRA)

Habitat: *Sinapis alba* is now widely distributed around the world, though it originally originated in the Mediterranean region. *Brassica nigra*is thought to be native to Europe's southern Mediterranean region, and probably South Asia, where it has been cultivated for millennia.

Climate: With some humidity, *Sinapis alba* prefers temperate climates. It is grown in the garden as well as commercially, and can withstand high temperatures, but during flowering and ripening, very hot days can reduce seed quality. *Brassica nigra*is often grown in the temperate zone, although it is primarily suitable for tropical areas, and mainly grown as a rainfed crop in low or moderate rainfed areas.

Soil: Mustard plants require high - nutrient soils with high nitrogen rates, but can be grown from light to heavy on a wide variety of soils, growing best on relatively heavy loamy sandy soils, not suited to very wet soils.

Growing: Sow seeds directly in the garden from early spring to late summer. Germination takes place in less than a week.Below are some tips in growing;

- Break up a section of the garden with a hoe in full or partial light. The afternoon shade helps the plant grow longer when your garden is in a particularly warm microclimate. For three years running the site should not have been used for other plants in Brassica broccoli, cabbage or sprouts in Brussels. The soil used for other crops from Cole can harbor diseases that may damage the mustard.
- Apply granular fertilizer with 10 - 20-10 nitrogen-phosphorus-potassium content to garden soil with a pH between 6.0 and 7.5. Sprinkle 1 cup of fertilizer over the soil at planting time per 10-foot row, and work it into the soil before sowing the mustard seed.
- Sow mustard seed directly in garden soil 1/2-inch-deep and 1 inch apart— at 45 to 85 degrees Fahrenheit soil temperature. U.S. sowing mustard prime time Agriculture Department planting zones 9 and 10 are for early spring crops from February to April, and fall crops from July to October.
- Hold well-watered before seeds germinate, which depending on temperature takes between four and seven days.
- Thin seedlings up to 6 inches to 18 inches apart depending on the cultivar 's mature size you are planting— when plants are 3 inches tall. Use the thinning to add salads with a peppery flavor.
- To avoid harm from flea beetles, whiteflies or aphids, cover the young plants with a floating row cover.
- Hold well-watered plants, soaking them good once a week if there is no rain. The leaves ' flavor becomes more pronounced when there is not enough water in the field and high temperatures.
- Pull out weeds by hand or with a hoe gently as they grow. Avoid damaging the mustard roots with the hoe, or crushing them.

- Apply an additional 1/2 cup of fertilizer to the base of the mustard plants along the wall. Scratch it softly to the ground with a hoe. The extra nitrogen helps the plants to grow their dark color.
- As required, harvest mustard leaves by cutting off the outer leaves at the base. The finer the texture makes them more tender. Plants are usually ready to harvest within 35 to 45 days of sowing.
- If you want to make mustard as a condiment with the plant seeds, cause the plants to flower and go to seed. After it bolts in warm weather, the plant produces a stalk from its center, flowers and forms seedpods. As the foliage of the plant begins to yellow, the seed stalks and pods are cut down and hanged in a cool, dry place to mature.

Harvesting: Mustard seeds are harvested just prior to the pods becoming ripe and bursting.

Preserving: Store dried seeds (whole or powdered) in an airtight container in a cool, dry place.

4.11 BLACK PEPPER (PIPER NIGRUM)

Habitat: *Piper nigrum* is native to south India, parts of Asia, and is extensively cultivated there and elsewhere in tropical regions.

Climate: Grows in humid, mountainous, tropical regions in partial shade, can also be grown as a houseplant in temperate regions.

Soil: Grows best in soil that is neither too dry nor susceptible to flooding, prefers moist, well-drained and rich in organic matter.

Growing: Plant peppercorn seeds indoors 1/2" deep. For proper germination keep seeds damp and remain at 50% humidity or higher and 75 to 85°F. After seeds have germinated, keep the soil moist and from drying out. Transplant in larger pots as the seedling grows. Below are some tips in growing;

- Fill individual pots with a soilless sterile potting mixture of 3 inches in diameter. Heat the mixture until it is fully evenly moist. Sow the 1/4-inch-deep Black Pepper seeds, planting two seeds per pot.
- Cover the pots with a clear plastic sheet, which helps keep the soil moist and provides a humid environment that helps germinate. Place the pots in a position that is Fahrenheit 70 to 75 degrees.

- Monitor the soil in the pots and lightly water it if it starts drying before the seeds sprout. Once the first seedlings break through the soil, remove the plastic completely.
- After the seeds germinate, transfer the pots to a 70-to 75-degree position that receives at least six hours of direct sunlight per day. When the top of the soil feels dry, water the seedlings but avoid wetting the foliage.
- Begin the application of the fertilizer two weeks after germination. Apply a soluble fertilizer for vegetable plants diluted to half the monthly rate recommended on the package. With this solution, water the "Black Pearl" peppers every two weeks.
- When the stronger plant produces its third set of leaves, pinch off the weaker seedling in every pot near the soil surface. When they are about eight weeks old, transplant the seedlings into a permanent 3-gallon planter or garden bed.
- Space the plants eight to 12 inches apart in all directions due to their sunlight requirement.

Harvesting: Black pepper is grown from the pepper plant's still-green unripe drupes. The drupes are briefly cooked in hot water to both clean and prepare for drying. Heating in hot water splits in the pepper's cell walls, and speeds the work of browning enzymes during drying. For several days, the drupes are dried in the sun or by machine, during which the pepper around the seed shrinks to a thin, wrinkled black layer, darkening. The spice is named black peppercorn once dried. White pepper consists of pepper plant seed alone, with the pepper fruit's darker-colored skin removed. This is usually achieved through a process called retting, where fully ripe red pepper berries are soaked in water for about a week, during which the pepper's flesh softens and decomposes. Rubbing then removes what's left of the fruit, and dry the exposed seed. Green pepper is made from the unripe drupes, similar to black. Dried green peppercorns are treated in a way that preserves the green color, such as sulfur dioxide injection, canning or freezing.

Preserving: Store dried peppercorns in an airtight container in a cool, dry place.

4.12 BORAGE (BORAGO OFFICINALIS)

Habitat: Borage is native to the Mediterranean, but is now naturalized in most parts of Europe, can also be found in the UK and in the United States.

Climate: Borage can grow in a wide variety of conditions, comes up like a weed in garbage dumps, roadsides, and near buildings.

Soil: Borage flourishes in ordinary soil, adapts to poor or rich soil.

Growing: Borage can be propagated by splitting rootstocks in spring and placing cuttings of shoots in sandy soil in a cold frame in summer and autumn, or from seeds sown in fairly good, light soil, from mid - March to May, in drills 18 inches apart, thinning the seedlings to about 15 inches apart in the rows. If left alone, Borage will openly seed itself and come up in the same place year after year. Even seeds may be sown in the fall. Those planted will then bloom in May, while those sown in the spring will not flower until June. Below are some tips in growing;

- Choose a place in your garden where the soil is well- drained. Even your planting site should be in partial shade to promote the germination and growth of your borage seeds.
- Turn the soil in your planting site with a shovel or garden fork to loosen the ground.
- Rake weeds, debris and rocks out of the soil in the planting site.
- Apply a layer of organic compost to the soil and rake it into the dirt.
- Water the soil until mildly humid. Borage seeds germinate faster when grown in moist soil.
- Sow 1/8-inch-thick seed borage. To make tiny holes, poke your finger into the soil and drop one or two borage seeds in each. The seeds are spaced 12 inches apart. You don't need extra soil to cover the seeds, because they need plenty of light to germinate.
- Water the seeds from the borage using a garden hose. Use a soft spray to keep the seeds from moving away.
- Throughout April and May, when soil temperature is about 70 degrees Fahrenheit, plant borage seeds appear. Sow seeds throughout shallow rows, spaced 12 inches apart in sunny areas. Cover seeds with about 1 cm of soil, and water until the surface of the soil is moist. Do not try to transplant borage; it forms a long taproot, so it does not transplant well. Thin seedlings start

crowding each other until the plants are spaced about 15 inches apart.

- Borage is a low - maintenance herb, requiring little treatment. Water and thin seedlings periodically before the ripening of the plants. During periods of prolonged drought plants grow in water. In the winter, push potted borage indoors prior to the first frost. Select the blue flowers regularly before they can set seed to prevent borage from self-seeding in the greenhouse.

Harvesting: Borage flowers can be harvested from June to August.

Preserving: Dry the flowers and leaves completely, then cut them into smaller pieces and store in airtight containers in a cool, dry place.

4.13 BURNET (SANGUISORBA MINOR AND S. OFFICINALIS)

Habitant: *Sanguisorba* is a genus of flowering plants in the *Rosaceae* family, native to the Northern Hemisphere temperate regions. Burnet is the common name. Burnets are perennial herbs widespread throughout Europe, Asia and other parts of the world.

Climate: They are fairly drought tolerant and require full sunlight. Although it prefers full sun it can grow in light shade.

Soil: Easily grown in average, medium, well-drained soils in full sun. Burnet prefers chalky moist fertile soil

Growing: Burnet's to grow easily. It is a perennial clumping about 50 cm across and 30 cm high. It is a delicate and tidy mounding plant, and also looks great in containers. It makes a good plant with an edging. Cutting back the blossoms will yield plenty of tender new leaves. Weight per thousand grains of seed: 5,5 g; seed drilling requirement: April 15,0 kg / ha; planting requirement: April 6 plants / m2; row distance: 30 – 50 cm. Cutting older leaves regularly to encourage the growth of young leaves that have the best taste. Self-seeds are healthy. Cut flower stalks immediately after bloom when needed for self-seeding. Evergreen in warm winter climates. Below are some tips in growing;

- Sow seeds indoors around 4 to 6 weeks before the last frost in early spring. Sow seed in flats or in single pots. Just gently cover the seed and just keep the seed starting mix moist. The seed will germinate within approximately 7 to 14 days.

- Transplant seedlings from the burnet salad into the garden about two weeks after the last spring frost. Carefully handle seedlings to prevent shock during transplantation. Large plants do not grow well.
- Cover lightly, with ⅛ inch of soil and keep the soil moist, until seeds germinate.
- Grow plants 12 to 15 inches apart. Space rows 18 inches apart.
- Salad burnet grows best in evenly moist soil. Keep the soil just moist.
- Early in the season sided dress salad burnet with aged compost; feed plants with compost tea or a diluted fish emulsion solution every 6 weeks.
- Mulch around salad burnet with aged compost to keep weeds down and add nutrients to the soil.
- Remove flower stalks to stimulate new leafy growth as they develop. Older leaves and plants are not as delicate and aromatic as young plants. Every year break the salad burnet into new plants. Rhizome spreads salad burnet; monitor the spread of plants.

Harvest: Young leaves, herb before flowering

4.14 CARAWAY (CARUMCARVI)

Habitat: Caraway is native to Western Asia, Europe and Northern Africa.

Climate: Caraway prefers warm, sunny locations.

Soil: Caraway grows best well-drained soil that is rich in organic matter.

Growing: Caraway plants do their best with seeds when started. Sow straight into well-drained soil and cover lightly in spring or fall with a thin layer of soil. By working up some compost into the top 1 - 2 inches you will prepare the soil. Caraway is a two-year-old. This means that it takes the plant two years to ripen, produce seeds and then die. You may not see any evidence of the plant during the winter even though the roots are getting ready for spring. So, carefully mark your position so you won't accidentally dig them up in the spring. Below are some tips in growing;

- Sow caraway indoors in biodegradable peat pots or cell packs 3 to 4 weeks before the last frost average for later transplantation of both pots and plants. Push seed into the surface of the potting soil and cover with vermiculite loosely. Seed germinates within four to fourteen days.

- Set caraway seedlings in the garden shortly after the last spring frost.
- In spring or fall, knit caraway outside. For an early start, sow caraway in spring about the date of the average last frost, as soon as the soil can be worked. Sow the seed. Sowing seeds in spring will not produce flowers or crops until the second summer. You can also sow seed in autumn for the next spring to germinate.
- Sow caraway seed ¼ inch deep.
- 8 inches apart from the space plants. Small productive plants 12 to 18 "apart when seedlings are about 4 inches tall. Space rows distance between 18 and 24 inches.
- Grow 6 caraway plants for cooking; grow 12 plants for preserving.
- Companion plants: Many vegetables help caraway grow well. Not developing well similar to fennel. Pea is a good companion. Caraway flowers attract beneficial insects that attack tomato hornworms including predatory wasps.
- Caraway requires regular watering, even till it is founded. Do not allow it to dry out seedlings. Once established, caraway between watering can go almost dry. Stop watering when flowers start to form; this will allow seeds to ripen faster.
- When seedlings are 3 inches tall, fertilize the caraway lightly and again when flower shoots appear. Give caraway at midseason a side dressing of aged compost.
- If the soil freezes in winter cover plants with leaf mold, straw, or aged compost to protect the roots.
- Eviting excessive pruning in the first year, or weakening the plant. Caraway may need winter protection in cold-winter regions for second year flowers and seeds to be produced. In the second-year pinch off flowers and caraway will grow for another season.
- Select a container at least 8 inches deep. Caraway has a taproot and does not grow very well in a container.

Harvesting: Harvest caraway fruit (seeds) when they have ripened but before they fall to the ground. Cut the flower heads and gently shake over a shallow pan.

Preserving: Dry the caraway fruit (seeds) and store it in an airtight container in a cool, dry place. You can even dry-roast the fruit (seeds) to enhance the flavor before using or storing.

4.15 CARDAMOM (ELETTARIACARDAMOMUM)

Habitat: Cardamom is native to southeastern Asia from India south to Sri Lanka and east to Malaysia and western Indonesia, where it grows in tropical rainforests.

Climate: Cardamom grows on the shady jungle floor of tropical rainforests; in cooler climates, it is grown indoors as a houseplant.

Soil: Prefers a rich, loamy, slightly acidic soil. For best results, fertilize or amend the soil with minerals phosphorus and potassium.

Growing: The smaller seeds should be sown in a light but dense soil buried approximately 1/8 "beneath the soil surface. Cardamom requires a steady supply of moisture and will not withstand drought. If growing in a greenhouse, it should be kept humid. Cardamom is not tolerant of cold and should be kept in a place with many hours of partial or indirect sunlight. Below are some tips in growing;

- Fill all-purpose potting soil with a 10-inch planter. Use a planter that has several holes in the drain. Place the planter above a saucer that is filled with pebbles. Spray water onto the pebbles.
- Swirl your finger over the middle of the soil and put the plant in the hole. Scoop up the dirt around the plant foundation.
- After planting, water the plant with distilled water, then once per day. Keep the soil moist but don't drip wet. To restore humidity, spray the cardamom with a mister. In winter, cut back on water; only heating any other day.
- Place the planter out of direct sunlight into a warm spot. Cardamom enjoys daylight. Choose a room that is temperature -controlled away from the windows and doors. For this plant, the ideal temperature is 80 degrees Fahrenheit but acceptable is 72 to 80 F.
- Fertilize the plant twice a month using a high-nitrogen, low-potassium houseplant fertilizer.
- If the leaf tips are brown, it means you're over-watering.
- Take cuttings from the root of a sturdy, established plant to

propagate.

Harvesting: Cardamom requires around 3 years of growth to produce seed-bearing capsules. They will slowly dry out after the flowers mature as capsules form. These can be collected when capsules start turning green and dried on screens in 6 - 7 days later. Turnabout regularly. When pods are dry and easy to break open the seeds can be collected. Place pods in a cup, and thresh them carefully by applying light pressure to split the dried seed pod. Separate seed from the chaff by winnowing with a small fan, or by putting it in a medium screen and shaking gently back and forth while slightly pushing through extraneous matter.

Preserving: Store seeds in a sealed container in a dry, cool location out of direct sunlight.

4.16 CASSIA (CINNAMOMUM CASSIA)

Habitat: *Cinnamomum cassia*, or cinnamon, originates in southern China, is widely cultivated there and elsewhere in southern and eastern Asia (India, Indonesia, Laos, Malaysia, Taiwan, Thailand, and Vietnam).

Climate: Cinnamon requires a low altitude and a hot, humid, tropical climate in partial shade.

Soil: Grows best in soils of loam or sandy loam. The best bark in sandy soil comes from trees while loam provides faster growth and higher yields.

Growing: Cinnamon is primarily propagated by seeds while cuttings may also be planted by a plantation. In a fertile soil, plant seeds in pots, and grow in a greenhouse or tropical climate. Once they are large enough to handle transplantation seedlings. The tree has two years to grow, after which it is cut down. Small shoots show up next year. These shoots have their thin inner bark stripped off. Below are some tips in growing;

- Grow cassia which is not susceptible to standing water in well-drained soil. Cassia tolerates the most moderate -quality garden soil and grows well. Space plants placed 3 to 4 feet away from each other and surrounding plants. In large pots filled with normal potting soil, Cassia often grows well.
- Provide cassia with a full six hours of sunlight or more per day. In sunny places where temperatures remain above freezing plants remain evergreen.
- Fill garden cassia with 1 to 2 inches of water roughly once a

week. Do this, after planting for the first year. Developed plants require irrigation only during dry periods when the soil starts to completely dry out. When the top 1 inch of soil feels dry, cassia is potted with water.

- Cover the soil with a 2-inch mulch layer in the garden bed, and replenish it every fall. Mulch retains soil moisture, and in winter protects plant roots. Leave a space between the stem of the cassia and the mulch so they do not strike.
- At any time during the growing season, take out dead branches or leaves. Take dead stems with pruning shears back into the nearest safe forest. Dispose of discarded leaves and stems.
- If frost is an issue, bring potted cassia plants indoors, and you want them to remain evergreen. Place them in a sunny window, and water just enough to keep the soil from drying out. Frost on outdoor plants will kill the foliage back. As new growth begins in spring, remove the dead leaves.
- Track cassia plants, which are more likely to affect indoor plants, for insects such as aphids and whiteflies. Inspect the underside of the leaves for aphid colonies, or gently shake the stems and test for tiny white flies fleeing the distressed plant. Spray the pests directly with neem oil containing an insecticidal soap or insecticide. Repeat applications on the spray label every three days or as guided until the pests are gone.
- Some forms of cassia, including *Senna pendula* var. In some regions Glabrata are invasive. Verify that your selected varieties are safe to grow in your field.
- Cassia plants used to belong to the Cassia genus but they were split into their own distinct senna genus. The botanical name Cassia may still be used in some nurseries.

Harvesting: The bark strips harvested are dried and become sticks to cinnamon. Cinnamon sticks are bent as they are because this is the natural way the bark dries away from the tree after being peeled. The dried sticks may be ground into a powder, too.

Preserving: Store cinnamon sticks or cinnamon powder in an airtight container in a cool, dry place.

4.17 CATNIP (NEPETA CATARIA)

Habitat: Catnip is native to Europe and to Central Asia in the southwest, and is commonly naturalized in many parts of the world.

Climate: Catnip grows partly to full sun in cultivated fields, in hedgerows, field boundaries and on dry banks and wasteland.

Soil: Catnip is easily grown in any garden soil, and does not require moisture in the same way as the other mints.

Growing: Catnip reproduces by dividing plants in the spring, or by sowing seed at the same time. Sow in rows, approximately 20 inches apart, thinning the seedlings down to about the same distance apart as the plants reach a considerable size. They need little attention, and if the field is kept free from weeds, it will last for several years. The seeds ' ability to germinate lasts five years. Below are some tips in growing;

- Plant catnip seeds 1/8-inch deep in draining 10-inch pots filled with a mixture of 1 gallon of each coarse river sand, moist sphagnum peat moss and sterilized loam, along with 4 tablespoons of dolomitic lime to raise the pH to between 6.1 and 7.8.
- After seeds are planted, water the growing medium before the water drains from the containers. Spray the growing medium with water from a spray bottle three or four times a day, keeping it moist at all times. Water with 1 inch of water every week after planting outside, if you receive less than 1 inch of weekly rainfall.
- Dig one hole for each plant, equal in width and depth to the 10-inch pots, outdoors in a sunny or partially shaded location, spacing each hole 12 to 18 inches apart, after the frost threat passes in spring.
- Place a 10-inch flat, draining pot in every hole that you dug. In the 10-inch pots that you buried; place the pots you planted the catnip inside. Growing catnip in one pot inside another, buried pot encourages removal when you take the plants inside.
- Thin the catnip seedlings so that after they reach 2 to 5 inches tall, only the healthiest one remains in each bowl. Grasp the seedlings by their stems and pull them to remove them from the containers.
- Place a 3-foot-tall, 3-foot-wide tomato cage around the catnip to prevent it from being disturbed or dug out of the ground by cats or other animals.
- Keep an eye out at the bottoms of the leaves for fine webbing, a

sign of spider mites. Without a magnifying glass, spider mites are hard to see so if you spot any, they will look like slowly moving white dots. If you have spider mites spray the leaves with a water hose.

- Pinch back the tips of the shoot to grow thick, well-shaped plants in spring. After each bloom, cut the main stem down to 1 inch above ground level using pruning shears for the job. The spent flower heads pinch off to prevent seed distribution.
- Take the leaves out of the plants after they have been cut back during bloom. Place the leaves 1 or 2 weeks into a paper bag to dry until use. You can pinch leaves off the plants too as you need them.
- Sanitize your pruning shears for 30 minutes in 1 cup of chlorine bleach and 9 cups of water, and let them air dry before using them. Wipe the blades between wounds with a cloth that is moistened with rubbing alcohol.

Harvesting: During the growing season, Harvest leaves flowers at any time when the plant is in full bloom.

Preserving: Dry the flowers and leaves, and then cut into smaller pieces. Store in an airtight container in a cool, dry place.

4.18 CHILI PEPPER (CAPSICUM SPECIES)

Habitat: *Capsicum annuum* is a domesticated species of the plant genus Capsicum native to southern North America and northern South America.

Climate: Chili pepper plants prefer full sun in a warm climate, these plants are mostly perennial in sub-tropical and tropical regions; however, they are usually grown as annuals in temperate climates.

Soil: Prefers warm, moist, nutrient-rich soil.

Growing: Start indoors and sow seeds ¼ inch deep, 8 to 12 weeks before the last frost. You can pre-soak seeds in warm water overnight prior to planting. After planting, keep moist and warm in a sunny location. Good results are also achieved by putting plastic cling wrap over the containers to create a hothouse environment. Remove when seedlings emerge.Below are some tips in growing;

- Not only does the amount of water you give your hot peppers affect their overall health during the growing season, but it can also "trick" them into developing the flavor and heat you prefer. Maintaining soil consistently moist when flowers appear on the plants and again when fruits form is especially important. As peppers reach maturity, flavor and pungency will be influenced by variations in the water. Hotness increases when you completely withhold water, as the peppers reach full size and mature color. The evening before you plan to harvest them, or the morning before an afternoon harvest, flood the pepper bed for extra heat.

- Monitor dirt changes to affect the hot pepper's taste and heat. Most hot peppers, in the range 5.5 to 5.8 pH, prefer slightly acid soil. If you have alkaline soil, use sulfur to make it acidic. In general, 1 pound of sulfur in a 100-square-foot garden can diminish ph by 1 point. Until planting, the more nitrogen - rich the soil, the milder your chili peppers get. For hotter peppers, prior to planting, restrict soil preparation to work a 2-inch compost layer in the top 6 to 8 inches of the pepper field. To improve the production of pepper, add a handful of Epsom salts rich in magnesium to the planting spot. Before setting in the seedling, cover the salt pile with a thin layer of soil, to prevent direct root contact.
- Chili peppers are lovers of the sun but it can be too much of a good thing— well, too much. Hot peppers need warm air and temperature in the soil as seedlings, without a doubt. Do not place transplants in the garden until at least two weeks after the last frost date in your town, when soil temperatures reach 65 degrees Fahrenheit or so. Do not postpone planting at the same time, because the seedlings must have passed the flowering stage before the air temperature reaches 85 degrees Fahrenheit. Otherwise, the plants will lose their blossoms, increasing pepper yield. Even after the peppers reach the process of blossoming, you will still have to keep an eye on the temperature gauge. Place shade cloth over pepper plants if it goes over 105 degrees, to protect them from sun damage.
- Keep track of where you plant chili peppers and related plants from year to year for optimum disease and pest-resistance. Many members of the "nightshade" family include bell peppers, tomatoes, eggplants and potatoes to which hot peppers belong. Nightshade plants are particularly vulnerable to fungal diseases, and these disease spores can winter in the soil. Do not grow them where any nightshade plants have flourished for at least three

years to reduce the chances of your hot peppers falling victim to any of those problems.

Harvesting: Harvest when the peppers are red for hottest flavor and after drying if you plan to crush or grind them. When it's green, you can harvest and use the pepper fresh but the flavor may not be as hot.

Preserving: Drying: Use a threaded sewing needle (can use thread or fishing line) to string hot peppers through stem. Hang in full sun.Freezing: First, wash thoroughly and let air dry or pat dry. Remove stems but leave the seeds in. The peppers can either be frozen in containers or plastic bags and they will taste "fresh" any time you eat them.

Crushed Hot Peppers (DIY): Easily make your own gourmet sweet, crushed peppers. Ovenpreheated at 250 degrees / broil. In a pan or on a sheet of tin foil, spot dried, hot peppers in the oven. Roast until peppers begin to dim (about 5 minutes). Let it cool down and put it in a blender. Crush to the size needed. All dried forms can be ground into powder too. Place dried peppers or powder in an airtight container, in a cool, dry place.

Caution: Caution should be taken when processing or handling this pepper as it can be hot and you would not want to touch your eyes, mucus membranes or similar sensitive areas after handling.

4.19 CELERY SEED (APIUMGRAVEOLENS, VARIETY DULCE)

Habitants: Celery is native to the Mediterranean. Wild celery was known to the Greeks and Romans. However, the cultivation of celery only began in Europe in the 17th century.

Climate: Celery grows best in a bright spot out of the direct sun, with temperatures of 70° to 75°F during the day and about 60°F at night. Provide plenty of water and good drainage and air circulation.

Soil: Celery grows best in moist, nutrient-rich soil.

Growing: Celery grows nearly year-round in mild areas. Start seeds indoors elsewhere in January, and bed seedlings in early spring after the danger of frost is over. Soak seeds prior to planting. Germination usually takes about ten days. Transplant at approximately three months, when seedlings are about 3 inches high. Plants around 6 inches apart in a vacuum. A copious amount of water. The juiciness of stalks depends on how much water the plants are receiving. Less ideal conditions result in thicker, heavier, bitterer stalks. Below are some tips in growing;

- Grow celery in compost-rich, moisture - retaining soil bordering saturated but still draining soil. Celery favors soil with a pH of 5.8 to 6.8. Celery has a low heat tolerance and prefers a calm, rainy location where temperatures range from 60 ° F to 70 ° F. Plant celery which offers 4 months of cool weather during the growing season.
- Celery is a cultivation of the cool weather. It takes 16 weeks of cool weather for the harvest to arrive. Start celery seed indoors for 8 to 10 weeks before spring's last frost. When seedlings have 5 to 6 leaves, set transplants in the garden 2 to 3 weeks before the average last frost date. (To prolong transplantation time and slow growth, cut seedlings down to 3 inches tall and then let them grow on.) Cold weather will inhibit growth as warm weather would. Temperatures below 50 ° F will cause celery to bolt for more than 12 hours. Celery plant in the early spring in cool spring and summer regions. In warm spring and summer regions, grow celery for harvest in late autumn or early winter in the late summer.
- Sow celery seed 1⁄4 to 1⁄2-inch-deep, 6 to 10 inches apart; 24 inches apart for spaceline. Transplant seedlings began indoors, set 6 to 10 inches apart, into trenches 3 to 4 inches deep. As plants grow, soil mounts up around the stems to blanch them. Self-blanching celery in blocks 6 to 12 inches apart; closer planting will give higher yield but slender stalks.
- Companion plants. Lettuce, spinach, English peas. Avoid pumpkins, cucumbers, and squash.
- They can grow celery in an 8-inch jar. Place the celery in large containers on 10-inch centers. Tie paper or cardboard cylinders around the stalks to blanch celery growing in a container.
- Hold the celery well-watered throughout all growth phases. Water shortages can slow growth, cause stalks to become stringy and allow plants to send up stalks of flowers. Celery is a big feeder. Before planting and side-dressing plants with compost at midseason add aged-compost to planting beds.
- Hold Weed - free celery planting beds to avoid competition for nutrients and moisture. Keep the cultivation shallow, so that

roots are not damaged. Blanch celery enhances its sweet flavor and white stalks. Celery not blanched could be bitter tasting. Blanching is accomplished by covering the stalks with cylinders of dirt, wool, or paper rolled up to the top of the stalks to shield them from the light, which allows them to develop chlorophyll and turn green. Celery blanch until 10 to 14 days before harvest. Upon blanching, celery that sits too long will become pithy and may rot.

- Celery usually doesn't encounter serious pest problems, but can be attacked by miner celery leaf and slugs (while blanching).
- Celery can be affected by pink rust, black heart and blight. Make sure that the soil contains enough magnesium and calcium to avoid these diseases.

Harvesting: Seeds of the celery are picked as they ripen. The primary umbels (flowers) of celery mature first, and are often broken before the secondary umbels are dry. It is possible to harvest the umbels by hand as they ripen. As the umbels are harvested the seeds can be easily separated and will be fairly clean. Screening or winnowing should remove any remaining dust and debris.

Preserving: Store dried celery seed in an airtight container in a cool, dry place.

4.20 CHERVIL (ANTHRISCUS CEREFOLIUM)

Habitat: Chervil is native to Southern Europe and the Caucasus but was spread by the Romans through most of Europe, where it is now naturalized.

Climate: Chervil grows sunny sides in forest garden; dappled shade; gloomy edges; hedgerows and cultivated beds. Plants don't like hot dry summers; best to provide a cool shade location for summer crops; however, winter crops require a sunny location.

Soil: Chervil prefers a well-drained moisture retentive soil.

Growing: Sow chervil seeds directly outdoors in the garden soil, because the long taproot can make transplantation difficult. Chervil bolts quickly (goes to flower and seed), so during the growing season, it may be necessary to re-so many times. Sow in succession from February through October for continuous harvesting. The seed usually germinates within two to three weeks. Below are some tips in growing;

Chervil seeds can be sown a couple of weeks before the last expected spring frost. Start chervil cultivation in single containers under fluorescent lights to be transplanted after the last spring frost. Use individual pots to avoid the tap root being disturbed during transplantation time. (The plant may bloom prematurely if the taproot is disturbed.) Chervil seed does not need to be covered for germination. It takes 7 to 14 days to germinate.

- Transplant chervil to the garden after the last frost in spring; be careful not to disturb the taproot.
- In spring, seed the chervil directly in the garden as soon as the soil can be worked. In cool summer regions, if you sow chervil in spring, you can do second sowing in mid- to late summer. Seeds can also be sown to germinate in autumn and grow in the following spring.
- Sow 1⁄4-inch-deep chervil seed and 3 inches apart. Sow seeds in a very fine soil bed, or sprinkle seeds over a fine vermiculite base.
- Space plants 6 inches apart; thin plants 8 to 12 inches apart when plants are 3 to 4 inches tall.
- Grow 6 chervil plants to cook over the course of the season; sow seed for continuous harvest at 2-week intervals. Grow 12 sustaining plants.
- Chervil is said to repel slugs and keep the lettuce free from ants and aphids. Chervil is said to make the radishes taste stronger.
- Keep the soil evenly moist; do not let the soil dry out when growing chervil.
- Prepare beds with aged compost to be deposited before planting. Side-dress chervil with aged compost during the growing season. There is no need for extra fertilizer.
- Mulch around plants in hot regions to 4 inches deep to keep the roots cool. Keep mulch away from plant base so crown rot and earwigs don't harm them.
- Cover chervil seed and young plants with floating row cover to keep young plants from consuming birds and rabbits. Pinching out flower stalks later in the season to prolong growth. To keep the plant 's energies channeled into leaf growth, cut down the flowering stems. Long floral stalks are a warning that the plant is close to the end of its life.

Harvesting: Harvest leaves throughout the growing season, regular harvesting also helps to prevent bolting.

Preserving: Dry the leaves well and cut into smaller pieces, store in an airtight container in a cool, dry place.

4.21 CHICORY (CICHORIUMINTYBUS)

Habitat: Chicory lives as a wild plant on roadsides in its native Europe, and in North America and Australia, where it has become naturalized.

Climate: Chicory lives on the roadsides as a wild plant, it is a cool -weather crop, and only tolerates mild summer temperatures. It requires well-distributed rainfall in drier areas, with good drainage or some irrigation.

Soil: Chicory grows on any type of soil, but, when cultivated grows best on mellow, deeply tilled, fertile soil or sandy loam.

Growing: Chicory requires a cool, firm seedbed. Chicory Spring seedings are usually successful. Summer seedings with proper irrigation were successful in some areas. After mid-August, you don't want to sow chicory flower seed as cooling temperatures and shortening day lengths in the fall hinder the production of chicory stands. When planting you can drill or broadcast Chicory flower seed, but drilling is preferred because it provides a more uniform planting depth. 1/4 "thick plant seeds. Below are some tips in growing;

- Plant chicory in a sunny bed if it grows for its seeds, or if it grows under partial shade to harvest the leaves. Choose a bed of soil that drains, and moderate fertility. Stop beds of clay - based soil as the roots aren't growing properly.
- In early spring, prepare the planting bed once the soil temperatures are warm to 65 F. Thoroughly weeding the room. Shake the soil with a fork to cultivate. Place a compost layer 4 inches thick into the bed to a depth of 8 inches.
- Dig furrows 1/16 "deep along the surface. The furrows space between 2 and 3 feet apart. Scatter the seeds of chicory into the furrows at a rate of two per inch. Cover a thin layer of soil over the seeds.
- Heavily spray the bed to settle the soil over the plants. Use a misting nozzle on a garden hose. Do not saturate the soil since the

seeds are buried very shallowly and will dislodge quickly under a strong stream of water.

- Wait for germination starting in seven days but don't be discouraged if the seeds sprout for up to three weeks. Thin the seedlings, if they grow for roots, to one per 6 inches, or leave them closely spaced if they grow for the leaves.
- Place a 2-inch layer of mulch between the rows to prevent weed growth and help keep soil moisture intact. Remove weeds, as nutrient and moisture rivalry tends to be reduced.
- Water once a week to a depth of 1 inch. Enable the soil to dry out slightly between watering to prevent the roots from rotting or growing too quickly, resulting in a bitter, fibrous crop.
- Side-dress the rows with a layer of well-composted manure 2 inches thick once the plants grow to 4 inches in height. In the middle to late summer, add more manure to promote the production of large, stable roots.
- Prune off the heads of the flower as they arise in the summer months. Watch for buds as early as May but prepare well into October to prune the plants. The pruned flower heads are discarded into a green waste bin.
- The leaves are picked before the middle rib thickens and becomes gritty if they are eaten raw. The mature leaves are harvested for cooking. In mid- to late autumn, dig up the roots for consumption as cool weather increases the size and consistency of the roots.

Harvesting: Chicory harvesting should take place as late as possible in the season as there is usually a marked increase in root size and weight during the cool weather. As they become known as weeds, care should be taken to remove all remaining pieces of roots. Roots are washed, cut into 2.5 cm square cubes, and dried over flames. The final process is to roast the dried chicory and grind it to a fineness suitable for mixing with the ground coffee.

Preserving: Can store dried chicory indefinitely. Ground chicory is usually packaged in bulk in waterproof barrels or containers, or in smaller household-use packets. Store in a dry , cool spot.

4.22 CHIVES (ALLIUM SCHOENOPRASUM)

Habitat: Allium schoenoprasum is native to Europe, Asia and North America. It's is the only species of Allium native to both the New and the Old Worlds.

Climate: Chives prefer full sun, commonly grown in cultivated beds, rock gardens, perennial beds and herb gardens.

Soil: Chives thrive in well-drained soil, rich in organic matter, with a pH of 6-7.

Growing: Chives can be propagated by seeds or through division. Sow seeds indoors in a rich mix of soil, in a greenhouse or the like. The seedlings can then be transplanted to the outdoor garden after about 4 weeks. Below are some tips in growing;

- Plant individual chive plants in a pot measuring 6 to 8 inches in diameter, or plant several plants 6 inches apart in a larger pot. To promote good drainage, using pots which have holes in the bottom.
- Plant individual chive plants in a pot measuring 6 to 8 inches in diameter, or plant multiple plants 6 inches apart in a larger pot. Using pots that have holes in the bottom to facilitate good drainage.
- Grow chives in a sunny spot where they will get at least six hours of direct sun per day.
- Upon planting, the water chives until water starts draining through the holes in the bottom of the pot, then allowing the soil to dry out slightly before watering again.
- Water the soil during the growing season, when the top 1 inch of the soil starts to dry out. An easy way to test the moisture in the soil is to put your finger into the soil near the edge of the bowl.
- Between spring to late summer fertilize every two weeks using natural, granular fertilizer, like a combination of 3-3-3. Use 1 dining cubicle. The pot is 6 to 8 inches in diameter. For larger pots, use 2 tablespoons, between 10 and 12 inches. Sprinkle the fertilizer onto the ground at the plant base. Water for release of the nutrients after each fertilizer application. Requirements for fertilizers differ by product; if a specific amount is indicated by the package label, follow the directives of the supplier. In winter, discontinue the fertilization of chives to allow the plants to rest.

- Chives from harvesting grow 6 inches tall. Use a sharp knife or tiny garden clippers to cut the green leaves 2 inches above the soil line. Harvesting chives consistently every two to three weeks through the growing season by removing the top growth when it gets 6 inches tall.
- Cut off the flower when it grows to encourage further development in the foliage.

Harvesting: The leaves are harvested by cutting down to the plant foundation. The plant must continuously grow new leaves during the growing season, allowing for a continuous harvest.

Preserving: If the leaves are not new to use, they can be dried and cut into smaller pieces. Place in a cool, dry place in an airtight container.

4.23 CICELY (MYRRHISODORATA)

Habitat: Chervil is native to Southern Europe and the Caucasus but was spread by the Romans through most of Europe, where it is now naturalized.

Climate: Chervil grows sunny sides in forest garden; dappled shade; gloomy edges; hedgerows and cultivated beds. Plants don't like hot dry summers; best to provide a cool shade location for summer crops; however, winter crops require a sunny location.

Soil: Chervil prefers a well-drained moisture retentive soil.

Growing: Sow chervil seeds directly outdoors in the garden soil, because the long taproot can make transplantation difficult. Chervil bolts easily (goes to flower and seed), so during the growing season, it may be necessary to re-sow several times. Sow in succession from February through October for continuous harvesting. The seed normally germinates within two to three weeks.Below are some tips in growing;

- In late spring, a fresh seed can start out indoors. Seeds need to be stratified or kept cold before germinating. Place the seeds in moistened peat moss three months before sowing in a covered container in the refrigerator. Germination will take about 30 days but can take up to eight months at times. Spring root division is the safest way to start cicely sweet.
- Sweet cicely has a long taproot and dislikes being moved. It is best to circularly transplant it when small before taproots grow long. Set off transplants in spring after the last freeze. Cicely transplant sweet into a sunny portion of the yard.

- Sow seeds for germination in spring or late summer the next year. Germination can be extremely slow. Sowing only germinates after a period of freezes and thaws.
- Sow seeds ½ inch deep. Set divisions at the same depth they were growing.
- Space plants 2 feet apart.
- How much to plant: Grow 6 plants for culinary use; grow 12 plants for preserving.
- Companion planting: Sweet cicely is rich in nectar; honeybees and beneficial insects are drawn to the garden. Stop cicely planting sweet parsley and carrots nearby; they attract the same insect pests.
- Sweet cicely grows best where the soil stays just moist; do not let the soil dry out.
- Add aged compost or industrial organic planting mix to planting beds before sweet cicely planting.
- Grow sweet cicely in a container at least 6 inches high and wide.
- In winter, sweet cicely may be grown indoors in a bowl. After the first autumn freezes the plant will die back to the ground.

Harvesting: Harvest leaves throughout the growing season, regular harvesting also helps to prevent bolting.

Preserving: Dry the leaves well and cut into smaller pieces, store in an airtight container in a cool, dry place.

4.24 CILANTRO/ CORIANDER (CORIANDRUM SATIVUM)

Habitat: Cilantro is native to regions spanning from southern Europe and North Africa to southwestern Asia.

Climate: Cilantro grows in full sun at a pH of 6.2 to 6.8; in the south and southwest where the sun is strong it will tolerate light shade.

Soil: Prefers well-drained, fertile soil.

Growing: In the garden where you can pick, Cilantro wants its own room and then let it go to seed. Plant it in a bed devoted to herbs, where it can reseed, or in a vegetable garden corner.Below are some tips in growing;

- Cover a well-drained garden bed with a 2 "compost layer. Choose a bed that receives about six hours of sunlight per day. Transform the compost into the top 6 inches of soil using a spade to help improve bed humidity and nutrition.
- Plant 1/4-inch-deep cilantro seeds, and space them in rows, 12 inches apart, and about 4 inches apart. Cilantro doesn't tolerate transplantation, and when sown directly in the greenhouse, does best.
- Mist the surface of the soil with water after sowing, to moisten it. Mist as needed until the seeds of the cilantro germinate, which can last up to a week. Reduce watering to once a week as needed to keep the top 6 inches of soil moisture, or water when the top 1/2 inch of soil feels dry.
- When they grow 3 to 4 inches tall, pinch off excess cilantro seedlings so that the remaining plants are spaced about 9 inches apart in the garden rows.
- Bright green yield, healthy plant leaves, as needed. Snip with small shears through the roots, or pinch them over a root. Alternatively, for a bigger harvest, cut the entire plant by up to half its height. Shearing can also prevent the cilantro from seeding.
- If you want coriander seeds, allow the cilantro to blossom. After the blooms wilt, pick the seed heads and the seed clusters dry. Separate the seeds from the stems and in an airtight jar store the dry seeds.

Harvesting: Continuous foliage of harvest cilantro in the cooler months of spring and fall, even in areas without hard freezes during winter. Harvest by cutting the leafy stems near to ground level; they will be between 6 and 12 inches long. Avoid cutting more than one-third of the leaves at once, or the plant can be damaged. Clipping the brown, round seed heads to remove the seeds; placing it upside down in a paper bag. The round husks will dry and break in two within a couple of days, dropping the edible seed within. Do not delay harvesting seeds, or the poor stems will fall over. Dry plants, and thoroughly leaves.

Preserving: Store dried cut leaves or seeds in an airtight container in a cool, dry place.

4.25 CINNAMON (CINNAMOMUMVERUM)

Habitat: *C.verum*, the true Ceylon cinnamon tree is native to South Asia (India and Sri Lanka), but had already been introduced pan-tropically in the 1700s.

Climate: Cinnamomum is suited to a wide variety of climates. Cinnamon needs a warm and humid climate with an annual well -distributed rainfall of about 2000-2500 mm and average temperatures of around 27 ° C. Wild cinnamon trees are adapted to the evergreen tropical rainforests. It grows best at low altitudes and is usually grown without rain, but being basically a forest tree, it tolerates light shade.

Soil: It grows well in the tropics on different soils but soil type has a marked effect on the quality of the bark. Fine gravelly sandy and laterite soils are stronger in Sri Lanka and India, rather than rocky and stony substrates, but more loamy soils are favored in the Seychelles and Madagascar. Cinnamon is considered susceptible to salinity, and the effect of waterlogged and marshy conditions is a bitter product.

Growing: Seed has low viability and is best sown when ripe. Remove the fruit pulp, as germination can be inhibited. Germination at 20 ° C will take 1 - 6 months. Stored seed should be sown to containers as soon as possible. Sowing seeds can be achieved in nurseries or directly on the ground. Greenhouse beds should be well-prepared and partially shaded with rich sandy soil free of roots and stones, with ample freshwater supply nearby. Clumps of seedlings can be transplanted into baskets after about 4 months, and are planted 4 or 5 months later. Alternatively, until the final transplantation, the seedlings may be left in the greenhouse, the shade gradually removed and the seedlings hardened. Semi-repeat side fire cuttings, 7 cm with a handle, June / July in a low-heat tub. Old rootstocks break. The benefit of splitting old rootstocks is that the stems are ready for cutting after planting at about 12 - 18 months, compared to as much as 3 years for seedlings.

Harvesting: Commercial cinnamon bark harvesting involves cutting down the stems low after an initial establishment period and harvesting the bushy regrowth stems at regular intervals afterward. Stems are split into 2 longitudinal stripes during the rainy season to promote the peeling of the bark. A first harvest may be obtained in Sri Lanka after 3 - 4 years, though with subsequent cuttings, quality and yield improve. Cinnamon is cut every 2 years on plantations, and the flush of straight shoots produced by coppicing yields the bark peeled for cinnamon quills. The estimated bark yield after 3 -

4 years is 56-67 kg / ha, and subsequently grows to 168-224 kg / ha. You can also get about 63 kg / ha of chips and about 2.5 tons / ha of undried leaves.

Preservation: Leaves are harvested after harvest; the brown stem is scraped off and the stems are cut to pieces of convenient length. The bark is torn longitudinally and peeled off using peeling knife on the same harvest day or after softening the next day. The cylindrical bark pieces dried 2 - 5 days in the sun and are wrapped in packets. The dried cinnamon bark (quills) is graded and traded according to color, taste, smell, and weight.

4.26 CLOVE (SYZYGIUMAROMATICUM)

Habitat: Cloves are native to the Maluku islands in Indonesia. Cloves are harvested primarily in Indonesia, India, Madagascar, Zanzibar, Pakistan, and Sri Lanka.

Climate: Clove trees need a warm, humid climate with an annual rainfall of 50 to 70 inches and a minimum temperature of 59 ° F, and a place in full sun or part shade. The majority of clove plantations are within 10 ° of the equators and close to the ocean.

Soil: Clove trees grow best in well-drained, fertile loam soil.

Growing: Sowing seeds in spring, or rooting cuttings in summer, will propagate trees. The first time the trees bloom when they are eight or nine years old. Trees can live for 100 years or longer, under favorable conditions. Below are some tips in growing;

- Purchase clove seed from a reputable source, and moist the soil prior to planting. Make sure the seeds you purchase are pollinated (which is why just planting a few buds out of the bottle in your spice rack won't do that). Maintain the soil where the clove grows wet but is not waterlogged.
- Place clove seeds directly onto the ground surface. The seeds need not be buried under the soil to lay down roots and begin to adjust to their new homes. When only a little time has passed, you will see the seedlings of the clove tree starting to sprout up. When they exceed one inch in height, move your clove seedlings into a larger pot.
- Make sure your clove plants are watered regularly, especially during periods of drought or overheat for your area. Clove trees like to grow in rich soil with good drainage, they are loamy and they are full of organic matter.

- More than anything, gardeners considering adding clove to their gardens should be mindful that growing clove trees is a long - term project requiring its share of long-term work — but it also comes along with long - term rewards. Do not expect to get any functional buds from your clove tree until the branches have begun to produce a full variety of flower clusters which will take at least six years before the first possible harvest, starting from the time you planted the seeds.
- When it comes to watering your clove trees a delicate balance must be achieved. Supported treatment options are light watering and regular misting to mimic the humidity of the natural tropical conditions of the clove plant if your climate does not quite suit.
- More than anything, gardeners considering adding clove to their gardens should be aware that growing clove trees is a long - term project requiring its share of long-term work — but it also comes along with long - term rewards. Do not expect to get any functional buds from your clove tree until the branches have started producing a full array of flower clusters, which will take at least six years before the first possible harvest, starting from the time you planted the seeds.

Harvesting: When they reach full size, the unopened flower buds are harvested by hand and are just turning pink. The buds are then snapped off the stalks of the flowers and placed on the mats of the leaves to dry in the sun. They'll have turned dark brown after three days and only weigh one-third of their weight when picked first.

Preserving: Store dried cloves (whole or powdered) in an airtight container in a cool, dry place.

4.27 COSTMARY (TANACETUMBALSAMITA)

Habitat: A broad perennial up to 3 feet in height that is native to West Asia and naturalized in Europe and parts of North America.

Climate: The coastal herbal plant is a hardy herb tolerating hot summer and cold winters. Full sun or part shade, but will not bloom unless in full sun; they are best in the part shade when grown for their leaves.

Soil: Prefers fertile loam (heavy feeder), but on the dry side. Needs to be replanted every few years as the center will die out.

Growing: Plants do contain very few seeds. Best propagated via rhizome. Costmary care is an easy task; once grown, the herb does not need any fertilizer and rarely needs water. Require for each plant to have at least 12 inches. Costmary profits from division every two to three years to keep the plant from becoming overgrown and tired. Dig the clump with your hands in spring or autumn, then pull the rhizomes apart, or pick them with a knife or shovel. Replant, or give away the divisions.Below are some tips in growing;

- Avoid the full shade. Costmary will not flower without sun.
- Plant costmary in humus-rich, well-drained soil; costmary prefers a soil pH of 6.0 to 6.7.
- Start costmary indoors in spring. Seeds germinate in 14 to 21 days.
- Transplant costmary to the garden mid- to late spring.
- Start costmary seed outdoors after the last spring frost or set out rhizome divisions.
- Sow seeds ¼ to ½ inch deep.
- Space costmary plants 2 to 3 feet apart.
- How much to grow: Grow 1 plant for culinary use. Use costmary in small amounts; it may be overpowering.
- Costmary is an aggressive grower; it has a root system of rhizomes that can crowd out other plants that grow nearby.
- Grow costmary in soil that is kept evenly moist.
- Costmary grows in loamy, humus-rich soil. There is no need for any extra fertilizer. Aged compost is added to the planting bed before planting.
- Costmary can look leggy, if not back-clipped. Discourage flowering for more foliage. Cutting costmary back to 3 or 4 inches above the ground when plant flowers; it will fill out in a few weeks. Plants break every 3 years.
- Costmary can be grown in a pot 12 inches wide and deep.
- In cold winter areas, costmary will die back to the ground; it will re-spring in the spring. Wintertime plants grow indoors in containers.

Harvesting: Clips leaves with scissors.Collect small quantities of leaves from each plant before they yellow. Harvest young costmary leaves as need.

The oils in the leaves are strongest before flowering.

Preservation: Refrigeration-Costmary leaves should remain in the refrigerator for a few days; cover the leaves in a damp paper towel and put them in a plastic perforated bag. Drying-Leaves can be dried out of direct sunlight on a screen or tray set out at a warm spot. Storage-Store dried leaves in an airtight jar.

4.28 BLACK CUMIN (NIGELLA SATIVA)

Habitat: *Nigella sativa* (black caraway, also known as black cumin, nigella, kalojeera, kalonji or kalanji) is an annual flowering plant of the *Ranunculaceae* family, native to a large area of the eastern Mediterranean, northern Africa, the Indian Subcontinent and western Asia.

Climate: The growth habits of the black cumin flower are that they love to grow in full sunlight when at all possible.

Soil: They also love rich, dark soil with a pH of between 6.0 and 7.0 to allow them to grow.

Growing: Black cumin develops extremely easily from seed and germinates very quickly, typically 10 - 14 days depending on temperatures. Where garden color is desired early in the season, it can be either directly sown or started in containers. Black cumin is at its finest when grown in groups or clumps to the best advantage when the feathery leaves and flowers can be used. Sowing seeds directly and consistently over the entire season gives the most color and best results for the longest flowering time.Below are some tips in growing;

- Check the pH of the soil with a home kit two to three months before the target planting date to check that the pH is slightly acidic with 6.0 to 7.0 values. Work ground rock sulfur inside the soil to reduce the pH if necessary. Follow the current pH-based package application rate. Let the soil rest for two to three months to allow the pH-change time to change.
- Apply 2 to 3 inches of organic compost to a depth of 8 to 10 inches inside the planting area. Compost raises the soil's nutrient value and is a natural fertilizer.
- Sprinkle gently over the planting area with *Nigella sativa* seeds. Sprinkle the seeds over about 1/4 to 1/2 inch of soil. Tap the soil

to keep the seeds in place, gently.

- Sprinkle lightly with water over the planting area to moisten the soil to a depth of 1 to 2 inches. Use caution to avoid seed runoff from heavy water streams. Keep the soil sufficiently moist, but not warm, to promote germination and growth of plants.
- Once the seedlings reach 2 to 3 inches tall plants thin out. Pinch off weak sprouts that grow too close to each other. Seek spacing the plants 4 to 6 inches apart. Hold the healthiest sprouts forever.
- Deadhead plants by cutting the flowers off when they start to fade. If you want them to turn into seed pods, leave the flowers onto the vine.

Harvesting: Once they have turned brown the seed pods can be cut off and harvested. Make sure that you do that before the pods open and scatter their seeds all over. Cutting off stems with scissors or garden snips and flipping upside down in a bowl or collecting tub to ensure no loss of seed.

Preserving: When you have a lot, use a rubber band to tie together, and leave to dry fully in a large paper grocery bag. Do not put it in a plastic bag, or leave it in a bucket or bowl as the heads of the seed may become moist and then moldy. Once the head of the seed is completely dry, it can either be shaken (when the pod has opened) or opened and gently rubbed to extract the seeds. Sieve or pick any no seed material, ensure the seeds are completely dry and store them for culinary use in an airtight container. If intended to be used for seeding the next crop, put in the refrigerator. Pods can also be left intact, and they are very attractive for dried flower decorations.

4.29 DILL (ANETHUMGRAVEOLENS)

Habitat: Dill is native to Asia Minor and the Mediterranean; it is now widely planted around the globe and has been naturalized in parts of Europe and North America.

Climate: Grow dill in full sun; plants are likely to fall over in shady locations.

Soil: Dill is best grown in rich, light, well-drained soils.

Growing: Sow seeds directly to the ground just before the last date of spring frost. Seedlings can be hard to transplant so seeds are not usually started indoors. For purposes of extending the time when fresh leaves can be harvested, additional seeds may be planted every two weeks until early summer. Dill can be grown in dark, large containers too.Below are some tips in growing;

- Dill grows best in bio-rich, well - drained soil. Until planting apply aged compost to the planting area. Dill favors a soil pH between 5.5 and 6.7.
- Dill seed can begin indoors for 4 to 6 weeks before being transplanted into the garden. Seedlings are forming taproots that transplant poorly so that dill is easiest to start and grow in place.
- Set out seedlings after the last frost in spring. Dill forms a taproot and is not easily transplanted.
- Sow dill in the garden after the frost threat has passed. Sow seed in sheets, or bands of 5 inches. Dill can readily reseed itself so it can be planted where it can grow for several years. Sow successive dill crops for a continuous, fresh harvest every 3 to 4 weeks.
- Direct sow seed in shallow trenches ¼ to ½ inch deep; thin successful seedlings from 8 to 12 inches apart.
- Space dill plants 10 to 12 inches apart. Space rows 2 to 3 feet apart. Dill is often grown in clumps, not rows.
- Grow 10 dill plants over the course of the season for cooking and culinary use; sow several successions two weeks apart. Grow 20 plants for preserving.
- Companion planting: plant dill with plants of the cabbage family and also tomatoes, chilies, sweet peppers, strawberries, and thyme trees. Dill boosts crop production for the cabbage family. Do not grow dill near carrots, or hybridize it with fennel. Dill attracts planting honeybees and beneficial insects. Dill can be used as a Tomato hornworm trap crop. It's said that the aroma of dill repels aphids and spider mites. Dill attracts the caterpillars which turn into black butterfly's swallowtails.
- Water dill regularly and evenly until established. Once the dill has been established it will grow best if the soil can almost dry between watering.
- Prepare aged composting beds for planting. Side dress dill with compost tea two to three times during the growing season.
- Keep the planting bed well weeded; weeds compete for nutrients and water

- It may benefit from staking on the prevailing wind in gardens. Keep weed-free on planting beds; weeds compete for humidity and nutrients. For sustained leaf growth, pinch out early flowers.
- Dill will grow easily in a container. Choose a container at least 12 inches deep as dill forms a taproot.
- Dill can be grown indoors in winter. Outdoors dill will likely die back to the ground after the first hard freeze.

Harvesting: Dill seed can be harvested by putting the seed heads in paper bags to dry out about 2 - 3 weeks after flowering (this process will also help prevent self-seeding in the garden). Leaves usually have the best flavor when the flowers first open; the leaves have by far the best flavor when fresh from the garden is harvested.

Preserving: For later usage, leaves or seeds may be dried or frozen; store dried dill leaves or seeds in a cool, dry place in an airtight container.

4.30 FENNEL (FOENICULUM VULGARE)

Habitat: Fennel is native to the Mediterranean, now naturalized throughout Europe, Asia and North America.

Climate: Fennel has become naturalized in many regions along dry, sunny roadsides, in pastures and other open sites. In Australia and the USA, it is considered an invasive species and a weed.

Soil: Fennel will grow in any ordinary garden soil, though it may have more yields on rich fertile or sandy soils.

Growing: Fennel is quickly propagated by seeds, sown in ordinary soil at a sunny place early in April. In some places it is considered to be invasive, so be careful in selecting the location as a fennel stand can last for several years.Below are some tips in growing;

- However, fennel can grow in all types of soil, in well-drained compost-rich soil.
- Fennel produces a taproot and is best planted. When started indoors, plant in individual peat pots so taproots are not disrupted during transplantation. Sow seed indoors 6 to 8 weeks prior to the final frost.
- Set fennel in the garden after the last frost in spring.
- Sow common fennel seed as early as 2 to 3 weeks before the last frost date average in spring. You can also sow fennel before the

first frost in late summer or early autumn for harvest. Fennel is semi-hardy and tolerates a mild frost. Enable succession plantings for harvest through autumn from late summer

- Sow fennel seed ⅛ to ½ inch deep. Seeds must be covered completely to germinate.
- Space fennel plants 10 to 12 inches apart. Space rows 2 to 3 feet apart. Fennel re-seeds itself readily; plant it where it can grow for several seasons.
- Grow 1 to 2 fennel plants for cooking; grow 4 to 5 plants for preserving.
- Companion planting: cultivate fennel to draw butterflies, colibris, and beneficial insects into the garden with sunflowers, calendulas, and nasturtiums. It is said that fennel interferes with the growth of beans, tomatoes, and kohlrabi. Do not plant fennel in the vicinity of dill or cilantro; they can pollinate.
- Give fennel regular, even watering until it is established. Once established, fennel can be kept on the dry side. Do not overwater fennel.
- Side dress fennel with aged compost at midseason.
- Mulch around fennel in summer to keep roots cool. To make bulb fennel tastier, mulch around the base of the plant to blanch the bulb and make it tender.
- Common fennel can grow 4 to 5 feet tall and may require staking or supports, especially if it is growing in a windy spot. Mound soil up around the base of Florence fennel to blanch the bulb and make it tender.
- Common fennel will grow easily in a container. Choose a container at least 12 inches deep; fennel forms a taproot so the container must be deep enough for the root.
- Fennel can be grown outdoors in mild winter regions.

Harvesting: Watch your fennel plant fade into light brown with yellow heads and the seeds begin to form in tight clusters. They will have a light green hue until they are ready to pick— when you see stripes begin to form on the individual seeds, you'll know it's time for harvesting. Snip off the heads as a whole and allow them to dry out at night, making the selection of seeds easier the next day. Thoroughly dry the seeds, use them whole or powdered.

Preserving: Store whole seeds or powder in an airtight container in a cool, dry place.

4.31 FENUGREEK (TRIGONELLA FOENUM-GRAECUM)

Habitat: Fenugreek is a native to India and southern Europe. For centuries it has grown wild in India, the Mediterranean and North Africa where it is mainly cultivated.

Climate: Fenugreek grows on sunny field verges, uncultivated ground, dry grasslands and hillsides.

Soil: Fenugreek prefers a well-drained loamy soil, but can grow in ordinary garden soil.

Growing: Pre-soak the seed in warm water for 12 hours, and then sow directly in the garden in spring. Below are some tips in growing;

- Mixing a good organic fertilizer into the soil before planting will yield better results when using fenugreek to improve the very poor soil.
- Companions growing with, Buckwheat, Beans and Crowder Peas.
- Spacing; Single Plants: 5" (15cm) each way (minimum); Rows: 5" (15cm) with 5" (15cm) row gap (minimum)
- Sow fenugreek from late spring to late summer, if more than eight weeks of the unused bed are left. Guide seed into cultivated soil so the seeds are roughly 2 inches (5 cm) apart and 1⁄2 inch (1 cm) wide. In all directions, thin to 6 "(15 cm) apart.

Harvesting: The leaves are harvested in the growing season and can be used fresh or dried. The seeds are harvested when fully ripe and dried for later use.

Preserving: Store dried whole or powdered seeds in an airtight container in a cool, dry place.

4.32 FILE (SASSAFRAS ALBIDUM)

Habitat: *Sassafras albidum* is native to eastern North America, from southern Maine and southern Ontario west to Iowa, and south to central Florida and eastern Texas.

Climate: Sassafras occurs at altitudes of sea level up to 1,500 meters all throughout the eastern deciduous forest habitat.

Soil: Sassafras trees prefer a rich, well-drained sandy loam (pH of 6–7), however, they will grow in any loose, moist soil.

Growing: Sow out in a cold frame as soon as the seed is ripe. Stored seed requires 4 months of cold stratification at 4 ° C. The seed shall be better sown as early as possible in the year. Transplant the seedlings as soon as possible into individual pots, and cultivate them in the greenhouse. Cultivate them on for their first winter in the greenhouse and then plant them out in early summer.Below are some tips in growing;

- Collect tree seeds from sassafras in fall, when the fruit turns dark blue. Hunt for tiny seeds, one inch or less. For each tree, you want to plant collect at least three seeds.
- Cut open the fruit to reveal the seeds. Remove the seeds, and water them. Dry the seeds with a towel, and place them in a sealable container in a shallow layer of dampened topsoil.
- Fall your seeds and close the container into the damp topsoil. Place the jar in a refrigerator and store for at least 120 days at 35 to 40 degrees Fahrenheit to approximate the normal dormancy time for the seed.
- Prepare a potting mixture for your seeds, using equal amounts of coarse sand, peat and perlite, during the week before the final spring frost. Using large plant pots, which are at least one foot deep, to support the development of tap root sassafras. To prevent the soil from becoming waterlogged after heavy rain, make sure that your plant pots have several drainage holes.
- Fill the growing plant pot with a potting mix until just one inch is between the soil and the pot rim. Plant three seeds at the center of each pot, one-quarter inch deep in the potting soil, leaving several inches of space between each seed. When you put your finger knuckle-deep into the soil, water the soil gently until it is moist to the touch.
- Place the plant pots outdoors and water them often enough without saturating the soil to maintain a moist growing environment. Place your plant pots in a partially shaded location outdoors until the seeds germinate. Wait two to three weeks after your seeds start sprouting and then pick the largest plant from each pot and

remove the other competing seedlings. Continue to water your seedlings until they reach one foot or so in height.

- Choose a transplant site that will receive full sunshine for part of the day. Make sure that the location you select has enough space for an adult sassafrass tree that can reach 60 feet in height with up to 40 feet of foliage spread.
- For each plant pot, prepare a planting hole that is several inches wider in diameter and equal in depth to your plant pots. Use a shovel to the bottom of the hole to aerate and remove the dirt. When you have a thick sand or clay soil, blend the compost or manure into the soil at the base of the pit. Remove your sassafras carefully from the plant pot and gently shake any soil that is attached to the roots.
- Place the sassafras upright in the roots of the hole and place so they protrude straight down or out of the tree without bending or pointing up. Gently cover the roots with the soil which you removed from the opening. Immediately after transplantation, water your sassafras gently to stabilize and moisten the soil.
- Irrigate the soil around your sassafras frequently enough so that the soil does not dry out. Spread the irrigation all over the area beneath the tree's leaves and branches to cover the entire root zone. Put a layer mulch around the base of your sassafras tree to prevent soil water loss and reduce competition from other plants.

Harvesting: Sassafras roots are usually harvested during the fall. Only the outer part or the "bark" of the root is used, dry for later use.

Preserving: Store dried sassafras root bark in an airtight container in a cool, dry place.

4.33 GINGER (ZINGIBER OFFICINALE)

Habitats: Ginger cultivation began in South Asia and has since spread to East Africa and the Caribbean.

Climate: Ginger prefers a tropical climate - sheltered spot, filtered sunlight, warm weather, high humidity.

Soil: Ginger prefers rich, moist soil that is never water-logged.

Growing: In late winter, or early spring, plant parts of ginger rhizome, each with growth buds. In the first place, every rhizome you plant will only grow a few leaves, at one spot. It will become a thick clump over time, and expand very slowly.Below are some tips in growing;

Select a healthy root of the ginger with multiple nodes of rhizome and visible eye buds. The buds can appear as small bumps, or close to potato eyes, on the surface of the rhizomes.

- Cut the ginger into 1-inch to 1 1/2-inch piece, with each cube having at least one eye bud. If there are not enough eyes on the ginger root, you can make the cubes larger if desired.
- Allow the ginger to dry out of direct sunlight in a cool location for two to three days following cutting. This drying period reduces the probability of rot or other damage as soon as the cuttings are planted.
- Plant the ginger cuttings in moist, well-drained soil, approximately 1 inch thick, amended with sand and compost or other organic material. Place the cuttings 15 inches apart to provide enough space for the creation of rhizomes.
- Water sparingly until the new ginger plant gets out of the soil and leaves to grow. Continue watering to keep the soil moist, as required, but avoid oversaturation.
- Ginger harvesting is done when plant growth is slowing and the leaves start turning yellow. Dig up the entire root and cut off the portion of the plant above ground. You can cut and plant unused ginger to create additional plants for gingering.

Harvesting: The best time to harvest ginger after the leaves have died down is anytime. This point usually takes eight to ten months to reach.

Preserving: Ginger root can be used fresh, or dried as cut pieces, powdered, or even crystallized/candied. Store dried ginger in an airtight container in a cool, dry place.

4.34 GRAINS OF PARADISE (AFRAMOMUMMEL-EGUETA)

Habitat: It is native to West Africa because of this commodity, which is sometimes called Pepper Coast (or Grain Coast). It is also an important cash crop in southern Ethiopia's Basketo district (Basketo special woreda)

Climate: It can be cultivated only in hot humid tropical climates.

Soil: Paradise grains grow well in a fertile, well - drained mix of soil, such as a combination of compost, manure and peat moss. Sand can also be mixed to improve drainage in the topsoil. For grains of paradise plants raised in containers, normal, well-drained potting soil can be used too. When planted the seeds should be covered with approximately 1/2 inch of soil.

Growing: Seeds have a higher chance of successful germination and if they start in a sunny location that is not crowded or shaded by other shrubs or trees the seedlings will be healthier. Southern or southeasterly exposure is ideal, but any area receiving direct sunlight is sufficient. The grains of paradise seeds can be sown indoors if the peppers are to be grown in a cooler climate in which they cannot thrive outdoors in the winter, or if grown within their hardiness range they can be sown directly into the greenhouse.

Harvesting: Inside the seeds produce red pods, with white jelly-like pulp inside. Once they ripen from green to red, the pods are ready for harvest. Only then will its flavor have produced by the seeds.

Preservation: The pods are collected, and dried in the sun for about a week. When dried, the pods are about 2 to 2 3/4 inches (5 to 7 cm) long by 1 1/4 inches (3 1/4 cm) wide.

4.35 HOLY BASIL (OCIMUMTENUIFLORUM)

Habitants: It is native to the Indian subcontinent and widespread as a cultivated plant throughout the Southeast Asian tropics.

Climate: Tulsi is a tropical plant, it needs warm temperatures to germinate, and should be kept at least 70 degrees in a place. Basil plants grow in areas receiving full sun. Though they can handle all-day sun, if they receive six hours each day, they will also grow well.

Soil: Basil plants need well-draining soil, but they also need moist soil.

Growing: Tulsi seeds should start indoors six to twelve weeks before the last frost. You can grow holy basil much like other herbs but it needs warm temperatures. Some herbs, including basil, grow in pots, providing not only good drainage but also versatility in positioning. You will be able to place basil pots in a sunny spot, even if it's on your stone patio, wooden deck or balcony... the areas where you wouldn't be able to dig your plants in the dirt. And with holy basil, it allows you to enjoy the clove-scented fragrance of their leaves by placing pots near where you sit. Just keep in mind the holy basil 's mature height of 24 inches when selecting a cup.

Harvesting: Encourage bushy growth by pinching off the heads of the flowers about a month before harvest. This also stops the plant from producing seeds, thereby extending the life span of the basil. Split the holy basil over two to three sets of leaves with a knife or pruning scissors above the bottom. It is necessary to cut the holy basil before flowers appear. If you want the flowers color in your dried holy basil blend, then wait for the opening of the flower heads. Before harvesting, do not allow the holy basil to go to seed.

Preservation: Holy basil makes a beautiful garden plant that draws bees to the garden. A unique sweet, fragrant herb doesn't last long after it's selected. It has to be dried and processed so that it can be used in midwinter.

4.36 HORSERADISH (ARMORACIA RUSTICANA)

Habitant: Horseradish is native to southeastern Europe and western Asia. It is now popular around the world.

Climate: Grows best in full sun, favor damp locations such as by streams, grown in cultivated gardens.

Soil: Easily grown in average, medium, well-drained soil in full sun. Soil rich in organic matter will produce the largest, most pungent roots.

Growing: Plants rarely produce viable seeds. Usually grown annually (i.e. in spring, plant root cuttings and in late fall harvest roots).Below are some tips in growing;

- Choose a sunny place outside the main garden plot or on one side where you don't have to bother the plants of horseradish as you expand the garden. Also, a little space makes keeping this rampant spreader from entering other areas of the garden easier. Horseradish is best grown in a warm, moist but well- drained soil rich in organic matter.
- Prepare the soil by digging and loosening it to 12-18 inches thick. For each root, you want to plant, add a shovelful of compost and thoroughly mix it in.
- Create a 4-inch-deep furrow with a hoe. For each root, you intend to plant make the furrow 18 to 24 inches deep.
- Place the roots in the furrow at a 45-degree angle with the tapered end pointing downwards. Adjust the depth to support the root at this angle, as appropriate.

- Fill in the furrow so that the top of the root is covered with 2 inches of soil. Press gently with your hands to remove air pockets.
- Apply 2 inches of mulch around the plants to help the soil conserve moisture and reduce the number of weeds.
- Water deeply when there is less than an inch of rainfall in a week.
- Weed the area regularly surrounding the horseradish plants. Weeds compete for moisture with the horseradish roots, and provide an insect breeding and hiding area.
- Work additional compost in soil when the plants start to grow in spring. High-nitrogen fertilizers foster lush top growth and spindly roots, thus resisting the urge to fertilize plants with regular garden fertilizer.
- When new spring growth is about 1 foot tall, pull back the soil around the crown. Cut off the little roots to allow the main roots to grow straight and long. Cut some of the sprouts from around the crown edges, limiting the number to three or four.

Harvesting: Usually harvested in the fall, but roots can be harvested at any time (they can remain in the soil all year round, but should be harvested no later than the next spring).

Preserving: Store dried horseradish root powder in an airtight container in a cool, dry place.

4.37 HYSSOP (HYSSOPUS OFFICINALIS)

Habitant: *Hyssopus officinalis* is native to Southern Europe, the Middle East, and the region surrounding the Caspian Sea.

Climate: The species as a whole is resistant to drought, and thrives in full sun and warm climates.

Soil: Hyssop is tolerant of chalky, sandy soils.

Growing: Sow seeds in a cold frame in spring, and cover only the grain. The seed is easy to germinate. When seedlings are large enough to handle, they may be transplanted into individual pots for their first winter to be grown in a greenhouse. In late spring or early summer, after the last expected frosts, they can be planted into their permanent locations. Hyssop plants may also spread by cuttings or splitting.Below are some tips in growing;

- Sow hyssop seed indoors just a week or so before the final frost in early spring. Start seeding under fluorescent light in flats. Germination takes about 14 days to complete.

- Transplant hyssop outdoors in mid- to late spring after the last frost.
- Hyssop can be cultivated from cuttings of seed, division, or base. Sow outdoors after the last frost of the year. Even seeds can be sown in fall. In late spring or early fall, start the plants from cuttings. Start plants in early spring or late summer, from the root divisions.
- Sow seeds ¼ inch deep. Cover seeds lightly.
- Space hyssop plants 12 to 18 inches apart.
- Grow 2 hyssop plants for cooking; grow 10 to 20 plants for tea and preserving.
- Plant your lavender hyssop, rosemary, garlic chives and catmint. Hyssop is repelling flea beetles and other pests; it lures cabbage moths so that it can be used as a trap. Hyssop attracts bees, butterflies, and cockroaches. Plant hyssop near beehives; the honey is flavored by the gathered nectar. Hyssop is said to elevate grapevine yield. Radishes are said to fail if they are planted next to hyssop.

Harvesting: Hyssop is harvested twice yearly, once at the end of spring and again at the start of fall. The plants are preferably harvested while blooming. It helps one to gather tips for the flowering. Once the stalks have been removed, the hyssop is collected and dried by either stacking on pallets (to allow draining), or hanging in a cool, dry place to dry. Once the leaves have been dried (about 6 days), they may be removed and finely chopped.

Preserving: Store dried hyssop in an airtight container in a cool, dry place.

4.38 LAVENDER (LAVANDULA SPECIES)

Habitant: *Lavandula x intermedia* is a hybrid species between English lavender (L. *angustifolia*) (also known as spike lavender) and Portuguese lavender (*L. latifolia*). Originally grown in France for use in potpourris and other aromatic crafts.

Climate: Grows on dry grassy slopes amongst rocks, also in exposed, usually parched, hot rocky situations in sun.

Soil: Grows best in neutral to alkaline soils that are not continuously damp. Does better in dry soil than a rich, moist soil.

Growing: Sow seeds in a greenhouse in the spring; they are to germinate within 1 - 3 months. Once the seedlings are big enough, plant out the next spring into individual pots.Below are some tips in growing;

- Lavender grows best in full sun; it does not mind a hot sunny location.
- Plant the lavender in soil that is well -drained. Clay soils or soil that is too rich in organic matter can allow the growth of foliage rather than blooms. Lavender favors a soil pH between 6.7 and 7.1. Lavender grows in moist, light and gravely soil.
- Lavender starts indoors 4 weeks before spring's last freeze. Seed should be refrigerated 4 to 6 weeks before sowing. Sow seed in fully sunny flats or under fluorescent lights. Seeds are to germinate within 14 to 21 days.
- Transplant lavender outdoors in mid to late spring.
- Sow lavender seed outdoors after the last frost in late spring or in fall. Fall seed planting will germinate the next year.
- Sow seeds ½ inch deep.
- Space lavender plants 24 to 36 inches apart.
- Grow 2 lavender plants for kitchen and cooking; grow 6 to 12 plants for tea and preserving.
- Echinaceous vine lavender, winter savory, germander, hyssop, yarrow, and boxwood. Lavender flowers attract nectar-seeking insects into the garden including beneficial insects.
- Water lavender daily until water is formed then when the soil is dry to a depth of one inch. Do not overstretch.
- Compost Tea side-dress plants every two months. Growing spring feed lavender with a slow - release organic fertilizer as growth begins.
- Place a light mulch of aged compost around lavender in hot summer regions.
- Blooms spent removing will inspire new blooms. Trim back foliage to shape the plant after it has bloomed. In spring, prune plants back to half; remove dead wood and shape the plant. If you prune old wood, it won't grow back again. Plants can weaken for

more than 5 years or so and in summer new ones can start from cuttings. Rejuvenate the soil, before planting new plants, with aged compost or manure.

- Choose a container at least 8 inches deep and wide; larger is better if you are growing lavender as a decorative perennial.
- Protect plants in cold winter regions by heaping straw around plants; in spring remove straw. Indoors, lavender grows in a glossy glass. Do not cultivate lavender over water in pots.

Harvesting: The best time to harvest lavender is when the buds have grown on the plant but the flowers have not yet opened. Hanging flower stems can be hung upside down to dry in bunches, once cut. Dried flowers can be separated from stalks and preserved entirely or in powder.

Preserving: Store dried lavender flowers and dried powder lavender in an airtight container in a cool, dry place.

4.39 LEMON BALM (MELISSA OFFICINALIS)

Habitant: Lemon balm, balm, common balm, or balm mint, is a perennial herbaceous plant in the *Lamiaceae* family of mint, native to south central Europe, the Mediterranean Basin, Iran, and Central Asia, but now naturalized in the Americas and other regions.

Climate: It is moderately tolerant of darkness, much more so than other plants. It grows best in partial shade, in dry climates. The lemon balm grows in clumps and spreads both by seed and vegetative. The plant stems die-off in warm temperate zones at the beginning of winter but shoot up again in spring.

Soil: In well-drained, sandy loam, cultivate the lemon balm. Lemon balm, however, can grow on nearly any soil but not very wet soil. He favors a pH of 6.7 to 7.3 in soil.

Growing: Sow seeds indoors after the last spring frost, about 2 months before transplanting lemon balm into the greenhouse. Seeds require light to germinate so do not cover them with fine soil or only cover them lightly. Germination is expected in about 14 days. Sow lemon balm in spring, about the last frost's average time. Seeds will germinate slowly. In late summer or fall, sow seed too. During the growing season, root divisions can be planted at any time but will be formed faster in cool weather. Cutting off from new growth in moist sand will begin. Grow the lemon balm with plants from the broccoli, cauliflower, and other family cabbage. The lemon balm scent helps deter insects that target crops from the cabbage family and also hides the cabbage smell. To stop the lemon balm from being invasive, put it in a bottomless jar in the garden, which will trap the roots there.

Harvesting: During the growing season pinch off and use leaves and sprigs as needed. Older, lower foliage has the best fragrance. Leaves for drying are best harvested before the plant flowers start yellowing in the summer, usually about the time lower leaves begin. Cut the plant back by half at midseason or in the autumn; it will regrow new leaves in about 4 weeks. Snip leaves and sprigs on a pruner in the yard. The leaves are easy to bruise so treat them carefully.

Preservation: Dry the lemon balm by cutting about two-thirds of the way down the stem and hanging upside down in small bunches (up to about 1 inch in diameter at the base).

4.40 LEMONGRASS (CYMBOPOGON CITRATUS)

Habitant: Lemongrassis a genus of Asian, African, Australian, and tropical island plants in the grass family.

Climate: While we have heard reports that it can survive the occasional light frost, it prefers a sunny location in a grassland habitat with abundant water when growing and a temperature no lower than about 10 ° C in the winter.

Soil: Grows best in moist, well-draining soil.

Growing: Lemongrass is not usually grown by seed, as the root cuttings or divisions make it too easy to grow. In the fall or spring split the plants and put new plants directly in the garden bed. Add in some compost or well-aged manure to help enrich the soil when digging the holes for the plants.Below are some tips in growing;

- Purchase a lemongrass starter plant at a nursery or garden center.
- Place between 2 to 4 inches of compost over the soil. Dig at least 4 to 6 inches of the compost into the soil.
- Find a space that is large enough to accommodate the lemongrass plant rootball. Allow 3 feet between plants, as mature lemongrass is a large plant growing to 2 to 3 feet wide and up to 6 feet high.
- Place the lemongrass plant inside the cavity. Make sure the top of the rootball is even with the soil surface, as planting too deep will cause the plant to suffocate and die.

- Pat soil firmly around the roots of the lemongrass.
- Immediately after planting Water the lemongrass. Water regularly thereafter to keep the soil even and sufficiently moist. Lemongrass requires damp yet not soggy soil.
- A general-purpose garden fertilizer is used to fertilize lemongrass regularly. Apply the fertilizer as stated on the package label.
- Lemongrass is suitable for cultivation in containers with a minimum diameter of 12 inches. Use the potting soil for general purposes. Place the lemongrass in bright sunlight, consistently keep the soil moist and fertilize the plant weekly.
- You can start a plant by dividing a small section of an existing plant if you have an established lemongrass plant, or if you have a friend who is willing to share a plant. Assure you have a clump of healthy roots in the section.
- In early autumn dig up a small part of lemongrass if you live in a place with frosty winters. Plant the section in a pot and carry it indoors for the winter, in a warm, sunny spot. Replant the lemongrass outdoors as frost danger passes through in spring.

Harvesting: Once the plant is at least a foot tall, you can cut leaves from the plant at any time. Use a sharp knife to slice each one off at ground level to collect whole stalks. Once you cut them, take the outer stalks first, and they should be at least 1/2 inch thick. Seek not to just break them off or the rest of the plant could be harmed. Before use, you may need to peel out the tougher outer leaves.

Preserving: Hold the whole stalk in the refrigerator with the leaves to keep it fresh for several days. For best results, keep it in a plastic bag with a damp paper towel. Chopped stalk pieces may be frozen or dried for later use. Store dried leaves in a cool, dry place, in an airtight container.

4.41 LEMON VERBENA (ALOYSIACITRODORA)

Habitant: Lemon verbena, also known as Vervain and Lemon Bee Bush, is a perennial shrub native to South America introduced by the Spanish into Europe in the 1600s. The plant possesses lemon-scented leaves, as its common name implies.

Climate: Needs a climate that is sunny, warm and humid. Grows well in forests, in roads and in open scrub. Lemon verbena is a tall, showy herb that is often cultivated in herb gardens. This plant is a good choice for pathways where you can brush the leaves and release their lemony fragrance.

Soil: Lemon verbena is best grown in moist, light, well-drained fertile loam soil.

Growing: Sow the seed in late spring, in a greenhouse. Just cover the seed and keep it in a light position, ensuring the compost doesn't dry out. Transplant seedlings into individual pots when large enough to handle, and grow them on in the greenhouse for at least their first winter. Plant them out in early summer, and for at least their first winter outdoors offer some shelter from the cold. Also, may take softwood cuttings in May or June. Grow on in a greenhouse for at least their first winter, and plant out after the last predicted frosts in late spring.

Harvesting: Harvest lemon verbena leaves while the plant is in flower, dry for later use.

Preserving: Store dried lemon verbena leaves in an airtight container in a cool, dry place.

Licorice (*Glycyrrhiza glabra*)

Habitant: *Glycyrrhiza glabra* is native to southern Europe and parts of Asia.

Climate: Licorice grows best in deep valleys with full sun.

Soil: Prefers well-drained soils.

Growing: Pre-soak the seed in warm water for 24 hours, and then plant in a greenhouse in spring or fall. Transplant the seedlings into single pots when they are large enough to handle, and grow them in a greenhouse for their first winter. Plant out in late spring or early summer when successful growth is in progress. Plants grow very slowly from seed so many farmers prefer division. Take root splits in spring or autumn. The growing division needs to have at least one bud of growth. Divisions can be replanted into the garden bed immediately.

Harvesting: Licorice root, harvested from second- or third-year plants in the autumn, is dried for later use— either as cut pieces, as root sticks or as ground powder.

Preserving: Store dried root pieces or dried leaves in an airtight container in a cool, dry place.

4.42 LOVAGE (LEVISTICUMOFFICINALE)

Habitant: Lovage is said to be native to a large part of Europe and southwest Asia. Other sources, however, state that it is native only to the eastern Mediterranean region in southeastern Europe and Southwestern Asia. Others still argue that in Iran and Afghanistan it is native to Southwest Asia, citing the European populations as naturalized.

Climate: Lovage can be found growing in sunny mountain pastures and hedgerows near streams, waste places, or in herb gardens.

Soil: Lovage prefers a rich moist but well-drained soil.

Growing: Sow seeds in a cold environment in spring or early autumn. The seed may germinate slowly so it is probably best sown when it is ripe. Transplant the seedlings into individual pots during the summer when they are large enough to handle them and plant them out to their permanent garden locations. May break in spring or autumn too.Below are some tips in growing;

- In well-drained, moist soil, grow lovage, apply aged compost or commercial organic planting mix to the soil before planting. Lovage requires a soil pH of between 6.0 and 7.0.
- In early spring, launch lovage seeds indoors under fluorescent lights in individual pots. Germination may be inconsistent if the seed is not new; optimal is fresh seed harvested in fall and planted immediately. It takes 10 to 14 days to germinate. Sow indoors in autumn in cold winter areas, and transplant out in spring.
- Place lovage seedlings in the garden when all frost threat is past and at least 40 ° night temperatures Lovage has a taproot so transplant seedlings into the garden before the taproots are well developed.
- Sow outdoor lovage seed in early spring as soon as the soil can be worked or the following spring seed can be sown for germination in the fall.
- Sow seeds outdoors ½ inch deep.
- Space plants 18 to 24 inches apart; lovage grows in clumps.
- Grow one lovage plant for culinary use; grow one plant for preserving.
- Plant lovage containing fennel, hyssop and catmint. Lovage can

be used as a trap crop to lure hornworms off tomatoes. The yellow-greenish flowers attract beneficial insects into the greenhouse.

- Keep the soil evenly moist.
- In the growing season, give plants compost tea or dilute the emulsion of fish a few times. Attach rotting compost or a commercial organic planting mix to replenish the soil every spring.
- Mulch lovage with ½ inch of compost in spring. Lovage dies back in winter; protect the roots with mulch.
- Clip off lovage flowers to encourage bushy growth and continuous leaf production.
- Grow lovage in a pot 12 inches wide and deep.
- In winter Lovage will die. In winter, containers may be brought indoors for winter harvest. Place the pots in a glittering place.

Harvesting: In the summer, Lovage flowers and seeds are picked. Once the plant reaches bloom, the leaves are harvested and either processed for their oil or dried for subsequent use. Lovage roots are harvestable from plants 3 years of age and older. They are harvested either in the early spring or in the autumn and used fresh or dried.

Preserving: Store dried lovage root in an airtight container in a cool, dry place.

4.43 MACE/ NUTMEG (MYRISTICAFRAGRANS)

Habitant: *Myristicafragrans* is an evergreen tree indigenous to the Banda Islands in the Moluccas (or Spice Islands) of Indonesia.

Climate: Nutmeg trees grow in warm, humid tropical climates in full to part sun.

Soil: Nutmeg trees prefer well-draining soil that is rich in organic matter, with a medium to heavy texture.

Growing: Seedling propagation of the nutmeg trees produces 50 percent unproductive male seedlings. Because determining plant sex before the sixth to eighth year is troublesome, and there are inconsistent yields from sexual propagation, growers conclude that grafting is the preferred method of propagation. Epicotyl grafting, approach grafting, and patch budding have proven successful— with the most widely adopted method being epicotyl grafting. Air-layering is an alternative method but it is not suitable due to its low success rate (35-40 percent).Below are some tips in growing;

Clear the field where weeds, rocks, plants and debris are grown. Use a rake to spread a 2-to 4-inch layer of rotted manure over the planting site. Blend the manure with a tiller into the top 8 to 12 inches of soil.

- Dig a hole with a shovel twice as wide and equal in size to the root ball of the nutmeg tree. Space the hole from other trees or buildings within 40 to 50 feet.
- Extract the plant from wrappings of its growing container or root ball. Cut several circular roots with a pair of pruning shears to facilitate the growth outwards. Prune all torn, mushy or dead roots free.
- Place the root ball of the tree in the center of the hole. Place the top of the root ball even at or 1 inch above the surrounding ground level, removing or adding soil as required to the bottom of the hole. Level the bottom of the planting hole if necessary, to ensure the tree stands upright vertically.
- Fill half the hole with soil and firmly seal it around the root ball. A garden hose fills the void with water. Allow for the water to drain down the soil.
- Fill the residual space with soil in the vacuum. Don't overfill the hole or plant the tree deeper than it used to rise. Firm the soil around the root ball use your hands to damp it down.
- Create a 4 "ring of soil around the tree to create a water reservoir. Place the 24 to 30-inch ring away from the trunk of the tree. Load water into the tank. Wait for the water to drain down the soil. Sprinkle loose soil over any depressions that occur from the drainage of the water in the planting area. Don't rub down the soil while it is damp.
- Place a 1-to 2-inch-deep mulch layer over the planting area to reduce the loss of soil moisture. Hold the mulch from the trunk of the tree 4 to 6 inches away to prevent the bark from becoming damp and rot from growing.
- When the top 2 inches of soil gets dry, water the nutmeg tree. Fill the reservoir with water once to twice, allowing the water to flow completely down the soil after each application. Don't spray so often that causes soggy soil.

Harvesting: When the nutmeg seeds are mature, mace is harvested; the outer red mace is removed from the nutmeg seed and then dried for later use. When young, mace's scarlet color and fleshiness can dry yellow and brittle.

Preserving: Store dried mace powder in an airtight container in a cool, dry place.

4.44 MARJORAM (ORIGANUMMAJORANA)

Habitant: The Marjoram plant is indigenous to Cyprus and southern Turkey.

Climate: Marjoram thrives on sunny dry slopes and rocky places, though occasionally it will grow in partial shade.

Soil: As for the soil type, this plant is not fussy although it is usually grown in rather dry, warm, well-drained conditions, thriving on chalky soils.

Growing: Sow seedsat temperatures of around 50 - 55 ° F in the early spring. Just cover the seed, and usually expect germination within 2 to 4 weeks. Transplant the seedlings into individual pots when large enough to treat, and plant them out in early summer in their permanent locations. Divide by March or October.Below are some tips in growing;

- Marjoram grows in loose, well-drained soil. Marjoram can thrive in poor, well-drained soil. Marjoram prefers a soil pH of between 6.7 and 7.0.
- To get a head start at the season, sow marjoram for transplantation after the last frost as early as 4 weeks before the average last frost date in spring. Marjoram sprouts slowly; the optimum germination temperature is 70 ° F.
- Place transplants on the last frost average date in spring. Marjoram can also be grown from root divisions taken in the fall, indoors overwintered and place out in the spring.
- Sow marjoram in the garden after the last frost in spring.
- Sow marjoram seeds ¼ inch deep.
- Space marjoram plants to 6 to 12 inches apart. Space rows 18 to 24 inches apart.
- Grow 2 to 4 plants for cooking and kitchen use; grow 6 to 10 plants for preserving.

- All vegetables and herbs grow well in Marjoram. Grow marjoram with chives made from garlic, basil, rosemary, and chamomile. Marjoram is said to enhance nearby herbal growth and flavor.
- Regularly give water to marjoram until it is formed. Once water marjoram has been developed allowing the soil to dry sparingly between watering. Moist soil makes marjoram less flavorful.
- Foliar feed marjoram by spraying with compost tea or liquid seaweed extract 2 to 3 times during the growing season.
- Mulch marjoram in warm weather to protect roots from too much heat.
- Keep marjoram pinched back to stimulate bushy growth and flavor. Prune in late spring to encourage vegetative growth before the plant flowers; prune again in mid - summer, so that it does not become woody. Marjoram can sprawl; either cut it back or give it room for growth. Replace the plants which are 3 or 4 years old.
- When flower buds appear, you cut marjoram back by one-third; new leaves will form to extend the harvest. (Dry and store seasoning clippings.) In cold-winter areas, split plants into sunny windows in the fall and overwinter to replant them in the spring.
- In containers, Marjoram grows easily as an annual. Select 6-inch-deep bin. Potted marjoram can be grown inside a bright, sunny window.
- Frost won't survive in Marjoram. Grow marjoram in winter as annual indoors. Grow marjoram or under fluorescent lights in a sunny window.

Harvesting: Harvest marjoram just as it begins flowering. Use either fresh or dried. If drying, do so slowly and in shade for best results.

Preserving: Store dried marjoram in an airtight container in a cool, dry place.

4.45 OREGANO (ORIGANUM VULGARE)

Habitant: Oregano is native to warm-temperate western and southwestern Eurasia and the Mediterranean region.

Climate: Oregano grows in meadows, dry grassy areas, scrubby areas and cultivated herb gardens in full sun.

Soil: Grow in well-drained soil. Take care not to let the roots encounterin wet, soggy soil.

Growing: In the fall, oregano is grown by seed or propagated by root division. Sowing seeds in early spring, germination typically occurs within 2 weeks. Transplant the seedlings when they are large enough, and then plant them in early summer garden beds.Below are some tips in growing;

- Plant the oregano in well-drained, loose soil. Oregano prefers 6.0 to 7.0 soil pH. Oregano can grow in low, well-drained soil.
- Sow indoor oregano as early as 4 weeks before the last frost date average. Oregano seed will germinate in 10 to 15 days; sow seed at 70 ° F for optimal germination. Seeds sprout best in the light; cover seeds with cheesecloth until seedlings start pushing up.
- After the last frost transplant seedlings out. Oregano can be grown from the root divisions taken in the fall, indoors overwintered and planted in the spring.
- Sow seed of oregano in the garden on the average spring frost date of last. The temperature of the soil should be 45 ° F or colder.
- Sow seed ¼ inch deep; thin successful seedlings or set transplants to 6 inches apart.
- Space oregano plants 12 inches apart. Space rows 18 to 24 inches apart.
- Grow 2 to 4 plants for cooking; grow 6 to 12 plants for preserving.
- With all vegetables and herbs, Oregano grows well. Butterflies and beneficial insects eat oregano flowers.
- Give the oregano even water regularly until it is formed. Once developed, water oregano allows the soil to dry sparingly between watering. Moist soil will result in less flavorful oregano.
- Oregano foliar feed by spraying with compost tea or extracting liquid seaweed 2 to 3 times during the growing season
- Mulch around oregano in hot weather with aged compost to keep the roots cool and reduce evaporation of moisture from the soil.
- Keep the oregano pinched back to induce bushy growth and for the best aroma and flavor; flowering will result in loss of leaf flavor. In summer, cut back plants almost to the ground once or twice; this will stimulate fresh growth. Every three years divide clumps and replant healthy, rooted segments.

- Oregano quickly grows in containers. Pick 6-inch-deep bin. Indoors, potted oregano can be grown in a bright, sunny window or under fluorescent light. When they become woody remove container plants every 3 years.
- For cold-winter areas, split plants indoors for re-planting for spring in the fall and overwinter. When you cover them outdoor overwinter plants with a thick mulch of leaves.

Harvesting: In late summer the plant can be used fresh or dried—harvest the entire plant (but not the roots) to dry and prepare for winter use.

Preserving: Store dried oregano in an airtight container in a cool, dry place.

4.46 PARSLEY (PETROSELINUM CRISPUM)

Habitant: Parsley is native to the central Mediterranean area (southern Italy, Algeria, and Tunisia), naturalized elsewhere in Europe, and widely cultivated as a herb, spice, and vegetable.

Climate: Parsley is a fairly hardy biennial (yearly in colder areas) though it requires some cold protection. It likes a sunny place for part of the day where it gets a bit of shade. If the parsley gets too much light then it's going to get yellow.

Soil: Parsley grows best in moist, well-drained soil.

Growing: Growing seed parsley— germination is slow, taking as long as four to six weeks, and is often challenging due to furanocoumarins in its seed cover. Germination time can be reduced by pre-soaking the seed in hot water for 12 hours, which is allowed to cool rapidly but be careful not to overdo the heat and cook the seed.Below are some tips in growing;

- Parsley is best grown in compost-rich, humid soil with a pH between 5.5 and 6.7. Prior to or at planting time, apply a slow - release fertilizer to the soil.
- Sow parsley indoors 4 to 6 weeks before transplanting into the garden in late winter or early spring. Soak seeds in warm water 24 hours before they are planted. Seeds will germinate within 5 to 6 weeks and plants will ripen in 70 to 90 days.
- Transplant parsley outdoors in mid to late spring after the soil has warmed to 50°F.

- Sow parsley seeds outdoors in spring after all frost danger has passed and the soil is at least 50 ° Parsley is slow to germinate and emerge from the soil — usually 5 to 6 weeks after seeding. For better germination, soak seed overnight in warm water before sowing or pour boiling water before sowing seed in the planting area. Sow outdoor seed from early winter until late spring in mild winter regions and hot summer regions.
- Sow parsley seeds ¼ inch deep.
- Space parsley 6 to 8 inches apart in all directions.
- Grow 6 plants for cooking; grow 10 to 20 plants for preserving.
- Grow the asparagus parsley, sweet corn, peppers, and tomatoes. Parsley is said to repel beetles from asparagus and reduce the number of rust flies in carrots. To prevent cross - pollination, it is best to grow parsley apart from family plants, including carrots, celery, and parsnips.
- Keep the soil evenly moist; do not overwater parsley.
- Parsley is a large feeder. During the growing season, feed parsley with liquid extracts of the seaweed two to three times. Side-dressed midseason parsley with aged compost.
- Where the weather gets hot, mulch around plants to keep the roots cool; make sure they don't cover the crown that can cause rot.
- Cut back the entire length of outer stems (cut at the base) to keep the parsley productive. Cut from flower stalks. (When flowers are allowed to stay, the leaves become bitter.) Sprinkle dead leaves as needed to keep the plant smooth. Cut back plants towards the end of the growing season to promote new growth.
- Parsley grows in a jar, and yields well. Choose a container with a minimum width and depth of 6 inches. Grow the parsley indoors in 6-inch pots in summer or winter.
- Bring indoor parsley for winter harvesting and avoid freezing damage, and start cut-and-comer harvesting. Peregrine leaves grown in the house will be better than the leaves grown outside.

Harvesting: Cut stalks of parsley close to the ground, beginning with stalks outside and working your way around. That will promote new growth. Choose early in the day for the best flavor, while it's still cool. You can cut the entire plant off at base at the end of the season. For best flavor use fresh, or hang up to dry for later use.

Preserving: Store dried parsley in an airtight container in a cool, dry place.

4.47 PEPPERMINT (MENTHA ×PIPERITA)

Habitant: Peppermint is native to Europe, and is now widely cultivated in all regions of the world.

Climate: Peppermint generally grows best in moist: sunny to part shade locations.

Soil: The peppermint plant prefers moist, but not waterlogged soil.

Growing: Peppermint rapidly and easily reproduces through the division of the underground stolon. It can be invasive, so carefully choose the location in the garden.Below are some tips in growing;

- Combine equal portions of finely milled perlite and sphagnum moss in a pan. Add water to the touch until the mixture feels moist. Periodically stir the mixture, and disperse the moisture. Let the water soak in the mix until the sphagnum moss gets plump.
- Pack the moss-perlite sphagnum mixture into 3-inch diameter starter pots and leave the top 1/2 inch of each pot free. Firm the moist mixture into each pot 's bottom to collapse all pockets of air. Use your fingertips to rough, or gently rake, the top 1/4 inch of the sphagnum moss-perlite mixture of each pot which is the rising mix.
- Sprinkle 5 to 6 peppermint seeds onto each pot 's growing mixture surface. Place the seeds apart by at least 1/4 inch. To anchor, gently press the seeds onto the growing mixture. Do not cover the seeds with the mixture, as light helps germinate the seed.
- Place the starter pots near a sunny window facing south or west, with no heavy curtains or outdoor shade exposure. Place a 10-inch fluorescent lamp above the pots to give additional light if necessary. Keep the light in during the seed germination process for about 14 hours each day.
- Place the pots on a heating mat to give the right temperature for germination. During the day, heat the pots to above 70 degrees Fahrenheit, and turn off the heating mat at night to mimic the normal soil heating and cooling process.

- Cover the pots with a propagation dome or a clear, plastic wrap plate. Any device prevents drying out the peppermint seeds. Every day, raise the dome or plastic wrap and test the moisture level of the mixture of the pots. Mist the peppermint seeds with water whenever the surface of the mixture barely feels warm.
- Note that the first peppermint seedlings appear two to four weeks after the seeds have been sown. Replace the propagation dome or plastic wrap and the heating pad after germination of the rest of the seeds and seedlings emerge. By snipping them off at their base with small scissors, replace the weakest peppermint seedlings, and allow the most robust seedlings to grow.
- If you replaced the natural light with such a lamp, keep the fluorescent lamp in place. The illumination from the lamp will prevent the seedlings from becoming leggy.
- Rotate the pots by increments of 180 degrees per week. The rotations prevent peppermint seedlings from leaning toward their source of light. Provide water when the ground under the surface dries out. Apply water directly to the soil, instead of misting whole plants. Doing so lowers the chances of fungal and bacterial development on the leaves of the plants.
- When they develop multiple sets of mature leaves, they transplant the peppermint plants. Put them in individual plantings filled with potting soil, or transplant them directly into the soil of an outdoor garden after all frost hazard has passed, spacing the plants 12 "apart.
- Peppermint could become invasive when it is planted outdoors. Keep it in a bowl, or add a barrier to the root if you raise the plant outside.
- Use plant pots with multiple bottom drainage holes, which allow excess soil moisture to drain.

Harvesting: Once the flowers begin to open and can be dried, the leaves and flowering tops are harvested.

Preserving: Dried peppermint should be stored in an airtight container in a cool, dry place.

4.48 POPPYSEED (PAPAVER SOMNIFERUM)

Habitant: *Papaver somniferum* is native to and grown east to Iran in the Mediterranean region; now cultivated in many tropical, subtropical, and warm temperate nations. Currently known for opium cultivation areIndia, Iran, Turkey, Yugoslavia, Macedonia, Bulgaria, China, Manchuria, and Asia, and for seeds in other parts of Europe and India. Cultivated in Japan, and also in Australia.

Climate: The poppy plant can be found growing in areas ranging from cool temperate wet steppe to subtropical dry to moist forest life zones. The plant does poorly in the humid tropics.

Soil: The poppy thrives in rich, well-manured soil.

Growing: *Somniferum on papaver*is propagated from seed. These seeds are not as temperature- sensitive as most poppy plants, but they do best germinate at 15 ° C. The seeds are planted in shallow furrows. In early autumn in some areas Poppyseed mixed with sand is spread over tilled fields. The fields are then weeded in the spring, the poppy usually reached a height of around 6 inches at that time. The plants are then thinned to a height of about two feet. They flower in the summer months of June and July and their capsules are ripe.Below are some tips in growing;

- Poppyseeds germinate best in cool, well-draining soils in a full-sun location.
- Dig and remove the top 6 inches of soil. To further improve its quality and provide trace nutrients to the growing poppy plants, mix a 2-inch layer of compost into the loose soil. Raising the top two inches of soil after alteration to ensure it is loose and crumbly.
- To help avoid over-seeding, mix one-part Poppyseeds with two parts of sand. Sprinkle the seed mixture evenly over the bed. Sow about one seed per square centimeter.
- Cover the seeds with a one-eighth-inch layer of loose soil or peat moss. Poppies require some light to germinate.
- Daily or as needed, mist the bed with water so that the soil remains moist near the surface. Where the soil temperature is about 55 degrees Fahrenheit, poppies normally germinate within 15 days.
- Thin the seedlings, and the remaining plants will be 6 to 10 inches apart until they grow to a height of 1 in.
- Poppies transplant poorly so it's best to start the seeds directly

in the garden. If you must start them indoors, sow them four to six weeks before the last spring frost in your area or four weeks before you plan to transplant them outdoors for a fall and winter garden.

Harvesting: Harvest the Poppyseeds once the seed pod itself has dried and the top has opened up.

Preserving: Store dried Poppyseeds in an airtight container in a cool, dry place.

4.49 ROSEMARY (ROSMARINUS OFFICINALIS)

Habitant: Rosemary is a perennial herb in the mint family that occurs naturally throughout the Mediterranean region and is widely cultivated elsewhere.

Climate: Rosemary is used as an ornamental plant in sunny gardens and for xeriscape landscaping (landscaping that requires little or no irrigation), especially Mediterranean climates. It is easily grown, considered pest-resistant and it is even reasonably hardy in cool climates. Rosemary can withstand droughts and can survive a severe lack of water for lengthy periods.

Soil: Rosemary grows best on friable loam soil with good drainage.

Growing: Rosemary can be propagated by cutting a shoot of a soft new growth from an existing plant. The clipping should be about 4 to 6 inches, and a few leaves should be removed from the bottom so that they can be planted directly into the soil. Rosemary may also be cultivated by seed, but germination is sluggish.Below are some tips in growing;

- Spread 2 inches of compost over a well-drained full-sun garden bed with a pH of 6.0 to 7.0 in soil. Sprinkle 1 tablespoon of 10-10-10 fertilizer over each square foot of the area under planting. In the top 6 to 8 inches of soil, mix the compost and fertilizer.
- Dig the rosemary planting holesbetween 12-14 inches apart and ensure that it is as flat as possible before placing the route. Place the roots in the cavity, so that the rosemary grows in its pot at the same depth. Place Rosemary plants in all directions, 3 feet apart.
- Cover the soil with a 2-inch mulch layer to separate the soil, to avoid weeds and to retain moisture. Pull the mulch back so it doesn't rest against the rosemary base, which can cause root or

crown rot.

- If the top 1 to 2 inches of the soil starts to feel dry, water the rosemary. Supply around 1 inch of water weekly by irrigation, which is adequate to humidify the soil to a depth of 6 inches, if there is not enough rain.
- During spring, sprinkle 1 tablespoon of 10-10-10 fertilizer around each rosemary plant when new growth occurs. Avoid applying the fertilizer directly to the stems or leaves, as it burns the plant. Upon application, pour the fertilizer into the soil, so that it can soak into the root zone.
- In the spring cut off any dead or damaged branches. Trim the remaining stems back to the desired shape and height. The individual stems can be trimmed for herbal use at any time during the spring or summer.
- Rosemary grows well as a potted plant if provided with good drainage. You can overwinter the pots indoors in cold climates.
- Propagate from stem cuttings of new plants. Take a cut from new spring or early summer growth and pot it in moist soil. The cutting roots usually within 4 to 8 weeks.

Harvesting: Rosemary leaves can be harvested in the spring or summer and used fresh, they can also be dried for later use.

Preserving: Store dried rosemary in an airtight container in a cool, dry place.

4.50 RUE (RUTAGRAVEOLENS)

Habitant: It is native to the Balkan Peninsula. It is now grown throughout the world in gardens, especially for its bluish leaves, and sometimes for its tolerance of hot and dry soil conditions.

Climate: It requires mild cold and tropical climates. It tolerates hot and dry conditions and prefers open spaces under bright sunlight ;Optimum temperature: 15-28^{0}C; Relative humidity: 60-75 %

Soil: Grows well in sandy-loam soils with medium fertility. Soils with well -decomposed FYM found best forgermination and better survival of seedlings & productivity. Soil pH: 6.0 to 8.0

Growing: Commonly commercially propagated by Seeds & Stem cuttings. Viable-2 - 3 years Sowing rate: 250 - 300 g / ha. You should continue cultivating on the beds. Transplantation occurs when the seedling is 20 - 25 cm tall. The seeds transplanted are spaced 40 by 40 cm away. Irrigation is done dairy on seedlings and is done once in a week or fortnight on established plants. Weed management is especially important following transplantation. NPK may be used in transplantation. The root rot should be regulated and the aphids.

Harvesting: Five to six-month-old plants can be harvested at full flowering stage due to the high percentage of essential oil, rutin and coumarin. The average yield is about 2.5 to 3 tons per hectare of dried herbs.

Preservation: The cut plants are dried.

4.51 SAFFRON (CROCUS SATIVUS)

Habitant: *Crocus sativus* is native to Greece or Southwest Asia and has been gradually propagated throughout much of Eurasia and later transported to parts of North Africa, North America and Oceania.

Climate: Saffron grows best in sunny, semi-arid lands. It can survive cold winters, tolerating frosts as low as −10 °C (14 °F) and can even endure short periods of snow cover. Irrigation is required if grown outside of moist environments.

Soil: Prefers friable, low-density, loose, well-watered, and well-drained clay soils with high organic content.

Growing: Since in reality they are sterile, purple flowers of the herb can produce no viable seeds. The reproduction of the plant requires human assistance by replanting the corms of the plant. Corms are underground, bulb-like, starch-storing organs— and for next season, they must be dug up, broken down, and replanted. A corm will live for one season, though it will produce up to ten "cormlets" via this vegetative division that can be planted in the next season and will grow into new plants. Below are some tips in growing;

- Prepare the early dropping soil by digging a 9-to 12-inch deep trench that tests the new bulb bed's length and width. Place the soil on a plastic tarp from the pit, and add a 2-to 4-inch organic mulch layer. Incorporate the mulch with the earth and replenish the trench with the modified soil.

- Creating a 3-to 6-inch deep hole in modified soil. Place the corm with the pointing side facing up into the cavity. Fill the saffron corm with the soil, and gently massage the dirt with your hand palm. Place 4 inches apart on each corm. Heat deeply into the bulbs.
- Mix or dissolve a soluble, organic garden fertilizer, using the water-to-fertilizer ratio recommended by the manufacturer. Directly after planting apply the mixture to the corms. When the soil is dry, water the bulb bed deeply. Keep the soil moist but not soggy, because the overly wet soil can cause rot.
- Harvest saffron stigmas as soon as the plant opens by gently pulling off the plant the bright red or orange stigma. Place the stigmas on a layer of paper towels, cover them with a paper towel and store them in a warm, dry place for about a week, until the stigmas get dry and brittle. Place the stigmas dried in a plastic or glass vial.
- Carefully dig the corms out of the ground before the ground has frozen but after the first two to three touches of frost. Place the corms and cover them with dry peat moss in a plastic tub or wooden crate. Position the tub or crate in a cool, dry place where the temperature ranges from 40-to 50-degrees Fahrenheit.
- Replant the saffron corms into well-drained, nutrient - rich soil in late summer to early fall. The bulbs are regularly fertilized with a soluble, organic garden fertilizer and water.
- The crocus saffron is sometimes confused for colchicums, which is a highly toxic plant. The best way to tell the two apart is by stamens number. The colchicum has six stamens whereas the crocus saffron has only three.

Harvesting: The stigmas and styles of the flower are harvested in the autumn when the plant is in flower and are dried for later use.

Preserving: Store saffron in an airtight container in a cool, dry place. Be aware it should be used as quickly as possible; the spice does not store well. It should be used (or replaced) within 12 months.

4.52 SAGE (SALVIA OFFICINALIS)

Habitant: Common sage is an evergreen herb in the mint family. Although it is native to the Mediterranean, sage is now naturalized and widely cultivated throughout the world.

Climate: Sage grows on sunny dry banks and stony places. Is usually found in limestone areas and often where there is very little soil.

Soil: Requires a very well-drained light sandy soil.

Growing: Sow seeds in March or April, in a greenhouse. Germination occurs usually within 2 weeks. Transplant the seedlings into individual pots in early summer when they are large enough to handle and plant them out. Below are some tips in growing;

- Sage develops in well-drained soil. Sandy loam is best but also wise in very average soil. Sage recommends a soil pH between 6.0 and 6.7.
- Start sage seed inside as early as 6 to 8 weeks before the last frost date average.
- After the last freeze, transplant seedlings out into the garden when the soil has warmed.
- Sow sage seed before or after the last frost in the garden in late spring. Sow a shallow crop.
- Sow sage seeds ¼ inch deep.
- Thin sage seedlings or transplants 20 inches apart or more in a vacuum. Vacuum rows distance between 20 and 24 inches.
- Grow one or two sage plants for cooking; grow 4 to 6 plants for preserving.
- Grow sage with calendula and chives, including cabbage, onions, strawberries, and tomatoes. Sage is said to deter pests in the cabbage family, such as introduced cabbage worms and root maggot flies. Flowers attract in the garden bees and other beneficial insects. Some claim that sage would inhibit cucumber growth and have a negative effect on the onions.
- Until it is well established, sage requires regular watering, even. Hold or support the sage on the dry side once formed. Overwatering or soggy soil quickly kills sage. Water shortages will enhance the sage's flavor, but avoid causing the plant to wilt

due to lack of moisture. Watering overhead can cause serious problems with the mildew.

- Sage grows best in sandy-loam soil but does not need nutrients regularly. Twice during the growing season, give the sage a side dressing of compost tea.
- Mulch around the sage in very hot regions with aged compost or chopped leaves; mulch can reduce evaporation of moisture from the soil. Mulch to protect the roots from freezing in severe winter regions too.
- To preserve vigor, split sage plants every three years. Prune away flower stems for the best flavor before they bloom. Back in autumn, trim or cut plants to regenerate foliage for the following season. After many years, Sage can grow woody and decline and should be replaced.
- Sage as an annual may be container-grown. Grow sage at least 8 inches wide and deep in a pot.
- Container-grown sage in a protected place, an unheated garage or patio during winter. Mulch over plants where winters are cold, to help sage survive through winter. If winters are wet or springs are hot and cold, sage doesn't grow well.

Harvesting: Before the plant comes into bloom, the leaves are better picked and then dried for later use.

Preserving: Store whole dried sage leaves in a cool, dry place, cut dried sage leaves, powdered sage leaves and other varieties in an airtight container.

4.53 SAVORY (SATUREJAHORTENSIS AND S. MONTANA)

Habitant: Savory is native to the Mediterranean but was used in Europe, North America and South America.

Climate: Summer savory grows in sunny cultivated gardens, alsocan be found on dry, gravelly, or stony slopes.

Soil: Summer savory prefers a rich, light soil with plenty of moisture.

Growing: Sow savory seed directly in the garden in April, and cover only. Seeds are about to germinate in 2 weeks. Do not transplant savory, as root disturbance resents strongly. An Autumn sowing will provide an earlier supply of leaves in areas with mild winters.Below are some tips in growing;

Savory summertime prefers rich, well - drained organic soil; savory winter prefers well-drained, sandy soil. Savory prefers a soil pH of between 6.7 and 7.3.

- Sow indoors as early as 6 to 8 weeks prior to the final frost. Seeds may germinate in 14 days or longer. Winter savory Germination can be erratic.
- Set seedlings in the garden after the last frost in spring.
- Sow savory in spring garden about the time of the last average date of frost. Savory both in summer and winter can also be started from cuttings and divisions.
- Sow savory ¼ inch deep. Savory will germinate with no soil cover.
- 12 to 18 inches apart on space plants. Space rows distance between 12 and 18 inches. Savory winter can require a more saltycondition than savory summer.
- Grow 2 to 4 plants savory for cooking both in summer and in winter; grow 6 to 8 plants each for preservation.
- Beans and tomatoes grow savory in the summer. It is said to make onions sweeter by planting savory near-onions. Other perennials including hyssop, lavender, thyme, and sage grow savory in winter. Honeybees are attracted to the flowers.
- Savory require watering up to established even regularly. It can be kept on the dry side once savory is established.
- Savory does not require extra feeding. Side dress plants with aged compost at midseason.
- Protect savory winter with a thick mulch of dried leaves or straw, from freezing temperatures. Remove the Spring Mulch.
- Summer savory is growing so fast that it can become top- heavy and need staking. Winter savory is a perennial; it should be cut every spring to a few inches tall and replanted every four to five years. Periodic trimming of plants to encourage new growth.
- Savory summer and winter can be grown in containers. Grow as annual, savory summer. Use a container that is at least 6 “wide and deep. Winter savory overwinter container-grown in an unheated garage or patio.

- Savory winters are hardy to about 10 ° F. In winter, cover plants with a dense mulch of chopped leaves or grass. Grow indoors in summer, savory in winter.

Harvesting: The leaves are harvested just before the plant comes into flower and maybe dried for later use.

Preserving: Store dried savory leaves in an airtight container in a cool, dry place.

4.54 SESAME (SESAMUM INDICUM)

Habitant: Sesamum indicum is native to Africa and India, and is now widely cultivated in many tropical areas.

Climate: Sesame grows in tropical sunny areas and warm temperate areas. However, sesame is typically grown in cultivated beds rather than in the wild, which is worth noting.

Soil: Sesame requires a well-drained porous soil.

Growing: In early spring sow sesame seed in a greenhouse. Transplant the seedlings into individual pots after the last predicted frosts, when they are large enough to handle them and plant them out. For warm temperate zones, where frost is very rare and light (or even non - existent), during the spring or autumn seasons, one can sow the seed directly in the greenhouse. Below are some tips in growing;

- Choose a place that will receive full sun all or most of the day. Sesame thrives when it can be in full sun for the whole season.
- Plant in loamy or sandy soil, well- drained and containing lots of organic matter. If you don't have fast-draining or fertile soil, make amends to improve conditions with sand or peat moss and compost.
- To build a 1/2-to 2-inch hole, tap your finger into the ground. Scatter about three seeds into the hole and gently cover with soil. The plants are spaced between 6 and 18 inches apart. Don't think too much about spacing, because, according to the Thomas Jefferson Agricultural Institute, sesame seedlings should be self-thin rather than crowding each other and vying for resources.
- If necessary, plant beside a wall, garden border or fence to help protect the plants from storms and strong wind. They are highly sensitive to wind and strong rain till the plants develop deep,

strong roots.

- Water to dry out just before the soil. Seedlings need regular growing water, while mature plants require dryer conditions. Avoid overwatering, since sesame does not tolerate long soggy conditions.
- When the seed pods start to look dried out, harvest the seeds, but before they start splitting open.

Harvesting: Since sesame is a small flat seed, make it hard to dry the seed after it has been picked; due to its size, air does not circulate well. The seeds are therefore best harvested when they are as dry as possible, and should then be stored at 6 percent or less of the moisture. If it is too hot, the seed can heat up and quickly become rancid. The seeds are usually washed and hulled until harvested and are sometimes sorted by color.

Preserving: Store dried sesame seeds in an airtight container in a cool dry place.

4.55 STAR ANISE (ILLICIUM VERUM)

Habitant: *Illicium verum* is thought most likely to be native to southern China and northeast Vietnam. It has been cultivated since about 2,000 BC, and it is unknown whether plants growing in these areas are wild or naturalized. Star anise is cultivated in China, Laos, Vietnam, Korea, Japan, Taiwan, Hainan and the Philippines.

Climate: Star anise is grown widely in tropical areas in warm, sunny climates.

Soil: Seedlings are planted out in a rich, evenly moist, well-manured field when they are three years old.

Growing: Plants are propagated from cuttings or seeds that nestle inside the star - shaped fruit points. Seeds are collected from high yield trees and sown for up to one year within three days of collection, or stored wet at 5 ° C (41 ° F).Below are some tips in growing;

- Buy potted star anise from your garden center or take an established plant 's summer cuttings as it grows quickly and easily. Use a sharp knife if you take cuttings, and cut 4-to 10-inch stalk pieces. Place the cuttings in a sterile medium, such as clay, perlite, and peat mixture. Keep the medium warm and moist, until the root of the cuttings.

- Choose your star anise place depending on how you wish it to develop. Choose a site in full sun, if you prefer a compact upright plant with yellow-olive leaves. Develop star anise in deep shade with darker leaves, for a comfortable habit.
- The first growing season after planting, water your star anise frequently to help establish a strong root system. After that, the desired results are to water. Water abundantly if you wish your shrub to grow quickly and spread; star anise will root its branches on moist sites and may become somewhat invasive. Occasionally water for slower growth.
- Apply an acid fertilizer to promote spring production, just before the annual growth of the shrub spurts.
- You can grow star anise as a hedge; it accepts an annual shearing.

Harvesting: The star -shaped fruits are harvested just before ripening.

Preserving: Store dried fruit, whole or ground, in a cool dry place.

Caution: Making sure the spice never gets confused or mixed with any other organisms is very important. For example, star anise *Illicium anisatum*, a Japanese plant, is almost identical to Chinese star anise but is highly toxic.

4.56 THYME (THYMUS VULGARIS)

Habitant: *Thymus vulgaris* is native to southern Europe from the western Mediterranean to southern Italy.

Climate: Thyme grows on dry slopes, rocks, rock walls, and in cultivated gardens in a sunny location.

Soil: Thyme also occurs on clay or calcareous soils. It succeeds in dry soils, poor soils and once it is formed it tolerates drought. Plants can be cultivated on old walls.

Growing: This hardy herb can be propagated from divisions of either seed, cuttings and / or root. Sow seeds indoors for best results keeping the temperature around 70 ° F for germination. Place snipped stem pieces in wet sand for propagation through cuttings with new growth. Roots will appear in two weeks' time. In the Fall, the root division works best. Ascertain that you carefully uproot a plant to conserve as much of its root oil as possible. Divide the root system into half or thirds, and replant the divisions during moist soil one foot apart.Below are some tips in growing;

Thyme grows best in slightly sandy, well-drained soil. Soil which is too rich in organic matter will produce large but less fragrant plants. Thyme prefers a soil pH of between 6.0 and 6.7.

- Sow thyme indoors as early as 3 to 4 weeks prior to the final spring frost. Start seed indoors under fluorescent lights at a ground temperature of 70 ° F. Sowing can take up to 30 days to germinate.
- Transplant seedlings for thyme into the garden once the last frost has passed. Thyme can also be grown from spring or early summer cuttings or divisions.
- As soon as the soil can be worked in spring, thyme can be sown outdoors, usually 3 or 4 weeks before the last frost. You can take cuttings in spring, or separate older plants.
- Planting and spacing. Sow thyme seeds ¼ inch deep.
- Tiny seedlings up to 12 inches apart if 2 to 3 inches tall. Rows 16 to 24 inches apart from each other.
- Grow 6 plants for cooking; grow 10 to 18 plants for preserving.
- All plants of the cabbage family, eggplants, potatoes, strawberries, and tomatoes are said to benefit from thyme. Plant thyme with hyssop, chives, garlic, and rosemary. The thyme fragrance is said to repel pest insects and mask the plant smell which attracts insects from the pest.
- Keep the soil uniformly moist until well-rooted plants do. Once established thyme requires little watering; once established thyme grows best in dry-side soil.
- Throughout the growing season, spray the foliage with compost tea 2 or 3 times. At the beginning of the season scrape a teaspoon of cottonseed or bone meal around the base of each plant.
- Protect plants from freezes with a mulch of chopped leaves, straw, or evergreen branches.
- Keep weed-free on planting beds. In order to prevent thyme from becoming woody, prune plants back in spring by one-third, and again after flowering in summer. Divide thyme every three years so that it doesn't become woody.
- Thyme quickly grows in containers. Choose a container with a

depth of a minimum of 6 inches of soil. Containers in a sheltered place over winter.

- Take thyme indoors for winter harvesting, or start indoor plants for winter growing. Cover outdoor plants in winter, under a layer of mulch— cut leaves or straw.

Harvesting: Leaves and flowers may be harvested just before they flower, dry for later use.

Preserving: Store dried thyme and powdered thyme in an airtight container in a cool, dry place.

4.57 TURMERIC (CURCUMA LONGA)

Habitant: *Curcuma longa* is native to tropical Tamilnadu, in southeast India.

Climate: Turmeric grows at warmer temperatures, from 20 ° C to 30 ° C (68 ° F to 86 ° F). The plant needs to thrive with a considerable amount of annual rainfall, and it prefers light shade.

Soil: Turmeric thrives best on loamy or alluvial fertile soils; it cannot tolerate soil that is waterlogged.

Growing: Turmeric plants are collected annually for their rhizomes, and then propagated for the following season from some of those rhizomes. During fall (September or October), the plant turmeric rhizomes into warm soil. The rhizomes should be planted 5-7 cm deep.Below are some tips in growing;

- The herb is a rhizome. Visit your local nursery to get the turmeric for planting. You can also take the roots from the turmeric sold at grocery stores and plant them in a plastic bag covered seed tray. When the roots start developing shoots, remove the tray from the plastic bag and plant the sprouts out of direct sunlight at a warm place. Plant the rhizomes in the soil about 2 to 3 inches deep, and ensure the soil is warm. Spring and fall are good times for most locations to plant the turmeric.
- Add a balanced fertilizer to help the turmeric get a good start in the soil during planting time. If your ground analysis indicates less than ideal soil conditions, take the time before planting to correct them. According to the Kew Royal Botanical Garden, rhizomes should be ready for harvest within 9 to 10 months after

plantation. To assist with water retention, apply mulch to the planting site, and introduce organic matter, such as compost, into the soil if the plants tend to be stunted.

- Weed the planting site regularly, and avoid waterlogged soil conditions to avoid turmeric problems. Keep the turmeric leaves clean, to avoid red spider mite infestation. Other common pests include shoot borers, rhizome scale, leaf rollers, nematodes from the root- knot, burrowing nematodes and nematodes from lesions. Fungal diseases such as rhizome rot, leaf spot, and leaf blotch can also present problems, especially if the plants are allowed to sit under waterlogged conditions.

Harvesting: After planting, the turmeric rhizomes are harvested for 9 to 10 months. When the lower leaves turn yellow, or when stems are dry and collapse over rhizomes, maturity has reached. If desired, some of the rhizomes can be dug carefully at the side of a clump and harvested as required, rather than collecting the entire clump. When rhizomes are not used fresh, they are boiled in hot ovens for about 30–45 minutes and then dried in. When dry, they 're ground into a powder with its vivid orange - yellow hue.

Preserving: Tumeric powder should be stored in an airtight container in a cool, dry place.

4.58 VANILLA (VANILLA PLANIFOLIA AND V. TAHITENSIS)

Habitant: *Vanilla planifolia* is a vanilla orchid plant that is native to Mexico and is also present in Central America and the West Indies. This is mostly grown in Mexico and Madagascar.

Climate: *Vanilla planifolia* prefers hot, wet, tropical rainforest climates in indirect sunlight.

Soil: Vanilla plants begin to grow in rich, fertile soil, but then they will climb and grow on nearby tropical trees and at some stage, the base of the plant will rot away and the plant will become completely epiphytic (a non-parasitic plant that grows on another).

Growing: Vanilla is readily propagated from cuttings grown in a warm, humid greenhouse. It will take three to four years for vanilla plants to bloom, grown from cuttings.Below are some tips in growing;

- Buy or ask a friend for a vanilla orchid cut which has at least 6 nodes, the points where the leaves meet the stems. To keep it moist, cover the cutting in a damp paper towel until you can plant it.
- Fill a flowerpot with either moist sphagnum moss or a combination of the bark, peat and perlite of equal parts orchid. Instead, if you live in zone 10 or higher, choose an outdoor planting spot close to a tree foundation. Spade the soil and apply a few pounds of compost if the field is not well -drained.
- Take off the cutting of the lowest two leaves. Insert the cutting base into either the potting medium or the underlying loose soil. Cover the two cutting bottom nodes with the potting medium or soil, densely packing it around them.
- Push a plant stake behind a potted orchid cut into the potting medium, and attach the cutting to it with plant ties. Alternatively, tie the cutting to the vine, if you start it outdoors, and mulch the soil around it.
- Hold the potting medium or soil moist until the cutting starts showing new growth within 4 to 6 weeks. Give a broader aid to an indoor cutting once it begins to vinify, like a wood slab or a trellis.
- Don't bring your vanilla orchid in full sunlight so it will burn. Giving it only partial sun, or indirect bright light.
- A vanilla orchid would start blooming at age about 3— maybe sooner if you used a long cutting.
- If the vine is allowed to grow on a tree it will blossom occasionally as long as it grows upward. Therefore, vines can grow up to 1.50 m and then be trained horizontally on the supporting branch and latercoiled around them. In this section of the plant, this induces further flower production
- Vanilla, being a climbing vine needs a growing support.
- You must hand-pollinate your vanilla orchid's flowers if you want to harvest any beans.
- Mulching is used on many of the species that have been studied and is applied differently. Vanilla mulching takes place at the earliest opportunity, after planting. It is advised to have a combination of grasses and leguminous plants.

- When the trees grow in support, they are pruned early to promote branching. Developing a paragliding shape for the trees is desirable to give the growing vines better shade and protection.

Harvesting: After about five months, the fruits ("pods," "beans") mature, where they are harvested and cured. The ferments cure and the pods dry. Curing reduces also the loss of essential oils. This portion of the plant obtains the vanilla extract.

Preserving: Store cured vanilla in an airtight container in a cool, dry place.

CHAPTER 5

Spices and Herbs by Regions

5.1 OVERVIEW

Regional climates can greatly differ, depending on the distance from the equator, land mass size, water proximity and hundreds of other factors. These differences in climate, as well as many other influences, mean different plants have evolved in different places around the globe. Of course, the herbs and spices we use to season and flavor our food are among those plants. For these reasons, some areas of the world have spices that occur significantly more naturally than others. Very few spices, for example, originate from Europe, although the region is known for a huge variety of herbs, while Asia is the birthplace of many of the most widely used spices we know today. But even within these regions, there is a multitude of smaller regions with their own microclimates which encourage the growth of various spices. This means that the local cooking methods evolved to match what was available, thanks to the evolution of different spices in specific regions. Let 's look at the one spice black pepper everybody uses. Black pepper originates from South Asia. Therefore, this prolific little berry, along with its other regional spices, cardamom and cinnamon (among many, many others!), has a prominent place in South Asian cuisine. South Asian cuisine makes use of these spices in many dishes, from savory to sweet, and gives a distinct flavor profile to the region's cooking. Nearby, a whole different variety of spices developed in Southeast Asia, such as ginger and cloves, giving the locals a completely different kind of flavor to work with in making their favorite regional dishes. And the same goes for this complex world, for just about anywhere.

5.2 GEOGRAPHIC SPICE INDEX

You will find spices in this chapter, arranged according to the region from which they probably originated. Because spice trade is almost as old as humanity itself, we cannot in any case recreate the natural occurrence of spice plants. We have included the most relevant spices used in local cuisine today for every region. This awareness cannot, of course, be exhaustive, in part because spice use may vary even in relatively small regions. You may find that this index is rather Asia-centric; while certainly true, I argue that this is not because of my personal interest in Indian and South - East Asian cooking, but rather because almost all the spices that are important in our day are of Asian origin (exclude allspice, vanilla and chili from this statement). Hence, splitting the Asian portion of this index into several sections seemed simple, while only one section deals with African or American spices, respectively. This index contains brief tips about more than 60 untreated herbs and spices on my pages. Some of these spices are very obscure, have highly specialized (often non-culinary) uses, are used only in a small area, or are of purely historical interest. Some others are very interesting and deserve a more comprehensive treatment.

5.2.1 Central and Northern Europe

Surprisingly few spices actually originate in Europe, though many were imported. The Romans brought many of their Mediterranean spices to the northern Alpine countries, and some of them found the climate acceptable and easy to grow; some even spread across the new habitat and became part of the local flora. The following plants are commonly thought to be of European origin, though in some literature you can find different opinions expressed.

- Bear's garlic (ramson) (*Allium ursinum*)
- Blue Fenugreek (*Trigonella caerulea*)
- Borage (*Borago officinalis*)
- Caraway (*Carumcarvi*)
- Celery (*Apiumgraveolens*)
- Chives (*Allium schoenoprasum*)
- Cicely (*Myrrhisodorata*)
- Gale (*Myrica gale*)
- Horseradish (*Armoracia rusticana*)

- Juniper (*Juniperuscommunis*)
- Mugwort (*Artemisia vulgaris*)
- Southernwood (*Artemisia abrotanum*)
- Watercress (*Nasturtium officinale*)

The Truffles (Truffle black or perigord, Tuber melanophore, and truffle white or alba, *Tuber magnatum*). They played an eminent role in 18th - century French cuisine, and, despite their high price, are still of great importance.

Angelica. It is spread across Northern Eurasia (Angelica *archangelica*, *Apiaceae*). All parts of the plant have a strong and penetrating odor and are sometimes used for cooking, particularly in Northern Europe (for fish soups, for example). Nevertheless, the plant is more essential for liqueur flavorings.

Asarabacca. (European ginger, *Asarumeuropaeum*, Aristolochiaceae / Aristolichiales / Magnoliidae) is a perennial forest herb in Europe except for the Mediterranean region. The fleshy rhizome contains a variable composition essential oil and has a pleasant aromatic flavor. A nephrotoxic compound called aristolochic acid has been found in the Chinese (*A. sieboldii*, *A. heterotropoides*) and the North American (*A. canadensis*, "wild ginger") relatives. However, as a wild vegetable and flavoring, both the European and an American species enjoy some success.

Calamus. (Sweet flag, *Acorus calamus*, Araceae / Arales / Arecidae), though originally from India, has now been naturalized throughout the Northern hemisphere. The rhizome is very aromatic and can be candied like ginger (hence the name "German ginger"), but is rarely used for food flavoring. It is quite bitter (which is why it often appears in liqueurs), and the high β-asarone content makes it fairly unsafe for regular use. Calamus traded in pharmacies almost always comes from low in β-asarone, American plants.

Elder. (*Sambucus nigra*, Caprifoliaceae / Dipsacales / Cornidae) bears highly scented flowers, used for dessert and beverage flavorings. The dark purple fruits have been used as a colorant for wine in times that were fortunately long past.

Garlic mustard (*Alliariapetiolata*, Brassicaceae) has leaves with a distinct flavor of garlic and pungent seeds, like mustard. It is occasionally used by peasants, particularly in Eastern Europe.

Ground ivy Lamiaceae (*Glechoma hederaceae*) is an extremely common weed during Central and Western Europe. The leaves, which have a distinctly mint-and thyme-like fragrance, are an interesting if seldom used, culinary spice; I have heard of Czech recipes using it. In the past, they were also used for brewing beer from which the word "alehoof" was used.

Hop (*Humulus lupulus*, Cannabaceae / Urticales / Dilleniidae) is of course very important for brewing beer but is hardly ever used to cook. Alcohol also has little use (other than wine) in the kitchen except, perhaps, to quench the cook 's thirst.

Poplar (*Populus alba*, Salicaceae / Salicales / Dilleniidae) produces characteristic aromatic leaf buds and young leaves; some sources indicate that they have been used as a flavoring in the past. They continue to be used as flavorings for local liquor types.

Reflexed stonecrop (*Sedum reflexum*, Crassulaceae / Rosales) has fleshy, fresh-flavored leaves which are mainly used as a garnish in Western Europe. Formerly, chopped stonecrop leaves were quite popular to add extra sensation to salads, but these are no longer fashionable in our days.

Salad burnet (*Sanguisorba minor*, Rosaceae) is an occasional cultivated wild plant in Western Europe. Rich in tannins, the leaves owe an astringent but nutty taste to it. Lettuce, salads and especially the Frankfurt Green Sauce (see borage) are spiced with the leaves.

Wild rosemary (Marsh tea, *Ledumpalustre*, Ericaceae / Ericales / Cornidae) is a wild plant of Northern hemisphere bogs and swamps. There are several subspecies, one of which is a popular tea-plant in Canada ("Labrador tea"). The European form was used for gruit beer, as was the ecologically similar gale, although it contains alcohol from narcotic sesquiterpene and is not completely harmless.

Sorrel (*Rumexacetosa*, Polygonaceae) is noted for its pungent, acidic leaves that contain oxalic acid. It is used every now and then, e. G., in Green Broth.

Tansy (*Tanacetum vulgare*, Asteraceae) grows throughout Europe but its culinary use, as far as I know, is restricted to Britain. The leaves have a heavy, not very pleasant, odor that is due mostly to the toxic thujone (see also southernwood).

Woodruff: In Western Europe forests (*Galiumodoratum*, Rubiaceae / Gentianales / Cornidae) grow wild. When the aerial parts are wilted, coumarin is released (see also tonka bean), which gives some typical flavored wines its incomparable flavor.

5.2.2 The Mediterranean Region

The area around the Mediterranean Sea has always been a cultural unit, partly belonging to Europe, Asia and Africa. Early spice trading routes lead to the Mediterranean Sea from China and India via the Arabian Peninsula, making the area an important place for cultural and culinary exchange. Many fragrant plants grew abundantly in the warm Mediterranean climate; and in the course of the millennia, merchants, refugees or immigrants from further east have brought even more. The following are generally regarded as Mediterranean native plants; however, some are open to dispute, e. G., cumin, or even an olive apparently "typical Mediterranean."

- Ajwain (*Trachyspermumcopticum*)
- Anise (*Pimpinella anisum*)
- Coriander (*Coriandrum sativum*)
- Cumin (*Cuminumcyminum*)
- Fennel (*Foeniculum vulgare*)
- Hyssop (*Hyssopus officinalis*)
- Garden cress (*Lepidium sativum*)
- Lavender (*Lavandula angustifolia*)
- Mahaleb cherry (*Prunus mahaleb*)
- Myrtle (*Myrtuscommunis*)
- Nigella (*Nigella sativa*)
- Oregano (*Origanum vulgare*)
- Olive (*Olea europaea*)
- Rocket (*Eruca sativa*)
- Rosemary (*Rosmarinus officinalis*)
- Rue (*Rutagraveolens*)
- Sage (*Salvia officinalis*)
- Saffron (*Crocus sativus*)
- Savory (*Saturejahortensis*)
- Sumac (*Rhuscoriaria*)
- Thyme (*Thyumus vulgaris*)

First, in the Hellenistic age,, Asian spices became popular in Europe. Spice trade later flourished in the late days of the Romans, several two thousand years ago; spice trade was dominated by the Arabs from the beginning. Apicius' De re coquinaria is one of the oldest cookbooks in Europe; it lists some tropical spices, the most important of which were long pepper. Black pepper, cloves, and Chinese cinnamon (cassia) are also prominently featured. The mysterious spice silphion (probably of northern African origin) was extinguished about 100 AD and replaced with asafetida (from central Asia). For five centuries the use of olive oil has been a spiritual staple in the Mediterranean. Today, the majority of Mediterranean Europe relies on its native or imported herbs. Basil (stemming originally from South or even South East Asia) today grows wild all-over South Europe and is commonly used, especially in Italian cuisine; the same holds for the native oregano. Garlic ranks more prominently than those in countries of Northern Europe. Regional saffron is used for specialties in fish or seafood, but the high price of this spice restricts its use. Several dishes across the area involve small amounts of chiles; however, fiery food is not common. Bouquet garni (see parsley) and Herbes de Provence (see lavender) are traditional spice mixtures of Southern Europe. Herbs tend to predominate in Asia Minor and West Asia. Coriander and cumin (from Persia but grown locally) are popular, and more popular is the use of pungent spices (mainly black pepper and chiles). The sumac tree berries are important for reproducing the astringent and sour taste found in many Turkish to Israeli dishes. Chilies take an important part in fiery stews and sauces in Northern Africa. Both coriander and cumin are widely used but African spices (paradise grains) are also common. Cinnamon and cloves find most use among tropical Asian spices. All these, and more, can appear in mixtures of Moroccan spices (ras el hanout, see cubeb pepper). While a large number of Mediterranean herbs are discussed here, the diagnosis isn't exhaustive: there are many more that sometimes find their way in the kitchen. There are sometimes wild relatives of herbs handled here that are gathered by experienced family members since their taste is considered superior to that of those grown commercially. This use is often quite local and hardly mentioned in cookbooks. This is especially true of mint family herbs, e. G., thyme, marjoram and in particular, oregano. Further interesting plants from the Mediterranean are:

Black lovage It is similar to lovage and celery (alexanders, *Smyrniumolusatrum*, Apiaceae), with aromatic roots, leaves and fruits. Today, that herb 's culinary value is low.

Mastic Mastiha is a resin derived from *Pistacia lentiscus var*. Chia (Anacardiaceae), a tree that grows only on the island of Chios in eastern Greece, although the associated species are harvested in lesser grades. It was a significant Middle Age commodity but is now only used in Greek cooking (see Mahaleb cherry for more).

Samphire (*Crithmummaritimum*, Apiaceae) extends along all of Europe's coastlines, from the Caribbean to the Black Sea. The leaves are succulent with a salty-aromatic taste, and have in the past been a popular flavoring for salads; samphire pickle, once eaten a lot in Britain, is still popular in the Mediterranean.

Pennyroyal, *Mentha pulegium* (Lamiaceae), is markedly different from the culinary mints. It has been used in Roman cuisine since antiquity (see silphion). It is a traditional herb in Britain, in spite of its moderate toxicity.

Calamint, *Calamintha nepeta*, is an aromatic herb used in Italian regional cuisine (nepitella). Its flavor is reminiscent of herbs related to Lamiaceae, e.g. thyme, savory mint or oregano.

Pine nuts (Pignoli) are seeds collected from the Mediterranean stone pine (*Pinus pinea*, Pinaceae / Pinales); related pine species are also used in temperate Asia. They have a marvelous ethereal-aromatic flavor and are especially important in Spanish and Italian cuisine, e. Pesto g. (See basil).

Purslane (*Portulaca oleracea*, Portulacaceae / Caryophyllales) is an annual herb which probably originates in the Himalayas, but is now naturalized in Western Asia and Southern Europe. The raw leaves and stems, though often eaten cooked as a vegetable, have a crispy texture and a salty, fresh taste that makes them a good garnish for cold Mediterranean foods, e. G., The appetizers of West Asia. The floral buds have a more pronounced flavor, and were used as a caper substitute.

5.2.3 West and Central Asia

Many important spices actually stem from West or Central Asia, even if some of them are in our days cultivated from Morocco to Vietnam:

- Almond (*Prunus dulcis*)
- Asafetida (*Ferulaassa-foetida*)
- Bay leaf (*Laurus nobilis*)
- Black mustard seed (*Brassica nigra*)
- Dill seed (*Anethumgraveolens*)
- Fenugreek (*Trigonella foenum-graecum*)
- Garden cress (*Lepidium sativum*)
- Garlic (*Allium sativum*)
- Lemon (*Citrus limon*)

- Marjoram (*Majorana hortensis*)
- Onion (*Allium cepa*)
- Poppy (*Papaver somniferum*)
- Rose (*Rosa damascena*)
- Tarragon (*Artemisia daracunculus*)

Maybe cumin and some of the other spices listed in the previous section actually originate in western Central Asia, spreading westward in prehistoric times by migrating peoples. Persian or Arabian cuisine today uses a multitude of spices, having easy access to ingredients from Indian or Southeast Asia. As an essential component of Arab-style coffee, cardamom is much appreciated. The Arabic peninsular cooking types have a preference for aromatic but fiery cuisine. Examples might be the Yemeni zhoug (see coriander), a spicy chili-laden paste and the Saudi Arabic baharat (see paprika). The Caucasus republics, situated between the Black Sea and the Caspian Sea, have developed a unique food style, although both Russian and Turkish influences are evident. Georgia has a mild yet spicy cuisine focused on dried herbal flavors (see the Georgian spice mix khmeli-suneli in blue fenugreek) and sour-fruity sauces made from fresh or preserved fruits. The fresh herbs are often spread over warm and cold dishes; Georgian cooking makes unique use of both parsley and coriander leaves in parallel; the latter is not used elsewhere in the country. In neighboring Azerbaijan and Iran, a common inkling of fruity flavors is found. Barberry, Berberis vulgaris (Berberidaceae / Ranunculales), called zereshk or sereshk in Farsi and k'ots'akhuri in Georgian, is a typical Iranian spice that is unfortunately absent from these spice pages; it is often used for flavoring ground meats or Persian rice (polo) dishes. Dried limes are another source of a sour taste in Iranian foods (see also khoreshteghorme sabzi fenugreek). Marigold (Tagetes erecta, Asteraceae) is a common herb typical of Georgian cooking, which appears in several recipes including the spice mix khmelisuneli (see blue fenugreek). In Georgian, it's simply called "yellow flower" (q'vit'eliq'vavili) or "imeretian saffron" (imerulizaprana), and sometimes just zaprana that can lead to confusion with the much different saffron. The marigold flowers are dried and ground to produce a yellow powder that has a mild, sweet fragrance. It may be better substituted by safflower. Sometimes, fresh sprigs are also used which have a distinct, much stronger flavor reminiscent of the huacatay of South America. Between the Caspian Sea and the Tianshan Mountains, the proper Central Asian region is a region quite devoid of local spices, although imported spices have been available since ancient times, as the ancient Spice Route running from China to the Mediterranean cuts through that region. Kazakhstan cookbooks often mention cress-like flavorings local herbs. It is very common to combine dried fruits with meats, where cooks often use local species of the genus Prunus (apricot, plum).

5.2.4 South Asia

South Asia, which includes the Himalayan peninsular Deccan and the southern slopes, has a number of native spice plants. Moreover, Southeast Asian spices have been available in India for millennia. Indian cuisine is thus one of the most aromatic and fragrant in the world. Long ago, a large number of spices originating in South Asia were either shipped to the West or the East. For example, we find spices of Indian origin in today's South East Asia that have no place in contemporary Indian cooking, e. G. Lemon grass or galangale inferior. The table below shows only those South Asian spices which flavor the contemporary South Asian cuisine.

- Basil (*Ocimumbasilicum*)
- Black cardamom (*Amomum subulatum*)
- Black cumin (*Buniumpersicum*)
- Black pepper (*Piper nigrum*)
- Cardamom (*Elettariacardamomum*)
- Cinnamon (*Cinnamomumzeylanicum*)
- Curry leaf (*Murrayakoenigii*)
- Indian bay leaf (*Cinnamomumtamala*)
- Long pepper (*Piper longum*)
- Mango (*Mangiferaindica*)
- Orange (*Citrus sinensis*)
- Screw pine (pandanus) flower (*Pandanus odoratissimus*)
- Turmeric (*Curcuma longa*)

Many more spices play an important part in Indian cuisine today. Chiles, brought by the Portuguese to Asia from the New World, are generously used, in particular in South India and Sri Lanka. Tamarind (from East Africa) is used to give a sour and tart taste to some Southern Indian curry dishes. Of the European and Central Asian spices, coriander, cumin, and garlic are now important to Indian food tastes. Cinnamon, originally grown on the island of Sri Lanka, is now prized throughout India and is commonly combined with Southeast Asian cloves. In South Asia, Arab presence is greatest in Afghanistan, Pakistan, and Northern India. Cooks prefer to use less chili but more fragrant spices (cloves, saffron, and cinnamon) in those areas. Indian spice mixtures are various but most of them have nothing in common with Western supermarket curry powder (see curry leaves). In addition, most of the mixtures are not powder but pastes, made from ground spices, garlic, ginger and oil, and are not stored or traded. Mixtures containing only dried spices include the Bengali panchphoron (see fenugreek), the North Indian garam masala and the more Southern sambar podi (see cumin and coriander for the latter two, respectively). Under coconut is listed the Southern Indian mixture (besebele powder). See the Northern Indian (Moghul style) cooking black cumin, and the spiced butter (tadka or tarka) ajwain. Also, see onion. See Indonesian bay-leaf for the aromatic Northern biriyani for a few traditional recipes, and tamarind for the fiery Southern vindaloo. Under cardamom Indian spiced tea (chai masala) is discussed. Nepali cuisine is similar in several respects to Indian cooking, and some preparations, e.g. pickles, are quite comparable. Nepalese food is usually milder than Indian food, both in terms of actual heat and aromatic spice use. This does not make Nepal 's food bland or uninteresting, as there are several additional fermentation-made flavorings due to Chinese influence: cheese, soy products and typical Nepali gundruk, dried fermented vegetable leaves. Noodles are another culinary signature left by neighboring China in different styles.

Finally, as it is now called, Burma, or Myanmar, is the meeting place and melting pot of Indian and Southeast Asian great cooking traditions. Noodles, shrimp paste, on one side soy sauce and sesame oil, and on the other side cardamom, cinnamon, turmeric and cumin witness the mixed heritage and give Burmese curries a distinct and very tasty character. In the Indian subcontinent, fish flavorings are rare; this is consistent with the observation that fermented products generally have only a small tradition. The only exception to this is dried fish in Sri Lanka (umbalakada), commonly referred to in English as the "Maldive cod." In North -Eastern India and the Chittagong Hill Tracts in Eastern Bangladesh, however, fermented preparation from water-living organisms plays an important role, indicating South East Asian presence (mostly Burmese). The Khasi people use a paste of fermented fish with different spices (tintap) as a condiment, the Chakma use shrimp paste (sidol) for cooking, and the Meitei have a dry-fermented fish called ngari in Manipur. Fermented soybean products are more rarely found in this region: Khasi tungrymbai, Manipuri hawaijar. Indian cuisine fascinates. Nevertheless, there are some Indian spices that little literature exists:

Kokam, kokum, *Garcinia indica*, Clusiaceae / Theales / Dilleniidae, a souring agent for the South West Indian fish curries, should be listed first. Hydroxycitric acid is the principal constituent responsible for the acidic taste. Goraka in Sri Lanka is a similar spice which is also rich in hydroxycitric acid, derived from *Garcinia cambogia*. It is used for many types of meat and especially for fish curries, giving the foods both a refreshing sour note and a brownish color.

A little-known spice radhuni or randhuni is only used in cuisine in Bengali and is practically unavailable outside of Bengal. *Trachyspermumroxburghianum* (Apiaceae, Hindi: ajmud) is its botanical name, but it often gets confused with ajwain. True radhuni has an aroma comparable to the fenugreek and celery. Truly authentic variants of the Bengali spice blend panchphoron include radhuni with black mustard seeds in the typical mixtures (see nigella for details). In Nepal, jimbu are the dried leaves of a local onion species (*Allium wallachii*, or *A. hypsistum*, or *A. przewalskianum*, Alliaceae). The plants grow in the higher Himalayan portion (district of Mustang) and are gathered from wild populations. It is a local spice mostly used by ethnic groups (especially Thakali, also Gurung) living in that area. When fried very early, a flavor very similar to asafetida is relieved that is not recognized in the region. The spice is known as jambu in the Indian union state of Uttarakhand, and used equivalently. For more usage notes see chives. In some European countries, Cockscomb (*Celosia argentea* var. *cristata*, Amaranthaceae / Caryophyllales) is an ornamental common. The bright red flowers (Hindi lalmurghka, Kashmiri moul) are used as food dyes in India.

Horseradish tree (Drumstick tree, *Moringa oleifera*, Moringaceae / Capparales) grows wild in North India (Himalayan Southern Foothills). The tree is very versatile: the young fruits (drumsticks) are eaten as vegetables, an interesting oil (ben oil) is extracted from the seeds, the leaves are used medicinally and the root and root bark are rich in glucosinolates that give them a pungent, horseradish-like flavor. This is one of the rarer flavors of Indian cooking, however.

The wild tree *Buchananialanzan* (Anacardiaceae) yields edible seeds (chironji) with almond flavour; they are used to make sweets in India, especially in the northeast. The herb *Saussurealappa* (syn. *S. costus*) (Asteraceae) is native to Kashmir and grows elsewhere; it is known as putchuk or kushtha in India, and the name costus is most common in the West. The dried root has a strong, perfumed smell, and is often used in perfumery; it was used as a culinary spice in Europe in the late antiquity and early Middle Ages.

Spikenard (*Nardostachysjatamansi*, Valerianaceae / Dipsacales / Cornidae, Sanskrit jatamansi or jatavati) is a similar case: it is also native to the Himalayan region and is very important for perfumery, but almost never used for cooking in our times. Since the Middle Ages, the related herb *Valerianaceltica* (Alpine valerian, speick) has often been used as a cheaper replacement for expensive spikenard. In general, Indian cuisines use very little herbs, in addition to coriander and to a lesser extent mint. Yet the North - Eastern edge cooking styles are an exception to this and employ a large number of herbs also found in Southeast Asia, mostly in the state of Manipur. There chameleon leaf, dill, long coriander and even Vietnamese coriander are used by cooks. One herb unique to this region is *Elsholtziablanda*, known as Manipuri lomba, Mizo lengser or lengmaser, Angami Naga niepfü or rünou, Lotha Naga zutsu and Ao Naga napa. The dried inflorescence, forming a thick spike, has an intense fresh lemon fragrance that is quite similar to lemon verbena, and is used to flavor pickled bamboo sprouts and soupy fish curries.

A closely related herb is sometimes called Shrubby Mint in English called *Elsholtziafruticosa*, which I found not in Nagaland but only in Manipur. It is known locally as kang-human. The flavor is quite similar to that of Vietnamese balm, which is closely botanically related and also looks very similar. Alkanet (Alkannatinctoria, Boraginaceae) is a Mediterranean plant grown in India for its red rhizome (Hindi ratan jot) which yields a dye. In the North West, the dried rhizome is sometimes used as a food coloring.

The so-called mango ginger (*Curcuma amada*) is a minor South Indian spice, rather strictly species. The plant looks much like turmeric, but the rhizome is pale yellow only; indeed, its scent closely mimics the fragrance of unripe mangos. Southern India's rather enigmatic "spice" is a dried lichen called kalpasi in Tamil, and dagorphul in Marathi (although the latter name occasionally stands for star anise). The material of the plant is rather tasteless but it is widely used.

5.2.5 Southeast Asia

Southeast Asia has a significant number of native aromatic plants due to its tropical climate, the majority of which are used fresh in local cuisines. In the late Middle Ages and the first centuries of modern times, the Moluccas, a group of small islands on the border between Asia and Australia and home to nutmeg and cloves, were the focus of European spice policy.

- Chameleon plant (*Houttuynia cordata*)
- Cloves (*Syzygiumaromaticum*)
- Coconut (*Cocos nucifera*)
- Cubeb pepper (*Piper cubeba*)
- Fingerroot (*Boesenbergiapandurata*)
- Ginger (*Zingigerofficinale*)
- Greater galanga (*Alpinia galanga*)
- Indonesian bay leaf (*Eugenia polyantha*)
- Indonesian cinnamon (*Cinnamomumburmannii*)
- Kaffir lime (*Citrus hystrix*)
- Lemon grass (*Cymbopogon citratus*)
- Lesser galanga (*Kaempferia galanga*)
- Lime (*Citrus aurantifolia*)
- Long pepper (*Piper retrofractum*)
- Mace and Nutmeg (*Myristicafragrans*)
- Perilla (*Perilla frutescens*)
- Rice paddy herb (*Limnophilaaromatica*)
- Screw pine (pandanus) leaf (*Pandanus amaryllifolius*)
- Vietnamese cinnamon (*Cinnamomumloureiroi*)
- Vietnamese coriander (*Persicariaodorata*)

Today, all these spices (with the exception of not - so-used cinnamon types, cloves, and nutmeg) are prominently present in at least some of the major South East Asian cuisines. Chiles, ginger, and garlic are also found throughout the region, as are coconut products: coconut milk, and coconut oil. Numerous independent culinary styles have developed in Southeast Asia; yet most of them prefer fresh spices (if available), and also fresh herbs (basil, coriander leaves and mint) are popular in Vietnam, Cambodia and Thailand as fragrant decoration. Pungent fish preparations are popular throughout the region: fish sauces (nampla in Thailand, nuoc mam in Vietnam), shrimp pastes (gapi in Burma, trassi in Malaysia and Indonesia), and special freshwater-fished Cambodian paste, prahok. Fish sauce is also popular in Southern China, where it is called yulu; but it is only a mild flavoring in

Chinese cuisine. Thai cooks use even more spices (e.g., kaffir lime leaves, lemongrass, and fingerroot) and other strong-smelling ingredients such as dried fish to achieve the typical scent of Thai dishes. Thai food is sometimes extremely hot and fiery, as they generously use chiles. See coconut for Curries in Thai. For more Thai recipes see basil and mint, too. Spice use is not that common in Cambodia and Vietnam, and the Philippines, too, cook very mildly. Besides garlic and ginger, the South American annatto seeds are used in the Philippine cuisine. The Spaniards introduced this spice to the Philippines, and it is hardly found in other Asian countries. Vietnamese cuisine is remarkable for its extensive use of fresh herbs, some of which are only rarely used outside Vietnam (Vietnamese coriander, long coriander), while others (rice paddy herb, chameleon herb) do not appear in any other style of cooking. Many very different regional cuisines have evolved on the many islands of Indonesia, to be explained by different conditions of life (jungle nomads, farmers or seafarers; village-bound or cosmopolitan urban cultures), food taboos due to different religions (Islam, Christianity, Hinduism, Buddhism, Animism), different climates (tropical jungle, mountain forests, highlands or even animism).

Many Indonesian cuisines do not use sweet spices, which is even more shocking because indigenous to Indonesia are cloves, nutmeg and the variety Sumatra cinnamon. Instead, ginger, onion, garlic and moderate quantities of chiles, in addition,galanga and turmeric are the most common spices. Indonesian dishes also include shrimp paste (trassi) and soy sauce (kecap), which is also used in a very sweet and thick variety (kecapmanis). Especially Jawanese dishes often contain large amounts of sugar and sweet-spicy taste while I enjoyed quite hot food in Sumatra, and Bali definitely displays the largest variety of various spices. Some highlights of Indonesian cooking will be discussed shortly under greater galangale (rendang, a buffalo stew from Western Sumatra), Sichuan pepper (sangsang, a spicy pork variety stew from Northern Sumatra), coconut (ayampa'piong, a Sulawesi chicken dish), mango (the pan-Indonesian fruit salad rujak) and lesser galangale (bebekbatulu, Balinese roast duck). In general, regarding Indonesian spice pastes (bumbu), see lemon grass, see Indonesian bay leaf for information about Balinese cuisine and see tamarind for Jawa cooking. Much more herbs and spices are used in the large region's many and varied culinary forms. In Vietnam in particular, there is a huge wealth of local herbs that are not commonly found in the West. The following are particularly worth noting:

- Torch ginger (*Etlingeraelatior*, Zingiberaceae) is a special spice: in Singa-pore and Malaysia (bungakantan), the inflorescence is used to flavor curries.

- The fragrant wild betel leaves are commonly used in Thailand (cha pluu) and Vietnam (la lot) to wrap rice or other foods into. These leaves derive from a member of the pepper genus (*Piper sarmentosum*, Piperaceae), which is closely related to the so-called betel pepper, *Piper betle*, as an indispensable part of the betel bits eaten in South East Asia and India (pan).
- Musk mallow (ambrette, *Abelmoschus moschatus*, Malvaceae / Malvales / Dilleniidae) is an aromatic seed plant that is closely related.
- Butterfly pea (*Clitoriaternatea*, Fabaceae) has huge, deep blue flowers that are used to give Thailand (anchan, anjan) and Malaysia (bungatelang) desserts a bluish hue. It is mostly substituted in our days by synthetic food dyes.
- Broadleaf thyme (*Cuban oregano*, Indian borage, Mexican mint, Plectranthusamboinicus, Coleus amboinicus, Lamiaceae), while introduced into the Caribbean, is a herb native to South East Asia. Because of the essential oil rich in carvacrol the leaves have a strong odor. The fresh herb is utilized as a garnish in Indonesia (daunjinten), but particularly in Vietnam (rau day tan las).
- Very rarely, it is believed that in South East Asia the pungent seeds of some members of the Araceae family (e.g., Giant Elephant's Head, *Colocasia gigantea*) are used as pepper surrogates.
- The fruits of the tree *Garcinia atroviridis* (Clusiaceae / Theales / Dilleniidae) are used as a source of acidity, especially in Malaysia (asamgelugur), similar to the use in South India and Sri Lanka of other Garcinia species.
- *Aleurites moluccana* (Euphorbiaceae / Euphorbiales / Dilleniidae) yields seeds ("candle nut," kemiri) which are very popular but bland components of Indonesian spice pastes. See also lemongrass over spice mixtures that include candlenuts.
- A quite interesting spice is extracted from *Pangiumedule* (Flacourtiaceae / Violales), an Indonesian pangi or kepayang herb. The seeds, known in Indonesian as kluak or kluwak, and in bahasatoraja as pamarassan, are a typical ingredient of some local Indonesian cuisines, e. g. In East Jawa, and Sulawesi Central. They provide a dark color, an intense nutty taste and a smooth texture, slightly oily. The seeds need a fermentation process for

the production of flavor and the removal of hydrocyanic acid, by which they change from a cream color to almost black.

- Sandalwood (*Santalum band*, Santalaceae / Santalales / Rosidae) is the core wood of a Lesser Sunda Islands native parasitic plant, possibly Timor. Much of it is cultivated in Southern India today and used for incenses. Although highly fragrant, never most has been used for cooking.

5.2.6 East Asia

The entire area of East Asia is dominated by Chinese culture. Chinese cuisine is highly diverse and sophisticated; it has inspired all East Asian cuisines, and is also an important addition to all South East Asian culinary styles.

- Chameleon plant (*Houttuynia cordata*)
- Chinese cinnamon (*cassia) (Cinnamomum cassia*)
- Ginger (*Zingigerofficinale*)
- Lesser galanga (*Kaempferia galanga*)
- Perilla (*Perilla frutescens*)
- Sichuan pepper (*Zanthoxylumpiperitum*)
- Star anise (*Illicium verum*)
- Wasabi (*Wasabia japonica*)
- Water pepper (*Persicaria hydropiper*)

Chinese cuisine derives its appeal not so much from different spices, but from a multitude of meat and vegetable ingredients with different flavors, shapes, colors and texture, and from a number of standardized cooking and frying methods; the only common spice mixture is the famous five -spice powder (wuxiang fen, see star anise), which is widely used to flavor fried meat in China. Soy sauce (jiang you) is the most important condiment in China, but other soy products, such as sweet bean paste (haixian jiang, better known by its Cantonese name hoisin jeung), hot bean paste (douban jiang), and fermented black beans (dou chi), are also required to prepare authentic Chinese foods.

Cantonese cuisine, which is native to Guangdong province, is the least spicy cooking type in China. The name "Cantonese" derives from the province capital Guangzhou that was formerly known as "Canton" in the West. Cantonese cuisine is famous for its exotic meat dishes made with dogs, cats, monkeys and snakes. It is also known for a variety of barbecued meats (siumei, Mandarin shao wei), such as spare ribs (cha siu, also spelled western char siu, Mandarin cha shao). A famous word for Cantonese food is dim sam (also spelt dim sum in English), which is not a dish but a light meal consisting of a variety of small dishes; the most popular choice is meat-stuffed dumplings made from pork, chicken or shrimp with light but subtle flavors. The term has mainly come to mean a variety of such steamed pasta outside of the Guangdong area. Though Cantonese in origin, dim sam is now enjoyed all over China (Mandarin dianxin). According to tradition, fiery food is rather rare in China, except in two Central Chinese provinces: Hunan and Sichuan (Szechwan), also known as Tian-fu ("heavenly province" or "place of abundance"). Chiles, garlic, and aromatic sesame oil are common in these two provinces but particularly in Sichuan. An important flavoring of Central Chinese cookery is red hot bean paste, doubanjiang made from fermented broad beans. Spicy Sichuan and Hunan foods have recently become available and popular in wider parts of China, due to domestic migration. By contrast, the mountainous Yunnan province's cuisine has not attracted much interest yet, although it is spicy and similar to Sichuan cuisine.

The North-Eastern Chinese cooking is usually termed the Shanghai style. It is particularly rich and often uses sweet flavors. A typical motive of Shanghai cooking is the use of rice wine (liaojiu). Red-braising (hongshao) is a cooking technique that originates in Shanghai, although it is today commonly found all over China. The fourth and last Great Cuisine is the Northern Beijing style, which has a large repertoire of baked foods (a Central Asian influence) and uses more wheat than rice due to climatic reasons. Two signature dishes are Beijing duck (beijingkaoya) and Mongolian hotpot (meng-guhuo-guo). Furthermore, sweet and sour dishes are popular: Fish or meat are battered, deep-fried and served with a sweet–sour sauce (tangcu"sugar and vinegar")

Only Sichuan pepper (more specifically, a closely related Japanese species) is used as a condiment, either alone or mixed with tangerine or orange peel and chilies in the form of the spice mixture shichimi tōgarashi. Japanese dishes, therefore, owe most of their flavor to their ingredients, whose freshness and skillful preparation are crucial, in addition to dried sea grass and kelp, various soy products (e.g., soy sauce shōyu) and other fermented crops (Miso). A pungent root, wasabi, is used as a green paste for raw fish (sashimi) and rice bits (sushi); several herbs (water pepper, perilla and young Sichuan pepper leaves) are used for flavorings and decoration purposes. In sharp contrast, Korea's cuisine, East Asia's most eastern country, is fiery and pungent, dominated by chiles, toasted sesame seeds, and garlic; pickled vegetables (kim chi), spicy as well as sour, are also very common. Soybean paste (den jang, also written doenjang or doinjang) similar to Japanese miso and bean-chile paste (gochujang, pronounced kochujang) are essential flavors. The fresh spring onions are a common garnish in both Korea and Japan. Some additional local herbs and spices are commonly used. Chinese cuisine, for example, uses many local onion species (Allium, see chives); for Sichuan, in particular, cookbooks mention "local Himalayan herbs" but don't give any precise identifiers. We should also note the following:

- Ginseng (*Panax ginseng*, Araliaceae / Araliales) is mainly known in traditional Chinese medicine as an expensive herb, and as an aroma for alcoholic beverages. It is however also used as a culinary spice, particularly in Korea.
- Camphor is an essential aromatic of old, although it has never been used much to cook. In China, however, in the past, camphor was used to spice frozen desserts, and even now it is sometimes a part of smoking mixtures, giving rise to specialties such as tea and camphor wood smoked duck (zhang cha yazi). There are two different products commonly referred to as "camphor": the better-known Chinese or Japanese camphor (from *Cinnamomumcamphora*, Lauraceae) is composed of 2-bornanone and generally considered to be much lower than the much more expensive Sumatra camphor or Baros camphor (from *Dryobalanopsaromatica*, Dipterocarpaceae/Malvales/ Dilleniidae) composed mostly of borneol.
- The fresh leaves of mitsuba (*Cryptotaenia japonica*, Apiaceae) are used as a culinary herb in Japanese cuisine. Fresh leaves are chopped over soups or salads and sprinkled. The herb is known as yaerqin, in Chinese).

5.2.7 Africa

In the West, few African spices ever became popular. I know only four, personally, of which the origin of sesame is uncertain.

- Grains of Paradise (*Melegueta Pepper*) (*Aframomummelegueta*)
- Negro Pepper (*Kani Pepper*) (*Xylopia aethiopica*)
- Sesame (*Sesamum indicum*)
- Silphion (*Silphium*)
- Tamarind (*Tamarindusindica*)

The former two (from western Asia) were traded as a cheap substitute for black pepper during the Age of Explorations unless the sea route to India was established. Users lost interest in them later, and are now almost forgotten (and hard to get). Silphion is the name of a legendary spice that was so popular in ancient Rome that it became extinct during the early Imperial period. The definition of botany is subject to debate. Tamarind is possibly from East Africa, but is cultivated throughout the world in a tropical climate in our days, and is an important ingredient in Asian or Latin American cuisine. Sesame is one of mankind's most significant oilseeds, yet little of the crop is used as a spice. Specialties that contain sesame can be found throughout the Old World, from Europe to Korea. African cooking today is dominated by Arab influences, predominantly in the North and East, where Islam prevails. There's a lot of colonial influence in the South, by both European colonists and Indian and Malaysian immigrants. East Africa has absorbed the techniques of Arabic and Indian cooking and developed a unique cuisine combining foreign influences with local tradition. Cooking has maintained the distinct character in West and Central Africa, and is hardly comparable with any other culinary form.

In the North, e. g. Food is often very pungent in Nigeria, Cameroon, Ghana and Benin, due to the use of extra hot chilies imported from the Caribbean. Other essential flavors are dried fish products, smoked meats and toasted peanuts; the traditional cooking medium is unrefined palm oil (from *Elaeisguineensis*) whose flavor also greatly contributes to the character of West African cooking. In addition, a number of local spices are used that are nevertheless hardly available outside the region (except paradise grains and, if one is very fortunate, negro pepper). Subtle spice mixtures based on cumin and coriander however dominate in North Africa, and aromatic Asian spices are popular. See the exceedingly complex mixture ras el hanout with cubeb pepper. Through spice mixtures such as Tunisian gâlat dagga (see grains of paradise), and Ethiopian berbere (see long pepper), Arabic or Indian influence is manifested. Quite a number of spices from other continents are grown in tropical Africa today, where they are mostly planted and exported as "cash crops." Nigeria, for example, is a large ginger producer. Eastern Africa's tiny but abundant islands are suppliers of several of the finest spices for European consumers: Réunion (formerly known as "Bourbon") produces vanilla and allspice, and Zanzibar has long outgrown Indonesia as the main producing country for cloves.

A number of African spices aren't well known. For example, various scented pelargoniums are native to South Africa; they are often referred to as "scented geraniums," but belong to the closely related but distinct (Geraniaceae / Geraniales) genus Pelargonium, not Geranium. Such herbs have an impressive spectrum of different flavors, usually lemon or rose-like floral, but there are also varieties of ginger, cinnamon and even nutmeg-like fragrances. Nevertheless, even though a few varieties are grown for the perfume industry, these amazing plants have still not found much application in the cooking. The potential of indigenous spices has not yet been exploited for West Africa, as well. In the rest of the world, most native West African spices are unavailable. In some instances, I don't even know the botanical name, like the akob bark and felom fruits (seeds). Some more West African spices are mentioned in the below list.

Many species of the Aframomum genus (Zingiberaceae) grow edible fruits and pungent seeds, e. G., *Aframomumdanielli*, *Aframomumcitratum* (' mbongo spice') See also paradise grains

- The related genus Amomum also has members from Senegal to Ethiopia rising in the tropic belt that are used locally. In the past, some of these were traded as cardamom adulterants or surrogates. See black cardamom too.
- Furthermore, there are African pepper species like *Piper clusii* (see cubeb pepper).
- Another source of pungent flavor may be present in the various indigenous species of Zanthoxylum (Rutaceae) found in tropical Africa, but there is scant literature (see Sichuan pepper regarding Asian relatives).
- "Calabash nutmeg "is the seed of Monodoramyristica (Annonaceae), a popular substitute for nutmeg in Europe in the 16th century; today the species is also cultivated in Jamaica.
- The oily seeds of the tree *Ricinodendronheudelotii* (Euphorbiaceae / Euphorbiales / Dilleniidae) have a distinctive, strong flavor and are used for sauces (local names njangsa, njasang) as spice and thickener.
- Wild mango or bush mango is the fruit of the *Irvingiagabonensis* jungle tree and of the related species I. Wonbolu (Irvingiaceae / Sapindales / Rosidae); the botanical relationship to mango is loose only. The dried and ground seeds are known as ogbono, and

give West African chicken stews ("sauces") a sticky texture and presumably some flavor.

- Koseret is the Amharic name for the herb *Lippiaadoensis* (Verbenaceae), used in Ethiopia as a culinary spice. It is notable in kitfo, raw ground beef flavored with spiced butter. Most Ethiopian cookbooks secretly substitute basil for this. See also long pepper about the berbere spice mix.
- Roselle (red sorrel, *Hibiscus sabdariffa*, Malvaceae / Malvales / Dilleniidae, *Arabic karkadi*) is the purple, dried calyx of a plant which is related to the popular ornamental hibiscus. In parts of Northern and Western Africa, a refreshing acidic beverage prepared from the calyces is quite popular; more rarely, one reads about rosella calyces used in salty food, e. G., Curries in India and Malaysia.

5.2.8 AMERICA

Unfortunately, the two Americas ' contribution to the spice list is rather short. This is not due to the lack of aromatic plants, but mostly due to the lack of knowledge about the European native American spices. Latin American spices are easier to get through in the USA because of globalization, but few of them have found a permanent place in the spice shelf. Of course, this one American nightshade plant revolutionized nearly every cuisine in the country.

- Allspice (*Pimentadioica*)
- Annatto seeds (*Bixa orellana*)
- Boldo leaves (*Peumus boldo*)
- Chile (*Capsicum frutescens, C. chinense, C. baccatum, C. pubescens, C. annuum*)
- Epazote (*Chenopodium ambrosioides*)
- Filè (sassafras) (*Sassafras albidum*)
- Lemon verbena (*Lippiacitriodora*)
- Mexican Pepper-leaf (*Piper auritum*)
- Mexican tarragon (*Tagetes lucida*)
- Nasturtium (*Tropaeolummajus*)
- Oilseed pumpkin (*Cucurbita pepo*)

- Paprika (bell pepper) (*Capsicum annuum*)
- Paracress (*Spilanthesacmella*)
- Pink pepper (*Schinus terebinthifolius*)
- Tonka beans (*Dipteryxodorata*)
- Vanilla (*Vanilla planifolia*)

Because the cooking style in North America (the US and Canada) is largely derived from and not very different from European cuisine, spice usage is generally rather low (exclude from this statement the Mexican-influenced cooking of the Southern states of the US). There is currently only one North American plant being treated on these pages: Sassafras (file) has great importance in New Orleans cooking albeit only regional. From the Caribbean islands Allspice was introduced to Europe; its alternative name, new spice, indicates its origin from the New World. Vanilla is native to Mexico and has been used since Aztec times to spice a chocolate-like drink.

An epazote is a culinary herb found in Mexico. Toasted pumpkin seeds are an ancient flavoring of Central American peoples, dating back to pre - Columbian times; but extraction of oil from toasted pumpkin seeds, as practiced in Central Europe, is a much more recent invention. The stem of annatto seeds from South America, much used locally, and pink pepper, a spice that has become popular in the modern cuisine over the past decades. Other South American spices are tonka beans and paracemas, but they found limited use outside of South America. Lemon verbena is usually underestimated as a further spice. However, the most important spice of both Americas are chiles and bell peppers, both thought to be native to the Amazon region but traded extensively as far north as the southern states of today's USA before the Europeans arrived. Today, they are highly valued in all tropical American, Asian and African countries. Some more interesting plants from North, Central and South America are:

- The Californian bay leaf tree (*Umbellulariacalifornica, Lauraceae*) has highly aromatic leaves that have, however, been largely replaced in recent years by Mediterranean bay leaves, even in the USA.
- The spicebush (*Lindera bezoin, Lauraceae*) is native to the Eastern USA. It has a heavy and friendly, spicy aroma in all parts of the plant. The fruits were used as replacements for allspice.
- Anise hyssop (*Agastache foeniculum, Lamiaceae*) is native to

North America; although it is a member of the same family, it is not closely related to hyssop. The large leaves are heavily scented, reminiscent of anise or licorice, but are hardly ever used in cooking.

- The early North American settlers knew of the winter-green aromatic leaves (*Gaultheria procumbens, Ericaceae / Ericales / Cornidae*), a North American dwarf shrub: winter-green tea was a popular drink. The scent is attributed to an essential oil that is almost entirely composed of methyl salicylate. The essential oil is still used in the USA and Canada for flavoring candies and confectionery.
- Bergamot (bee balm, *Monarda didyma* and relatives, *Lamiaceae*), in the days of the pioneers, was, similar to the former, a common tea herb that reduced reliance on expensive, imported black tea from Asia (Boston Tea Party). It has a lemony but also a thyme-like, spicy taste that is well suited to cooking. See lemon balm, too.
- A sour beverage from the fruits of several species of American sumac was distilled, e. G., smooth sumac, Rhus glabra (*Anacardiaceae*), but I have no experience of culinary use comparable to that of Mediterranean sumac.
- West Indian bay leaves (also bay rum or Caribbean bay leaves) derive from the *Pimentaracemosa* (*Myrtaceae*) tree, or from the closely related allspice tree, according to some other sources. They have a strong aroma of the clove and are especially used in Jamaican and Cuban cooking.
- Mexican bay-leaves are quite different, derived from *Litseaglaucescens* (*Lauraceae*), a tree closely linked to the Mediterranean laurel.
- The so-called white cinnamon (wild cinnamon, *Canella winterana, Canellaceae / Magnoliales*) is of the Caribbean and Florida descent. Occasionally its aromatic bark is used as an alternative to true cinnamon, but I find its herbal-pungent taste more akin to sweet flag or galanga.
- There is a lot of use of herbs called "organo" or "marjoram" in cookbooks in Central America, but I assume that they are meant to be native herbs in the first place. The "Mexican oregano" is classified as at least two different herbs: *Poliominthalongiflora*

(*Lamiaceae*) and *Lippiagraveolens* (*Verbenaceae).* In addition, the genera Lippia, Coleus and *Plectranthus* (both *Lamiaceae*) have found many more aromatic plants that have found local culinary applications.

- Aztec herb (*Lippiadulcis*) is an aromatic and intensely spicy, leafy shrub. It is moderately common among Western herb lovers (mainly for infusions) despite its toxicity (because of Campher), while I do not know anything about indigenous use.
- The *Crotalaria longirostrata* (*Fabaceae*) herb is native to Mexico and used in Oaxacan cuisine, where its fresh leaves give soups and tamales a mild bean flavor. Native names such as chepil, chipil and chipilín are referred to here.
- In Tabasco and Veracruz, the word herba de conejo ("rabbit herb") refers to aherb used to flavor bean dishes. Different sources give the *Tridaxcoronpifolia* (*Asteraceae*) and *Castilleja lanata* (Scrophulariaceae) botanical identity
- *Peperomia pseudoalpina* (*Piperaceae*) heart - shaped leaves are a natural flavoring of some provinces of Central Mexico (Oaxaca, Veracruz, Puebla). The plant is known as "coriander of the forests" as tequelite or cilantro silvestre, where the latter name refers to both the coriander-like taste and wild occurrence.
- The *Porophyllumtagetoides* (*Asteraceae*) with the local names pepicha, pipitza, tepicha and chepiche is another Mexican herb with a coriander flavor.
- Peruvian coriander (*Porophyllumruderale*, Asteraceae) in Mexico is called papalo, or papaloquelite, and in Peru and Bolivia killi or quillquiña. The "intermediate flavor between coriander leaves and rocket" is mentioned, but I don't find it too close to those. This herb is used mostly for salsa (Bolivia: sarsa), including coriander leaves.
- Mint marigold (*Tagetes minuta* and *Tagetes elliptica*, Asteraceae) is an important herb in the Bolivian and Peruvian Andean cuisines. In cookbooks, Quechua huacatay (Aymara wacataya) is mostly known by its name. The herb has an amazing, spicy-fresh flavor and should only be used in fresh condition, although it can be made from a pesto-like concoction (Salsa de Huacatay, black mint sauce), which retains much of the original taste.
- The peanut (groundnut,*Arachis hypogaea*, Fabaceae) originates

from South America, but is widely cultivated worldwide as a source of protein and oil. In many West African and South East Asian cuisines toasted peanuts are an important flavoring.

- Heliotrope (Heliotropiumarborescens, Boraginaceae) is a popular ornamental of Peruvian origin; its vanilla-scented flowers have been used as cake, fudge-like desserts and sherbets flavorings in Europe.
- Sweet honey leaf (*Stevia rebaudiana*, Asteraceae) is native to Paraguay's highland and is traditionally used as a sweetener by indigenous peoples, especially for local mate tea. Thanks to several diterpene glycosides (steviol, stevioside), fresh and dried leaves have an extremely sweet taste. The plant is currently grown commercially in Japan, where stevia extract plays an important role as an artificial sweetener. However, it has not been so popular in other countries as yet.
- The members of the Ocotea genus are trees with aromatic calyces of leaves, bark, and fruit; in the two species *O. pretiosa*, *O. native*of the Amazon basin, quixos the fragrance comes close to cinnamon. A major Spanish expedition perished in the 16th century, almost to the last guy, when searching the origin of this spice (see annatto). Yet "American cinnamon" gained no culinary importance after its discovery, but the closely related species O. In Brazil sassafras is grown commercially as a source of safrole.

5.2.9 AUSTRALIA

Few Australian plants have ever gained economic importance, the leading example being macadamia nuts (*Macadamia integrifolia* and *M. tetraphylla*, Proteaceae / Proteales / Rosidae). However, there are plenty of aromatic plants, some of which may grow in importance in the coming cuisines.

- Lemon myrtle (*Backhousiacitriodora*)
- Tasmanian pepper (*Tasmannialanceolata*)

Both spices are currently barely known (less used) outside Australia, but such things may change rapidly in our global world. Notice that there are more indigenous flavorings in Australia that can be classified as spices: the dried, tiny bush tomato berries (Solanum centrale, Solanaceae) have a complex taste that is not entirely different from Italian sun-dried tomatoes, though less fruity and spicier. Another nominee is the so-called seeds of wattle, roasted and roasted seeds of different Acacia species, e. G., Victoriae Acacia, A. Sophorae, A. Murrayana (The Mimosae / Fabales). Such plants have a long track record of Indigenous use.

5.3 HERBS AND SPICE TRADE

Luckily for our taste buds, people from different regions began traveling and trading somewhere along the road-and discovering that their neighbors near and far had some delicious flavors in their food which they had not encountered. Since humans love variety, these traders began trading their locally grown spices, taking them far and wide. Once people in Europe got wind from these exciting new spices, whole expeditions were sent out to set up routes eastwards, going deeper into Asia, while Indian traders were busily setting up the Spice Route to bring their wares westward. Along the way, as trade routes were formed between all the continents, even more spices and herbs, as well as diverse cooking methods, were introduced to various places and even more complex flavor profiles were created as people combined these spices in new and interesting ways. The Spice Route past is a long-and often hostile and even violent-one, but as the world is getting closer and easier to navigate, ordinary folk like us have gained the advantage of easy access to a vast array of spices and herbs. This means you and I can enjoy a Mexican - origin meal on Monday, Thai food on Tuesday, Hungarian on Wednesday or any of a multitude of regional specialty dishes on any day of the week.

There's a wide world market for imported spices and culinary herbs, estimated at just over US$ 2.3 billion. LDCs like Madagascar, Comoros, and the United Republic of Tanzania receive a large portion of their foreign exchange from spice exports. Vanilla, cloves, chilies, cardamoms, nutmeg and mace are the principal spices exported by LDC countries. Since 1995, the LDCs have generated more than half of the world's vanilla imports and more than one-fifth of the world's clove demand. Other important exports of LDCs include black pepper, paprika, coriander, cumin, cinnamon, ginger, turmeric, etc. Annual world spice imports averaged 500,000 tons from 1995 to 1999, increasing at a rate of 8.5 percent a year. This high growth rate is a good indicator of an improvement in spice consumption. The LDCs have immense potential to extend their exports to the markets of both developed and developing countries. Those are fast -growing high-value markets, and the LDCs have the natural capital to deal with. They are likely to benefit from the rapid market growth, may take market share from their developing world competitors, add to the variety of spices and herbs they provide and increase value through preliminary processing. But with spice exports, the LDCs have not performed well in recent years, given growing world imports. Total exports from LDCs dropped by 8.5 percent in quantity and 4 percent in value over the period from 1995 to 1999. There were two major explanations for this: the drastic decline in world prices of vanilla following an over-supply situation in Indonesia and the poor climate conditions in many African producing countries in 1998 and 1999.Strategies to realize the full potential offered by the export of spices and culinary herbs need to take account of the following five types of opportunity:

Diversifying the product range. Products such as coriander, cumin, bay, oregano and thyme offer sizeable and attractive export opportunities.

- Diversifying export markets.
- Increased market share and improved positioning. Improving quality performance and production timelines are key requirements.
- Make the most of niche opportunities. While relatively small at present (slightly more than 1 percent of total spice demand), the spice market's organic segment is growing, driven by a desire to consume real, wholesome goods that drive overall spice consumption. The trend towards a greater variety of ethnic cuisine offers every country's prospect of creating a niche for spices and admixtures that will be used in their own cooking.
- Packaging and shipping. In the developing countries, there is a movement towards greater processing and packaging of spices. Spice admixtures often gain ground in response to a greater desire for convenience.
- The critical success factor: increasing the supply capacity and competitiveness of the LDCs and helping them leverage their enhanced capabilities in international markets through information, access to inputs, support services, utilities and infrastructure.

Under current business conditions, LDC exporters may concentrate on supplying high-quality, whole spices to large processors and packers. We may aim to supply large, industrial users with high-quality ground spices in bulk, who are more likely to be able to purchase directly from source and use more than one supplier. The aim should be to draw investments by processors with market awareness and distribution capabilities in the customer countries in the longer term. Smaller processors and traders in the producing countries will look for alliances and joint ventures with processors.

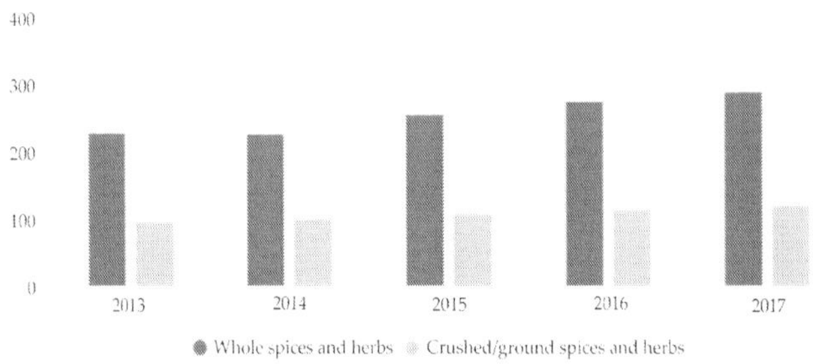

Figure 23: Developing countries supply of crushed/ground spices and herbs between 2013 and 2017

Source: Eurostat, 2018

REFERENCES

1. Black CD, Herring MP, Hurley DJ, O'Connor PJ. Ginger (Zingiber officinale) reduces muscle pain caused by eccentric exercise. J Pain. 2010;11(9):894-903. doi:10.1016/j.jpain.2009.12.013
2. Bouzembrak, Y., Camenzuli, L., Janssen, E., & Van der Fels-Klerx, H. J. (2018). Application of Bayesian Networks in the development of herbs and spices sampling monitoring system. Food Control, 83, 38-44.
3. Nakatani, N. (2000). Phenolic antioxidants from herbs and spices. Biofactors, 13(1-4), 141-146.
4. Tapsell, L. C., Hemphill, I., Cobiac, L., Sullivan, D. R., Fenech, M., Patch, C. S., ... & Fazio, V. A. (2006). Health benefits of herbs and spices: the past, the present, the future.
5. Bower, A., Marquez, S., & de Mejia, E. G. (2016). The health benefits of selected culinary herbs and spices found in the traditional Mediterranean diet. Critical reviews in food science and nutrition, 56(16), 2728-2746.
6. Eurostat 2017 Spices and herbs https://ec.europa.eu/eurostat/web/products-datasets/-/prc_hicp_manr
7. https://www.herbco.com/c-2-bulk-herbs-spices.aspx
8. http://theepicentre.com/get-spice/
9. https://www.britannica.com/quiz/season-to-taste-fact-or-fiction
10. https://www.thehomesteadgarden.com/category/gardening-tips/

11. Herbs, F. A. O. (2005). Spices and Essential Oils: Post-harvest Operations in Developing Countries. Douglas, J. Heyes and B. Smallfield, Rome.
12. Hlebowicz J, Darwiche G, Björgell O, Almér LO. Effect of cinnamon on postprandial blood glucose, gastric emptying, and satiety in healthy subjects. Am J Clin Nutr. 2007;85(6):1552-1556. doi:10.1093/ajcn/85.6.1552
13. Jaffee, S. (2005). Delivering and taking the heat: Indian spices and evolving product and process standards. The World Bank.

INDEX

B

C

H

I

L

M

NUTRITION AND THE DEVELOPING BRAIN

Editor

Rubem Dalca

scitus academics

Nutrition and the Developing Brain

Edited by **Rubem Dalca**

ISBN: 978-1-68117-722-9

Library of Congress Control Number: 2016958850

SCITUS Academics LLC,

616, Corporate Way, Suite 2, 4766,

Valley Cottage, NY 10989

www.scitusacademics.com

Notice

Printed in United States of America on Acid Free Paper

PREFACE

The brain is a specific tissue in which functionality depends on the generation and conduction of electrical potentials over long axonal components of cell bodies and the synaptic gaps between these cell bodies. Nutrients and growth factors regulate the development of the brain during fetal and early postnatal life. The rapidly developing brain is much more susceptible to nutrient inadequacy but demonstrates its highest plasticity as well. All nutrients are important for the growth and development of neuronal cells, but some appear to have greater effects in late fetal and neonatal times. Protein, choline, vitamin A, selenium, iodine, copper, zinc, iron, certain fats, energy and folate are included.

This book, Nutrition and the Developing Brain, is intended to examine evidence from human and animal research, emphasizing the influence of p articular nutrients on cognitive development and brain function. The effects of under-nutrition on the brain of young children and infants can be harmful and persistent. All life processes are regulated by the influence and ultimately interplay of nurturing and biological factors. Brain growth and development, as well as their practical results, behavior, are no different. In embryonic, fetal and early postnatal life, genetic determinants specify the destiny of neuronal progenitors and their migration to the brain regions. These genetic determinants also modulate the transmission of synaptic signals and help establish and maintain the main nervous system. Green determinants play an equally important role in shaping the neural configuration through postnatal synaptic blooming and cutting, which incorporates ongoing experiences in the development of the human brain's synaptic architecture. Nutrition is an environmental factor because access to information from the environment is represented by it, but contrary to many other environmental resources such as medical care, experience or education, nutrition can directly modify the gene structure and mediate the expression of genetics by providing specific molecules that allow genes to exercise their potential or possibly targeted effects on brain.

TABLE OF CONTENTS

1

DHA EFFECTS IN BRAIN DEVELOPMENT AND FUNCTION

Lotte Lauritzen [1,*], Paolo Brambilla [2,3], Alessandra Mazzocchi [4], Laurine B. S. Harsløf [1], Valentina Ciappolino [2] and Carlo Agostoni [4]

[1]Department of Nutrition Exercise and Sports, University of Copenhagen, Rolighedsvej 26, 1958 Frederiksberg C, Denmark

[2]Psychiatric Clinic, Department of Neurosciences and Mental Health, Fondazione IRCCS Ospedale Cà Granda-Ospedale Maggiore Policlinico, University of Milan, 20121 Milan, Italy

[3]Department of Psychiatry and Behavioural Neurosciences, University of Texas at Houston, 2800 South Macgregor Way, Houston, TX 77021, USA

[4]Pediatric Clinic, Fondazione IRCCS Ospedale Cà Granda-Ospedale Maggiore Policlinico, Department of Clinical Sciences and Community Health, University of Milan, 20121 Milan, Italy

ABSTRACT

Docosahexaenoic acid (DHA) is a structural constituent of membranes specifically in the central nervous system. Its accumulation in the fetal brain takes place mainly during the last trimester of pregnancy and continues at very high rates up to the end of the second year of life. Since the endogenous formation of DHA seems to be relatively low, DHA intake may contribute to optimal conditions for brain development. We performed a narrative review on research on the associations between DHA levels and brain development and function throughout the lifespan. Data from cell and animal studies justify the indication of DHA in relation to brain function for neuronal cell growth and differentiation as well as in relation to neuronal signaling. Most data from human studies concern the contribution of DHA to optimal visual acuity development. Accumulating data indicate that DHA may have effects on the brain in infancy, and recent studies indicate that the effect of DHA may depend on gender and genotype of genes involved in the endogenous synthesis of DHA. While DHA levels may affect early development, potential effects are also increasingly recognized during childhood and adult life, suggesting a role of DHA in cognitive decline and in relation to major psychiatric disorders.

Keywords: docosahexaenoic acid; brain development; desaturases; psychiatric disorders

1. INTRODUCTION

Long chain polyunsaturated fatty acid (LC-PUFA), including docosahexaenoic acid (DHA) and arachidonic acid (AA), are incorporated into membrane phospholipids and, apart from their structural role in these membranes, they also act as precursors of autocoid signaling molecules (e.g., docosanoids) and as potent activators of a number of gene transcription factors (e.g., peroxisome proliferator activated receptors). The essentiality of n-3 LC-PUFA is generally mainly contributed to the incorporation of DHA in uniquely high levels in the central nervous system—although DHA is incorporated in most other tissues where it may also have important functional effects.

Overall, membrane PUFA composition (the principal components of which are linoleic acid (LA), AA and DHA) seems to be more responsive to DHA in the diet than to intake of LA and AA [1]. Animal studies have demonstrated that an increase in dietary α-linolenic acid (ALA) is almost completely reflected in membrane n-3/n-6 PUFA-ratios at LA/ALA intakes of <10, whereas the dietary balance between ALA and LA has little influence at higher ALA intakes, and a similar biphasic response is also seen in diets that contain LC-PUFA [2]. These results show a high sensitivity of tissue membranes to dietary variations in the PUFA-supply within the normal range, strongly favoring incorporation of n-3 LC-PUFA over LA and AA. In the case of a dietary deficiency of n-3 PUFA, there is a trend for DHA to be replaced with the nearest n-6 PUFA equivalents, whereas few changes are seen for the reciprocal lack of dietary of n-6 PUFA [3,4]. Thus, n-3 PUFA seem to be the main determinant of membrane PUFA composition and unsaturation. Membrane DHA incorporation in different tissues, e.g., erythrocytes (RBC), has been shown to depend on diet, mainly fish intake and in infants also breastfeeding, but is also to some extent supported by DHA formed endogenously by desaturation and elongation of ALA. This conversion is limited by the delta-6 desaturase enzymatic step, which generally has a low efficiency, but the rate conversion has been shown to be affected by genetic setup in the fatty acid desaturase (*FADS*) gene cluster and to vary depending on age and circulating levels of sex hormones.

The present paper will give an update of the current literature and try to answer the following questions: (1) Does the high rate of DHA accumulation in the brain have any functional importance? (2) If yes, is the endogenous synthesis of DHA high enough to support optimal functional levels of DHA in the brain? Finally, we will also address whether DHA may contribute to normal brain functioning later in life.

2. BRAIN DHA ACCUMULATION DURING DEVELOPMENT DEPENDING ON DIET

The accumulation of DHA in the brain takes place during the brain growth spurt in the intrauterine and neonatal period up to two years of age and the high levels of DHA in the brain are maintained throughout life [5]. Due to the lack of *de novo* PUFA synthesis, the rate of membrane DHA incorporation in early life—in the brain as well as in other tissues—depends on maternal transfer, dietary supply (*i.e.*, breastfeeding) and endogenous LC-PUFA production. The DHA accumulation in the brain during the third trimester of pregnancy is substantially higher (in % of fatty acids (FA%)) than the overall body deposition rates, whereas brain incorporation of AA is more in line with that which occurs in other tissues [6]. Fetal LC-PUFAs accumulation occurs mainly during the last trimester, in which weight increase becomes more rapid and growth is accompanied by a deposition of fat tissue, which begins around the 30th week of gestation [7]. Fetal fat tissues contain relatively low levels of DHA and AA [8,9] compared to the large relative amounts of LC-PUFAs that are deposited in the brain [8,10]. However, the absolute amount of DHA in fetal adipose tissue exceeds that in the brain [7]. Based on *post-mortem* studies it has been calculated that whole-body DHA accretion during the third trimester amounts to around 50 mg/day while the accretion of AA is approximately twice as high (100 mg/day) [8]. It has been estimated that this fetal LC-PUFA accumulation is supported by a supply of approximately 50 mg/(kg × day) of n-3 LC-PUFA and 400 mg/(kg × day) of n-6 LC-PUFA [6].

The intrauterine PUFA supply occurs via transfer of non-esterified PUFA mainly derived from the maternal circulation across the placenta [10,11]. The overall fat concentration in maternal plasma increases throughout pregnancy [7], and placental fat transport is driven by a concentration gradient as the fetus has substantially lower fat concentrations [12], including the concentration of DHA and AA [7,13]. The relative proportion of DHA and AA is, however, consistently higher in circulating lipids of the neonate [14], whereas the concentrations of LA and ALA differ much less from that in the maternal blood [15,16], indicating a preferential transfer of LC-PUFA. The exact mechanisms involved in placental PUFA transfer remain unclear, but is generally considered to involve proteins with some specificity for LC-PUFA, especially DHA, over PUFAs with shorter chain length [17,18,19]. Additionally, DHA has been shown to be incorporated into triacylglycerol in human placental cells, whereas AA is primarily esterified in phospholipids [20,21], and this differential esterification may contribute to the preferential transport of DHA and accumulation of AA in the placenta itself. AA has been shown to be taken up from the maternal blood by the placenta at higher rates than DHA, while DHA accumulates in the fetal blood stream at a three-fold higher rate than AA [20]. Although other interpretations are plausible, the specificity for placental transfer of DHA over AA could be interpreted as a specific retention of AA on the maternal side possibly for prostaglandin production in relation to the initiation of delivery. Maternal dietary n-3 LC-PUFA has a slight gestation prolonging effect, which may be explained by a dampening of the AA-derived eicosanoid response [22], which results in an increase birth weight and intrauterine LC-PUFA accretion. In infants born preterm the progressive accumulation of LC-PUFA in fetal tissues is truncated at the end of pregnancy and accumulation is also strongly limited in growth-retarded fetuses [23].

Post-natal accumulation of LC-PUFA in infant tissues is supported by maternal transfer of PUFA through breastmilk, and blood levels of LC-PUFA in breast-fed infants remain higher than maternal levels for some time postnatally [24,25]. In neonate baboons, dietary DHA has been shown to consistently support greater brain DHA incorporation and maintenance of cortex DHA concentration, while brain AA is unaffected by dietary supply and decreases with age [26]. Moreover, brain autopsies from human infants have shown an around 25% higher mean FA% of DHA in cortical phospholipid of breast-fed (9.7%) compared to age-matched formula-fed infants (7.6%) [27]. The overall percentage of LC-PUFA was maintained in formula-fed infants by a compensatory increase in the incorporation of n-6 LC-PUFA, which however was incomplete in formula-fed preterm infants with the lowest concentration of cortical DHA, where an increase in the n-9 series PUFA was also detected [27]. A second autopsy study also showed an increase in cortex DHA with age in breast-fed but not in formula-fed infants, whereas the percentage of AA in the brain increased with age irrespective of diet [28] just as in the infant baboons. Similarly, the RBC DHA content of breast-fed infants has been found to be higher than that of formula-fed infants [28]. Breastmilk has been shown to be a main contributor to the DHA content in infant RBC [29], and infant RBC DHA has been shown to be associated with maternal n-3 LC-PUFA intake and RBC DHA status during lactation [30]. RBC DHA decrease after infancy as complementary feeding usually supplies less DHA [31]. The intake of n-3 LC-PUFA has been shown to be low in a number of studies in children [32] and European children have been shown to have whole blood n-3 LC-PUFA levels consistently below 2.5 FA% between 3 and 8 years of age [33].

The phenomenon of increasing LC-PUFA in fetal and infant blood and tissues relative to that of their mother has been described as "bio-magnification" [34], but could also be interpreted as a natural consequence of a dual liver system *i.e.*, the combined PUFA metabolism and conversion of LA and ALA to AA and DHA in both the mother and the fetus/infant. Both term and preterm infants have been shown to convert stable isotope labeled LA and ALA to AA and DHA, respectively [35,36,37], and the synthesis has been shown to decrease with post-conceptional age [38]. The desaturase capacity has been estimated to be in the order of 40 mg/(kg × day) of AA and 13 mg/(kg × day) of DHA in neonates born in the 32nd week of gestation, but to decrease to around 14 and 3 mg/(kg × day) at 1 month past expected term [39]. This synthetic rate may still provide a substantial contribution to fulfill infant needs, which, based on maintenance of plasma DHA homeostasis, have been estimated to be around 5 mg/(kg × day) of DHA [40]. However, this does not exclude that exogenous sources of DHA are needed in the diet to fulfill the requirements of the growing infant.

3. EFFECTS OF FADS POLYMORPHISMS ON LC-PUFA LEVELS

Overall, data suggest that n-6 PUFAs in breastmilk, plasma and RBC membranes across all ages are more affected by single nucleotide polymorphisms (SNPs) in the *FADS* gene cluster than n-3 PUFAs, typically with an increase in LA and a decrease in AA levels in minor allele carriers [29,41,42,43,44,45]. Minor allele homozygotes of various *FADS* SNPs have also been found to have lower blood (RBC and plasma) levels of AA and higher levels of LA and ALA during pregnancy [41,42]. *FADS* polymorphisms have been estimated to explain as much as 29% of the variation in serum AA contents in adults, in whom serum DHA concentrations are determined primarily by the dietary supply of preformed DHA [43]. Colostrum AA and DHA levels have been found to be decreased in minor allele carriers of a number of *FADS* SNPs [46], but studies in mature breastmilk have shown that the concentration of AA is influenced to a larger extent than that of DHA [41,47,48]. Findings in plasma from both mothers and neonates have shown strong inverse associations between the minor allele for two *FADS* SNPs and the concentrations of DHA and eicosapentaenoic acid (EPA) as well as AA in the newborn infants, thus confirming that synthesis of DHA provides a relevant contribution to status [49]. Curiously, a study of 2000 cord blood samples found that minor allele *FADS* SNPs in the mother gave rise to increased levels of n-6 PUFA before the delta-5 desaturation step (LA and di-homo-γ-linoleic acid), whereas minor allele SNPs in the child resulted in decreased levels of AA and other n-6 LC-PUFA beyond this point in the metabolic pathway [50]. More data on the biochemical effects of *FADS* polymorphisms are needed to derive a biologically plausible interpretation of their potential functional effects. Furthermore, both AA and DHA needs to be considered together since apart from the main determinants of their levels, either endogenous or exogenous, their balance may be critical for the functional outcomes in infancy and beyond.

We have recently found that some *FADS* polymorphisms may substantially contribute to RBC DHA levels in late infancy (to the same extent as breastfeeding) [29]. Some SNP minor alleles (rs1535 and rs3834458) were even found to dose-dependently up-regulate DHA status [29], whereas minor alleles of all the investigated SNPs lowered AA in a consistent way [51]. Interestingly, identical analyses did not reveal any effect of these SNPs on RBC DHA at 3 years of age [29], which could be explained by increased residual variation in the model due to a more diverse fish intake or could be interpreted as a decline in the endogenous DHA biosynthesis, consistent with other findings [39]. Furthermore, a longitudinal study of serum phospholipid fatty acid composition at 2 and 6 years of age in 331 children found higher tracking in n-3 LC-PUFA levels in children who were major allele carriers [52]. Instead tracking of n-6 LC-PUFA was lower in major allele homozygotes of various *FADS* SNPs compared to tracking in carriers of at least onc minor allele [52]. More longitudinal outcome data may suggest plausible biological interpretations. However, although DHA may mainly be determined by variation in intake, mainly of preformed DHA, the genetic patterns also appear to be of relevance for tissue DHA levels in the perinatal phases, although probably less later in life as the rate of endogenous synthesis declines, thus increasing the importance of exogenous DHA.

4. DIETARY DHA AND POSTNATAL DEVELOPMENT

The majority of the randomized controlled trials investigating the effect of dietary LC-PUFA supplementation in term infants have added both DHA and AA, and only few have investigated the effect of varying DHA intakes at a constant intake of AA. With respect to the functional effects of LC-PUFA supplementation in infancy, the most accepted developmental effect is an increased rate of visual acuity development [53]. This effect seems to be explained solely by DHA, as a meta-regression analysis found that variability in the effects on visual acuity between studies was explained by the dose of DHA [54]. However, little is known regarding the persistency of this effect on vision and the potential effects that this early visual deficit may have on cognitive development.

Overall, meta-analyses of the randomized controlled trials that have investigated the effect of LC-PUFA supplementation on neurodevelopmental outcomes throughout the first two years of life have not shown any clear benefit of LC-PUFA addition to infant formula on development of term or preterm infants [55,56,57]. However, a meta-analysis that combined all LC-PUFA formula supplementation trials in both term and preterm infants found a trend for an effect on the Bayley scale Mental Developmental Index at around 12 months of age, which were not affected by the maturity of the infant at birth [57]. This meta-analysis did not find any effect of LC-PUFA dose, although there was a trend towards an effect of the DHA dose, but no such trend for AA [57]. The studies that have supplemented the infants with DHA indirectly via n-3 LC-PUFA supplementation of their pregnant or lactating mothers, generally provide a more clean way to study effects of the early DHA supply as this has little effect on the AA supply to the infant. A meta-analysis of randomized trials that supplemented lactating mothers with n-3 LC-PUFA showed that infants of supplemented mothers had larger heads at 2 years of age [58]. Furthermore, the meta-analyses looking at the developmental effects of maternal n-3 LC-PUFA supplements in pregnancy and lactation have suggested some effects on neurodevelopment based on a few studies [58,59]. However, at the current stage, this does not provide any definite proof that an increase in the early DHA supply improves the mental development of infants.

So far, few studies have shown that the effect of perinatal n-3 LC-PUFA supplementation may be affected by the gender of the child. In two large investigations, the DINO and DOMInO trials [60,61], an increased early DHA supply was associated with different effects on cognitive outcomes in girls and boys. A gender-treatment interaction on cognitive outcomes was also observed in a small Danish trial of maternal fish oil supplementation during lactation [62], although no clear effects were observed when the children were followed up at 7 years of age [63]. The different effects of increased DHA supply on various outcomes in girls and boys all appear to counteract the normally observed gender differences in behavior. It is not clear if these effects should be interpreted as beneficial in one gender and adverse in the other or if it is due to some other effect of DHA that diminish the cultural gender differences which we have come to perceive as normal biological differences. Interestingly, in the Danish maternal fish oil supplementation trial treatment-gender interactions were found also on blood pressure at 7 years of age [64]. Blood pressure is not normally defined as cognitive outcome, but is nevertheless affected by the central nervous system in response to anxiety. As was the case with cognitive outcomes, boys and girls in the fish oil group were found to have comparable diastolic and mean arterial blood pressures, whereas girls had higher blood pressures than boys in the control group [64]. The intervention was also found to level out gender differences on energy intake and physical activity at 7 years of age [64]. Accordingly, these results indicate that early DHA intake could also have long-term health consequences, which might be mediated effects in the brain and lifestyle choices.

Many of the available studies on the effects of maternal or, more commonly, infant n-3 LC-PUFA supplementations on neurodevelopmental outcome during infancy have several limitations, which become more and more evident as our knowledge on the physiology of LC-PUFA, and DHA in particular, progresses. The vast majority of the studies, whether on cognition of other functional outcomes or if they provide the supplement during pregnancy, lactation or to the infant in various types of formula, show a great heterogeneity with respect to LC-PUFA sources, doses of DHA (and AA) and durations of interventions. It should be noted that the methodologies for primary outcome assessment as well as age of effect examination differed between trials, and the effects in the first few years of life and potential long-term effects may be quite different. For outcomes such as neurological and cognitive development, there may be a necessity to use different tests at different ages to accommodate changes in age and maturity level. However, many trials have investigated effects on numerous outcome measures, which are often internally inconsistent, or show no apparent pattern over time. In addition, studies often have low power in terms of the number of participants and sometimes also high rates of dropouts as well as lack of intention-to-treat analysis and a sufficient description of allocation concealment. Although baseline demographic characteristics are constantly reported, often baseline n-3 PUFA intake or status is not included in the characterization. This omission is critical for the interpretation, since baseline n-3 PUFA status will likely affects the response to changes in n-3 PUFA intake—both with respect to acute and persistent functional effect. Finally, as mentioned above, the effects of early n-3 LC-PUFA supply may vary in boys and girls, and this is not taken into account in the older studies. The emerging knowledge indicates that it is critical to take these aspects into account and that the variation in these aspects complicate attempts to combine data in meta-analyses to achieve conclusions with respect to the functional consequences of the addition of LC-PUFA, even beyond the single, specific effects of DHA.

5. EFFECTS OF FADS POLYMORPHISMS ON COGNITION AND NEUROBEHAVIORAL OUTCOMES

Current knowledge about the functional effects of *FADS* polymorphism is limited and although the most clear effect on PUFA metabolism as mentioned is a decrease in AA production, functional associations with*FADS* genotype cannot be interpreted as a consequence of a reduction in AA. The influence of *FADS*polymorphisms on LC-PUFA status—and specifically the observed variations between specific SNPs and specific LC-PUFA over time—introduces new variables to be considered in the evaluation of the effects of*FADS* genotype on development and health of young children.

Several studies have showed that infant *FADS* genotype, examined by use of different individual SNPs, modifies the effect of breastfeeding on IQ-like neurodevelopmental outcomes in childhood [46,65,66], while other studies did not find any significant interaction [67,68] (Figure 1).

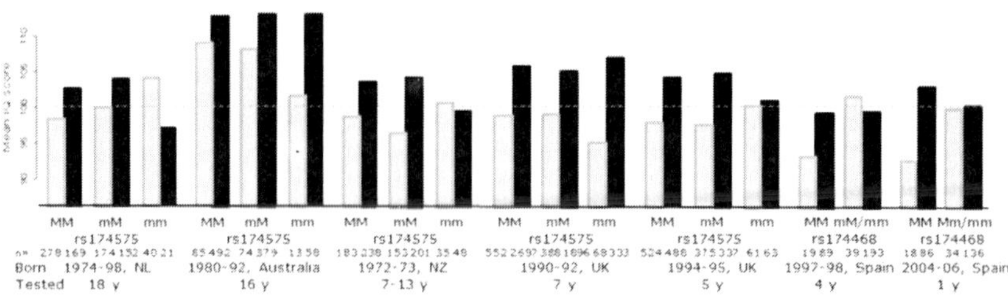

Figure 1. Results from studies examining the potential modifying effect of single nucleotide polymorphisms in the fatty acid desaturase gene cluster on the effect of breastfeeding on IQ-like neurodevelopmental outcomes in children. The figure is based on data from [46,65,66,67,68] and gives the average IQ in the SNP×feeding groups (breast-fed in black and formula-fed in light gray). The grey stippled line is the reference line for mean normal IQ.

As expected, based on the observed differences between breast-fed and formula-fed infants, all the studies have higher scores in breast-fed compared to formula-fed major allele carriers, but with no apparent differences between homozygotes and heterozygotes, which might be expected based on the additive effects of number of major alleles that is expected according to the observed effects on LC-PUFA. Thus, an interaction is dependent on a different pattern among the minor allele homozygotes (or the minor allele carriers in the Spanish study in which these were pooled with the heterozygotes). In the studies that found an interaction this is based on an equal "IQ" in breast-fed and formula-fed in the two cohorts in the Caspi study [66] and the two Spanish cohorts [46], whereas the largest of the studies found an even bigger difference between breast-fed and formula-fed among the minor allele homozygotes [65]. However, in all of the studies there were only few formula-fed minor allele carriers, and thus the largest variation in this group was likely skewed because of the scores of few children were at a high risk of chance effects. The studies differ with respect to breastfeeding frequency as well as the definition of breastfeeding, which in the large UK study was defined as >1 months [65] and ever having been breast-fed in the Dutch study [67], but was not clearly defined in other studies [46,66]. The Australian study tried to examine the effect of breastfeeding duration (not apparent in the figure in which we have pooled all the breast-fed groups), but they did not have the power to judge this due to a lack of a statistical (although visual indicated) dose-response between duration of breastfeeding and IQ [68]. The Dutch study found that the effect appeared to vary—although not significantly—between different cognitive functions and testing ages [67]. Furthermore, given the variation in the year of birth of the subjects in the studies, it is also reasonable to assume that there could have been differences in the PUFA composition of the formulas and presumably also in the maternal fish intake, and lifestyle in general, and thus in the DHA content of the breastmilk of the study populations. Little is known regarding interactions between the *FADS* polymorphisms and intake of AA and DHA from breastmilk and infant formula or the dietary ratio between the precursors, LA and ALA, but it is reasonable to suspect that this might have an influence on the functional response. Due to the increasing availability of micro-invasive methods for determination of blood fatty acid status, future Mendelian randomization studies should now be able to study effects of these potential sources of heterogeneity.

Additionally, the studies on interactions between breastfeeding and IQ used different *FADS* SNPs (mainly rs174575, rs1535 and rs174468), but, as indicated by the aforementioned study from Harsløf and coworkers, they may not all down-regulate the endogenous DHA synthesis in the infants [29]. This could contribute to the observed variable associations, and it is important to consider potential disequilibrium with other SNPs in the interpretation of the results from the *FADS* SNP studies. Interestingly, Steer *et al.* reported opposing effects of rs174574 and rs3834458 in the modulation of the association between breastfeeding and IQ [69]. Opposing effects of rs1535 and rs174448 have also been observed in problem solving and communication skills assessed by the Ages and Stages Questionnaire (ASQ) in a recent study among 3 year-olds [51]. None of the *FADS* SNP-breastfeeding interaction studies have considered whether the effect might differ between boys and girls. As in the DINO trials [60], the Mendelian randomization of *FADS* SNPs *versus* 3-year ASQ outcomes found that the effect of DHA increasing SNPs appeared to be negative in girls and positive in boys [51]. Due to the previously mentioned opposing effects of these SNPs on DHA in early life and the lack of opposing effects on AA plus the lack of association between *FADS* SNPs and DHA status at 3 years of age, these results might be interpreted as proof of a *programming effect* specifically of early DHA dietary intakes. However, the lack of effect of the *FADS* SNPs on DHA status later in life could be due to a blurring effect of a more diverse dietary intake of DHA from fish. Therefore, it is not possible to rule out that DHA supply contrary to the current hypothesis of an early window of vulnerability might have an effect on brain function at all stages of life.

6. NEUROBEHAVIOURAL OUTCOMES IN OLDER CHILDREN

Brain DHA accretion continues into childhood, and although the accretion rate declines, the incorporation of DHA is still high at least during the preschool years. Once high levels of DHA are achieved in the brain these are maintained during later life, and this presumably also depends on an optimal dietary supply, as dietary intake of DHA from fish in adults has been shown to be the dominant determinant of DHA levels in various lipid pools [70]. However, to our knowledge no studies have examined the dietary requirements in order to achieve optimal brain DHA maintenance. Few studies have investigated the effect of *FADS* SNPs or n-3 LC-PUFA supplementation on cognitive development, emotions and behavior in toddlers and later in childhood or even in healthy adults.

A single study pooling data from three trials that randomized to LC-PUFA formulas immediately after birth or after breastfeeding for 6 weeks or 4–6 months, respectively and continued supplementation throughout the first year of life, found significant beneficial effects on problem solving at 9 months of age only in the two studies that started intervention early [71]. However, one study that examined the effects of DHA-enriched baby food also found an apparent improvement of cognitive outcomes [72]. Furthermore, a trial that provided a teaspoonful of cod liver oil (free of vitamin A and D) from 9 to 12 month of age found an increase in voluntary attention in a free play test after the intervention, especially in boys, compared with un-supplemented children [73].

Results from studies in schoolchildren in low-income countries have shown relatively convincing cognitive effects of fish oil supplementation. The effects have been shown to be stronger in children with low socioeconomic status or malnutrition-related health problems and a low consumption of fish and very little n-3 PUFA [74,75]. However, no overall cognitive effects were found after fish oil supplementation of 6–11 year-old South African children with poor iron and n-3 LC-PUFA status [75], but paradoxically an adverse effect of fish oil was observed on memory mainly in girls and specifically those with iron deficiency anemia. Little research has been performed on the effects of n-3 LC-PUFA on brain functions in school-aged children from high-income countries. One functional magnetic resonance imaging study showed that DHA supplementation was associated with increased activation of the prefrontal cortex and better reaction time during sustained attention in healthy 8–10 year-old boys [76]. In a cross-over intervention trial with more than 800 schoolchildren we have recently found that healthy school meals rich in fish improve school performance [77]. Some observational studies have also observed a positive association between n-3 LC-PUFA intake and cognitive performance. A study of 4000 American children found that the association between n-3 LC-PUFA intake and cognitive performance was stronger in girls than in boys [78]—again, an example of gender-related nutrition. Four randomized trials have supplied schoolchildren from high-income countries specifically with n-3 LC-PUFA [79,80,81,82]. Three of these studies found some beneficial effects on cognition or school performance of 0.4–1 g/day of n-3 LC-PUFA, while the study that did not find any effect supplied only around 0.2 g/day [79]. This dose-response effect is however not always consistent, as a three-armed study in 90 British 10–12 year-old children found a beneficial effect of 0.4 g/day of DHA on word recognition, but poorer performance in children who had 1 g/day [80].

In the last mentioned study, all the DHA supplemented children had a more relaxed mood compared to controls [80], which is consistent with another trial that found an apparent effect on mood, *i.e.*, a reduction in impulsivity and anti-social behavior, in 450 healthy 8–10 year-old children supplemented with fish oil *versus*olive oil [81]. Similar behavioral effects were also observed in one of the South African studies, which showed a decrease in physical activity during school hours, less oppositional behavior, inattention and lower scores on a rating scale of traits of attention-deficit hyperactivity disorder (ADHD) after fish oil supplementation [82]. Comparable behavioral effects have also been indicated, although not firmly proven, in children with ADHD [83] and in addition, one study has found an association between *FADS* SNPs and the development of ADHD, specifically in the context of prenatal alcohol exposure [84]. It is difficult to draw any firm conclusions based on the results of these trials and observational studies in schoolchildren due to differences in dose, duration and most of all the tested outcomes. Furthermore, the studies on behavioral conditions may be biased due to methodological flaws such as limited sample size and the large number of neurological tests that were performed in most of the studies (out of which only a few showed significant effects). More well-conducted studies, adjusted for multiple test administrations, are therefore needed in order to provide more convincing evidence for an effect of n-3 LC-PUFA intake on cognitive, behavioral and emotional effects in children.

So far, the effect of gender has not been given much attention in intervention trials with n-3 LC-PUFA in preschool and school-aged children. However, as was the case in the studies on the effects of DHA in the perinatal period, a gender-treatment effect has been observed on mean arterial blood pressure after fish oil supplementation from 9 to 18 month of age in healthy Danish infants, which just as in the previously mentioned maternal fish oil supplementation study was mostly affected in boys [85]. In this case, blood pressure was reduced in the boys, which however was still counteracting the observed gender difference in the control group, resulting in an almost similar mean arterial blood pressure in the girls and boys of the fish oil supplemented group [85]. A similar gender-equalizing effect was observed on the systolic blood pressure later in infancy in a study that compared fish oil *versus* no supplement during the complementary feeding period [86]. In that study, the observed changes in systolic blood pressure were found to correlate with the previously mentioned changes in free play attention [73], which could indicate a common emotional component. Furthermore, a recent randomized controlled cross-over trial in young adults also observed a gender-specific effect of fish oil supplementation on the sensation of appetite that abolished gender differences observed after a three week intervention in the soy oil control period [87].

7. NEUROBEHAVIOURAL OUTCOMES BEYOND CHILDHOOD

Only a few studies have examined if fish oil supplementation can affect brain functions in healthy young adults, but some studies indicate that DHA may be important for cognition and behavior during late adulthood. DHA supplementation improved memory in healthy, young adults whose habitual diets were low in DHA, and the response was still modulated by sex [88] suggesting consistence with the effects found in late infancy with the achievement of gross motor milestones [89]. An observational study conducted in 6158 individuals of >65 years found that high fish consumption, but not dietary n-3 LC-PUFA intake, had a protective effect on cognitive decline [90,91]. A systematic review and meta-analysis from 2006 gathered all available evidence from observational, preclinical and clinical studies to assess the effects of n-3 LC-PUFA on cognitive protection [92]. Four of the trials have shown a protective effect of n-3 LC-PUFA only among those with mild cognitive impairment conditions [93]. In another trial with 485 subjects with mild memory complaints, an improvement of memory was demonstrated after 0.9 g/day of DHA for 24 weeks [94]. A recent meta-analysis of all randomized trials that have investigated the effect of fish oil on cognitive decline also indicated a potential beneficial effect, but only in trials that had supplied >1 g/day of DHA in subjects who at the beginning of the trail exhibited some signs of cognitive decline [90]. There are nine separate observational studies that have suggest a possible link between increased fish consumption and reduced risk of Alzheimer's disease [95,96]. Furthermore, analysis of human cadaver brains has shown that people with Alzheimer disease have less DHA in their frontal lobe and hippocampus compared with unaffected individuals [97]. In addition, studies in mice provide support for the protective role of n-3 LC-PUFA, showing that a dietary intake of DHA induces an increase in DHA levels in the hippocampus with subsequent improvement of memory performances [98].

DHA has also been suggested to be effective in major psychiatric disorders. Most of the conducted studies have used n-3 LC-PUFA levels in RBC membranes as a measure of exposure, due to the objectivity of this measure and their high correlation with habitual dietary intake [99] in addition to their presumed reflection of brain LC-PUFA levels. Such studies have shown a significant correlation between DHA deficits and schizophrenia [100]. Life style in schizophrenia is characterized by heavy smoking, drinking, high-caloric diets, low physical activity and use of drugs that cause oxidative stress in the body. However, a recent study found no reduction of either DHA or AA in large groups of un-medicated Indian and Malaysian patients suffering from schizophrenia [101]. There is a tendency for RBC membrane levels of DHA and AA to diminish during storage, and this may happen abnormally rapidly in schizophrenia [102], possibly because of an increased oxidative stress [103]. Interestingly, several studies reported a better outcome in psychotic patients supplemented with n-3 LC-PUFA, either EPA or DHA [104].

Accumulating evidence also suggests that n-3 LC-PUFA supplementation may be efficacious for the treatment of positive and negative symptoms in patients with schizophrenia or at ultra-high risk for psychosis [105]. There is also some evidence that n-3 LC-PUFA may be relevant in relation to the pathophysiology of depression [106]. Cross-national studies indicate that higher intake of fish/seafood is correlated with lower lifetime prevalence rates of unipolar and bipolar depression [107]. In fact, depression may present with an increased production of pro-inflammatory cytokines and elevations in plasma homocysteine levels [108], and n-3 LC-PUFA have in randomized controlled trials been shown to be able to reduce both [109]. Thus, it can be speculated that n-3 PUFAs produce a positive effect on mood, partly because of the high brain content of DHA and its involvement in neurogenesis and neuroplasticity and partly due to their anti-inflammatory properties [110] as well as their effect on carbon metabolism, which is known to be of importance in relation to the metabolism of mono-aminergic neurotransmitters [111]. Some epidemiological studies have in the same way found that lower n-3 LC-PUFA intake is linked to an increased risk for emerging depressive symptoms [112]. Therefore, higher habitual dietary n-3 LC-PUFA intake may be protective against mood swings or even ultimately prevent mood dysregulation [113]. There is however a need for large well-performed randomized controlled trials in this area in order to confirm such effects.

8. CONCLUSIONS

The effects of DHA on brain and cognitive development have been extensively investigated in the last years. Its functional effects have been progressively, but not entirely, separated from those of AA. Clinical trials on maternal and infant dietary intakes are not entirely clear and consistent, but seem to indicate a complex interaction between the genotype pattern of *FADS*, gender, dietary intakes and lifestyle. For these reasons it is difficult to disentangle the effects of dietary DHA from the results of the randomized supplementation trials. In future studies an appropriate sample size should be calculated in order to adjust for the different variables. Mendelian trials provide a new tool to investigate the effects of LC-PUFA on cognitive development, but the interpretation of results from such trials requires an improved understanding of the biochemical effects of individual *FADS* SNPs and also needs to consider the potential differences between boys and girls.

Thus, our two questions, (1) Does the high rate of DHA accumulation in the brain have any functional importance? (2) If yes, is the endogenous synthesis of DHA high enough to support optimal functional levels of DHA in the brain? Regarding the first, there is clear evidence that DHA contributes to the visual development of infants, as also concluded by EFSA [53], but the associations with cognitive development are still not clearly interpreted [114,115], and one of the main problems could be that the effects differ between boys and girls, which needs to be considered in future trials. Due to the proposed early window of vulnerability, so far few studies have focused on the potential effects of n-3 LC-PUFA intake on cognitive and behavioral outcomes in children and young adults, but available studies indicate that the hypothesis might be worth challenging. Finally, there is preliminary evidence that DHA may ameliorate cognitive decline and affect behavioral symptoms in major neuropsychiatric disorders such as dementia, schizophrenia and depression. There is an extremely poor availability of trials on the effect of DHA supplementations that have investigated the changes in the fatty acid status as a function of the *FADS*polymorphisms. Most evidence indicates that the DHA accumulation is mainly affected by dietary intake, specifically of preformed DHA. However new studies indicate that the genetic make-up in the *FADS* gene cluster may contribute substantially to the current understanding, but that the effects may be SNP-specific and may even vary with age, or at least are most evident in the perinatal period, where the endogenous synthesis of LC-PUFA is upregulated and diet may be more easily controlled for, especially during lactation (or formula feeding).

AUTHOR CONTRIBUTION

All the Authors contributed significantly to the paper. Lotte Lauritzen drafted the two sections about*FADS* and Neurobehavioural Outcomes in Older Children with assistance from Laurine BS Harsløf, the section on Dietary DHA and Postnatal Development and the background section on Brain DHA Accumulation during Development Depending on Diet was drafted by Lotte Lauritzen and Carlo Agostoni, and the section on Neurobehavioural Outcomes beyond Childhood was drafted by Paolo Brambilla and Valentina Ciappolino. Alessandra Mazzocchi and Valentina Ciappolino performed the literature search, commented on the manuscript, proof read and ordered the references.

CONFLICTS OF INTEREST

The authors declare no conflict of interest. PB was partially supported by grants from the Italian Ministry of Health (GR-2010-2317873).

REFERENCES

1. Vlaardingerbroek, H.; Hornstra, G.; de Koning, T.J.; Smeitink, J.A.; Bakker, H.D.; de Klerk, H.B.; Rubio-Gozalbo, M.E. Essential polyunsaturated fatty acids in plasma and erythrocytes of children with inborn errors of amino acid metabolism. Mol. Genet. Metab. 2006, 88, 159–165.
2. Abbott, S.K.; Else, P.L.; Atkins, T.A.; Hulbert, A.J. Fatty acid composition of membrane bilayers: Importance of diet polyunsaturated fat balance. Biochim. Biophys. Acta 2012, 1818, 1309–1317.
3. Neuringer, M.; Connor, W.E.; Lin, D.S.; Barstad, L.; Luck, S. Biochemical and functional effects of prenatal and postnatal omega 3 fatty acid deficiency on retina and brain in rhesus monkeys. Proc. Natl. Acad. Sci. USA 1986, 83, 4021–4025.
4. Carrie, I.; Clement, M.; de Javel, D.; Frances, H.; Bourre, J.M. Specific phospholipid fatty acid composition of brain regions in mice: Effects of n-3 polyunsaturated fatty acid deficiency and phospholipid supplementation. J. Lipid Res. 2000, 41, 465–472.
5. Carver, J.D.; Benford, V.J.; Han, B.; Cantor, A.B. The relationship between age and the fatty acid composition of cerebral cortex and erythrocytes in human subjects. Brain Res. Bull. 2001, 56, 79–85.

6. Cetin, I.; Alvino, G.; Cardellicchio, M. Long chain fatty acids and dietary fats in fetal nutrition. J. Physiol. 2009, 587, 3441–3451.

7. Haggarty, P. Effect of placental function on fatty acid requirements during pregnancy. Eur. J. Clin. Nutr. 2004, 58, 1559–1570.

8. Kuipers, R.S.; Luxwolda, M.F.; Offringa, P.J.; Boersma, E.R.; Dijck-Brouwer, D.A.J.; Muskiet, F.A.J. Fetal intrauterine whole body linoleic, arachidonic and docosahexaenoic acid contents and accretion rates. Prostaglandins Leukot. Essent. Fat. Acids 2012, 86, 13–20.

9. Kuipers, R.S.; Luxwolda, M.F.; Offringa, P.J.; Boersma, E.R.; Dijck-Brouwer, D.A.J.; Muskiet, F.A.J. Gestational age dependent changes of the fetal brain, liver and adipose tissue fatty acid compositions in a population with high fish intakes. Prostaglandins Leukot. Essent. Fat. Acids 2012, 86, 189–199.

10. Clandinin, M.T.; Chappell, J.E.; Leong, S.; Heim, T.; Swyer, P.R.; Chance, G.W. Intrauterine fatty acid accretion rates in human brain: Implications for fatty acid requirements. Early Hum. Dev. 1980, 4, 121–129.

11. Herrera, E. Implications of dietary fatty acids during pregnancy on placental, fetal and postnatal development—A review. Placenta 2002, 23, S9–S19.

12. Campbell, F.M.; Gordon, M.J.; Dutta-Roy, A.K. Placental membrane fatty acid-binding protein preferentially binds arachidonic and docosahexaenoic acids. Life Sci. 1998, 63, 235–240.

13. Haggarty, P. Fatty acid supply to the human fetus. Annu. Rev. Nutr. 2010, 30, 237–255.

14. Montes, R.; Chisaguano, A.M.; Castellote, A.I.; Morales, E.; Sunyer, J.; Lopez-Sabater, M.C. Fatty acid composition of maternal and umbilical cord plasma and early childhood atopic eczema in a Spanish cohort. Eur. J. Clin. Nutr. 2013, 67, 658–663.

15. Agostoni, C.; Galli, C.; Riva, E.; Rise, P.; Colombo, C.; Giovannini, M.; Marangoni, F. Whole blood fatty acid composition at birth: From the maternal compartment to the infant. Clin. Nutr. 2011, 30, 503–505.

16. Oliveira, O.R.C.; Santana, M.G.; Santos, F.S.; Conceicao, F.D.; Sardinha, F.L.C.; Veiga, G.V.; do Carmo, M.G.T. Composition of fatty acids in the maternal and umbilical cord plasma of adolescent and adult mothers: Relationship with anthropometric parameters of newborn. Lipids Health Dis. 2012, 11, 157.

17. Gil-Sanchez, A.; Larque, E.; Demmelmair, H.; Acien, M.I.; Faber, F.L.; Parrilla, J.J.; Koletzko, B. Maternal-fetal in vivo transfer of

(13C)docosahexaenoic and other fatty acids across the human placenta 12 h after maternal oral intake. Am. J. Clin. Nutr. 2010, 92, 115–122.

18. Campbell, F.M.; Dutta-Roy, A.K. Plasma-membrane fatty acid binding protein (FABPpm) is exclusively located in the maternal facing membranes of the human placenta. FEBS Lett. 1995, 375, 227–230.

19. Dutta-Roy, A.K. Transfer of long-chain polyunsaturated fatty acids across the human placenta. Prenat. Neonat. Med. 1997, 2, 101–107.

20. Crabtree, J.T.; Gordon, M.J.; Campbell, F.M.; Dutta-Roy, A.K. Differential distribution and metabolism of arachidonic acid and docosahexaenoic acid by human placental choriocarcinoma (BeWo) cells. Mol. Cell. Biochem. 1998, 185, 191–198.

21. Campbell, F.M.; Clohessy, A.M.; Gordon, M.J.; Page, K.R.; Dutta-Roy, A.K. Uptake of long chain fatty acids by human placental choriocarcinoma (BeWo) cells: Role of plasma membrane fatty acid binding protein. J. Lipid Res. 1997, 38, 2558–2568.

22. Allen, K.G.D.; Harris, M.A. The role of n-3 fatty acids in gestation and parturition. Exp. Biol. Med. 2001,226, 498–506.

23. Cetin, I.; Giovannini, N.; Alvino, G.; Agostoni, C.; Riva, E.; Giovannini, M.; Pardi, G. Intrauterine growth restriction is associated with changes in polyunsaturated fatty acid fetal-maternal relationships. Pediatr. Res. 2002, 52, 750–755.

24. Jørgensen, M.H.; Nielsen, P.K.; Michaelsen, K.F.; Lund, P.; Lauritzen, L. The composition of polyunsaturated fatty acids in erythrocytes of lactating mothers and their infants. Mater. Child Nutr.2006, 2, 29–39.

25. Koletzko, B.; Schmidt, E.; Bremer, H.J.; Haug, M.; Harzer, G. Effects of dietary long-chain polyunsaturated fatty acids on the essential fatty acid status of premature infants. Eur. J. Pediatr. 1989,148, 669–675.

26. Hsieh, A.T.; Brenna, J.T. Dietary docosahexaenoic acid but not arachidonic acid influences central nervous system fatty acid status in baboon neonates. Prostaglandins Leukot. Essent. Fat. Acids 2009, 81, 105–110.

27. Farquharson, J.; Cockburn, F.; Patrick, W.A.; Jamieson, E.C.; Logan, R.W. Infant cerebral cortex phospholipid fatty-acid composition and diet. Lancet 1992, 340, 810–813.

28. Makrides, M.; Neumann, M.A.; Byard, R.W.; Simmer, K.; Gibson, R.A. Fatty acid composition of brain, retina, and erythrocytes in breast- and formula-fed infants. Am. J. Clin. Nutr. 1994, 60, 189–194.

29. Harsløf, L.B.S.; Larsen, L.H.; Ritz, C.; Hellgren, L.I.; Michaelsen, K.F.; Vogel, U.; Lauritzen, L. FADSgenotype and diet are important determinants of DHA status: A cross-sectional study in Danish infants. Am. J. Clin. Nutr. 2013, 97, 1403–1410.

30. Lauritzen, L.; Carlson, S.E. Maternal fatty acid status during pregnancy and lactation and relation to newborn and infant status. Mater. Child Nutr. 2011, 7, 41–58.

31. Rise, P.; Tragni, E.; Ghezzi, S.; Agostoni, C.; Marangoni, F.; Poli, A.; Catapano, A.L.; Siani, A.; Iacoviello, L.; Galli, C. Different patterns characterize omega-6 and omega-3 long chain polyunsaturated fatty acid levels in blood from Italian infants, children, adults and elderly.Prostaglandins Leukot. Essent. Fat. Acids 2013, 89, 215–220.

32. Rahmawaty, S.; Charlton, K.; Lyons-Wall, P.; Meyer, B.J. Dietary intake and food sources of EPA, DPA and DHA in Australian children. Lipids 2013, 48, 869–877.

33. Wolters, M.; Schlenz, H.; Foraita, R.; Galli, C.; Rise, P.; Moreno, L.A.; Molnar, D.; Russo, P.; Veidebaum, T.; Tornaritis, M.; et al. Reference values of whole-blood fatty acids by age and gender from European children aged 3–8 years. Int. J. Obes. 2014, 38, S86–S98.

34. Crawford, M.A.; Hassam, G.A.; Williams, G.; Whitehouse, W.L. Essential fatty acids and fetal brain growth. Lancet 1976, 1, 452–453.

35. Sauerwald, T.U.; Hachey, D.L.; Jensen, C.L.; Chen, H.; Anderson, R.E.; Heird, W.C. Intermediates in endogenous synthesis of C22:6 omega-3 and C20:4 omega-6 by term and preterm infants. Pediatr. Res.1997, 41, 183–187.

36. Salem, N., Jr.; Wegher, B.; Mena, P.; Uauy, R. Arachidonic and docosahexaenoic acids are biosynthesized from their 18-carbon precursors in human infants. Proc. Natl. Acad. Sci. USA 1996, 93, 49–54.

37. Szitanyi, P.; Koletzko, B.; Mydlilova, A.; Demmelmair, H. Metabolism of C-13-labeled linoleic acid in newborn infants during the first week of life. Pediatr. Res. 1999, 45, 669–673.

38. Uauy, R.D.; Mena, P.; Wegher, B.; Nieto, S.; Salem, N., Jr. Long chain polyunsaturated fatty acid formation in neonates: Effect of gestational age and intrauterine growth. Pediatr. Res. 2000, 47, 127–135.

39. Carnielli, V.P.; Simonato, M.; Verlato, G.; Luijendijk, I.; de Curtis, M.; Sauer, P.J.; Cogo, P.E. Synthesis of long-chain polyunsaturated fatty

acids in preterm newborns fed formula with long-chain polyunsaturated fatty acids. Am. J. Clin. Nutr. 2007, 86, 1323–1330.

40. Lin, Y.H.; Llanos, A.; Mena, P.; Uauy, R.; Salem, N., Jr.; Pawlosky, R.J. Compartmental analyses of H-2(5)-alpha-linolenic acid and C-13-U-eicosapentaenoic acid toward synthesis of plasma labeled 22:6 n-3 in newborn term infants. Am. J. Clin. Nutr. 2010, 92, 284–293.

41. Xie, L.; Innis, S.M. Genetic Variants of the FADS1 FADS2 gene cluster are associated with altered n-6 and n-3 essential fatty acids in plasma and erythrocyte phospholipids in women during pregnancy and in breast milk during lactation. J. Nutr. 2008, 138, 2222–2228.

42. Koletzko, B.; Lattka, E.; Zeilinger, S.; Illig, T.; Steer, C. Genetic variants of the fatty acid desaturase gene cluster predict amounts of red blood cell docosahexaenoic and other polyunsaturated fatty acids in pregnant women: Findings from the Avon Longitudinal Study of Parents and Children. Am. J. Clin. Nutr. 2011, 93, 211–219.

43. Schaeffer, L.; Gohlke, H.; Muller, M.; Heid, I.M.; Palmer, L.J.; Kompauer, I.; Demmelmair, H.; Illig, T.; Koletzko, B.; Heinrich, J. Common genetic variants of the FADS1 FADS2 gene cluster and their reconstructed haplotypes are associated with the fatty acid composition in phospholipids. Hum. Mol. Genet. 2006, 15, 1745–1756.

44. Mathias, R.A.; Vergara, C.; Gao, L.; Rafaels, N.; Hand, T.; Campbell, M.; Bickel, C.; Ivester, P.; Sergeant, S.; Barnes, K.C.; et al. FADS genetic variants and omega-6 polyunsaturated fatty acid metabolism in a homogeneous island population. J. Lipid Res. 2010, 51, 2766–2774.

45. Bokor, S.; Dumont, J.; Spinneker, A.; Gonzalez-Gross, M.; Nova, E.; Widhalm, K.; Moschonis, G.; Stehle, P.; Amouyel, P.; de Henauw, S.; et al. Single nucleotide polymorphisms in the FADS gene cluster are associated with delta-5 and delta-6 desaturase activities estimated by serum fatty acid ratios. J. Lipid Res. 2010, 51, 2325–2333.

46. Morales, E.; Bustamante, M.; Gonzalez, J.R.; Guxens, M.; Torrent, M.; Mendez, M.; Garcia-Esteban, R.; Julvez, J.; Forns, J.; Vrijheid, M.; et al. Genetic variants of the FADS gene cluster and ELOVL gene family, colostrum LCPUFA levels, breastfeeding, and child cognition. PLoS ONE 2011, 6, e17181.

47. Moltó-Puigmartí, C.; Plat, J.; Mensink, R.P.; Müller, A.; Jansen, E.; Zeegers, M.P.; Thijs, C. FADS1 FADS2 gene variants modify the association between fish intake and the docosahexaenoic acid proportions in human milk. Am. J. Clin. Nutr. 2010, 91, 1368–1376.

48. Lattka, E.; Rzehak, P.; Szabo, E.; Jakobik, V.; Weck, M.; Weyermann, M.; Grallert, H.; Rothenbacher, D.; Heinrich, J.; Brenner, H.; et al.

Genetic variants in the FADS gene cluster are associated with arachidonic acid concentrations of human breast milk at 1.5 and 6 mo postpartum and influence the course of milk dodecanoic, tetracosenoic, and trans-9-octadecenoic acid concentrations over the duration of lactation. Am. J. Clin. Nutr. 2011, 93, 382–391.

49. Steer, C.D.; Hibbeln, J.R.; Golding, J.; Smith, G.D. Polyunsaturated fatty acid levels in blood during pregnancy, at birth and at 7 years: Their associations with two common FADS2 polymorphisms. Hum. Mol. Genet. 2012, 21, 1504–1512.

50. Lattka, E.; Koletzko, B.; Zeilinger, S.; Hibbeln, J.R.; Klopp, N.; Ring, S.M.; Steer, C.D. Umbilical cord PUFA are determined by maternal and child fatty acid desaturase (FADS) genetic variants in the Avon Longitudinal Study of Parents and Children (ALSPAC). Br. J. Nutr. 2013, 109, 1196–1210.

51. Jensen, H.A.R.; Harsløf, L.B.S.; Nielsen, M.S.; Christensen, L.B.; Ritz, C.; Michaelsen, K.F.; Vogel, U.; Lauritzen, L. FADS single-nucleotide polymorphisms are associated with behavioral outcomes in children, and the effect varies between genders and is dependent on PPAR genotype. Am. J. Clin. Nutr.2014, 100, 826–832.

52. Glaser, C.; Rzehak, P.; Demmelmair, H.; Klopp, N.; Heinrich, J.; Koletzko, B. LISA study group. Influence of FADS polymorphisms on tracking of serum glycerophospholipid fatty acid concentrations and percentage composition in children. PLoS ONE 2011, 6, e21933.

53. European Food Safety Authority. The Panel on Dietetic Products, Nutrition and Allergies on a request from Mead Johnson Nutritionals. Scientific opinion on DHA and ARA and visual development. EFSA J. 2009, 941, 1–14.

54. Uauy, R.; Hoffman, D.R.; Mena, P.; Llanos, A.; Birch, E.E. Term infant studies of DHA and ARA supplementation on neurodevelopment: Results of randomized controlled trials. J. Pediatr. 2003, 143, S17–S25.

55. Simmer, K.; Patole, S.K.; Rao, S.C. Long-chain polyunsaturated fatty acid supplementation in infants born at term. Cochrane Database Syst. Rev. 2011, 12.

56. Qawasmi, A.; Landeros-Weisenberger, A.; Leckman, J.F.; Bloch, M.H. Meta-analysis of long-chain polyunsaturated fatty acid supplementation of formula and infant cognition. Pediatrics 2012, 129, 1141–1149.

57. Schulzke, S.M.; Patole, S.K.; Simmer, K. Long-chain polyunsaturated fatty acid supplementation in preterm infants. Cochrane Database Syst. Rev. 2011, 2.

58. Delgado-Noguera, M.F.; Calvache, J.A.; Cosp, X.B. Supplementation with long chain polyunsaturated fatty acids (LCPUFA) to breastfeeding mothers for improving child growth and development.Cochrane Database Syst. Rev. 2010, 12.

59. Gould, J.F.; Smithers, L.G.; Makrides, M. The effect of maternal omega-3 LCPUFA supplementation during pregnancy on early childhood cognitive and visual development: A systematic review and meta-analysis of randomized controlled trials. Am. J. Clin. Nutr. 2013, 97, 531–544.

60. Makrides, M.; Gibson, R.A.; McPhee, A.J.; Collins, C.T.; Davis, P.G.; Doyle, L.W.; Simmer, K.; Colditz, P.B.; Morris, S.; Smithers, L.G.; et al. Neurodevelopmental outcomes of preterm infants fed high-dose docosahexaenoic acid: A randomized controlled trial. JAMA 2009, 301, 175–178.

61. Makrides, M.; Gibson, R.A.; McPhee, A.J. Effect of DHA supplementation during pregnancy on maternal depression and neurodevelopment of young children: A randomized controlled trial. JAMA2010, 304, 1675–1683.

62. Lauritzen, L.; Jørgensen, M.H.; Olsen, S.F.; Straarup, E.M.; Michaelsen, K.F. Maternal fish oil supplementation in lactation: Effect on developmental outcome in breast-fed infants. Reprod. Nutr. Dev. 2005, 45, 535–547.

63. Cheatham, C.L.; Nerhammer, A.S.; Asserhøj, M.; Michaelsen, K.F.; Lauritzen, L. Fish oil supplementation during lactation: Effects on cognition and behavior at 7 years of age. Lipids 2011, 46, 637–645.

64. Asserhøj, M.; Nehammer, S.; Matthiessen, J.; Michaelsen, K.F.; Lauritzen, L. Maternal fish oil supplementation during lactation may adversely affect long-term blood pressure, energy intake, and physical activity of 7-year-old boys. J. Nutr. 2009, 139, 298–304.

65. Steer, C.D.; Smith, G.D.; Emmett, P.M.; Hibbeln, J.R.; Golding, J. FADS2 polymorphisms modify the effect of breastfeeding on child IQ. PLoS ONE 2010, 5.

66. Caspi, A.; Williams, B.; Kim-Cohen, J.; Craig, I.W.; Milne, B.J.; Poulton, R.; Schalkwyk, L.C.; Taylor, A.; Werts, H.; Moffitt, T.E. Moderation of breastfeeding effects on the IQ by genetic variation in fatty acid metabolism. Proc. Natl. Acad. Sci. USA 2007, 104, 18860–18865.

67. Groen-Blokhuis, M.M.; Franic, S.; van Beijsterveldt, C.E.M.; de Geus, E.; Bartels, M.; Davies, G.E.; Ehli, E.A.; Xiao, X.J.; Scheet, P.A.; Althoff, R.; et al. A prospective study of the effects of breastfeeding andFADS2 polymorphisms on cognition and hyperactivity/attention problems. Am. J. Med. Genet. B. Neuropsychiatr. Genet. 2013, 162, 457–465.

68. Martin, N.W.; Benyamin, B.; Hansell, N.K.; Montgomery, G.W.; Martin, N.G.; Wright, M.J.; Bates, T.C. Cognitive function in adolescence: Testing for interactions between breast-feeding and FADS2polymorphisms. J. Am. Acad. Child Adolesc. Psychiatry 2011, 50, 55–62.

69. Steer, C.D.; Lattka, E.; Koletzko, B.; Golding, J.; Hibbeln, J.R. Maternal fatty acids in pregnancy, FADSpolymorphisms, and child intelligence quotient at 8 y of age. Am. J. Clin. Nutr. 2013, 98, 1575–1582.

70. Browning, L.M.; Walker, C.G.; Mander, A.P.; West, A.L.; Madden, J.; Gambell, J.M.; Young, S.; Wang, L.; Jebb, S.A.; Calder, P.C. Incorporation of eicosapentaenoic and docosahexaenoic acids into lipid pools when given as supplements providing doses equivalent to typical intakes of oily fish. Am. J. Clin. Nutr. 2012, 96, 748–758.

71. Drover, J.; Hoffman, D.R.; Castaneda, Y.S.; Morale, S.E.; Birch, E.E. Three randomized controlled trials of early long-chain polyunsaturated fatty acid supplementation on means-end problem solving in 9-month-olds. Child Dev. 2009, 80, 1376–1384.

72. Hoffman, D.R.; Theuer, R.C.; Castaneda, Y.S.; Wheaton, D.H.; Bosworth, R.G.; O'Connor, A.R.; Morale, S.E.; Wiedemann, L.E.; Birch, E.E. Maturation of visual acuity is accelerated in breast-fed term infants fed baby food containing DHA-enriched egg yolk. J. Nutr. 2004, 134, 2307–2313.

73. Harbild, H.L.; Harsløf, L.B.S.; Christensen, J.H.; Kannass, K.N.; Lauritzen, L. Fish oil-supplementation from 9 to 12 months of age affects infant attention in a free-play test and is related to change in blood pressure. Prostaglandins Leukot. Essent. Fat. Acids 2013, 89, 327–333.

74. Parletta, N.; Cooper, P.; Gent, D.N.; Petkov, J.; O'Dea, K. Effects of fish oil supplementation on learning and behaviour of children from Australian Indigenous remote community schools: A randomised controlled trial. Prostaglandins Leukot. Essent. Fat. Acids 2013, 89, 71–79.

75. Baumgartner, J.; Smuts, C.M.; Malan, L.; Kvalsvig, J.; van Stuijvenberg, M.E.; Hurrell, R.F.; Zimmermann, M.B. Effects of iron and n-3 fatty acid supplementation, alone and in combination, on cognition in school children: A randomized, double-blind, placebo-controlled intervention in South Africa. Am. J. Clin. Nutr. 2012, 96, 1327–1338.

76. McNamara, R.K.; Able, J.; Jandacek, R.; Rider, T.; Tso, P.; Eliassen, J.C.; Alfieri, D.; Weber, W.; Jarvis, K.; DelBello, M.P.; et al. Docosahexaenoic acid supplementation increases prefrontal cortex

activation during sustained attention in healthy boys: A placebo-controlled, dose-ranging, functional magnetic resonance imaging study. Am. J. Clin. Nutr. 2010, 91, 1060–1067.

77. Sørensen, L.B.; Dyssegaard, C.B.; Damsgaard, C.T.; Petersen, R.A.; Dalskov, S.M.; Hjorth, M.F.; Andersen, R.; Tetens, I.; Ritz, C.; Astrup, A.; et al. The effect of Nordic school meals on concentration and school performance in 8 to 11 year-old children in the OPUS School Meal Study: A cluster-randomized controlled cross-over trial. Br. J. Nutr. 2015, 113, 1280–1291.

78. Eilander, A.; Hundscheid, D.C.; Osendarp, S.J.; Transler, C.; Zock, P.L. Effects of n-3 long chain polyunsaturated fatty acid supplementation on visual and cognitive development throughout childhood: A review of human studies. Prostaglandins Leukot. Essent. Fat. Acids 2007, 76, 189–203.

79. Osendarp, S.J.M.; Baghurst, K.I.; Bryan, J.; Calvaresi, E.; Hughes, D.; Hussaini, M.; Karyadi, S.J.M.; van Klinken, B.J.W.; van der Knaap, H.C.M.; Lukito, W.; et al. Effect of a 12-mo micronutrient intervention on learning and memory in well-nourished and marginally nourished school-aged children: 2 parallel, randomized, placebo-controlled studies in Australia and Indonesia. Am. J. Clin. Nutr. 2007,86, 1082–1093.

80. Kennedy, D.O.; Jackson, P.A.; Elliott, J.M.; Scholey, A.B.; Robertson, B.C.; Greer, J.; Tiplady, B.; Buchanan, T.; Haskell, C.F. Cognitive and mood effects of 8 weeks' supplementation with 400 mg or 1000 mg of the omega-3 essential fatty acid docosahexaenoic acid (DHA) in healthy children aged 10–12 years. Nutr. Neurosci. 2009, 12, 48–56.

81. Kirby, A.; Woodward, A.; Jackson, S.; Wang, Y.; Crawford, M.A. A double-blind, placebo-controlled study investigating the effects of omega-3 supplementation in children aged 8–10 years from a mainstream school population. Res. Dev. Disabil. 2010, 31, 718–730.

82. Smuts, C.M.; Greeff, J.; Kvalsvig, J.; Zimmermann, M.; Baumgartner, J. Long-chain n-3 polyunsaturated fatty acid supplementation decrease physical activity during class in iron deficient South African school children. Br. J. Nutr. 2015, 113, 212–224.

83. Rytter, M.J.H.; Andersen, L.B.B.; Houmann, T.; Bilenberg, N.; Hvolby, A.; Mølgaard, C.; Michaelsen, K.F.; Lauritzen, L. Diet in the treatment of ADHD in children-A systematic review of the literature.Nord. J. Psychiatry 2015, 69, 1–18.

84. Brookes, K.J.; Chen, W.; Xu, X.; Taylor, E.; Asherson, P. Association of fatty acid desaturase genes with attention-deficit/hyperactivity disorder. Biol. Psychiatry 2006, 60, 1053–1061.

85. Harsløf, L.B.; Damsgaard, C.; Hellgren, L.; Andersen, A.; Vogel, U.; Lauritzen, L. Effects on metabolic markers are modified by PPARG2 and COX2 polymorphisms in infants randomized to fish oil. Genes Nutr. 2014, 9, 1–11.

86. Damsgaard, C.T.; Schack-Nielsen, L.; Michaelsen, K.F.; Fruekilde, M.B.; Hels, O.; Lauritzen, L. Fish oil affects blood pressure and the plasma lipid profile in healthy Danish infants. J. Nutr. 2006, 136, 94–99.

87. Damsbo-Svendsen, S.; Rønsholdt, M.D.; Lauritzen, L. Fish oil-supplementation increases appetite in healthy adults. A randomized controlled cross-over trial. Appetite 2013, 66, 62–66.

88. Stonehouse, W.; Conlon, C.A.; Podd, J.; Hill, S.R.; Minihane, A.M.; Haskell, C.; Kennedy, D. DHA supplementation improved both memory and reaction time in healthy young adults: A randomized controlled trial. Am. J. Clin. Nutr. 2013, 97, 1134–1143.

89. Agostoni, C.; Zuccotti, G.V.; Radaelli, G.; Besana, R.; Podesta, A.; Sterpa, A.; Rottoli, A.; Riva, E.; Giovannini, M. Docosahexaenoic acid supplementation and time at achievement of gross motor milestones in healthy infants: A randomized, prospective, double-blind, placebo-controlled trial. Am. J. Clin. Nutr. 2009, 89, 64–70.

90. Yurko-Mauro, K.; Alexander, D.D.; van Elswyk, M.E. Docosahexaenoic acid and adult memory: A systematic review and meta-analysis. PLoS ONE 2015, 10.

91. Morris, M.C.; Evans, D.A.; Tangney, C.C.; Bienias, J.L.; Wilson, R.S. Fish consumption and cognitive decline with age in a large community study. Arch. Neurol. 2005, 62, 1849–1853.

92. Lim, W.S.; Gammack, J.K.; van Niekerk, J.; Dangour, A.D. Omega-3 fatty acid for the prevention of dementia. Cochrane Database Syst. Rev. 2006, 1.

93. Chiu, C.C.; Su, K.P.; Cheng, T.C.; Liu, H.C.; Chang, C.J.; Dewey, M.E.; Stewart, R.; Huang, S.Y. The effects of omega-3 fatty acids monotherapy in Alzheimer's disease and mild cognitive impairment: A preliminary randomized double-blind placebo-controlled study. Prog. Neuropsychopharmacol. Biol. Psychiatry 2008, 32, 1538–1544.

94. Yurko-Mauro, K.; McCarthy, D.; Rom, D.; Nelson, E.B.; Ryan, A.S.; Blackwell, A.; Salem, N., Jr.; Stedman, M. Beneficial effects of

docosahexaenoic acid on cognition in age-related cognitive decline.Alzheimers Dement. 2010, 6, 456–464.

95. Astarita, G.; Jung, K.M.; Berchtold, N.C.; Nguyen, V.Q.; Gillen, D.L.; Head, E.; Cotman, C.W.; Piomelli, D. Deficient liver biosynthesis of docosahexaenoic acid correlates with cognitive impairment in Alzheimer's disease. PLoS ONE 2010, 5, e12538.

96. Bazan, N.G.; Molina, M.F.; Gordon, W.C. Docosahexaenoic acid Signal lipidomics in nutrition: Significance in aging, neuroinflammation, macular degeneration, Alzheimer's, and other neurodegenerative diseases. Annu. Rev. Nutr. 2011, 31, 321–351.

97. De Mel, D.; Suphioglu, C. Fishy business: Effect of omega-3 fatty acids on zinc transporters and free zinc availability in human neuronal cells. Nutrients 2014, 6, 3245–3258.

98. Labrousse, V.F.; Nadjar, A.; Joffre, C.; Costes, L.; Aubert, A.; Gregoire, S.; Bretillon, L.; Laye, S. Short-term long chain omega3 diet protects from neuroinflammatory processes and memory impairment in aged mice. PLoS ONE 2012, 7, e36861.

99. Sands, S.A.; Windsor, S.L.; Reid, K.J.; Harris, W.S. The impact of age, body mass index, and fish intake on the EPA and DHA content of human erythrocytes. Lipids 2005, 40, 343–347.

100. Hoen, W.P.; Lijmer, J.G.; Duran, M.; Wanders, R.J.; van Beveren, N.J.; de Haan, L. Red blood cell polyunsaturated fatty acids measured in red blood cells and schizophrenia: A meta-analysis.Psychiatry Res. 2013, 207, 1–12.

101. Peet, M.; Shah, S.; Selvam, K.; Ramchand, C.N. Polyunsaturated fatty acid levels in red cell membranes of unmedicated schizophrenic patients. World J. Biol. Psychiatry 2004, 5, 92–99.

102. Khan, M.M.; Evans, D.R.; Gunna, V.; Scheffer, R.E.; Parikh, V.V.; Mahadik, S.P. Reduced erythrocyte membrane essential fatty acids and increased lipid peroxides in schizophrenia at the never-medicated first-episode of psychosis and after years of treatment with antipsychotics. Schizophr. Res. 2002, 58, 1–10.

103. Reyazuddin, M.; Azmi, S.A.; Islam, N.; Rizvi, A. Oxidative stress and level of antioxidant enzymes in drug-naive schizophrenics. Indian J. Psychiatry 2014, 56, 344–349.

104. Hashimoto, M.; Maekawa, M.; Katakura, M.; Hamazaki, K.; Matsuoka, Y. Possibility of polyunsaturated fatty acids for the prevention and treatment of neuropsychiatric illnesses. J. Pharmacol. Sci. 2014, 124, 294–300.

105. Amminger, G.P.; Schafer, M.R.; Papageorgiou, K.; Klier, C.M.; Cotton, S.M.; Harrigan, S.M.; Mackinnon, A.; McGorry, P.D.; Berger, G.E.

Long-chain omega-3 fatty acids for indicated prevention of psychotic disorders: A randomized, placebo-controlled trial. Arch. Gen. Psychiatry 2010, 67, 146–154.

106. Machado-Vieira, R.; Mallinger, A.G. Abnormal function of monoamine oxidase-A in comorbid major depressive disorder and cardiovascular disease: Pathophysiological and therapeutic implications. Mol. Med. Rep. 2012, 6, 915–922.

107. Weissman, M.M.; Bland, R.C.; Canino, G.J.; Faravelli, C.; Greenwald, S.; Hwu, H.G.; Joyce, P.R.; Karam, E.G.; Lee, C.K.; Lellouch, J.; et al. Cross-national epidemiology of major depression and bipolar disorder. JAMA 1996, 276, 293–299.

108. Severus, W.E.; Littman, A.B.; Stoll, A.L. Omega-3 fatty acids, homocysteine, and the increased risk of cardiovascular mortality in major depressive disorder. Harv. Rev. Psychiatry 2001, 9, 280–293.

109. Huang, T.; Zheng, J.; Chen, Y.; Yang, B.; Wahlqvist, M.L.; Li, D. High consumption of omega-3 polyunsaturated fatty acids decrease plasma homocysteine: A meta-analysis of randomized, placebo-controlled trials. Nutrition 2011, 27, 863–867.

110. Bourre, J.M. Roles of unsaturated fatty acids (especially omega-3 fatty acids) in the brain at various ages and during ageing. J. Nutr. Health Aging 2004, 8, 163–174.

111. Assies, J.; Mocking, R.J.T.; Lok, A.; Ruhé, H.G.; Pouwer, F.; Schene, A.H. Effects of oxidative stress on fatty acid- and one-carbon-metabolism in psychiatric and cardiovascular disease comorbidity. Acta Psychiatr. Scand. 2014, 130, 163–180.

112. Lin, P.Y.; Huang, S.Y.; Su, K.P. A meta-analytic review of polyunsaturated fatty acid compositions in patients with depression. Biol. Psychiatry 2010, 68, 140–147.

113. McNamara, R.K.; Vannest, J.J.; Valentine, J.C. Role of perinatal long-chain omega-3 fatty acids in cortical circuit maturation: Mechanisms and implications for psychopathology. World J. Psychiatry2015, 5, 15–34.

114. European Food Safety Authority. The Panel on Dietetic Products, Nutrition and Allergies on a request from Mead Johnson Nutritionals. Scientific Opinion on DHA and ARA and brain development. EFSA J. 2009, 1000, 1–13.

115. European Food Safety Authority. The Panel on Dietetic Products, Nutrition and Allergies. Scientific Opinion on the substantiation of a health claim related to DHA and contribution to normal brain development pursuant to Article 14 of Regulation (EC) No 1924/2006. EFSA J. 1924, 12.

2

DOES VITAMIN C DEFICIENCY AFFECT COGNITIVE DEVELOPMENT AND FUNCTION?

Stine Normann Hansen, Pernille Tveden-Nyborg and Jens Lykkesfeldt

Department of Veterinary Disease Biology, Faculty of Health and Medical Sciences, University of Copenhagen, Ridebanevej 9, DK-1870 Frederiksberg C, Copenhagen, Denmark

ABSTRACT

Vitamin C is a pivotal antioxidant in the brain and has been reported to have numerous functions, including reactive oxygen species scavenging, neuromodulation, and involvement in angiogenesis. Absence of vitamin C in the brain has been shown to be detrimental to survival in newborn SVCT2(−/−) mice and perinatal deficiency have shown to reduce hippocampal volume and neuron number and cause decreased spatial cognition in guinea pigs, suggesting that maternal vitamin C deficiency could have severe consequences for the offspring. Furthermore, vitamin C deficiency has been proposed to play a role in age-related cognitive decline and in stroke risk and severity. The present review discusses the available literature on effects of vitamin C deficiency on the developing and aging brain with particular focus on *in vivo* experimentation and clinical studies.

Keywords: vitamin C; cognition; oxidative stress; ROS; aging; development; stroke

1. INTRODUCTION

Increasing evidence is pointing to vitamin C (VitC) as an important redox homeostatic factor in the central nervous system, linking an inadequate dietary supply of VitC to negative effects on cognitive performance [1,2,3]. The brain displays comparatively high concentrations of VitC in particular during deficiency when most other organs are depleted [4,5,6], underlining an essential role of VitC in the brain. This has been further supported by findings of perinatal mortality and cerebral hemorrhages in newborn mice devoid the VitC transporter—thus, exposed to VitC depletion in the brain [7,8]—and by data showing reduced hippocampal volume and impaired spatial memory in guinea pig models of dietary induced VitC deficiency [1,9].

The only known clinical condition directly associated with lack of VitC is scurvy, representing the terminal and lethal collapse following prolonged and severe VitC deficiency (depletion) [10,11]. More recently, increased attention has been devoted the potential chronic effects of a suboptimal VitC status such as hypovitaminosis C—in humans defined as a plasma concentration below 23 μmol/L [12]—in disease development, e.g., impaired brain development [1,9], in multifactorial complexes of life-style associated diseases [13], and in neurodegenerative disorders [14,15,16,17].

Reports from cross-sectional population surveys have consistently estimated that at least 10%–15% of the adult population in the Western world suffers from suboptimal VitC levels/hypovitaminosis C [10,18]. This prevalence has been reported to be substantially increased in subgroups of particular risk, such as developing countries, communities of low socio-economic status, smokers, elderly, pregnant women, and children with poor nutritional status [19,20,21,22,23]. Thus, effects of VitC deficiency potentially affect millions on a global scale.

The present review outlines the functions of VitC in the brain and discusses putative effects of deficiency on cognitive ability based on findings from studies in animals and humans.

2. VITAMIN C

VitC is a water soluble vitamin contributing as electron donor in several important biological reactions in the body. The active form of the vitamin, L-ascorbic acid, primarily exists as the monoanion ascorbate at physiological pH [24]. Most species are able to biosynthesize the micronutrient from glucose in the liver but higher-order primates including humans, as well as guinea pigs, and some bat, fish, and bird species are all dependent on an adequate dietary supply of VitC due to an evolutionary buildup of mutations and deletions in the gene encoding for L-gulono-1,4-lactone oxidase causing it to become non-functional [11]. Being unable to catalyze the final step in VitC biosynthesis, this renders the affected species entirely dependent on a dietary supply [25,26,27,28].

The uptake and distribution of VitC in the body is under close homeostatic control and primarily regulated by tissue specific sodium dependent vitamin C co-transporters (SVCT) 1 and 2, actively transporting VitC in exchange of sodium [29,30,31,32]. This results in a saturable plasma concentration (around 70 μM in humans [33]) that can only be increased via parenteral administration of VitC. Thus, the route of administration should clearly be considered when evaluating the potential effects of VitC intervention. In healthy individuals, the amount of the two-electron oxidation product of ascorbate, dehydroascorbic acid (DHA), represents only a small fraction of the total VitC pool due to efficient intracellular recycling to ascorbate and is generally considered to be of little importance in the overall homeostasis [34]. Within the body, VitC displays complex non-linear pharmacokinetics, as well as differential tissue distribution [31]. This includes the brain that is able to preferentially retain VitC at the expense of other tissues during chronic states of severe deficiency and to uphold concentrations 100-fold higher than, e.g., liver and kidney, which are readily depleted [5,6].

The brain depends on the SVCT2 receptor to govern active VitC transport across the choroid plexus to the brain extracellular fluid and further on to neuronal cells [4]. Interestingly, neuroglia does not express the SVCT2 transporter, thus, are thought to rely on passive facilitated diffusion through GLUT-transporters for DHA transport; the DHA subsequently being reduced to ascorbate intracellularly [35,36]. In the brain, VitC is also regionally distributed with the hippocampus and the frontal and occipital cortex displaying high concentrations [36,37]. However, these apparent regional differences might—at least in part—be due to differences in neuronal density. Neurons have particularly high VitC concentrations (about 10 mM) compared to glia (about 1 mM) and, e.g., frontal-cortex and hippocampus both display high levels of neurons compared to other brain areas [38]. The SVCT2 expression also correlates with a 10-fold higher metabolism and, hence, reactive oxygen species (ROS) formation and, thus, antioxidant requirement in neurons [39,40,41,42]. Proposed causes and consequences of VitC deficiency in the brain are depicted inFigure 1.

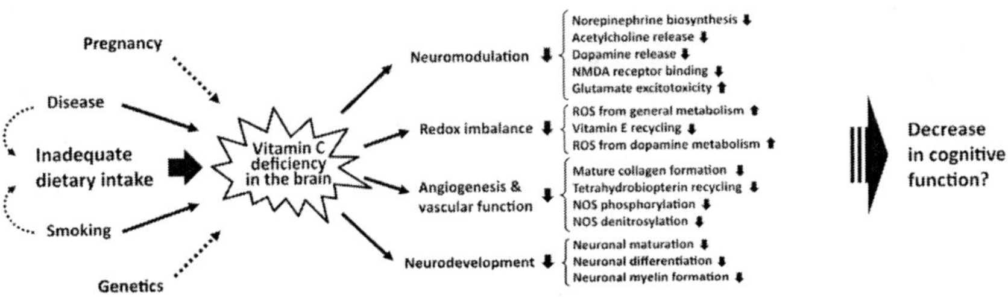

Figure 1. Proposed causes and consequences of VitC deficiency in the brain. Several risk factors for VitC deficiency have been identified, including disease, smoking, and inadequate dietary intake, but also pregnancy and genetics have been shown to affect VitC levels. Based on VitC's involvement in important processes in the brain, there is reason to believe that these could be adversely affected by a deficiency. The functions of VitC are both related to its antioxidant function of upholding redox balance in the brain but also other important functions. These include modulation of the cholinergic, catecholinergic, and glutamergic systems of the brain, as well as the general development of neurons through maturation, differentiation and myelin formation. VitC is involved in several processes in the vascular system and hereby help maintain integrity and function of, e.g., nitric oxide synthase, which regulates vessel relaxation through production of nitric oxide. Abbreviations: NOS, nitric oxide synthase; ROS, reactive oxygen species; NMDA, *N*-methyl-D-aspartate.

3. THE FUNCTIONS OF VITAMIN C IN THE BRAIN

3.1. The Antioxidant Role of Vitamin C in the Brain

Due to a high level of poly-unsaturated fatty acids (PUFA), combined with high rates of cellular metabolism, the brain is particularly vulnerable to oxidative damage [43,44,45,46,47]. Several *in vitro* and *in vivo* experiments have supported a crucial role for VitC in the brain, both as a powerful antioxidant and scavenger of ROS, as well as a key factor in the recycling of other brain antioxidants, e.g., vitamin E (VitE) [5,6,48]. Mice born without functional SVCT2 transporters do not survive birth and display increased oxidative stress in the brain, as well as cerebral hemorrhages [7,8]. Redox-imbalances in the brain have been associated with ischemia-induced neurodegeneration [49,50,51,52] and in chronic diseases such as Alzheimer's syndrome and Huntington's chorea [14,15,16,17,53,54,55] as well as psychological disorders such as schizophrenia [56,57,58]. Collectively, these observations have led to the hypothesis that VitC plays a pivotal role in maintaining redox balance in the brain and subsequently that VitC deficiency leads to neuronal damage through processes involving increased ROS and oxidative stress [13].

Additionally, VitC has also been found to induce the expression of brain-derived-neurotrophic-factor—a component of several survival pathways—and may, thereby, contribute to the defense mechanisms of the brain [43].

3.2. Vitamin C as a Neuromodulator

VitC is known to participate in neuronal maturation and myelin formation, and also be involved in central nervous system signal transduction through neurotransmitters [59,60]. The dopamine (DA) receptor is involved in several different brain processes including pleasure, reward, motor control, and memory [61,62]. VitC supplies electrons for the dopamine-β-hydroxylase catalyzing the formation of norepinephrine from DA, and may provide neuroprotection from ROS and quinones generated by DA metabolism [63,64,65,66,67]. Thus, DA receptor activation has been shown to cause release of VitC in the brain and SVCT2 knock-out mice have a deficient adrenal catecholamine system and show increased adrenal cell apoptosis supporting an important role of VitC in DA homeostasis [68,69]. VitC has also been shown to induce the release of acetylcholine (ACh) and norepinephrine from synaptic vesicles of neurons, linking VitC to neuronal signal-transmission [70].

Another neuromodulatory role of VitC appears to be its involvement in presynaptic re-uptake of glutamate [2,71]. Here, VitC prevents excitotoxic damage caused by excess extracellular glutamate otherwise leading to hyperpolarization of the *N*-methyl-D-aspartate (NMDA) receptor and subsequent neuronal damage [40,71]. VitC has also been shown to inhibit the binding of glutamate to the NMDA receptor, in this way exhibiting a direct effect in the prevention of neuronal over-stimulation by glutamate [72,73].

Collectively, these findings imply that VitC is an important neuromodulator in the brain, and that depletion may have serious consequences for neuronal function and integrity.

3.3. Vitamin C in Angiogenesis

VitC's function in collagen maturation is well recognized. Collagen is a primary component of supportive tissue and constitutes the basal membrane of blood vessels [7]. The final steps in the formation of mature triple helix collagen depend on VitC as electron donor in the hydroxylation of procollagen prolyl and lysyl residues [74,75,76]. VitC deficiency disrupts this collagen maturation leading to an impaired integrity of the vascular wall and ultimately resulting in hemorrhage—a cardinal symptom of scurvy—supported by cerebral bleedings in the SVCT2 knock-out mouse [7,8]. However, findings of unaffected levels of hydroxyproline in SVCT2 knock-out mice and in guinea pigs subjected to scorbutic VitC depletion suggests that VitC may even act as a *de novo* stimulator of collagen synthesis pointing towards more advanced functions besides that as an electron donor [8,77,78].

VitC's angiogenic properties are further underlined by its apparent function as an electron donor for Fe2+-2-oxyglutarate-dependent dioxygenases catalyzing the hydroxylation of hypoxia inducible factor 1α (HIF-1α) [79,80]. Importantly, HIF-1α is involved in neuronal development, oxygen homeostasis and angiogenesis including vascular endothelial growth factor and erythropoetin [79,80,81,82]. Regulation of HIF is associated with VitC dependent hydroxylation and subsequent degradation, thus deficiency may increase HIF-1α levels hereby disturbing normal vascular development, likely to be particularly important in the growing fetus and in cases of regeneration following brain injury [51,52,83,84,85].

4. VITAMIN C DEFICIENCY AND COGNITIVE DYSFUNCTION

The versatile roles of VitC related to the structural and functional integrity of the brain fuels the question of whether VitC deficiency may propagate cognitive dysfunction as suggested by findings in experimental animal models and reports from population surveys [1,3,15]. Of particular interests are the effects of VitC on the developing brain, age related neuronal degeneration, and induced brain injury (Figure 2).

4.1. Putative Consequences of Vitamin C Deficiency in the Developing Brain

The developing brain is particularly vulnerable to redox imbalances due to an undeveloped ROS defense system and a high metabolic rate, as well as having a high concentration of PUFA prone to oxidation [38,40,86]. While the establishment of the SVCT2 transporter as essential for VitC supply to the fetal brain and perinatal survival in mice demonstrates VitC as a key factor in brain development, the absolute lethality unfortunately also prohibits further assessment of functional consequences [7,8]. However, guinea pigs exposed to chronic but non-scorbutic VitC deficiency in early life showed significantly impaired spatial memory

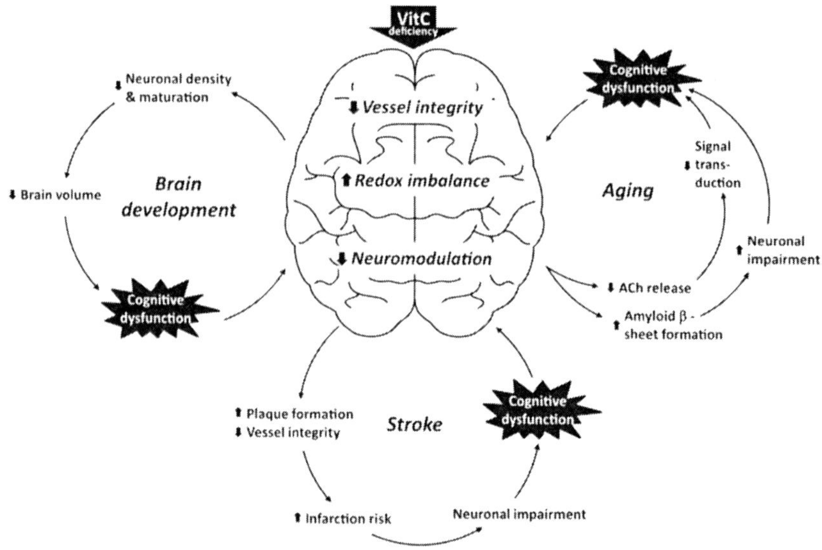

Figure 2. Vitamin C deficiency on cognitive function. The involvement of vitamin C in vessel integrity, redox balance and neuromodulation in the brain has prompted investigations into the effect of the vitamin on the developing brain, in aging and in stroke. In the developing brain, neuronal density and maturation is compromised by VitC deficiency, giving rise to decreased brain volume. In the aging brain deficiency affects ACh release and may impair cognitive function through reduced signal transduction but also through amyloid β deposition resulting in generation of reactive oxygen species and increased neuronal impairment in people suffering from Alzheimer's disease. In stroke, VitC deficiency may result in decreased vessel integrity through, e.g., decreased NOS generation and impaired synthesis of mature collagen; potentially leading to increased plaque formation and incidence of stroke. Furthermore, an increase in infarct area may result from redox imbalance causing increased neuronal death.

compared to sufficient counterparts and significant reductions in neuronal number in both the dentate gyrus and cornu ammonia of the hippocampus [1]. A later study of the effects of pre- and postnatal VitC deficiency in guinea pigs showed that 10%–15% reductions in hippocampal volume persisted into early adulthood (two months of age) and was unaffected by repletion [9]. None of the animals (sufficient or deficient) performed adequately in the Morris Water Maze to allow for conclusions to be drawn on cognitive performance, however, swimming parameters in the cued trial showed no apparent difference in locomotor ability between deficient and controls [9]. In gulo(−/−) mice unable to synthesize VitC, no effect on brain function could be detected following behavioral testing in the Y-maze, Morris Water Maze, and explorative activity monitor [87]. The apparent discrepancy between findings in mice and guinea pigs may be due to variations in trial conditions and/or specific responses to the imposed deficiency as well as species variations, e.g., prenatal *vs.* postnatal exposure and altricial *vs.* precocial offspring. However, a negative effect of chronic VitC deficiency on the hippocampus seems to be a consistent finding, at least in the guinea pig model, possibly leading to functional consequences, such as reductions in spatial memory [88,89].

Studies of young children exposed to intrauterine growth restriction (IUGR) or having a very low birth-weight have shown deleterious effects on cognition [90,91,92]. IUGR leads to a generalized malnutrition and, thus, does not asses the isolated effects of VitC deficiency. However, considering the high levels of VitC found in the young brain and the high metabolic rate during growth, a negative consequence of deficiency is not unlikely [93,94,95]. Tolsa *et al.* [92] reported that preterm babies (n = 14; gestational age: 32.5 ± 1.9 weeks) suffering from IUGR (birth weight: 1246 ± 299 g) displayed reduced brain volumes, particularly in the gray matter when compared to normal-weight preterm babies. At the calculated term date, the maintained significant volume reductions and, thus, no "catch-up" growth was observed. Furthermore, the reduction in brain volume was accompanied by a lower maturity by the Assessment of Preterm Infants' Behavior-scoring. In a follow-up study of children born prematurely (n = 11, gestational age: median = 28 weeks) and with very low birth weights (median = 998 g; range: 840–1490 g), Isaacs *et al.* [91] reported long term effects on cognition at age 13.5 years (median age). When compared to controls, magnetic resonance-scans of prenatal children showed consistent atrophy of hippocampus and also atrophy of perioccipital white matter and corpus callosum as well as enlarged third and lateral ventricles. The neurological examination revealed that the study group had significantly impaired everyday memory by the Rivermead Behavioral Memory test, as well as lower numeracy skills in the Wechsler Objective Numerical Dimensions test [91].

The underlying pathogenesis of the anatomical and cognitive changes seen in IUGR is not completely understood, but it has been proposed that IUGR newborns have increased levels of oxidative stress. Biri *et al.*[96] measured oxidative stress markers in preterm IUGR babies (n = 13; gestational age: 33–34 weeks) and their mothers. Maternal blood, umbilical cord blood and placental samples were used to measure superoxide dismutase (SOD), glutathione peroxidase (GSH-Px), malondialdehyde (MDA), antioxidant potential (AOP), adenosine deaminase (ADA), catalase (CAT) and xanthine oxidase (XO). The umbilical cord blood from preterm babies showed significantly elevated oxidative stress in all markers, except GSH-Px and AOP, while the IUGR mothers differed significantly in all markers other than CAT, when compared to controls. The placental samples were also significantly changed in all aspects, except SOD and ADA [96]. In 29 preterm babies, Berger and coworkers [97] also reported high levels of F_2-isoprostanes, but these were accompanied by high levels of VitC and did not correlate with potentially redox-active iron. Collectively, the above results suggest that the IUGR fetus is exposed to increased oxidative stress and tissue damage including the brain, which may partly explain the decreased cognitive functions and reduced brain volumes observed in these children later in life.

Another potential source of oxidative stress in the prenatal period is pre-eclampsia [98]. Several studies have concluded that maternal pre-eclampsia results in decreased cognitive performance in the affected children, when compared to children from healthy pregnancies [99,100,101,102]. Women with pre-eclampsia have been shown to have reduced levels of VitC and several studies have investigated a potential effect of VitC supplementation [103]. In a double-blind study, Chappell *et al.* [104] compared markers of oxidative stress in 79 high-risk women supplemented with VitC (1000 mg/day) and VitE (400 IU/day); 81 high-risk women offered placebo and 32 low-risk women not taking any supplements. VitC, plasminogen activator inhibitor (PAI)-2, and placenta growth factor concentrations were decreased in placebo group; and 8-epi-prostaglandin F2α, leptin, and the PAI-1/-2 ratio were increased in the placebo group compared with the low-risk group, whereas the vitamin supplement group displayed VitC, 8-epi-prostaglandin F2α, leptin, and PAI-1/-2 values similar to the low risk group. Thus, supplementation of high-risk women was associated with improvement in the biochemical indices of preeclampsia. However, studies have also reported no elaborate effects of supplementation with VitC and VitE in combination (1000 mg/day and 400 IU/day respectively) on the incidence of pre-eclampsia in high-risk women [105,106]. This might be due to the absence of VitC status as a defined inclusion criterion, hereby allowing for variation in the degree of plasma saturation and subsequent differential outcomes of supplementation [13]. Table 1 gives an overview of some of the studies done on oxidative stress and/or VitC in the developing brain, both *in vivo* and in clinical studies.

4.2. Vitamin C and Aging

According to "The free radical theory of aging", aging is the accumulated consequence of a lifetime of free radical assaults on the cells and macromolecules of the body [108]. In the aging brain, neurodegeneration has been associated with increased oxidative stress either through loss of electrons from the respiratory chain, inflammatory response or peroxide generation from β-amyloid [44,109,110].

Table 1. Vitamin C, oxidative stress and brain development.

	Species	Intervention	Measurements	Outcome	Reference
	In Vivo Studies				
	SVCT2(−/−) mice (ED: 18.5–19.5).	Dams: 0.33 g/L VitC in drinking water.	VitC content, MDA, F_2-isoprostanes and F_4-neuroprostanes in brain (cortex). Additional IHC.	The SVCT2(−/−) fetuses: Increased F_2-isoprostanes ($p < 0.001$), F_4-neuroprostanesincreased ($p < 0.05$), isoketal staining, apoptotic cells and hemorrhages. Decreased collagen-IV staining and VitC content ($p < 0.001$).	[7]
	Dunkin Hartley guinea pigs (6/7 to 60/61days). Postnatal deficiency	VitC in diet: 923 mg/kg or 100 mg/kg feed.	Asc, DHA, glutathione, MDA and SOD in brain. Quantitation of hippocampal neurons. Functional assessment in MWM.	Decreased performance in MWM ($p < 0.05$) and reduced number of neurons in hippocampus ($p < 0.05$) in VitC deficient animals.	[1]
	Dunkin Hartley guinea pigs (GD: 18). Prenatal deficiency.	VitC prenatal: 900 mg/kg diet or 100 mg/kg diet. Postnatal: 750 mg/kg or 100 mg/kg.	Asc, DHA and MDA in brain. Hippocampal neurogenesis and volume. Functional assessment in MWM.	Significant and persistent lower hippocampal volume ($p < 0.001$).	[9]
	Dunkin Hartley guinea pigs (2 days to 3 weeks).	VitC in diet: 1036 mg/kg or 36 mg/kg	Asc, DHA, glutathione, SOD, MDA, α- and γ-tocopherol, protein carbonyls, 8-oxo-deoxyguanosine and base excision repair in brain.	VitC deficiency caused significant reductions in Asc ($p < 0.001$), DHA ($p = 0.034$), MDA ($p < 0.001$) and protein carbonyls ($p = 0.003$) and an increase in base excision repair ($p = 0.014$).	[5]
	Design and Subjects	**Intervention**	**Measurements**	**Outcome**	**Reference**
	Clinical Studies				
Cohort Studies	92 preterm children (7.86 ± 0.7 years, birth weight: 1475.13 ± 556.44 g) 40 age-matched controls.		Cognitive testing: Spatial pattern/Recognition, Intradimensional/Extra Dimensional Set-Shifting task, Tower of London task, Spatial Working Memory task, Spatial Memory Span task, and a Psychomotor screening.	Preterm children had decreased performance in Psychomotor test ($p < 0.01$), Recognition test ($p < 0.01$), Spatial Memory Span ($p < 0.01$), and Spatial Working Memory ($p < 0.001$)	[90]
Cohort Studies	13 IUGR preterm infants (gestational age 33–34 weeks) and 12 controls.		Maternal blood, umbilical cord blood and placental samples: SOD, GSH-Px, MDA, AOP, ADA, CAT and XO.	All markers, except GSH-Px and AOP were elevated in umbilical cord blood. IUGR mothers differed significantly in all markers other than CAT. Placental samples were significantly changed in all markers, except SOD and ADA ($p < 0.01$ or less).	[96]
	Design and Subjects	**Intervention**	**Measurements**	**Outcome**	**Reference**
	Clinical Studies				
Controlled trials	Randomized clinical trial: 160 women in high risk for pre-eclampsia (16–22 weeks pregnant) and 32 controls	VitC (1000 mg/day) and VitE (400 IU/day) or placebo	Plasma VitC, PAI-2, placenta growth factor, 8-epi-prostaglandin F2α, leptin, PAI-1/2 ratio.	Vitamin supplemented: VitC, 8-epi-prostaglandin F2α, leptin, and PAI-1/-2 equal to controls; whereas placebo-treated displayed decreased VitC, PAI-2, and placenta growth factor and increased 8-epi-prostaglandin F2α, leptin, and PAI 1/-2 ratio	[104]
Controlled trials	Randomized clinical trial: 283 women with high risk of pre-eclampsia (16–22 weeks pregnant).	VitC (1000 mg/day) and VitE (400 IU/day) or placebo.	PAI-1 and -2 measured every month until delivery. Pre-eclampsia assessed by the development of proteinuric hypertension.	VitC + E supplementation was associated with a decrease in the PAI-1/PAI-2 ratio ($p = 0.015$) and a significantly decreased risk of pre-eclampsia ($p = 0.002$).	[107]
Controlled trials	Double blind randomized clinical trial: 100 women in high risk of pre-eclampsia (14–20 weeks pregnant).	VitC (1000 mg/day) and VitE (400 IU/day) or placebo.	Incidence of pre-eclampsia.	No significant effect of VitC + E treatment.	[105]
Controlled trials	Double blind randomized clinical trial: 1365 women in high risk for pre-eclampsia (14–22 weeks pregnant).	VitC (1000 mg/day) and VitE (400 IU/day) or placebo.	Occurrence of pre-eclampsia defined as hypertension and onset of proteinuria.	Supplementation with VitC + E did not reduce risk of pre-eclampsia	[106]

Abbreviations: VitC, vitamin C; VitE, vitamin E; Asc, ascorbate; DHA, dehydroascorbic acid; PAI, plasminogen activator inhibitor; MDA, malondialdehyde; GSH-Px, glutathione peroxidase; AOP, antioxidant potential; ADA, adenosine deaminase; CAT, catalase; XO, xanthine oxidase; SOD, superoxide dismutase; MVM, Morris Water Maze; IUGR, intrauterine growth restriction; IHC, immunohistochemistry; ED, embryonic day; GD, gestational day; SVCT, sodium-dependent vitamin C transporter.

VitC treatment (60–120 mg/kg intra-peritoneal; three to eight days) has been shown to attenuate reduced performance in elevated plus maze and passive avoidance test in seven-month-old Swiss mice, pointing towards VitC as an important factor in age-related cognitive decline [111]. In humans, meno-pause is associated with some degree of decline in cognition [112,113]. In ovariectomized rat-models of human meno-pause supplementation with VitC and VitE prevented deficits, hereby linking antioxidant status to the protection of cognitive function [114]. Notably, as both rat and mouse models are capable of synthesizing VitC, results should be interpreted with caution as the findings may not translate well to humans. In aging guinea pigs subjected to long term, non-scorbutic VitC deficiency (100 mg VitC/kg feed), no significant effects of age on biochemical markers in the brain were detected compared to controls (323 mg VitC/kg feed), and it was concluded that the age-related change in VitC status observed in several species is more likely related to maturation rather than aging *per se* [115,116].

In aging humans (n = 137; age: 66–90 years), plasma concentrations of VitC has been reported to be positively correlated with cognitive performance and plasma levels of VitC were significantly reduced in elderly suffering from different kinds of dementia [117,118]. This is supported by findings from a prospective cohort study from Department of Health/Medical Research Council Nutritional Programme in which participants (n = 921; ≥65 years) with the lowest VitC status displayed the poorest cognitive function, a finding which persisted when corrected for age, illness, social class, or other dietary variables [3]. In the Nurses' Health study, long-term VitC and VitE supplementation prior to cognitive testing was significantly associated with better cognitive performance (p = 0.03) in women aged 70–79 years with a trend towards increasing performance with increased duration of use (p = 0.04) [118]. A similar effect on cognitive performance was found by Masaki *et al.* [119] in the Honolulu-Asia Aging study, where VitC or VitE supplementation was associated with a higher cognitive performance (OR: 1.25, 95% CI, 1.04–1.50). Furthermore, a protective effect of supplementation on vascular dementia (OR: 0.12, 95% CI, 0.02–0.88) and mixed/other dementia (OR: 0.31, 95% CI, 0.11–0.89) was found. However, other studies have not found an effect of self-reported VitC and/or VitE supplementation on age-related dementia [120].

The effect of VitC on cognition associated to aging has been extensively studied with regards to Alzheimer's disease (AD) [121]—a disease already affecting many and predicted to increase in global prevalence [122]. Though the etiology of the disease is not completely elucidated, ROS and oxidative stress has been linked to disease progression [123]. As AD patients have been reported to have decreased levels of plasma VitC, investigations into the role of antioxidants including VitC in AD pathogenesis has been conducted [124,125]. In animal models of AD, VitC supplementation has been reported to reduce some of the cognitive dysfunction seen in control animals [126,127]. In a study of APP/PSEN1 mice, acute VitC administration (125 mg/kg intra-peritoneal, one hour prior to testing) significantly improved cognitive performance in both Y-maze and Morris Water Maze, albeit no significant effect was shown on either amyloid load, acetylcholine esterase (AChE) or oxidative stress markers. Likewise, an acute memory-enhancing effect of VitC has also been reported in other studies of age-dependent cognitive decline as well as in cognitive sound animal models [111,128].

In another set-up, 6 months old APP/PSEN1 mice exposed to four months of dietary VitC (1 g/kg diet) alone or in combination with high (750 IU/kg diet) or low (400 IU/kg diet) VitE showed decreased levels of F_4-neuroprostanes and MDA—markers of lipid oxidation—in supplemented *vs.* controls while amyloid deposition was unaffected

[126]. The low VitE + C treatment attenuated spatial memory deficits in APP/PSEN1 mice and improved performance in wild-type mice in the water maze. Interestingly, the high VitE + C treatment apparently impaired spatial memory compared to controls [126]. Likewise, VitC supplementation of drinking water (1333 mg/kg/day) to AβPP mice (six months old) for six months had no significant effect on the amyloid plaque load; however, an effect of VitC on cognitive function was noted [129].

Collectively, these data support the hypothesis that VitC can reduce cognitive decline in animal models while the exact mechanisms are yet to be disclosed. Studies of VitC supplementation during scopolamine-induced amnesia in mice have shown induction of AChE in the medial forebrain, pointing towards a role of VitC in the intricate regulation of cholinergic neurotransmission [130]. One potential mode of action has been suggested to be through the involvement of BH4:BH2 in the metabolism of mono-amine neurotransmitters—DA, norepinephrine and serotonin—as VitC maintains reduced biopterin status and may thereby indirectly regulate levels of neurotransmitters, known to be reduced in the memory deficits reported for AD patients [131].

In humans, VitC content of plasma and cerebrospinal fluid (CSF) in relation to cognitive decline in patients with mild to moderate AD (n = 32; mean age: 71 ± 7 years) was investigated [15]. A poor CSF/plasma VitC ratio was found to be a predictor of cognitive decline in AD in that each unit increase in CSF/plasma VitC ratio was associated with 1.1 units less point loss on Mini-Mental-Stale Examination and 2.7 units less loss on Alzheimer's Disease Assessment Scale—Cognitive Section in 12 months [15]. Several studies have found that supplementation with VitC with or without VitE reduces the risk of AD [16,132], while others have failed to find this association [120,133]. In a study by Gray *et al.* [133] 2969 participants (≥65 years) were followed for a mean of 5.5 years and no significant effect of either VitC, VitE or multivitamin was associated with a decreased incidence of AD or dementia. In contrast, Engelhart *et al.* [16] found that supplementation lowered the risk of AD based on findings in 5395 participants (≥55 years) followed from 1990/93 to 1997/99. The reason for inconsistent findings in human studies may be found in the considerable variation in inclusion criteria including those of, e.g., plasma VitC status and definitions of "supplement user" between studies. Since ROS is thought to be a crucial part of the AD disease progression, it can be speculated that a consistently high VitC status acts in a preventive manner, while VitC supplementation *per se* is not a treatment for clinical AD. Thus, infrequent supplement users may not achieve the same benefits as individuals with consistent intake of adequate VitC. To more reliably investigate the possible preventive effect of VitC supplementation on AD development, well-designed randomized, controlled trials with adequate sample size and appropriate inclusion criteria are necessary [134,135]. In Table 2 is presented *in vivo* and clinical studies done on VitC and aging.

Table 2. Vitamin C and aging.

Species	Intervention	Measurement	Outcome	Reference
		In vivo studies		
APP/PSEN1 and B6C3F1/J mice (6–10 months).	VitC in diet (1 g/kg) and high or low dose VitE (750/400 IU/kg).	Functional assessment, amyloid, F_4-neuroprostanes and MDA.	Supplementation with VitC and low VitE decreased markers of oxidative stress in transgenic mice ($p < 0.05$). Improvement of MWM performance was seen in low VitE group ($0.05 > p < 0.001$).	[127]
Swiss mice (3 and 7 months).	IP injection of 60 and 120 mg/kg VitC for three or eight consecutive days.	Elevated plus maze, passive avoidance test.	Treatment improved performance in young animals ($p < 0.05$) and reversed performance deficits in old animals ($p < 0.05$).	[111]
Dunkin Hartley guinea pig (3–9 months and 36–42 months).	Diet containing 325 mg VitC/kg or 100 mg VitC/kg.	VitC, MDA, glutathione, 8-oxodG and SOD in brain. SVCT2 mRNA expression in brain.	Deficiency did not cause significant changes in oxidative stress markers but aging *per se* showed a significant effect ($p < 0.05$). No detectable effect on SVCT mRNA expression in deficient.	[116]
AβPP mice (6–12 months).	1333 mg/kg/day VitC in drinking water.	IHC for anti-Aβ. Western blot, Aβ-ELISA, OxyBlot, glutathione, functional assessment by MVM and elevated plus maze.	VitC prevented some behavioral abnormalities in AβPP mice ($0.05 > p < 0.02$), down-regulated amyloid ($p < 0.05$), significant difference of Aβ42/Aβ40 ratio ($p < 0.02$) and increased in synaptophysin ($p < 0.05$). Phosphorylated tau was decreased ($p < 0.05$).	[129]
Female ovariectomized Wistar rats (80 days).	VitE (40 mg/kg) and VitC (100 mg/kg) IP once daily for 30 days.	MWM, open field test.	Vitamin C + E treatment prevented deficits in reference memory in MWM ($0.01 > p < 0.05$).	[114]
Swiss mice (3 months).	VitC (60 mg/kg) IP injection of for three consecutive days.	Elevated plus maze and passive avoidance.	VitC injection reversed amnesia induced by scopolamine (0.4 mg/kg) and diazepam (1 mg/kg) ($p < 0.05$).	[111]
B6C3F1/J mice (12 weeks).	VitC (125 mg/kg) IP.	Behavioral testing, MDA and Asc content in cortex, AChE activity, brain glutathione.	VitC treatment reversed some of the memory deficits induced by scopolamine (1 mg/kg IP) ($0.05 > p < 0.001$) and increased medial forebrain AChE acticity ($p < 0.001$).	[130]

Species	Intervention	Measurement	Outcome	Reference
		In vivo studies		
CD1 mice (16 months).	Oxiracetam (62.5/125/250 mg/kg). VitC (50/100/200 mg/kg). VitC (125 mg/kg) + oxiracetam (100 mg/kg) IP for three consecutive days.	Light-dark aversion test.	VitC alone or in combination with oxiracetam significantly reduced scopolamine-induced (0.25 mg/kg IP, day 4) amnesia ($p < 0.01$).	[136]
Dunkin Hartley guinea pigs (3–9 and 8–14 months).	Diet with 325 mg/kg VitC	VitC in brain and CSF.	Concentrations of VitC significantly increased in CSF with age ($p < 0.05$). Elevated Asc oxidation ratio in young compared to old animals ($p < 0.05$).	[115]
Dunkin Hartley guinea pigs (3–9 and 36–42 months).	Diet with 325 mg VitC/kg or 100 mg VitC/kg.	VitC in CSF and 8-oxodG, MDA, glutathione and SOD in brain.	No effect was observed besides on VitC concentration in brain and CSF in deficient animals.	[116]

	Design and Subjects	Intervention	Measurements	Outcome	Reference
			Clinical Studies		
Cohort studies	12 AD or dementia patients (71 ± 11 years) and healthy controls (35 ± 5 years).		Blood samples of VitC and DHA.	Dementia and AD patients had significantly lower Asc and DHA levels ($p < 0.001$).	[137]
	Patients ($n = 134$) (AD, vascular dementia or Parkinson's) and 58 matching controls.		Plasma content of: α-carotene, β-carotene, lycopene, VitA, VitC, VitE and TAC.	VitC was significantly lower in AD ($p < 0.001$), vascular dementia ($p < 0.001$) and Parkinson's disease with dementia ($p < 0.01$).	[117]

	Design and Subjects	Intervention	Measurements	Outcome	Reference
			Clinical Studies		
Cohort studies	Prospective cohort study: 633 participants age ≥65 years.	Direct inspection of ingested supplements (two weeks of base-line). Participants were followed for a mean of 4.3 years.		None of the VitE or VitC users developed AD despite a predicted incidence of 3.9 and 3.2, respectively ($p = 0.04$).	[132]
	Nurses' Health study: 14.968 women age 70–79 in 1995–2000.	Semi-quantitative questionnaire on lifestyle, supplemental use and medical history biennially from 1980.	TICS, 10-word list, immediate and delayed recall, verbal fluency, digit span backwards test.	Long-term VitC + E supplementation was associated with better cognitive function ($p = 0.03$) and a trend toward better performance ($p = 0.04$)	[118]
	The Honolulu-Asia Aging Study: 3385 men age 71–93 years.	Questionnaires on vitamin supplementation in 1982/1988.	Assessment of cognitive performance by CASI in 1991–1993.	VitC and/or VitE supplementation decreased the incidence of vascular (OR: 0.12) and mixed/other type dementia (OR: 0.31) and was associated with a higher cognitive performance (OR: 1.25).	[119]
	The Rotterdam Prospective Study: 5395 participants age ≥55 years in 1990–1993	Interview of dietary intake of VitC, VitE, β-carotene, supplements, educational level, *etc.*	Clinical examination and MMSE, GMS, CAMDEX in 1993–1994 and 1997–1999.	High dietary intake of VitC and VitE may lower the risk of Alzheimer's disease. RR = −0.82/standard deviation increase in VitC intake.	[16]
	Prospective cohort study. 32 patients with mild to moderate AD age 71 ± 7 years.	Physical examination.	ADAS-cog, MMSE, CDR and geriatric depression. CSF and blood samples at baseline.	CSF/Plasma VitC content predicted cognitive decline partially due to a compromised blood brain barrier integrity.	[15]
	Nurses' Health Study: 16.010 women age ≥70 years in 1995–2000.	Food-questionnaire in 1980 and expanded version in 1984, 1986 and every four years thereafter. FRAP assessment.	TICS scores and ten word list, global composite scores. East Boston Memory test on three occations.	No significant association between FRAP scores and cognitive function, when adjusted for confounders	[138]

Table 2. *Cont.*

	Design and Subjects	Intervention	Measurements	Outcome	Reference
			Clinical Studies		
	Adult Changes in Thought Prospective Study: 2969 participants age ≥65 years.	Self-reported VitC, VitE or multivitamin supplement. Participants were followed for a mean of 5.5 years.	Health and lifestyle parameters (e.g., BMI, smoking and alcohol consumption) CASI score every second year.	Neither VitC, VitE nor multivitamin use was associated with a decreased incidence of AD or dementia.	[133]
Cohort studies	Prospective cohort study: 137 elderly age 66–90 years.	Nutritional data collected in 1980 and 1986.	Cognitive evaluation in Logical Memory, Abstraction and Visual Reproduction trials in 1986.	Plasma concentrations of VitC were positively correlated with Rey-Osterrieth Copy test performance and Visual Reproduction ($p < 0.05$).	[139]
	Prospective cohort study: 921 elderly age ≥65 years.	A one week food diary or interviews to quantify consumer habits. Participants were followed for 20 years.	Medical examination including Hodkinson Abbreviated Mental test.	Participants with the lowest dietary/plasma VitC status had the poorest cognitive function (OR: 1.6).	[3]
Clinical trials	Randomized open-label clinical trial: 23 AD patients receiving cholinergic treatment.	400 IU VitE and 1000 mg VitC per day or no vitamin treatment. CSF samples at baseline, one month and twelve months.	Clinical and neuropsychological assessment.	Significant increases in VitC content in CSF and decreases in autoxidation ($p < 0.05$). No neuropsychological differences	[140]

Abbreviations: VitC, vitamin C; VitE, vitamin E; DHA, dehydroascorbic acid; Asc, ascorbate; VitA, vitamin A; AchE, Acethylcholine esterase; TAC, total antioxidant capacity; SOD, superoxide dismutase; MDA, malondialdehyde; SVCT, sodium-dependent vitamin C transporter; Aβ, beta-amyloid; AD, Alzheimer's disease; CSF, cerebrospinal fluid; BMI, body mass index; ELISA, enzyme-linked immunosorbent assay; IP, intra-peritoneal; IHC, immunohistochemistry; FRAP, ferric reducing antioxidant capacity; MMSE, Mini Mental State Examination; CAMDEX, Cambridge Mental Disorders of the Elderly Examination; ADAS-cog, Alzheimer's Disease Assessment Scale; TICS, Telephone Interview of Cognitive Status; CASI, Cognitive Abilities Screening Instrument; GMS, Geriatric Mental State; CDR, Clinical Dementia Rating; RR, relative risk; OR, odds ratio; SD, standard deviation.

4.3. Vitamin C and Stroke

Another disease that frequently leads to negative consequences on cognitive ability is stroke. Animal model studies have shown that brain VitC concentrations increase during ischemia, leading to speculations that VitC may play a neuroprotective role in these events [141]. In rodents and primates, VitC supplementation has been shown to limit the infarct area produced by middle cerebral artery occlusion (MCAO) both with and without subsequent reperfusion [51,52,142,143]. C57BL/6J mice subjected to transient or permanent MCAO and treated with administration of either ascorbate (250/500 mg/kg intravenous (IV)) or DHA (40/250/500 mg/kg IV) immediately before, 15 min after or 3 h after MCAO suggests a beneficial and dose-dependent effect of DHA on cerebral blood flow with subsequent reduced infarction-size and mortality, whereas ascorbate did not result in comparable effects [51].

In spontaneously hypertensive and stroke prone rats (SHR-SP), the expression of proteins such as glutathione S-transferase glutathione peroxidase are reportedly decreased together with brain total antioxidant capacity, while MDA is increased compared to spontaneously hypertensive rat (SHR) that are not stroke prone, indicating that increased oxidative stress may be a risk factor in the progression of stroke-related disease [52]. To further examine this, 24 h-MCAO was performed following four weeks of ascorbate 200 mg/kg) and VitE (100 mg/kg) supplementation PO once daily in both SHR-SP and SHR rats. The mean infarct area of SHR-SP was significantly larger than SHR with areas of 31.6% ± 5.4% and 23.0% ± 3.3%, respectively ($p = 0.004$) [52]. Treatment of SHR-SP rats with VitC and VitE decreased MDA, increased total antioxidant capacity and glutathione peroxidase activity and decreased the infarct area significantly, suggesting protective means of antioxidant treatment on oxidative stress and ischemia. Moreover, treatment of Macacca radiata monkeys with ascorbate (500 mg/kg; maximum of 2 grams, IV) before MCAO significantly reduced infarct size compared to control animals (7.3% ± 2.7% *vs.* 22.1% ± 6.7%; $p = 0.0003$), supporting a protective role of VitC in the amelioration of stroke induced damage [142].

Table 3. Vitamin C and stroke.

	Species	Intervention	Measurement	Outcome	Reference
			In vivo studies		
	C57BL/6J mice.	DHA (40/250/500 mg/kg) or Asc (250/500 mg/kg) IV on three time points following MCAO.	Cortical cerebral blood flow, infarct volume, neurological assessment, mortality.	DHA improved cerebral blood flow dose-dependent. Decreased infarct size and mortality ($p < 0.05$). Asc did not show these effects.	[51]
	SHR and SHR-SP rats (4–5 months old).	VitC (200 mg/kg) and VitE (100 mg/kg) PO once daily for 4 weeks MCAO of 24 h duration	2D Western blot of antioxidative protein expression, TAC, GSH-Px and MDA in brain. Cerebral infarct area.	VitC + E treatment significantly reduced oxidative stress and infarct area in SHR-SP ($p < 0.01$).	[52]
	Male Sprague-Dawley rats (4 weeks old) with or without STZ-induced diabetes for six weeks	VitC (100 mg/kg) PO once daily for 2 weeks following MCAO/Re	Infarct volume and edema, neurological score.	VitC treatment significantly reduced infarct area, edema and neurological score in both non-diabetic and diabetic animals compared to untreated controls ($p < 0.01$).	[143]
	Maccaca radiata monkey.	Ascorbate (500 mg/kg up to 2 g IV) immediately before MCAO of 4 h duration.	Cerebral infarct area.	VitC treatment significantly reduced infarct area ($p = 0.0003$).	[142]
	Design and Subjects	**Intervention**	**Measurements**	**Outcome**	**Reference**
			Clinical Studies		
Cohort studies	Department of Health and Social Security nutritional survey: 730 participants age ≥65 years.	Food diary and interviews. Participants were followed for 20 years.	Plasma VitC, physical examination.	Participants in the highest third of VitC intake had a RR = 0.5, when compared with the lowest third.	[146]
Cohort studies	The Nurses' Health Study: 85,118 participants age 30–55 years.	Semi-quantitative questionnaire on lifestyle, supplemental use and medical history. The participants were followed for 16 years		VitC supplemental use is significantly associated with lower risk of coronary heart disease (RR = 0.72).	[147]
Cohort studies	Cancer-Norfolk prospective study: 20,649 participants age 40–79 years.	Health and lifestyle questionnaire, socioeconomic data	Physical examination, plasma VitC content.	Plasma VitC was inversely related to risk of stroke. Participants in top quantile had a RR = 0.58.	[145]
	Design and Subjects	**Intervention**	**Measurements**	**Outcome**	**Reference**
			Clinical Studies		
Cohort studies	Basel prospective study 2974 men	The participants were followed for 12 years.	Baseline values of VitC and β-carotene in plasma.	Low levels of VitC and β-carotene were related to an increased risk of dying from ischemic heart disease or stroke.	[50]
Clinical trials	Double-blind randomized clinical trial: 40 patients (0–2 years after cardiac transplant).	500 mg VitC and 400 IU VitE twice daily for one year or placebo.	Plasma VitC and VitE content. Average intimal index, coronary endothelium-dependent vasoreactivity.	Supplementation with VitC + E caused retardation of early signs of atherosclerosis associated with heart transplantation ($p = 0.008$)	[148]
Clinical trials	Sixty ischemic stroke patients (72.8 ± 10.4 years). VitC *vs.* non-VitC group and 20 controls (69.8 ± 10.5 years).	500 mg/day VitC IV in addition to standard stroke treatment for ten days starting the day after stroke	NIHSS neurological status and bilirubin, creatinine, uric acid, and TAC day one, three, five and ten of treatment. NIHSS three months after stroke.	No difference in clinical status of patients during the ten day treatment or after the three months follow-up.	[144]

Abbreviations: VitC, vitamin C; VitE, vitamin E; DHA, dehydroascorbic acid; Asc, Ascorbate; TAC, total antioxidant capacity; MDA, malondialdehyde; GSH-Px, glutathione peroxidase; MCAO, middle cerebral artery occlusion; IV, intra-venous; PO, per os; NIHSS, National Institutes of Health Stroke Scale; SHR, spontanous hypertensive rat; SHR-SP, spontaneously hypertensive rat stroke prone; RR, relative risk; STZ, streptozotocin.

Unfortunately, results from human studies have not been as consistent as those from animal studies. In a study of ischemic stroke patients receiving standard stroke treatment with or without VitC supplementation (500 mg/day IV) for ten days, no significant effect of VitC on National Institute of Health-Stroke Scale neurological status ten days or three months after the stroke incident could be detected [144]. However, compared to experimental studies in animals, human treatments are commenced at a much later stage following stroke-diagnosis, and may thus fail in targeting a potential narrow therapeutic window of VitC intervention. Another reason may be that the effect of VitC is more pronounced in the prevention of stroke in humans. Several large-scale human epidemiological studies have found an inverse relationship between plasma VitC and incidence of stroke suggesting that deficiency could be an important contributor to the development of disease [50,145,146,147]. An overview of studies on VitC and stroke in both experimental animal models and in humans is presented in Table 3.

5. CONCLUSIONS

Collectively, controlled experimental animal studies support VitC as a key factor in the prevention of cognitive decline following both aging associated alteration as well as neurodegenerative disorders. Data also supports a direct effect of VitC deficiency on brain function particularly during development and/or regeneration following traumatic brain injury such as ischemic insults. Reports from large population surveys in humans point to VitC deficiency as contributing factor in disease propagation, however, randomized controlled clinical trials have not been able to confirm the putative beneficial effects of VitC supplementation and/or intervention. A possible contributing reason for this apparent discrepancy may include differences in inclusion criteria, typically in recording of VitC status prior to study start, admitting individuals already saturated in VitC and thus unlikely to experience any effects of further supplementation. Moreover, the common use of multivitamins or combinations of vitamins in the intervention studies often precludes the ability to isolate the effects of the individual substances. Consequently, further randomized controlled trials using VitC as a single substance are required to elaborate on whether findings from experimental models translate into effects in humans, possibly focusing on specific subgroups with increased prevalence of VitC deficiency, as well as the identification of potential therapeutic windows.

ACKNOWLEDGMENTS

Jens Lykkesfeldt is supported by the LifePharm Centre for *In Vivo* Pharmacology.

AUTHOR CONTRIBUTIONS

The manuscript was conceived by all authors. Stine Normann Hansen wrote the draft manuscript which was subsequently edited by all authors.

CONFLICTS OF INTEREST

The authors declare no conflict of interest.

REFERENCES

1. Tveden-Nyborg, P.; Johansen, L.K.; Raida, Z.; Villumsen, C.K.; Larsen, J.O.; Lykkesfeldt, J. Vitamin C deficiency in early postnatal life impairs spatial memory and reduces the number of hippocampal neurons in guinea pigs. Am. J. Clin. Nutr. 2009, 90, 540–546.
2. Tveden-Nyborg, P.; Lykkesfeldt, J. Does vitamin C deficiency result in impaired brain development in infants? Redox Rep. 2009, 14, 1–6.
3. Gale, C.R.; Martyn, C.N.; Cooper, C. Cognitive impairment and mortality in a cohort of elderly people. Br. Med. J. 1996, 312, 608–611.
4. Rice, M.E. Ascorbate regulation and its neuroprotective role in the brain. Trends Neurosci. 2000, 23, 209–216.
5. Lykkesfeldt, J.; Trueba, G.P.; Poulsen, H.E.; Christen, S. Vitamin C deficiency in weanling guinea pigs: Differential expression of oxidative stress and DNA repair in liver and brain. Br. J. Nutr. 2007, 98, 1116–1119.
6. Hughes, R.E.; Hurley, R.J.; Jones, P.R. Retention of ascorbic acid by guinea pig tissues. Br. J. Nutr.1971, 26, 433–438.
7. Harrison, F.E.; Dawes, S.M.; Meredith, M.E.; Babaev, V.R.; Li, L.; May, J.M. Low vitamin C and increased oxidative stress and cell death in mice that lack the sodium-dependent vitamin C transporter SVCT2. Free Radic. Biol. Med. 2010, 49, 821–829.
8. Sotiriou, S.; Gispert, S.; Cheng, J.; Wang, Y.H.; Chen, A.; Hoogstraten-Miller, S.; Miller, G.F.; Kwon, O.; Levine, M.; Guttentag, S.H.; et al. Ascorbic acid transporter Slc23a1 is essential for vitamin C transport into the brain and for perinatal survival. Nat. Med. 2002, 8, 514–517.

9. Tveden-Nyborg, P.; Vogt, L.; Schjoldager, J.G.; Jeannet, N.; Hasselholt, S.; Paidi, M.D.; Christen, S.; Lykkesfeldt, J. Maternal vitamin C deficiency during pregnancy persistently impairs hippocampal neurogenesis in offspring of guinea pigs. PLoS One 2012, 7, e48488.

10. Hampl, J.S.; Taylor, C.A.; Johnston, C.S. Vitamin C deficiency and depletion in the United States: The Third National Health and Nutrition Examination Survey, 1988 to 1994. Am. J. Public Health 2004, 94, 870–875.

11. Lykkesfeldt, J.; Michels, A.J.; Frei, B. Vitamin C. Adv. Nutr. 2014, 5, 16–18.

12. Smith, J.L.; Hodges, R.E. Serum levels of vitamin C in relation to dietary and supplemental intake of vitamin C in smokers and nonsmokers. Ann. N. Y. Acad. Sci. 1987, 498, 144–152.

13. Tveden-Nyborg, P.; Lykkesfeldt, J. Does vitamin C deficiency increase lifestyle associated vascular disease progression? Evidence based on experimental and clinical studies. Antioxid. Redox Signal.2013, 19, 2084–2104.

14. Rebec, G.V.; Barton, S.J.; Marseilles, A.M.; Collins, K. Ascorbate treatment attenuates the Huntington behavioral phenotype in mice. Neuroreport 2003, 14, 1263–1265.

15. Bowman, G.L.; Dodge, H.; Frei, B.; Calabrese, C.; Oken, B.S.; Kaye, J.A.; Quinn, J.F. Ascorbic acid and rates of cognitive decline in Alzheimer's disease. J. Alzheimers Dis. 2009, 16, 93–98.

16. Engelhart, M.J.; Geerlings, M.I.; Ruitenberg, A.; van Swieten, J.C.; Holman, A.; Witteman, J.C.M.; Breteler, M.M.B. Dietary intake of antioxidants and risk of Alzheimer disease. J. Am. Med. Assoc. 2002,287, 3223–3229.

17. Valencia, A.; Sapp, E.; Kimm, J.S.; McClory, H.; Reeves, P.B.; Alexander, J.; Ansong, K.A.; Masso, N.; Frosch, M.P.; Kegel, K.B.; et al. Elevated NADPH oxidase activity contributes to oxidative stress and cell death in Huntington's disease. Hum. Mol. Genet. 2013, 22, 1112–1131.

18. Hercberg, S.; Galan, P.; Preziosi, P.; Bertrais, S.; Mennen, L.; Malvy, D.; Roussel, A.M.; Favier, P.; Briancon, S. The Su.Vi.Max study—A randomized, placebo-controlled trial of the health effects of antioxidant vitamins and minerals. Arch. Int. Med. 2004, 164, 2335–2342.

19. Mosdol, A.; Erens, B.; Brunner, E.J. Estimated prevalence and predictors of vitamin C deficiency within UK's low-income population. J. Public Health 2008, 30, 456–460.

20. Wrieden, W.L.; Hannah, M.K.; Bolton-Smith, C.; Tavendale, R.; Morrison, C.; Tunstall-Pedoe, H. Plasma vitamin C and food choice in the third Glasgow Monica population survey. J. Epidemiol. Community Health 2000, 54, 355–360.

21. De Oliveira, A.M.; Rondo, P.H.D.; Barros, S.B.D. Concentrations of ascorbic acid in the plasma of pregnant smokers and nonsmokers and their newborns. Int. J. Vitam. Nutr. Res. 2004, 74, 193–198.

22. Ortega, R.M.; Lopez-Sobaler, A.M.; Quintas, M.E.; Martinez, R.M.; Andres, P. The influence of smoking on vitamin C status during the third trimester of pregnancy and on vitamin C levels in maternal milk. J. Am. Coll. Nutr. 1998, 17, 379–384.

23. Johnston, C.S.; Thompson, L.L. Vitamin C status of an outpatient population. J. Am. Coll. Nutr. 1998,17, 366–370.

24. Dhariwal, K.R.; Hartzell, W.O.; Levine, M. Ascorbic acid and dehydroascorbic acid measurements in human plasma and serum. Am. J. Clin. Nutr. 1991, 54, 712–716.

25. Nishikimi, M.; Kawai, T.; Yagi, K. Guinea pigs possess a highly mutated gene for l-gulono-gamma-lactone oxidase, the key ezyme for L-ascorbic acid biosynthesis missing in this species. J. Biol. Chem.1992, 267, 21967–21972.

26. Nishikimi, M.; Fukuyama, R.; Minoshima, S.; Shimizu, N.; Yagi, K. Cloning and chromosomal mapping of the human nonfunctional gene for l-gulono-gamma-lactone oxidase, the enzyme for L-ascorbic acid biosynthesis missing in man. J. Biol. Chem. 1994, 269, 13685–13688.

27. Chatterjee, I. Evolution and biosynthesis of ascorbic acid. Science 1973, 182, 1271–1272.

28. Nandi, A.; Mukhopadhyay, C.K.; Ghosh, M.K.; Chattopadhyay, D.J.; Chatterjee, I.B. Evolutionary significance of vitamin C biosynthesis in terrestrial vertebrates. Free Radic. Biol. Med. 1997, 22, 1047–1054.

29. Fischer, H.; Schwarzer, C.; Illek, B. Vitamin C controls the cystic fibrosis transmembrane conductance regulator chloride channel. Proc. Natl. Acad. Sci. USA 2004, 101, 3691–3696.

30. Corti, A.; Casini, A.F.; Pompella, A. Cellular pathways for transport and efflux of ascorbate and dehydroascorbate. Arch. Biochem. Biophys. 2010, 500, 107–115.

31. Lindblad, M.; Tveden-Nyborg, P.; Lykkesfeldt, J. Regulation of vitamin C homeostasis during deficiency. Nutrients 2013, 5, 2860–2879.

32. Tsukaguchi, H.; Tokui, T.; Mackenzie, B.; Berger, U.V.; Chen, X.Z.; Wang, Y.X.; Brubaker, R.F.; Hediger, M.A. A family of mammalian Na+-dependent L-ascorbic acid transporters. Nature 1999, 399, 70–75.

33. Lykkesfeldt, J.; Loft, S.; Nielsen, J.B.; Poulsen, H.E. Ascorbic acid and dehydroascorbic acid as biomarkers of oxidative stress caused by smoking. Am. J. Clin. Nutr. 1997, 65, 959–963.

34. Lykkesfeldt, J.; Viscovich, M.; Poulsen, H.E. Ascorbic acid recycling in human erythrocytes is induced by smoking in vivo. Free Radic. Biol. Med. 2003, 35, 1439–1447.

35. Berger, U.V.; Hediger, M.A. The vitamin C transporter SVCT2 is expressed by astrocytes in culture but not in situ. Neuroreport 2000, 11, 1395–1399.

36. Harrison, F.E.; Green, R.J.; Dawes, S.M.; May, J.M. Vitamin C distribution and retention in the mouse brain. Brain Res. 2010, 1348, 181–186.

37. Milby, K.; Oke, A.; Adams, R.N. Detailed mapping of ascorbate distribution in rat brain. Neurosci. Lett.1982, 28, 15–20.

38. Dobbing, J. The later growth of the brain and its vulnerability. Pediatrics 1974, 53, 2–6.

39. Rice, M.E.; Russo-Menna, I. Differential compartmentalization of brain ascorbate and glutathione between neurons and glia. Neuroscience 1998, 82, 1213–1223.

40. Erecinska, M.; Cherian, S.; Silver, I.A. Energy metabolism in mammalian brian during development.Prog. Neurobiol. 2004, 73, 397–445.

41. Mun, G.H.; Kim, M.J.; Lee, J.H.; Kim, H.J.; Chung, Y.H.; Chung, Y.B.; Kang, J.S.; Hwang, Y.I.; Oh, S.H.; Kim, J.G.; et al. Immunohistochemical study of the distribution of sodium-dependent vitamin C transporters in adult rat brain. J. Neurosci. Res. 2006, 83, 919–928.

42. Mefford, I.N.; Oke, A.F.; Adams, R.N. Regional distribution of ascorbate in human brain. Brain Res.1981, 212, 223–226.

43. Grant, M.M.; Barber, V.S.; Griffiths, H.R. The presence of ascorbate induces expression of brain derived neurotrophic factor in SH-SY5Y neuroblastorna cells after peroxide insult, which is associated with increased survival. Proteomics 2005, 5, 534–540.

44. Halliwell, B. Reactive oxygen species and the central nervous system. J. Neurochem. 1992, 59, 1609–1623.

45. Bendich, A.; Machlin, L.J.; Scandurra, O.; Burton, G.W.; Wayner, D.D.M. The antioxidant role of vitamin C. Adv. Free Radic. Biol. Med. 1986, 2, 419–444.

46. Seregi, A.; Schaefer, A.; Komlos, M. Protective role of brain ascorbic acid content against lipid peroxidation. Experientia 1978, 34, 1056–1057.

47. Qiu, S.; Li, L.; Weeber, E.J.; May, J.M. Ascorbate transport by primary cultured neurons and its role in neuronal function and protection against excitotoxicity. J. Neurosci. Res. 2007, 85, 1046–1056.

48. Buettner, G.R.S.; Schafer, F.Q. Ascorbate as an antioxidant. In Vitamin C—Function and Biochemistry in Animals and Plants; Asard, H., May, J., Smirnoff, N., Eds.; Garland Science/BIOS Scientific Publishers: Bodmin, UK, 2004; pp. 173–188.

49. Abramov, A.Y.; Scorziello, A.; Duchen, M.R. Three distinct mechanisms generate oxygen free radicals in neurons and contribute to cell death during anoxia and reoxygenation. J. Neurosci. 2007, 27, 1129–1138.

50. Gey, K.F.; Stahelin, H.B.; Eichholzer, M. Poor plasma status of carotene and vitamin C is associated with higher mortality from ischaemic heart disease and stroke: Basel Prospective Study. Clin. Investig.1993, 71, 3–6.

51. Huang, J.; Agus, D.B.; Winfree, C.J.; Kiss, S.; Mack, W.J.; McTaggart, R.A.; Choudhri, T.F.; Kim, L.J.; Mocco, J.; Pinsky, D.J.; et al. Dehydroascorbic acid, a blood-brain barrier transportable form of vitamin C, mediates potent cerebroprotection in experimental stroke. Proc. Natl. Acad. Sci. USA 2001, 98, 11720–11724.

52. Zhang, X.H.; Lei, H.; Liu, A.J.; Zou, Y.X.; Shen, F.M.; Su, D.F. Increased oxidative stress is responsible for severer cerebral infarction in stroke-prone spontaneously hypertensive rats. CNS Neurosci. Ther.2011, 17, 590–598.

53. Bogdanov, M.B.; Andreassen, O.A.; Dedeoglu, A.; Ferrante, R.J.; Beal, M.F. Increased oxidative damage to DNA in a transgenic mouse model of Huntington's disease. J. Neurochem. 2001, 79, 1246–1249.

54. Heo, J.-H.; Hyon, L.; Lee, K.-M. The possible role of antioxidant vitamin C in Alzheimer's disease treatment and prevention. Am. J. Alzheimers Dis. Dement. 2013, 28, 120–125.

55. Browne, S.E.; Bowling, A.C.; MacGarvey, U.; Baik, M.J.; Berger, S.C.; Muqit, M.M.K.; Bird, E.D.; Beal, M.F. Oxidative damage and metabolic dysfunction in Huntington's disease: Selective vulnerability of the basal ganglia. Ann. Neurol. 1997, 41, 646–653.

56. Do, K.Q.; Trabesinger, A.H.; Kirsten-Kruger, M.; Lauer, C.J.; Dydak, U.; Hell, D.; Holsboer, F.; Boesiger, P.; Cuenod, M. Schizophrenia: Glutathione deficit in cerebrospinal fluid and prefrontal cortex in vivo. Eur. J. Neurosci. 2000, 12, 3721–3728.

57. Akyol, O.; Herken, H.; Uz, E.; Fadillioglu, E.; Unal, S.; Sogut, S.; Ozyurt, H.; Savas, H.A. The indices of endogenous oxidative and antioxidative processes in plasma from schizophrenic patients the possible role of oxidant/antioxidant imbalance. Prog. Neuro-Psychopharmacol. Biol. Psychiatry 2002, 26, 995–1005.

58. Kulak, A.; Cuenod, M.; Do, K.Q. Behavioral phenotyping of glutathione-deficient mice: Relevance to schizophrenia and bipolar disorder. Behav. Brain Res. 2012, 226, 563–570.

59. Lee, J.Y.; Chang, M.Y.; Park, C.H.; Kim, H.Y.; Kim, J.H.; Son, H.; Lee, Y.S.; Lee, S.H. Ascorbate-induced differentiation of embryonic cortical precursors into neurons and astrocytes. J. Neurosci. Res. 2003, 73, 156–165.

60. Eldridge, C.F.; Bunge, M.B.; Bunge, R.P.; Wood, P.M. Differentiation of axon-related schwann-cells in vitro. Ascorbic acid regulates basal lamina assembly and myelin formation. J. Cell Biol. 1987, 105, 1023–1034.

61. Jaber, M.; Robinson, S.W.; Missale, C.; Caron, M.G. Dopamine receptors and brain function.Neuropharmacology 1996, 35, 1503–1519.

62. Beaulieu, J.M.; Gainetdinov, R.R. The physiology, signaling, and pharmacology of dopamine receptors. Mol. Cell. Biol. 2014, 34, 182–217.

63. Diliberto, E.J.; Allen, P.L. Semidehydroascorbate as a product of the enzymic conversion of dopamine to norepinephrine—Coupling of semidehydroascorbate reductase to dopamine-β-hydroxylase. Mol. Pharmacol. 1980, 17, 421–426.

64. Levine, M.; Asher, A.; Pollard, H.; Zinder, O. Ascorbic acid and catecholamine secretion from cultured chromaffin cells. J. Biol. Chem. 1983, 258, 3111–3115.

65. Desole, M.S.; Miele, M.; Enrico, P.; Esposito, G.; Fresu, L.; Denatale, G.; Miele, E. Investigations into the relationship between the dopaminergic system and ascorbic acid in rat striatum. Neurosci. Lett.1991, 127, 34–38.

66. Berman, S.B.; Zigmond, M.J.; Hastings, T.G. Modification of dopamine transporter function: Effect of reactive oxygen species and dopamine. J. Neurochem. 1996, 67, 593–600.

67. Hastings, T.G.; Lewis, D.A.; Zigmond, M.J. Role of oxidation in the neurotoxic effects of intrastriatal dopamine injections. Proc. Natl. Acad. Sci. USA 1996, 93, 1956–1961.

68. Pierce, R.C.; Rebec, G.V. Stimulation of both D1-dopamine and D2-dopamine receptors increases behavioral activation and ascorbate release in the neostriatum of freely moving rats. Eur. J. Pharmacol.1990, 191, 295–302.

69. Bornstein, S.R.; Yoshida-Hiroi, M.; Sotiriou, S.; Levine, M.; Hartwig, H.G.; Nussbaum, R.L.; Eisenhofer, G. Impaired adrenal catecholamine system function in mice with deficiency of the ascorbic acid transporter (SVCT2). FASEB J. 2003, 17, 1928–1930.

70. Kuo, C.H.; Hata, F.; Yoshida, H.; Yamatodani, A.; Wada, H. Effect of ascorbic acid on release of acetylcholine from synaptic vesicles prepared from different species of animals and release of noradrenaline from synaptic vesicles of rat brain. Life Sci. 1979, 24, 911–916.

71. Sandstrom, M.I.; Rebec, G.V. Extracellular ascorbate modulates glutamate dynamics: Role of behavioral activation. BMC Neurosci. 2007, 8, 32.

72. Majewska, M.D.; Bell, J.A.; London, E.D. Regulation of the NMDA receptor by redox phenomena—Inhibitory role of ascorbate. Brain Res. 1990, 537, 328–332.

73. Yusa, T. Increased extracellular ascorbate release reflects glutamate re-uptake during the early stage of reperfusion after forebrain ischemia in rats. Brain Res. 2001, 897, 104–113.

74. Hara, K.; Akiyama, Y. Collagen-related abnormalities, reduction in bone quality, and effects of menatetrenone in rats with a congenital ascorbic acid deficiency. J. Bone Miner. Metab. 2009, 27, 324–332.

75. Telang, S.; Clem, A.L.; Eaton, J.W.; Chesney, J. Depletion of ascorbic acid restricts angiogenesis and retards tumor growth in a mouse model. Neoplasia 2007, 9, 47–56.

76. Peterkofsky, B. Ascorbate requirement for hydroxylation and secretion of procollagen—Relationship to inhibition of collagen synthesis in scurvy. Am. J. Clin. Nutr. 1991, 54, S1135–S1140.

77. May, J.M.; Harrison, F.E. Role of vitamin C in the function of the vascular endothelium. Antioxid. Redox Signal. 2013, 19, 2068–2083.

78. Barnes, M.J.; Constabl, B.J.; Kodicek, E. Excretion of hydroxyproline and other amino acids in scorbutic guinea pigs. Biochim. Biophys. Acta 1969, 184, 358–365.

79. Schofield, C.J.; Ratcliffe, P.J. Oxygen sensing by HIF hydroxylases. Nat. Rev. Mol. Cell Biol. 2004, 5, 343–354.

80. Flashman, E.; Davies, S.L.; Yeoh, K.K.; Schofield, C.J. Investigating the dependence of the hypoxia-inducible factor hydroxylases (factor inhibiting HIF and prolyl hydroxylase domain 2) on ascorbate and other reducing agents. Biochem. J. 2010, 427, 135–142.

81. Tomita, S.; Ueno, M.; Sakamoto, M.; Kitahama, Y.; Ueki, M.; Maekawa, N.; Sakamoto, H.; Gassmann, M.; Kageyama, R.; Ueda, N.; et al. Defective brain development in mice lacking the HIF-1 alpha gene in neural cells. Mol. Cell. Biol. 2003, 23, 6739–6749.

82. Iyer, N.V.; Kotch, L.E.; Agani, F.; Leung, S.W.; Laughner, E.; Wenger, R.H.; Gassmann, M.; Gearhart, J.D.; Lawler, A.M.; Yu, A.Y.; et al. Cellular and developmental control of O2 homeostasis by hypoxia-inducible factor 1 alpha. Genes Dev. 1998, 12, 149–162.

83. Pate, S.K.; Lukert, B.P.; Kipp, D.E. Tissue vitamin C levels of guinea pig offspring are influenced by maternal vitamin C intake during pregnancy. J. Nutr. Biochem. 1996, 7, 524–528.

84. Schjoldager, J.G.; Tveden-Nyborg, P.; Lykkesfeldt, J. Prolonged maternal vitamin C deficiency overrides preferential fetal ascorbate transport but does not influence perinatal survival in guinea pigs. Br. J. Nutr. 2013, 110, 1573–1579.

85. Scaife, A.R.; McNeill, G.; Campbell, D.M.; Martindale, S.; Devereux, G.; Seaton, A. Maternal intake of antioxidant vitamins in pregnancy in relation to maternal and fetal plasma levels at delivery. Br. J. Nutr. 2006, 95, 771–778.

86. Ikonomidou, C.; Kaindl, A.M. Neuronal death and oxidative stress in the developing brain. Antioxid. Redox Signal. 2011, 14, 1535–1550.

87. Harrison, F.E.; Yu, S.S.; van den Bossche, K.L.; Li, L.; May, J.M.; McDonald, M.P. Elevated oxidative stress and sensorimotor deficits but normal cognition in mice that cannot synthesize ascorbic acid. J. Neurochem. 2008, 106, 1198–1208.

88. Burgess, N.; Maguire, E.A.; O'Keefe, J. The human hippocampus and spatial and episodic memory.Neuron 2002, 35, 625–641.

89. Giap, B.T.; Jong, C.N.; Ricker, J.H.; Cullen, N.K.; Zafonte, R.D. The hippocampus: Anatomy, pathophysiology, and regenerative capacity. J. Head Trauma Rehabil. 2000, 15, 875–894.

90. Luciana, M.; Lindeke, L.; Georgieff, M.; Mills, M.; Nelson, C.A. Neurobehavioral evidence for working memory deficits in school-aged children with histories of prematurity. Dev. Med. Child. Neurol. 1999,41, 521–533.

91. Isaacs, E.B.; Lucas, A.; Chong, W.K.; Wood, S.J.; Johnson, C.L.; Marshall, C.; Vargha-Khadem, F.; Gadian, D.G. Hippocampal volume and everyday memory in children of very low birth weight.Pediatr. Res. 2000, 47, 713–720.

92. Tolsa, C.B.; Zimine, S.; Warfield, S.K.; Freschi, M.; Rossignol, A.S.; Lazeyras, F.; Hanquinet, S.; Pfizenmaier, M.; Hüppi, P.S. Early alteration of structural and functional brain development in premature infants born with intrauterine growth restriction. Pediatr. Res. 2004, 56, 132–138.

93. Kratzing, C.C.; Kelly, J.D.; Kratzing, J.E. Ascorbic acid in fetal rat brain. J. Neurochem. 1985, 44, 1623–1624.

94. Adlard, B.P.F.; Desouza, S.W.; Moon, S. Ascorbic acid in fetal human brain. Arch. Dis. Child. 1974, 49, 278–282.

95. Zalani, S.; Rajalakshmi, R.; Parekh, L.J. Ascorbic acid concentration of human fetal tissues in relation to fetal size and gestational age. Br. J. Nutr. 1989, 61, 601–606.

96. Biri, A.; Bozkurt, N.; Turp, A.; Kavutcu, M.; Himmetoglu, O.; Durak, I. Role of oxidative stress in intrauterine growth restriction. Gynecol. Obstet. Investig. 2007, 64, 187–192.

97. Berger, T.M.; Polidori, M.C.; Dabbagh, A.; Evans, P.J.; Halliwell, B.; Morrow, J.D.; Roberts, L.J.; Frei, B. Antioxidant activity of vitamin C in iron-overloaded human plasma. J. Biol. Chem. 1997, 272, 15656–15660.

98. Redman, C.W.; Sargent, I.L. Latest advances in understanding preeclampsia. Science 2005, 308, 1592–1594.

99. Ehrenstein, V.; Rothman, K.J.; Pedersen, L.; Hatch, E.E.; Sorensen, H.T. Pregnancy-associated hypertensive disorders and adult cognitive function among Danish conscripts. Am. J. Epidemiol. 2009,170, 1025–1031.

100. Tuovinen, S.; Raikkonen, K.; Kajantie, E.; Leskinen, J.T.; Henriksson, M.; Pesonen, A.K.; Heinonen, K.; Osmond, C.; Barker, D.; Eriksson, J.G. Hypertensive disorders in pregnancy and intellectual abilities in the offspring in young adulthood: The Helsinki Birth Cohort Study. Ann. Med. 2012, 44, 394–403.

101. Cheng, S.W.; Chou, H.C.; Tsou, K.I.; Fang, L.J.; Tsao, P.N. Delivery before 32 weeks of gestation for maternal preeclampsia: Neonatal outcome and 2-year developmental outcome. Early Hum. Dev. 2004,76, 39–46.

102. Many, A.; Fattal, A.; Leitner, Y.; Kupferminc, M.J.; Harel, S.; Jaffa, A. Neurodevelopmental and cognitive assessment of children born

growth restricted to mothers with and without preeclampsia.Hypertens. Pregnancy 2003, 22, 25–29.

103. Mikhail, M.S.; Anyaegbunam, A.; Garfinkel, D.; Palan, P.R.; Basu, J.; Romney, S.L. Preeclampsia and antioxidant nutrients—Decreased plasma levels of reduced ascorbic acid, alpha-tocopherol and beta-carotene in women with preeclampsia. Am. J. Obstet. Gynecol. 1994, 171, 150–157.

104. Chappell, L.C.; Seed, P.T.; Kelly, F.J.; Briley, A.; Hunt, B.J.; Charnock-Jones, D.S.; Mallet, A.; Poston, L. Vitamin C and E supplementation in women at risk of preeclampsia is associated with changes in indices of oxidative stress and placental function. Am. J. Obstet. Gynecol. 2002, 187, 777–784.

105. Beazley, D.; Ahokas, R.; Livingston, J.; Griggs, M.; Sibai, B.M. Vitamin C and E supplementation in women at high risk for preeclampsia: A double-blind, placebo-controlled trial. Am. J. Obstet. Gynecol.2005, 192, 520–521.

106. Villar, J.; Purwar, M.; Merialdi, M.; Zavaleta, N.; Ngoc, N.T.N.; Anthony, J.; de Greeff, A.; Poston, L.; Shennan, A.; WHO vitamin C and vitamin E trial group. World Health Organisation multicentre randomised trial of supplementation with vitamins C and E among pregnant women at high risk for preeclampsia in populations of low nutritional status from developing countries. BJOG 2009, 116, 780–788.

107. Chappell, L.C.; Seed, P.T.; Briley, A.L.; Kelly, F.J.; Lee, R.; Hunt, B.J.; Parmar, K.; Bewley, S.J.; Shennan, A.H.; Steer, P.J.; et al. Effect of antioxidants on the occurrence of preeclampsia in women at increased risk: A randomised trial. Lancet 1999, 354, 810–816.

108. Harman, D. The aging process. Proc. Natl. Acad. Sci. USA 1981, 78, 7124–7128.

109. Mecocci, P.; Macgarvey, U.; Kaufman, A.E.; Koontz, D.; Shoffner, J.M.; Wallace, D.C.; Beal, M.F. Oxidative damage to mitochondrial-DNA shows marked age-dependent increases in human brain.Ann. Neurol. 1993, 34, 609–616.

110. Bowling, A.C.; Mutisya, E.M.; Walker, L.C.; Price, D.L.; Cork, L.C.; Beal, M.F. Age-dependend impairment of mitochondrial-function in primate brain. J. Neurochem. 1993, 60, 1964–1967.

111. Parle, M.; Dhingra, D. Ascorbic acid: A promising memory enhancer in mice. J. Pharmacol. Sci. 2003,93, 129–135.

112. Matthews, K.; Cauley, J.; Yaffe, K.; Zmuda, J.M. Estrogen replacement therapy and cognitive decline in older community women. J. Am. Geriatr. Soc. 1999, 47, 518–523.

113. Jacobs, D.M.; Tang, M.X.; Stern, Y.; Sano, M.; Marder, K.; Bell, K.L.; Schofield, P.; Dooneief, G.; Gurland, B.; Mayeux, R. Cognitive function in nondemented older women who took estrogen after menopause. Neurology 1998, 50, 368–373.

114. Monteiro, S.C.; Matté, C.; Bavaresco, C.S.; Netto, C.A.; Wyse, A.T.S. Vitamins E and C pretreatment prevents ovariectomy-induced memory deficits in water maze. Neurobiol. Learn. Mem. 2005, 84, 192–199.

115. Lykkesfeldt, J.; Moos, T. Age-dependent change in vitamin C status: A phenomenon of maturation rather than of ageing. Mech. Ageing Dev. 2005, 126, 892–898.

116. Tveden-Nyborg, P.; Hasselholt, S.; Miyashita, N.; Moos, T.; Poulsen, H.E.; Lykkesfeldt, J. Chronic vitamin C deficiency does not accelerate oxidative stress in ageing brains of guinea pigs. Basic Clin. Pharmacol. Toxicol. 2012, 110, 524–529.

117. Foy, C.J.; Passmore, A.P.; Vahidassr, M.D.; Young, I.S.; Lawson, J.T. Plasma chain-breaking antioxidants in Alzheimer's disease, vascular dementia and Parkinson's disease. QJM Mon. J. Assoc. Physicians 1999, 92, 39–45.

118. Grodstein, F.; Chen, J.; Willett, W.C. High-dose antioxidant supplements and cognitive function in community-dwelling elderly women. Am. J. Clin. Nutr. 2003, 77, 975–984.

119. Masaki, K.H.; Losonczy, K.G.; Izmirlian, G.; Foley, D.J.; Ross, G.W.; Petrovitch, H.; Havlik, R.; White, L.R. Association of vitamin E and C supplement use with cognitive function and dementia in elderly men. Neurology 2000, 54, 1265–1272.

120. Fillenbaum, G.G.; Kuchibhatla, M.N.; Hanlon, J.T.; Artz, M.B.; Pieper, C.F.; Schmader, K.E.; Dysken, M.W.; Gray, S.L. Dementia and Alzheimer's disease in community-dwelling elders taking vitamin C and/or vitamin E. Ann. Pharmacother. 2005, 39, 2009–2014.

121. Harrison, F.E.; Bowman, G.L.; Polidori, M.C. Ascorbic acid and the brain: Rationale for the use against cognitive decline. Nutrients 2014, 6, 1752–1781.

122. Brookmeyer, R.; Johnson, E.; Ziegler-Graham, K.; Arrighi, H.M. Forecasting the global burden of Alzheimer's disease. Alzheimers Dement. 2007, 3, 186–191.

123. Bowman, G.L. Ascorbic acid, cognitive function, and Alzheimer's disease: A current review and future direction. Biofactors 2012, 38, 114–122.

124. Polidori, M.C.; Mecocci, P. Plasma susceptibility to free radical-induced antioxidant consumption and lipid peroxidation is

increased in very old subjects with Alzheimer's disease. J. Alzheimers Dis. 2002, 4, 517–522.

125. Riviere, S.; Birlouez-Aragon, I.; Nourhashemi, F.; Vellas, B. Low plasma vitamin C in Alzheimer patients despite an adequate diet. Int. J. Geriatr. Psychiatr. 1998, 13, 749–754.

126. Harrison, F.E.; Allard, J.; Bixler, R.; Usoh, C.; Li, L.; May, J.M.; McDonald, M.P. Antioxidants and cognitive training interact to affect oxidative stress and memory in APP/PSEN1 mice. Nutr. Neurosci.2009, 12, 203–218.

127. Harrison, F.E.; Hosseini, A.H.; McDonald, M.P.; May, J.M. Vitamin C reduces spatial learning deficits in middle-aged and very old APP/PSEN1 transgenic and wild-type mice. Pharmacol. Biochem. Behav.2009, 93, 443–450.

128. Shahidi, S.; Komaki, A.; Mahmoodi, M.; Atrvash, N.; Ghodrati, M. Ascorbic acid supplementation could affect passive avoidance learning and memory in rat. Brain Res. Bull. 2008, 76, 109–113.

129. Murakami, K.; Murata, N.; Ozawa, Y.; Kinoshita, N.; Irie, K.; Shirasawa, T.; Shimizu, T. Vitamin C restores behavioral deficits and amyloid-beta oligomerization without affecting plaque formation in a mouse model of Alzheimer's disease. J. Alzheimers Dis. 2011, 26, 7–18.

130. Harrison, F.E.; Hosseini, A.H.; Dawes, S.M.; Weaver, S.; May, J.M. Ascorbic acid attenuates scopolamine-induced spatial learning deficits in the water maze. Behav. Brain Res. 2009, 205, 550–558.

131. Ward, M.S.; Lamb, J.; May, J.M.; Harrison, F.E. Behavioral and monoamine changes following severe vitamin C deficiency. J. Neurochem. 2013, 124, 363–375.

132. Morris, M.C.; Beckett, L.A.; Scherr, P.A.; Hebert, L.E.; Bennett, D.A.; Field, T.S.; Evans, D.A. Vitamin E and vitamin C supplement use and risk of incident Alzheimer's disease. Alzheimer Dis. Assoc. Dis.1998, 12, 121–126.

133. Gray, S.L.; Anderson, M.L.; Crane, P.K.; Breitner, J.C.S.; McCormick, W.; Bowen, J.D.; Teri, L.; Larson, E. Antioxidant vitamin supplement use and risk of dementia or Alzheimer's disease in older adults. J. Am. Geriatr. Soc. 2008, 56, 291–295.

134. Lykkesfeldt, J.; Poulsen, H.E. Is vitamin C supplementation beneficial? Lessons learned from randomised controlled trials. Br. J. Nutr. 2010, 103, 1251–1259.

135. Frei, B.; Birlouez-Aragon, I.; Lykkesfeldt, J. Authors' perspective: What is the optimum intake of vitamin C in humans? Crit. Rev. Food Sci. Nutr. 2012, 52, 815–829.

136. DeAngelis, L.; Furlan, C. The effects of ascorbic acid and oxiracetam on scopolamine-induced amnesia in a habituation test in aged mice. Neurobiol. Learn. Mem. 1995, 64, 119–124.

137. Barabás, J.; Nagy, E.; Degrell, I. Ascorbic acid in cerebrospinal fluid—A possible protection against free radicals in the brain. Arch. Gerontol. Geriatr. 1995, 21, 43–48.

138. Devore, E.E.; Kang, J.H.; Stampfer, M.J.; Grodstein, F. Total antioxidant capacity of diet in relation to cognitive function and decline. Am. J. Clin. Nutr. 2010, 92, 1157–1164.

139. LaRue, A.; Koehler, K.M.; Wayne, S.J.; Chiulli, S.J.; Haaland, K.Y.; Garry, P.J. Nutritional status and cognitive functioning in a normally aging sample: A 6-year reassessment. Am. J. Clin. Nutr. 1997, 65, 20–29.

140. Arlt, S.; Mueller-Thomsen, T.; Beisiegel, U.; Kontush, A. Effect of one-year vitamin C and E supplementation on cerebrospinal fluid oxidation parameters and clinical course in Alzheimer's disease. Neurochem. Res. 2012, 37, 2706–2714.

141. Landolt, H.; Lutz, T.W.; Langemann, H.; Stauble, D.; Mendelowitsch, A.; Gratzl, O.; Honegger, C.G. Extracellular antioxidants and amino acids in the cortex of the rat: Monitoring by microdialysis of early ischemic changes. J. Cereb. Blood Flow Metab. 1992, 12, 96–102.

142. Henry, P.T.; Chandy, M.J. Effect of ascorbic acid on infarct size in experimental focal cerebral ischaemia and reperfusion in a primate model. Acta Neurochir. 1998, 140, 977–980.

143. Iwata, N.; Okazaki, M.; Xuan, M.Y.; Kamiuchi, S.; Matsuzaki, H.; Hibino, Y. Orally administrated ascorbic acid suppresses neuronal damage and modifies expression of SVCT2 and GLUT1 in the brain of diabetic rats with cerebral ischemia-reperfusion. Nutrients 2014, 6, 1554–1577.

144. Lagowska-Lenard, M.; Stelmasiak, Z.; Bartosik-Psujek, H. Influence of vitamin C on markers of oxidative stress in the earliest period of ischemic stroke. Pharmacol. Rep. 2010, 62, 751–756.

145. Myint, P.K.; Luben, R.N.; Welch, A.A.; Bingham, S.A.; Wareham, N.J.; Khaw, K.T. Plasma vitamin C concentrations predict risk of incident stroke over 10-year in 20,649 participants of the European prospective investigation into Cancer-Norfolk Prospective Population Study. Am. J. Clin. Nutr. 2008,87, 64–69.

146. Gale, C.R.; Martyn, C.N.; Winter, P.D.; Cooper, C. Vitamin C and risk of death from stroke and coronary heart disease in cohort of elderly people. Br. Med. J. 1995, 310, 1563–1566.

147. Osganian, S.K.; Stampfer, M.J.; Rimm, E.; Spiegelman, D.; Hu, F.B.; Manson, J.E.; Willett, W.C. Vitamin C and risk of coronary heart disease in women. J. Am. Coll. Cardiol. 2003, 42, 246–252.

148. Fang, J.C.; Kinlay, S.; Beltrame, J.; Hikiti, H.; Wainstein, M.; Behrendt, D.; Suh, J.; Frei, B.; Mudge, G.H.; Selwyn, A.P.; et al. Effect of vitamins C and E on progression of transplant-associated arteriosclerosis: A randomised trial. Lancet 2002, 359, 1108–1113.

3

THE ROLE OF GANGLIOSIDES IN NEURODEVELOPMENT

Kate Palmano[1], Angela Rowan[2,*], Rozey Guillermo[3], Jian Guan[3] and Paul Mc Jarrow[2]

[1]33 Centennial Dr, Whitianga 3510, New Zealand

[2]Fonterra Co-operative Group Ltd., Private Bag 11029, Palmerston North 4442, New Zealand

[3]Centre for Brain Research, Auckland University, Private Bag 92019, Auckland 1142, New Zealand

ABSTRACT

Gangliosides are important components of neuronal cell membranes and it is widely accepted that they play a critical role in neuronal and brain development. They are functionally involved in neurotransmission and are thought to support the formation and stabilization of functional synapses and neural circuits required as the structural basis of memory and learning. Available evidence, as reviewed herein, suggests that dietary gangliosides may impact positively on cognitive functions, particularly in the early postnatal period when the brain is still growing. Further, new evidence suggests that the mechanism of action may be through an effect on the neuroplasticity of the brain, mediated through enhanced synaptic plasticity in the hippocampus and nigro-striatal dopaminergic pathway.

Keywords: gangliosides; brain; cognition; nutrition; neurodevelopment; neuroplasticity

1. INTRODUCTION

Gangliosides are sialylated glycosphingolipids which are widely distributed throughout body tissues, principally as components of cell membranes [1]. They are expressed more predominantly in nervous tissue and are particularly abundant in brain, where they constitute 10%–12% of the lipid matter of the neuronal membrane and are largely concentrated in the grey matter [2,3,4,5,6]. Gangliosides are situated in the outer leaflet of plasma membranes [7] where they are anchored by their ceramide lipid moiety with the glycans extending into the extracellular space. They are largely localised in membrane microdomains, also known as lipid rafts, in association with other sphingolipids and cholesterol [5,8,9]. Lateral interactions within these lipid rafts and with membrane proteins play an important part in cellular processes including cytokine and adhesion signal transduction and membrane protein regulation [8,10,11]. Gangliosides contribute significantly to the cell surface glycans in neuronal cells [6,12] and constitute 75% of the conjugated sialic acid in brain [9]. They mediate functions such as cell-cell recognition through their specific sialoglycan components [10], and cell adhesion, motility and growth through glycosynaptic microdomains [13].

In humans, brain growth occurs from 12 weeks of gestation through the first three years of infancy. Adequate nutrition to support the rapid growth and development of the brain during this period is paramount [14,15]. Nutritional deprivation results in impaired development, fewer neuronal connections, impaired synaptic connectivity and irreversible consequences for cognitive function throughout life [16]. It is widely accepted that gangliosides play a critical role in neuronal function and brain development, affecting such processes as neurotransmission, neurogenesis, synaptogenesis, modulating synaptic transmission, cell proliferation, and neuronal differentiation [6,12,17,18,19,20]. Many of these processes are fundamental to the so-called neuroplasticity of the brain i.e. the ability of the brain to undergo the activity-dependent functional and morphological remodeling which subserves learning and memory [21]. While prenatal brain development mainly involves neurogenesis and migration, postnatal brain development largely involves remodeling, specifically vascular remodeling, white matter (myelin) development and synaptic development and pruning [22,23]. The most dramatic brain remodeling takes place during the first postnatal year but remodeling continues throughout life such that the mature adult brain retains considerable functional plasticity [23,24].

Further, there is mounting evidence that diet can influence brain plasticity throughout life [25].

With the growing recognition of the importance of specific nutrients in the development and maintenance of the cognitive functions of the brain, gangliosides have become a new target for infant nutrition. This review briefly summarizes the role of gangliosides in brain development, function and neuroplasticity, with emphasis on the link between dietary intake of gangliosides and effect on cognitive functioning, and special focus on the role of ganglioside supplementation in neonates. New insights into mechanisms through which dietary gangliosides may exert their effect in both young and old brain are discussed.

2. STRUCTURE OF GANGLIOSIDES

Gangliosides consist of a ceramide core to which glycans are attached through a single glycosidic linkage at the 1-hydroxyl position [6]. The glycan chains, based on a combination of glucose, galactose and *N*-acetylgalactosamine, contain between 1–4 (and unusually up to 7) sialic residues [9] and are highly heterogeneous, resulting in high structural and functional diversity. The sialic acid ligand in brain gangliosides is exclusively of the Neu5Ac type, but this can be further modified by acetylation at the 4, 7 or 9 hydroxyl groups, with major effects on bioactivity and molecular recognition [26]. Additional heterogeneity in the ceramide moiety adds further complexity but in mammalian brain C18- and C20- sphingosine are the predominant sphingoid bases while the fatty acid ligand is typically C18 [6,9,27]. The long chain saturated fatty acid contributes to structural rigidity and increased lateral self-association of gangliosides within the outer leaflet of the membrane [9].

The commonly applied Svennerholm system of ganglioside nomenclature [28] is used herein. Thus G denotes ganglioside, M/D/T/Q (mono-/di-/tri-/tetra- *etc.*) indicate the number of sialic acid residues, with an assigned number being originally based on the migration order of the gangliosides on thin layer chromatography (e.g.GM3 > GM2 > GM1; see Figure 1 for structures). Gangliosides having 0, 1, 2 or 3 sialic acid residue(s) linked to the inner galactose residue are further classified into asialo, a-, b- and c-series, respectively (see Figure 1). Subclass definition of gangliosides (and glycosphingolipids) is based on the neutral carbohydrate core attached to the ceramide (for greater detail see [6,9,20,29,30]).

3. GANGLIOSIDES IN HUMAN BRAIN

The adult brain contains ganglioside of higher complexity than other tissues [9]. During development of the brain, the pattern and expression levels of gangliosides undergo marked change, from the simple GM3 and GD3 forms predominant in the embryonic brain, to the more complex and closely related forms GM1a, GD1a, GD1b and GT1b, for which GM3 and GD3 are precursors [20,31,32]. These four complex gangliosides are the dominant species in the human brain, comprising up to 90% of gangliosides present in the adult brain [33]. Other brain gangliosides include 9-*O*-acetyl GD3 as well as the complex forms GM2, GT1a and GQ1b [18,20].

Changes in brain ganglioside expression are highly region-specific [18], appear to be tightly regulated and correlate with neurodevelopmental milestones including neural tube formation, neuritogenesis, synaptogenesis, axonogenesis and myelination [4,20,34]. A dramatic increase in brain ganglioside accretion occurs in the 10th week of gestation and continues through the first five postnatal years [35], coinciding with the most active period of myelination [36]. In particular, GM1a and GD1a are increased 12–15 fold during this period with the most active accretion being at term, coinciding with dendrite arborisation, axonal outgrowth and synaptogenesis [35]. The appearance of GM4, the third most abundant ganglioside in human white matter, occurs after the beginning of myelination [35]. Content and composition of gangliosides in the brain also change during ageing, with as much as 64% of total ganglioside being lost over 6 decades, as well as an increase in GQ1b, GT1b, and GD1b with concomitant decrease in GM1a and GD1a [37].

The change in expression levels and patterns of gangliosides during brain development is largely attributed to spatiotemporally-regulated developmental changes in the expression levels and patterns of endogenous brain glycosyltransferases. These glycosyltransferases are fundamental to ganglioside biosynthesis and in brain, as in other tissues, a series of specific glycosyltransferases catalyse the sequential addition of carbohydrate moieties to a glycosyl ceramide precursor, with complex gangliosides being synthesized sequentially from simple gangliosides in different pathways, see Figure 1, [38,39,40,41]. GM3 synthase occurs at a pivotal point in this pathway, producing the simple ganglioside GM3, which serves as a precursor to complex gangliosides of the a- and b-series that are common in the brain [40]. Conversely, ganglioside catabolism occurs stepwise via glycan-specific exoglycosidases (reviewed in [38,42,43]), with the action of sialidases providing a mechanism for functional remodeling *in vivo* [5,12].

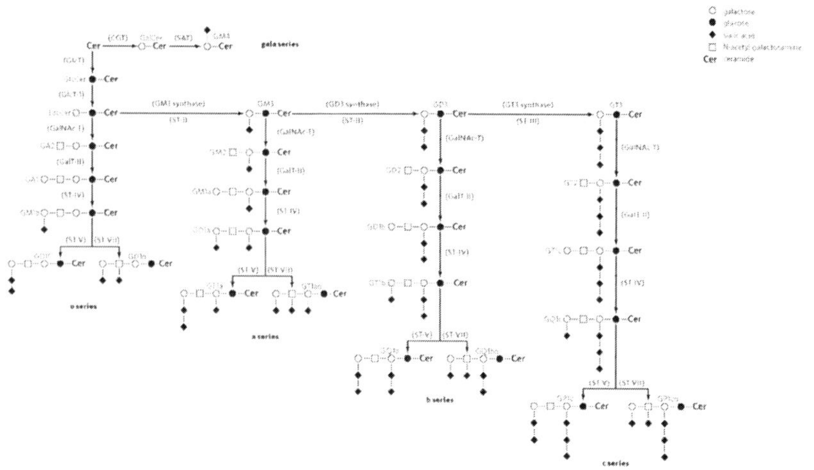

Figure 1. Biosynthetic pathways and nomenclature of major brain gangliosides. The gangliosides are synthesised sequentially by a series of glycosyl and sialyl (S) transferases (T).

4. ROLE OF GANGLIOSIDES IN NEURODEVELOPMENT AND NEURO-MAINTENANCE

Early studies on isolated neurons indicated an enrichment of gangliosides in terminal axons and synaptic endings, suggesting a role in formation, development and maintenance of the nervous system [44]. The importance of gangliosides in normal functioning and integrity of the nervous system and in neurodevelopment and neural regeneration has since been demonstrated by studies with mice genetically engineered to be deficient in gangliosides expressed in both developing and mature brain (GM3, GD3, GM1a, GD1a, GD1b, GT1b). The range of structural and functional defects resulting from selective knock-out of ganglioside synthase genes included altered neural development, neuronal degeneration, loss of sensory and motor functions, demyelination, axonal deterioration, behavioral abnormalities and learning and memory loss [9,20,45,46,47]. In particular, mice expressing primarily GM3 and devoid of complex brain-type gangliosides (GM1a, GD1a, GD1b, GT1b) suffered weight loss, progressive motor and sensory dysfunction, and deterioration in spatial learning and memory with aging [48,49]. Similarly, postnatal mice with a deficiency in GD3 and the downstream b-series gangliosides were shown to have impaired brain neurogenesis in the hippocampus resulting from a progressive loss in neural stem cell population, and exhibited depression-like behavior [46]. In humans, a defect in the GM3 synthase gene resulting in absence of GM3 and key complex gangliosides in the brain is an underlying cause of an inherited form of epilepsy which manifests not only in seizure onset in infancy, but neurological decline and blindness [50]. Many other disturbances of ganglioside metabolism in humans caused by deficiency of enzymes are associated with neurological disease, leading to uncontrolled accumulation in the brain often with severe consequences (reviewed elsewhere [9,30,51]). In addition to the well-established spectrum of disorders known as gangliosidoses that includes Tay-Sachs disease, Sandhoff disease, and GM1a gangliosidoses, GM1a and GM2 in the brain are also involved in other neurological conditions including Alzheimer's, Huntington's and Parkinson's diseases [52]. There are also implications of a role of gangliosides in other conditions including infection, inflammation, insulin metabolism and cancer [52,53,54]. From a therapeutic point of view, clinical trials have shown that injected ganglioside GM1a was able to remediate symptoms of Parkinson's disease [55] and acute ischaemic stroke [56], and also enhance neurological repair after spinal cord injury [57].

At the cellular level, the ability of exogenously applied gangliosides to potentiate neurite outgrowth in primary neurons, sensory ganglia and neuroblastoma cell lines, and axonal sprouting in regenerating axons has long been established [58,59,60,61,62,63,64]. More recently, a multifactorial mechanism for neurite outgrowth involving GM1a clustering in lipid rafts has been described [65]. Additionally, GD3 is thought to play a crucial role in maintaining the self-renewal capacity of neural stem cells, and hence neurogenesis, by sustaining EFG-induced EFGR signaling through recruitment of, and interaction with, EFGR in lipid raft micro domains [66]. The ganglioside 9-*O*-acetyl GD3 has been implicated in glial-guided neuronal migration and neurite outgrowth in both the developing and adult rat nervous system [67,68], and in facilitation of axonal growth and Schwann cell-induced myelination during both development and regeneration of mouse peripheral nerve [47]. Studies in developing human brain (gestational week 6–15) indicated GM1a involvement in glia-neuronal contacts during neuroblast migration, and GM3 localisation in cell proliferation zones [69]. Additionally, gangliosides can enhance or regulate cell survival. They have been shown to be neuroprotective against toxic insult [70], anti-apoptotic during neural cell differentiation [71], and pro-apoptotic during neural cell proliferation [72].

On another level, gangliosides have been shown to be essential for stabilising the architecture and functioning of lipid rafts, and therefore for maintenance of integrity in nerve tissue [73,74,75]. In this context, GD1a and GT1b are thought to mediate axon-myelin interactions through association with myelin-associated glycoprotein, thus helping to ensure long-term axon-myelin stabilization, axonal protection and regeneration [76,77,78].

5. GANGLIOSIDES IN NEUROPLASTICITY AND MEMORY FORMATION

Gangliosides play a pivotal role in modulating ion channel function and receptor signaling, both of which are key functions in neuronal excitability and synaptic transmission (reviewed in [12,79]). In particular, GM1a has been implicated in maintaining neuronal viability, conduction velocity and excitability through regulation of Na^+ channels, and in combination with GD1a, neuronal Ca^{2+} homeostasis [12,80]. In regard to synaptic transmission, exogenous GM1a and GQ1b were both shown to evoke neurotransmitter release in isolated mouse brain cortical synaptosomes through enhanced depolarisation-induced Ca^{2+} influx into the synaptosomes [81], while GQ1b was shown to be essential for synapse formation and synaptic activity in cultured rat cortical neurons [82]. A well-established model for synaptic transmission involving the Ca^{2+}-binding sialic acid component of gangliosides is described elsewhere [41]. In this regard, interactions between brain gangliosides, calcium, and functional membrane proteins (ion channels, ion pumps, receptors, kinases) are thought to play a crucial modulatory role in the transmission and storage of information involved in memory [34], which forms the basis for learned behaviours and higher order cognitive processing [83]. In particular, gangliosides support the formation and stabilization of functional synapses and neural circuits required as the structural basis of memory, and in some studies, mostly employing GM1a, they have been shown to protect against age- or lesion-induced memory deficit in animals when administered exogenously (for reviews and schema see [34,41]). Synapse strength can be modified over time according to level of stimulation and it this synaptic plasticity, along with structural remodeling of neural networks activated during learning that is thought to be the underlying mechanism for learning and memory (reviewed in [21,83,84]). These cognitive functions are primarily associated with the hippocampal region of the mammalian brain and notably, there is a 30% accretion of ganglioside in the hippocampus between weeks 16 and 22 of gestation [3]. Gangliosides GM1a and GD1b have been shown to improve synaptic plasticity in the CA3 hippocampal region, a region most closely associated with spatial learning and memory [85], via prolonged post-tetanic potentiation [86]. Further, GQ1b has been functionally implicated in long term potentiation (LTP), one of the key types of synaptic plasticity in which the expression of long-lasting activity-dependent synaptic modifications occur in response to periods of high frequency stimulation [21,25,83,84,87]. In particular, GQ1b was shown to participate in activity-dependent LTP in CA1 neurons of guinea pig hippocampal slices, through modulation of NMDA receptors/Ca^{2+}channels [88], while transgenic mice with decreased levels of GQ1b exhibited learning impairment and an altered synaptic plasticity, as evidenced by an attenuation of the induction of LTP in hippocampal CA1 neurons [89]. In mice, inhibition of the biosynthesis of b-pathway gangliosides, including GQ1b, also resulted in impairment of LTP induction in hippocampal CA1 neurons, accompanied by failure to learn in a simple memory task [90]. Moreover, intraperitoneal administration of ganglioside was able to reverse the lead-induced impairment of LTP activity in rat brain [91]. *In vivo*, the relationship between both short term and long term stimulation (behavioural training) and increased expression of brain gangliosides, particularly in the hippocampus, has long been known [92,93,94]. Although exogenous administration of gangliosides into normal or functionally compromised animals has resulted in inconsistent effects on memory, behaviour and learning, possibly due to compositional differences in the ganglioside, mode of administration and different age models (reviewed in [95,96]), some of the findings nevertheless support a positive impact of gangliosides on cognitive functions. Intracerebroventricular injection of GQ1b resulted in improved spatial learning and memory in rats [97] while high doses of bovine brain ganglioside (principally GM1a, GD1a, GD1b, GT1b) administered intraperitoneally improved learning ability and memory retention in neonatal, developing and aged rats [98]. Further, exogenous GM1a improved spatial learning and memory deficits in a rat model of Alzheimer's disease [99]. From another perspective, decrease in functional brain gangliosides induced in rats by intraventricular injection of brain ganglioside antiserum resulted in retrograde amnesia and learning deficit [100,101].

Overall, these findings support the view that the highly regulated differential expression of gangliosides appears to be closely involved in not only shaping the neural architecture of the developing brain and stabilising that architecture throughout life, but also ensuring optimum functioning and adaptation of the neural circuitry involved in neurotransmission, memory and learning.

6. INFLUENCES OF DIET ON BRAIN GANGLIOSIDES

The vital role of gangliosides in pre- and early postnatal development of the brain, and indeed in normal functioning and maintenance of the brain, raises the question of diet and the role it plays in providing adequate nutritional background to support appropriate expression and distribution of brain gangliosides, especially during crucial periods of brain growth, learning and assimilation. There is much evidence to support the notion that diet modulates brain structure and function, in particular brain plasticity (neurogenesis, synaptic formation and function), exerting its influence throughout the lifespan of an organism ([14,25] for review). Currently there is much interest in amelioration of age-related cognitive decline through dietary modification and in this respect, longitudinal studies have shown that frequent dairy intake is associated with better cognitive performance [102], in particular improvement in spatial working memory [103]. In neonates it has long been recognised that maternal diet throughout pregnancy and lactation can affect growth and development of the offspring, and undernutrition can have profound effects on learning and behaviour throughout life [14,15,16,104].

In rats, neonatal undernutrition induced by feeding mothers a low protein diet during lactation resulted in decreased incorporation of exogenous sialic acid into brain gangliosides and glycoprotein sialic acid, accompanied by a learning deficit which persisted into adulthood [105]. Significantly, 80% of the sialic acid injected during the brain growth period was incorporated into the synaptosomes [106]. Short term environmental stimulation in the undernourished pups resulted in improved learning coincident with improved brain ganglioside concentration, but neither parameter could be remediated to those levels found in the well-nourished control group [107]. Along with decreased ganglioside levels, postnatal undernutrition was also associated with structural deficits such as decreased brain glial cell population and reduced capacity to form myelin [108]. Lack of essential fatty acids in the maternal diet during gestation and lactation depressed both brain ganglioside and glycoprotein sialic acid concentrations and irreversibly impaired learning behaviour in the offspring [109]. Although no direct causality can be established from these studies, they nonetheless indicated that dietary-induced modifications in brain ganglioside content might impact upon cognitive functioning. Of note is another study in which similar postnatal undernutrition in rats resulted in a 55% decrease in brain gangliosides at 21 days, with a selective decrease in mono- and disialo-gangliosides [110].

The first evidence in humans that neonatal brain ganglioside concentration could be influenced by diet was provided in a study of infants who died of sudden infant death syndrome, in which it was found that those babies who had been largely breastfed had a 32% higher ganglioside content (measured as ganglioside-bound sialic acid) in their frontal cortices than formula fed babies [111]. Human milk naturally contains gangliosides, and although cow's milk, which provides the basis for most infant formula (IF), contains similar amounts of ganglioside to human breast milk, ganglioside concentration can be substantially lower in unsupplemented IF and indeed, gangliosides may be completely absent from soy-based IF [30,112,113,114]. This implies that ganglioside intake is generally lower in IF-fed infants than in breast-fed infants. There are also some compositional differences. In human milk, GD3 is predominant in the early stages of lactation while GM3 becomes more predominant in later stages [115,116,117,118,119,120]. Minor amounts of GM1a and highly polar gangliosides are also present [121]. In bovine milk and bovine milk-based IF, GD3 is the predominant species, while GM3, GT3, GM2, GD1a and GD1b essentially comprise the remainder [116,117,122]. In both human milk and bovine milk, ganglioside levels are highest in colostrum and decline markedly through transitional to mature milk [115,116,118,123,124]. Irrespective of neonatal *de novo* contribution, this changing content and distribution of gangliosides during lactation implies an essential role of dietary ganglioside in development, with the demand being seemingly greatest during the early postnatal period.

Both observational studies [125,126,127] and clinical trial data have indicated that breast-fed infants have better cognitive performance than formula-fed babies [128,129]. This has provided the springboard for a closer examination of the micronutrient status of IF in relation to brain development and cognition, and the need to supplement IF so as to more closely match human breast milk and ensure optimum nutrition during crucial brain growth periods. Although other dietary components such as ω-3-long chain polyunsaturated fatty acids have been implicated in improved neonatal cognitive development [130,131,132], and docosahexaenoic acid (DHA) is now regularly supplemented into IF [30], the notion that dietary gangliosides may have beneficial effects in brain development and cognitive functions nevertheless remains persuasive in the light of their critical involvement in neural development and function.

An implied assumption of dietary modulation is that the dietary component(s) of interest will survive digestion, be absorbed from the gut and thence distributed to the target or effector site. It has been shown in neonatal rats that dietary ganglioside (principally GD3 in bovine milk complex lipid) can be absorbed from the intestine directly into the microdomains of enterocyte basolateral membranes, and be further distributed such as to increase total ganglioside levels in plasma, brain and retina, with increased GD3 in enterocytes and retina, but no alteration in ganglioside profile in plasma or brain [53,122,133]; and that GM1a exogenously administered to pregnant rats can distribute widely into tissues including brain, and also cross the placenta into fetal tissues [134]. Furthermore, these studies indicated that gangliosides can be remodeled post-absorption according to tissue biochemical signature and demand, and notably in brain where levels of polysialogangliosides, particularly GD1a, were increased as a result of GM1a administration [134]. This was also evidenced by studies where dietary ganglioside administered to weanling rats brought about changes in GQ1b and GM4 profiles in brain myelin and synaptomes, respectively [135]. *Ex vivo* studies have confirmed that milk-derived gangliosides GM3 and GD3 can both be transferred across the human placenta [136]. These studies provide not only evidence of the bioavailability of supplemented dietary gangliosides to both the pre-natal and postnatal brain, but also evidence that dietary gangliosides can modify tissue ganglioside composition. Further, the suggestion that the increase of GD3 in the retina resulting from dietary GD3 supplementation may stimulate neonatal retinal maturation and development [133] provides a functional link.

7. DIETARY GANGLIOSIDES AND COGNITIVE FUNCTIONS

Both animal experiments and human clinical trials have provided some evidence that dietary gangliosides may improve cognitive functions. Rats supplemented with ganglioside-containing bovine complex milk lipid (CML; at dose rates of 0.2% and 1% of the total food intake) from an early age, through weaning and on to young adulthood showed improved cognitive measures of novelty recognition and spatial memory although brain ganglioside concentrations were not different from controls [137]. In a follow-up trial where pregnant rats were supplemented with CML at much lower doses (0.01% and 0.05% of the total food intake), GM1a, GD1a, GD1b and GT1b brain ganglioside concentrations in offspring were increased transiently around birth in the higher dose animals relative to the controls and brain weight was increased, but there were no changes in long-term behavioural functions as measured by standard behavioural tests [138]. However, the authors pointed to limitations in the trial, such as the scheduled feeding regimen which might have affected motivated learning through dopamine-induced changes in neural plasticity, and also the non-deficit model employed. In another study, ganglioside supplementation of the pre-weaning diet of artificially reared rats with CML (0.2% w/v of diet) had no effect on spatial short-term memory although the authors found the results inconclusive due to methodological considerations and in particular, lack of adequate controls. Notably, however, brain GD1b + GQ ganglioside content was significantly higher in naturally suckled pups [96]. Clearly, there are many variables to consider in design of animal trials which will adequately address the hypothesis, among them the developmental window within which gangliosides are likely to have the most impact, dose and duration of intervention and relevance in humans. In humans, the brain growth spurt begins in the last trimester of gestation and continues through to 18 months post-natally [139], whereas in rats this critical period is purely post-natal with highest rate of brain growth occurring between days 6–11 [140].

In a recent preclinical trial using neonatal piglets, in which brain growth and development is similar to humans, dietary phospholipids and gangliosides supplied as a phospholipid-rich milk protein concentrate from postnatal day 2 to day 28 improved spatial learning, moderately increased brain weight, and resulted in greater volumes in multiple brain areas along with more grey and white matter [141]. However, as brain ganglioside levels were not measured, a direct role of dietary ganglioside in the observed outcomes could not be assigned.

To date there have been few human clinical trials involving the impact of orally administered gangliosides on cognitive and other neurological functions but the outcomes have shown promise. In a large scale trial, biomedically manufactured gangliosides GM1a, GD1a, GD1b and GT1b were administered orally for 3 months or more to 2230 children suffering from cerebral palsy [142]. The authors reported that oral ganglioside treatment improved the neurological symptoms associated with cerebral palsy, in particular muscle tension, limb function, language ability and intelligence. Faster improvement was observed with younger children (0–3 years). In a small clinical study conducted to assess the impact of IF supplemented with gangliosides from CML on cognitive functions of normal healthy infants aged 0–6 months, supplementation to breast milk levels resulted in not only increased serum ganglioside levels but also improved cognitive performance on the Griffith's scale (hand-eye co-ordination and performance IQ) when compared with an unsupplemented IF control group. Further, serum ganglioside concentration and cognitive measures in the supplemented group were very similar to a breast milk-fed reference group [143]. GM3 is the predominant ganglioside in human enterocytes and serum [144,145], and also in human breast milk after the first 3 months of lactation [42]. Although dietary gangliosides were provided principally as GD3 both in the conventional and supplemented IF, the serum ratio of GM3 to GD3 remained similar in all groups, indicating that in accordance with data from animal experiments [133], dietary ganglioside can be assimilated and remodeled following uptake into intestinal cells.

CML is a naturally rich source of gangliosides but being derived from bovine milkfat globule membrane material (see [138]) it also contains other components such as phospholipids and fatty acids which might contribute to cognitive development and memory. Cognitive improvements induced by CML supplementation cannot therefore be exclusively assigned to gangliosides. However, observed increases in serum and brain ganglioside levels following CML supplementation [133,143], coupled with lack of effect of supplementation on brain total phospholipid content [133] and provision by CML of less than effective doses of known effectors DHA and choline [138] strongly suggest the involvement of gangliosides. Moreover, [143] provide persuasive argument that the ganglioside component of CML was the likely causative agent of improved cognitive effects in their human clinical trial.

Results from other non-dietary human trials reinforce the notion that interventional gangliosides can influence neurological development and cognitive measures, as well as remediate symptoms of disease. In a small controlled trial it was found that intravenous administration of GM1a to low birth weight infants resulted in significant improvement in neurobehaviour at 6 and 12 months, in particular gross and fine movement, as assessed by the Gesell scale [146] while non-dietary GM1a was shown to diminish neurodevelopmental pathologies resulting from foetal alcohol syndrome [147,148,149]. Intracerebroventricularly administered GM1a was reported to improve cognitive function in patients with Alzheimer's disease [150].

8. MECHANISTIC STUDIES OF DIETARY GANGLIOSIDE EFFECT ON COGNITIVE FUNCTIONS

In a recent study designed to probe the possible mechanisms through which ganglioside/CML supplementation might improve spatial learning and memory in postnatal rats, as observed in an earlier trial [137], our group found that CML supplementation (1% w/w of diet) may enhance neuroplasticity in the CA3 hippocampal region of young normal rats during development, possibly through enhanced synaptogenesis [151]. Our data also indicated that improved cognitive function of CML supplementation did not associate with either vascular remodeling or glutamatergic neuroplasticity in the hippocampus, striatum, and frontal cortex of young normal rats but may be associated with changes in the nigral-striatal dopaminergic pathway, which also plays a key role in learning and memory. We found an increased dopamine output in the striatum of CML-supplemented rats that was related to nigral dopamine expression, and we hypothesise that CML supplementation may result in increased synaptic transmission in the striatal dopamine terminals via an increase in the number of synaptic projections (Figure 2). The hippocampus, visual and auditory cortices and striatum undergo rapid development during early neonatal life, characterised by morphogenesis and synaptogenesis [24]. It is possible that increased synaptogenesis induced by dietary ganglioside during this rapid period of growth results in strengthened synapses and hence improved functionality. This view is supported by earlier work which showed that exogenous application of ganglioside GM1a to lesioned nigral-dopamine neurons favoured the recovery of dopaminergic synaptic function in rat striatum through increased density of striatal nerve terminal networks in unlesioned cells, via collateral sprouting [152].

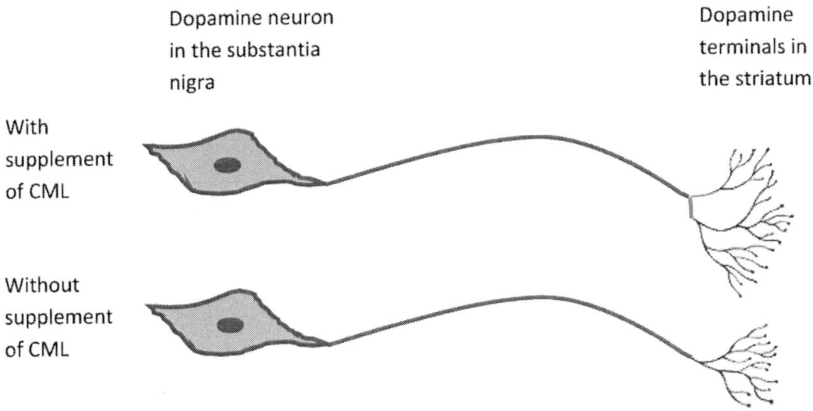

Figure 2. Potential mechanism for the effect of dietary ganglioside-containing complex milk lipid on learning and memory in postnatal rats. The diagram shows the nigra-striatal pathway in which the dopamine neurons project from the nigra to the striatum. Supplementation may increase the synaptic connectivity in the striatal dopamine terminals, depicted as amplification in the number of synaptic processes emerging from the terminal, and hence increase efficiency of dopamine trafficking.

In earlier work it had been found that intraperitoneal injection of GM1a improved spatial learning and memory in cognitively impaired aged rats [153], improved memory deficits in senescent rats [154] and improved spatial learning and memory deficits in a rat model of Alzheimer's disease [99]. In our recent studies on aged rats, we found that long term dietary supplementation of CML also resulted in improved spatial learning and memory and that these benefits were associated with improved vascular density, dopamine output, and neuroplasticity in the brain regions associated with memory [155]. We speculated that CML supplementation promoted slower, but more stable, neuroplasticity in the hippocampus, leading to a more durable memory. As yet, there is little known about either the role (if any) of gangliosides in vascular remodeling or the role of vascular remodeling in learning and memory functions. The contribution of vascular remodeling to the learning and memory outcomes in the aged rats therefore remains unclear.

Glutamate is the most abundant excitatory neurotransmitter in the mammalian central nervous system and is regarded as an initiator and important regulator of synaptic plasticity, particularly through modulation of LTP (reviewed in [156]). However, dopamine, another key neurotransmitter in the brain, plays a critical role in cognitive functioning and is known to improve learning and memory formation

[157,158]. The neurophysiological basis for these effects is also thought to reside in a prolonging effect on neuroplasticity, *i.e.*, through LTP [158]. Given that gangliosides have been functionally implicated in LTP, it is tempting to speculate that improvements in cognitive functions arising from the dietary ganglioside component might also be potentiated through dopamine effects on LTP neuroplasticity in aged rats.

9. CONCLUSIONS

Overall, the current body of research supports the idea that dietary gangliosides are important in supporting brain development early in life and further, that they might impact upon cognitive functions throughout life and crucially during infancy when there is high nutrient demand as the brain undergoes rapid remodeling. Although more research is clearly required, the elucidation of potential mechanisms through which dietary gangliosides might exert their effects on learning and memory is an interesting new development, and lends credence to the notion of nutritional benefit.

Human breast milk is a natural source of gangliosides for the neonate. Since the ganglioside content of IF is generally lower than that of breast milk, it is likely that in those cases where substitute feeding with IF is required, dietary ganglioside supply may be sub-optimal. In this case, routine supplementation of IF with gangliosides to breast milk levels to ensure optimal cognitive outcomes would seem warranted. Also, since nutrition during pregnancy is of prime importance to the developmental and cognitive outcomes of the foetus, and dietary or exogenous gangliosides are able to cross the placenta and assimilate in brain tissues, the concept of supplementing the maternal diet to ensure optimum supply of ganglioside during gestation, as has been suggested elsewhere [159], is an attractive one. Although gangliosides are found in other food products, milk products provide a convenient, natural and relatively rich source for supplementation (see [30]). In particular, bovine milk-derived CML is a food-acceptable and appropriate substrate for ganglioside fortification of IF, and safety of usage for infants has been confirmed in two recent clinical trials [143,160].

ACKNOWLEDGMENTS

The authors are grateful to Claire Woodhall for editorial assistance.

AUTHOR CONTRIBUTIONS

RG contributed background review material as part of an MSc project, JG and PM provided supervision and as well as background expertise on neurodevelopment and milk lipids, AR provided nutrition expertise, KP contributed lipids expertise and led the preparation of the manuscript.

CONFLICTS OF INTEREST

PM and AR are employees of Fonterra Co-operative Group Ltd. Financial support for the background research, literature review and manuscript preparation was received from the New Zealand Primary Growth Partnership (PGP) program, funded by Fonterra Co-operative group and the NZ Ministry for Primary Industries (MPI).

REFERENCES

1. Lucki, N.C.; Sewer, M.B. Nuclear sphingolipid metabolism. Ann. Rev. Physiol. 2012, 74, 131–151.
2. Ledeen, R.W.; Yu, R.K. Gangliosides-structure, isolation, and analysis. Methods Enzymol. 1982, 83, 139–191.
3. Kracun, I.; Rosner, H.; Drnovsek, V.; Vukelic, Z.; Cosovic, C.; Trbojeviccepe, M.; Kubat, M. Ganglioisides in the human brain-development and aging. Neurochem. Int. 1992, 20, 421–431.
4. Yu, R.K.; Nakatani, Y.; Yanagisawa, M. The role of glycosphingolipid metabolism in the developing brain. J. Lipid. Res. 2009, 50, S440–S445.
5. De Chaves, E.P.; Sipione, S. Sphingolipids and gangliosides of the nervous system in membrane function and dysfunction. FEBS Lett. 2010, 584, 1748–1759.
6. Kolter, T. Ganglioside biochemistry. ISRN Biochem. 2012, 2012, 1–36.
7. Ledeen, R. Gangliosides of the neuron. Trends Neurosci. 1985, 8, 169–174.
8. Prinetti, A.; Loberto, N.; Chigorno, V.; Sonnino, S. Glycosphingolipid behaviour in complex membranes. Biochim. Biophy. Acta-Biomembranes 2009, 1788, 184–193.
9. Schnaar, R.L.; Gerardy-Schahn, R.; Hildebrandt, H. Sialic acid in the brain: Gangliosides and polysialic acid in nervous system

development, stability, disease, and regeneration. Physiol. Rev. 2014, 94, 461–518.

10. Lopez, P.H.H.; Schnaar, R.L. Gangliosides in cell recognition and membrane protein regulation. Curr. Opin. Struct. Biol. 2009, 19, 549–557.

11. Inokuchi, J.-I. Membrane microdomains and insulin resistance. FEBS Lett. 2010, 584, 1864–1871.

12. Nowycky, M.C.; Wu, G.; Ledeen, R.W. Glycobiology of ion transport in the nervous system. Adv. Neurobiol. 2014, 9, 321–342.

13. Todeschini, A.R.; Hakomori, S.-I. Functional role of glycosphingolipids and gangliosides in control of cell adhesion, motility, and growth, through glycosynaptic microdomains. Biochim. Biophys. Acta-Gen. Subj. 2008, 1780, 421–433.

14. Morgan, B.L.G. Nutritional requirements for normative development of the brain and behavior. Ann. NY. Acad. Sci. 1990, 602, 127–132.

15. Georgieff, M.K. Nutrition and the developing brain: Nutrient priorities and measurement. Amer. J. Clin. Nutr. 2007, 85, 614S–620S.

16. Uauy, R.; Mena, P.; Peirano, P. Mechanisms for nutrient effects on brain development and cognition.Nestle Nutr. Workshop Ser. Clin. Perform. Program. 2001, 5, 41–70.

17. Chiavegatto, S.; Sun, J.; Nelson, R.J.; Schnaar, R.L. A functional role for complex gangliosides: Motor deficits in GM2/GD2 synthase knockout mice. Exp. Neurol. 2000, 166, 227–234.

18. Kracun, I.; Rosner, H.; Drnovsek, V.; Heffer-Lauc, M.; Cosovic, C.; Lauc, G. Human brain gangliosides in development, aging and disease. Int. J. Devel. Biol. 1991, 35, 289–295.

19. Wu, G.S.; Lu, Z.H.; Wang, J.F.; Wang, Y.; Xie, X.; Meyenhofer, M.F.; Ledeen, R.W. Enhanced susceptibility to kainate-induced seizures, neuronal apoptosis, and death in mice lacking gangliotetraose gangliosides: Protection with LIGA 20, a membrane-permeant analog of GM1. J. Neurosci. 2005, 25, 11014–11022.

20. Yu, R.K.; Tsai, Y.-T.; Ariga, T. Functional roles of gangliosides in neurodevelopment: An overview of recent advances. Neurochem. Res. 2012, 37, 1230–1244.

21. Bruel-Jungerman, E.; Davis, S.; Laroche, S. Brain plasticity mechanisms and memory: A party of four.Neuroscientist 2007, 13, 492–505.

22. Tam, S.J.; Watts, R.J. Connecting vascular and nervous system development: Angiogenesis and the blood-brain barrier. Ann. Rev. Neurosci. 2010, 33, 379–408.

23. Kolb, B.; Gibb, R. Brain plasticity and behaviour in the developing brain. J. Can. Acad. Child. Adolesc. Psychiatry 2011, 20, 265–276.

24. Thompson, R.A.; Nelson, C.A. Developmental science and the media: Early brain development. Am. Psychol. 2001, 56, 5–15.

25. Murphy, T.; Dias, G.P.; Thuret, S. Effects of diet on brain plasticity in animal and human studies: Mind the gap. Neural Plast. 2014, 2014, 563160–563160.

26. Schauer, R. Sialic acids: Fascinating sugars in higher animals and man. Zool. 2004, 107, 49–64.

27. Rosenber, A.; Stern, N. Changes in sphingosine and fatty acid components of gangliosides in developing rat and human brain. J. Lipid. Res. 1966, 7, 122–131.

28. Svennerholm, L. Chromatographic separation of human brain gangliosides. J. Neurochem. 1963, 10, 613–623.

29. Yu, R.K.; Tsai, Y.T.; Ariga, T.; Yanagisawa, M. Structures, biosynthesis, and functions of gangliosides—An overview. J. Oleo. Sci. 2011, 60, 537–544.

30. Mendez-Otero, R.; Pimentel, P.M.; Ukrainstev, S.; McJarrow, P. Role of gangliosides in neurological developmen. In Nutrition in infancy; Waton, R.R., Grimble, G., Preedy, V.R., Zibadi, S., Eds.; Humana Press: New York, NY, USA, 2013; Volume 2, pp. 105–118.

31. Ngamukote, S.; Yanagisawa, M.; Ariga, T.; Ando, S.; Yu, R.K. Developmental changes of glycosphingolipids and expression of glycogenes in mouse brains. J. Neurochem. 2007, 103, 2327–2341.

32. Yu, R.K.; Macala, L.J.; Taki, T.; Weinfeld, H.M.; Yu, F.S. Developmental-changes in ganglioside composition and synthesis in embryonic rat-brain. J. Neurochem. 1988, 50, 1825–1829.

33. Ando, S.; Chang, N.C.; Yu, R.K. High-performance thin-layer chromatography and densitometric determination of brain ganglioside compositions of several species. Anal. Biochem. 1978, 89, 437–450.

34. Rahmann, H. Brain gangliosides and memory formation. Behav. Brain Res. 1995, 66, 105–116.

35. Svennerholm, L.; Bostrom, K.; Fredman, P.; Mansson, J.E.; Rosengren, B.; Rynmark, B.M. Human brain gangliosides-developmental changes from early fetal stage to advanced age. Biochim. Biophys. Acta1989, 1005, 109–117.

36. Kinney, H.C. Human myelination and perinatal white matter disorders. J. Neurolog. Sci. 2005, 228, 190–192.

37. Segler-Stahl, K.; Webster, J.C.; Brunngraber, E.G. Changes in the concentration and composition of human brain gangliosides with aging. Gerontology 1983, 29, 161–168.

38. Sandhoff, K.; Kolter, T. Biosynthesis and degradation of mammalian glycosphingolipids. Philos Trans. R. Soc. Lond. B Biol. Sci. 2003, 358, 847–861.

39. Maccioni, H.J.F. Glycosylation of glycolipids in the golgi complex. J. Neurochem. 2007, 103, 81–90.

40. Proia, R.L. Gangliosides help stabilize the brain. Nat. Genet. 2004, 36, 1147–1148.

41. Wang, B. Molecular mechanism underlying sialic acid as an essential nutrient for brain development and cognition. Adv. Nutr. 2012, 3, 465S–472S.

42. Rueda, R.; Gil, A. Role of gangliosides in infant nutrition. In Lipids in infant nutrition; Huang, Y.S., Sinclair, A.J., Eds.; AOAC Press: Champaign, IL, USA, 1998; pp. 213–234.

43. Sandhoff, K.; Harzer, K. Gangliosides and gangliosidoses: Principles of molecular and metabolic pathogenesis. J. Neurosci. 2013, 33, 10195–10208.

44. Derry, D.M.; Wolfe, L.S. Gangliosides in isolated neurons and glial cells. Science 1967, 158, 1450–1452.

45. Allende, M.L.; Proia, R.L. Simplifying complexity: Genetically resculpting glycosphingolipid synthesis pathways in mice to reveal function. Glycoconj. J. 2014, 31, 613–622.

46. Wang, J.; Cheng, A.; Wakade, C.; Yu, R.K. Ganglioside GD3 is required for neurogenesis and long-term maintenance of neural stem cells in the postnatal mouse brain. J. Neurosci. 2014, 34, 13790–13800.

47. Ribeiro-Resende, V.T.; Gomes, T.A.; de Lima, S.; Nascimento-Lima, M.; Bargas-Rega, M.; Santiago, M.F.; de Melo Reis, R.A.; de Mello, F.G. Mice lacking GD3 synthase display morphological abnormalities in the sciatic nerve and neuronal disturbances during peripheral nerve regeneration.PLoS One 2014, 9, e108919.

48. Tajima, O.; Egashira, N.; Ohmi, Y.; Fukue, Y.; Mishima, K.; Iwasaki, K.; Fujiwara, M.; Inokuchi, J.; Sugiura, Y.; Furukawa, K.; et al. Reduced motor and sensory functions and emotional response in GM3-only mice: Emergence from early stage of life and exacerbation with aging. Behav. Brain Res.2009, 198, 74–82.

49. Tajima, O.; Egashira, N.; Ohmi, Y.; Fukue, Y.; Mishima, K.; Iwasaki, K.; Fujiwara, M.; Sugiura, Y.; Furukawa, K.; Furukawa, K. Dysfunction of muscarinic acetylcholine receptors as a substantial basis for progressive neurological deterioration in GM3-only mice. Behav. Brain Res. 2010, 206, 101–108.

50. Simpson, M.A.; Cross, H.; Proukakis, C.; Priestman, D.A.; Neville, D.C.A.; Reinkensmeier, G.; Wang, H.; Wiznitzer, M.; Gurtz, K.; Verganelaki, A.; et al. Infantile-onset symptomatic epilepsy syndrome caused by a homozygous loss-of-function mutation of GM3 synthase. Nat. Genet. 2004, 36, 1225–1229.

51. Utz, J.R.J.; Crutcher, T.; Schneider, J.; Sorgen, P.; Whitley, C.B. Biomarkers of central nervous system inflammation in infantile and juvenile gangliosidoses. Mol. Gen. Metab. 2015, 114, 274–280.

52. Bisel, B.; Pavone, F.S.; Calamai, M. GM1 and GM2 gangliosides: Recent developments. Biomol. Concepts2014, 5, 87–93.

53. Park, E.J.; Suh, M.; Thomson, B.; Thomson, A.B.R.; Ramanujam, K.S.; Clandinin, M.T. Dietary ganglioside decreases cholesterol content, caveolin expression and inflammatory mediators in rat intestinal microdomains. Glycobiology 2005, 15, 935–942.

54. Rueda, R.; Sabatel, J.L.; Maldonaldo, J.; Molina-Font, J.A.; Gil, A. Addition of gangliosides to an adapted milk formula modifies levels of fecal escherichia coli in preterm newborn infants. J. Pediatr.1998, 133, 90–94.

55. Schneider, J.S.; Gollomp, S.M.; Sendek, S.; Colcher, A.; Cambi, F.; Du, W. A randomized, controlled, delayed start trial of GM1 ganglioside in treated parkinson's disease patients. J. Neurol. Sci. 2013, 324, 140–148.

56. Argentino, C.; Sacchetti, M.L.; Toni, D. GM1 ganglioside therapy in acute ischemic stroke-response.Stroke 1990, 21, 825–825.

57. Geisler, F.H.; Coleman, W.P.; Grieco, G.; Poonian, D.; Sygen Study, G. The sygen® multicenter acute spinal cord injury study. Spine 2001, 26, S87–S98.

58. Roisen, F.J.; Bartfeld, H.; Nagele, R.; Yorke, G. Ganglioside stimulation of axonal sprouting in vitro.Science 1981, 214, 577–578.

59. Byrne, M.C.; Ledeen, R.W.; Roisen, F.J.; Yorke, G.; Sclafani, J.R. Ganglioside-induced neuritogenesis: Verification that gangliosides are the active agents, and comparison of molecular species. J. Neurochem. 1983, 41, 1214–1222.

60. Tsuji, S.; Arita, M.; Nagai, Y. GQ1b, a bioactive ganglioside that exhibits novel nerve growth factor (NGF)-like activities in the two neuroblastoma cell lines. J. Biochem. 1983, 94, 303–306.

61. Gorio, A. Ganglioside enhancement of neuronal differentation, plasticity, and repair. CRC Crit. Rev. Clin. Neurobiol. 1986, 2, 241–296.

62. Ferreira, A.; Busciglio, J.; Landa, C.; Caceres, A. Ganglioside-enhanced neurite growth: Evidence for a selective induction of high-molecular-weight MAP-2. J. Neurosci. 1990, 10, 293–302.

63. Doherty, P.; Ashton, S.V.; Skaper, S.D.; Leon, A.; Walsh, F.S. Ganglioside modulation of neural cell-adhesion molecule and N-cadherin-dependent neurite outgrowth. J. Cell Biol. 1992, 117, 1093–1099.

64. Wu, G.; Fang, Y.; Lu, Z.H.; Ledeen, R.W. Induction of axon-like and dendrite-like processes in neuroblastoma cells. J. Neurocytol. 1998, 27, 1–14.

65. Ichikawa, N.; Iwabuchi, K.; Kurihara, H.; Ishii, K.; Kobayashi, T.; Sasaki, T.; Hattori, N.; Mizuno, Y.; Hozumi, K.; Yamada, Y.; et al. Binding of laminin-1 to monosialoganglioside GM1 in lipid rafts is crucial for neurite outgrowth. J. Cell Sci. 2009, 122, 289–299.

66. Wang, J.; Yu, R.K. Interaction of ganglioside GD3 with an EGF receptor sustains the self-renewal ability of mouse neural stem cells in vitro. Proc. Nat. Acad. Sci. 2013, 110, 19137–19142.

67. Mendez-Otero, R.; Santiago, M.F. Functional role of a specific ganglioside in neuronal migration and neurite outgrowth. Brazil. J. Med. Biol. Res. 2003, 36, 1003–1013.

68. Santiago, M.F.; Liour, S.S.; Mendez-Otero, R.; Yu, R.K. Glial-guided neuronal migration in P19 embryonal carcinoma stem cell aggregates. J. Neurosci. Res. 2005, 81, 9–20.

69. Stojiljkovic, M.; Blagojevic, T.; Vukosavic, S.; Zvezdina, N.; Pekovic, S.; Nikezic, G.; Rakic, L. Ganglioside GM1 and GM3 in early brain development: An immunocytochemical study. Int. J. Dev. Neurosci. 1996, 14, 35–44.

70. Mochetti. Exogenous gangliosides, neuronal plasticity and repair, and the neurotrophins. Cell Mol. Life Sci. 2005, 62, 2283–2294.

71. Bieberich, E.; MacKinnon, S.; Silva, J.; Yu, R.K. Regulation of apoptosis during neuronal differentiation by ceramide and b-series complex gangliosides. J. Biol. Chem. 2001, 276, 44396–44404.

72. Nakatsuji, Y.; Miller, R.H. Selective cell-cycle arrest and induction of apoptosis in proliferating neural cells by ganglioside GM3. Exp. Neurol. 2001, 168, 290–299.

73. Ohmi, Y.; Tajima, O.; Ohkawa, Y.; Furukawa, K.; Furukawa, K. Gangliosides play pivotal roles in the regulation of complement systems and in the maintenance of integrity in nerve tissues. J. Neurochem.2010, 113, 21.

74. Ohmi, Y.; Ohkawa, Y.; Yamauchi, Y.; Tajima, O.; Furukawa, K.; Furukawa, K. Essential roles of gangliosides in the formation and maintenance of membrane microdomains in brain tissues.Neurochem. Res. 2012, 37, 1185–1191.

75. Ohmi, Y.; Tajima, O.; Ohkawa, Y.; Yamauchi, Y.; Sugiura, Y.; Furukawa, K.; Furukawa, K. Gangliosides are essential in the protection of inflammation and neurodegeneration via maintenance of lipid rafts: Elucidation by a series of ganglioside-deficient mutant mice. J. Neurochem. 2011, 116, 926–935.

76. Yang, L.J.S.; Zeller, C.B.; Shaper, N.L.; Kiso, M.; Hasegawa, A.; Shapiro, R.E.; Schnaar, R.L. Gangliosides are neuronal ligands for myelin-associated glycoprotein. Proc. Nat. Acad. Sci. 1996, 93, 814–818.

77. Sheikh, K.A.; Sun, J.; Liu, Y.J.; Kawai, H.; Crawford, T.O.; Proia, R.L.; Griffin, J.W.; Schnaar, R.L. Mice lacking complex gangliosides develop wallerian degeneration and myelination defects. Proc. Nat. Acad. Sci. 1999, 96, 7532–7537.

78. Vajn, K.; Viljetic, B.; Degmecic, I.V.; Schnaar, R.L.; Heffer, M. Differential distribution of major brain gangliosides in the adult mouse central nervous system. PLoS One 2013, 8, e75720.

79. Ledeen, R.W.; Wu, G.S. Ganglioside function in calcium homeostasis and signaling. Neurochem. Res.2002, 27, 637–647.

80. Ledeen, R.W.; Wu, G. Nuclear sphingolipids: Metabolism and signaling. J. Lipid Res. 2008, 49, 1176–1186.

81. Tanaka, Y.; Waki, H.; Kon, K.; Ando, S. Gangliosides enhance KCl-induced Ca2+ influx and acetylcholine release in brain synaptosomes. Neuroreport 1997, 8, 2203–2207.

82. Mizutani, A.; Kuroda, Y.; Muramoto, K.; Kobayashi, K.; Yamagishi, K.; Inokuchi, J. Effects of glucosylceramide synthase inhibitor and ganglioside GQ1b on synchronous oscillations of intracellular Ca2+ in cultured cortical neurons. Biochem. Biophys. Res. Comm. 1996, 222, 494–498.

83. Cooke, S.F.; Bliss, T.V.P. Plasticity in the human central nervous system. Brain 2006, 129, 1659–1673.

84. Nonaka, M.; Fujii, H.; Kim, R.; Kawashima, T.; Okuno, H.; Bito, H. Untangling the two-way signalling route from synapses to the nucleus, and from the nucleus back to the synapses. Philos. Trans. R. Soc. Lond. B Biol. Sci. 2014, 369.

85. Kesner, R.P. Behavioral functions of the CA3 subregion of the hippocampus. Learn. Mem. 2007, 14, 771–781.

86. Egorushkina, N.V.; Ratushnyak, A.S.; Egorushkin, I.V. The influence of exogenous gangliosides on the dynamics of the development of prolonged posttetanic potentiation. Neurosci. Behav. Phys. 1993, 23, 435–438.

87. Bliss, T.V.P.; Collingridge, G.L. A synaptic model of memory-long term potentiation in the hippocampus. Nature 1993, 361, 31–39.

88. Fujii, S.; Igarashi, K.; Sasaki, H.; Furuse, H.; Ito, K.; Kaneko, K.; Kato, H.; Inokuchi, J.; Waki, H.; Ando, S. Effects of the mono- and tetrasialogangliosides GM1 and GQ1b on ATP-induced long-term potentiation in hippocampal CA1 neurons. Glycobiology 2002, 12, 339–344.

89. Ikarashi, K.; Fujiwara, H.; Yamazaki, Y.; Goto, J.-I.; Kaneko, K.; Kato, H.; Fujii, S.; Sasaki, H.; Fukumoto, S.; Furukawa, K.; et al. Impaired hippocampal long-term potentiation and failure of learning in beta 1,4-N-acetylgalactosaminyltransferase gene transgenic mice. Glycobiology 2011, 21, 1373–1381.

90. Fujiwara, H.; Ikarashi, K.; Yamazaki, Y.; Goto, J.-I.; Kaneko, K.; Sugita, M.; Kato, H.; Sasaki, H.; Inokuchi, J.-I.; Furukawa, K.; et al. Impairment of hippocampal long-term potentiation and failure of learning in mice treated with D-threo-1-phenyl-2-decanoylamino-3-morpholino-1-propanol. Biomed. Res. 2012, 33, 265–271.

91. She, J.Q.; Wang, M.; Zhu, D.M.; Sun, L.G.; Ruan, D.Y. Effect of ganglioside on synaptic plasticity of hippocampus in lead-exposed rats in vivo. Brain Res. 2005, 1060, 162–169.

92. Dunn, A.J.; Hogan, E.L. Brain gangliosides: Increased incorporation of (1–3 h) glucosamine during training. Pharmacol. Biochem. Behav. 1975, 3, 605–612.

93. Irwin, L.N.; Samson, F.E., Jr. Content and turnover of gangliosides in rat brain following behavioural stimulation. J. Neurochem. 1971, 18, 203–211.

94. Savaki, H.E.; Levis, G.M. Changes in rat brain gangliosides following active avoidance conditioning.Pharmacol. Biochem. Behav. 1977, 7, 7–12.

95. McJarrow, P.; Schnell, N.; Jumpsen, J.; Clandinin, T. Influence of dietary gangliosides on neonatal brain development. Nutr. Rev. 2009, 67, 451–463.

96. Wainwright, P.E.; Lomanowska, A.M.; McCutcheon, D.; Park, E.J.; Clandinin, M.T.; Ramanujam, K.S. Postnatal dietary supplementation with either gangliosides or choline: Effects on spatial short-term memory in artificially-reared rats. Nutr. Neurosci. 2007, 10, 67–77.

97. Jung, W.R.; Kim, H.G.; Kim, K.L. Ganglioside GQ1b improves spatial learning and memory of rats as measured by the y-maze and the morris water maze tests. Neurosci. Lett. 2008, 439, 220–225.

98. Mei, Z.T.; Zheng, J.Z. Effects of exogenous gangliosides on learning and memory in rats. Jpn. J. Physiol.1993, 43, S295–S299.

99. Yang, R.; Wang, Q.; Min, L.; Sui, R.; Li, J.; Liu, X. Monosialoanglioside improves memory deficits and relieves oxidative stress in the hippocampus of rat model of alzheimer's disease. Neurolog. Sci. 2013,34, 1447–1451.

100. Karpiak, S.; Graf, L.; Rapport, M. Passive avoidance learning is inhibited by antiserum to brain ganglioside. Neurosci. Abstr. 1976, 2, 443.

101. Karpiak, S.E.; Rapport, M.M. Inhibition of consolidation and retrieval stages of passive-avoidance learning by antibodies to gangliosides. Behav. Neural. Biol. 1979, 27, 146–156.

102. Crichton, G.E.; Elias, M.F.; Dore, G.A.; Robbins, M.A. Relation between dairy food intake and cognitive function: The maine-syracuse longitudinal study. Int. Dairy J. 2012, 22, 15–23.

103. Crichton, G.E.; Murphy, K.J.; Howe, P.R.C.; Buckley, J.D.; Bryan, J. Dairy consumption and working memory performance in overweight and obese adults. Appetite 2012, 59, 34–40.

104. Landon, J.; Davison, M.; Krageloh, C.U.; Thompson, N.M.; Miles, J.L.; Vickers, N.H.; Fraser, M.; Breier, B.H. Global undernutrition during gestation influences learning during adult fife. Learn. Behav. 2007,35, 79–86.

105. Morgan, B.L.; Winick, M. Effects of administration of N-acetylneuraminic acid (NANA) on brain nana content and behavior. J. Nutr. 1980, 110, 416–424.

106. Morgan, B.L.; Winick, M. The subcellular localization of administered N-acetylneuraminic acid in the brains of well-fed and protein restricted rats. Brit. J. Nutr. 1981, 46, 231–238.

107. Morgan, B.L.; Winick, M. Effects of environmental stimulation on brain N-acetylneuraminic acid content and behavior. J. Nutr. 1980, 110, 425–432.

108. Morgan, B.L.; Naismith, D.J. The effect of early postnatal undernutrition on the growth and development of the rat brain. Brit. J. Nutr. 1982, 48, 15–23.

109. Morgan, B.L.; Oppenheimer, J.; Winick, M. Effects of essential fatty acid deficiency during late gestation on brain N-acetylneuraminic acid metabolism and behaviour in the progeny. Brit. J. Nutr.1981, 46, 223–230.

110. Vaswani, K.K.; Sharma, M. Effect of neonatal undernutrition on rat brain gangliosides. Int. J. Vit. Nutr. Res. 1985, 55, 323–329.

111. Wang, B.; McVeagh, P.; Petocz, P.; Brand-Miller, J. Brain ganglioside and glycoprotein sialic acid in breastfed compared with formula-fed infants. Am. J. Clin. Nutr. 2003, 78, 1024–1029.

112. Lacomba, R.; Salcedo, J.; Alegria, A.; Barbera, R.; Hueso, P.; Matencio, E.; Jesus Lagarda, M. Sialic acid (N-acetyl and N-glycolylneuraminic acid) and ganglioside in whey protein concentrates and infant formulae. Int. Dairy J. 2011, 21, 887–895.

113. Rueda, R. Gangliosides, immunity, infection and inflammation. In Diet, Immunity and Inflammation; Calder, P.C., Yaqoob, P., Eds.; Woodhead Publishing Ltd.: Cambridge, UK, 2013; pp. 341–354.

114. Zhang, J.; Ren, Y.; Huang, B.; Tao, B.; Pedersen, M.R.; Li, D. Determination of disialoganglioside GD3 and monosialoganglioside GM3 in infant formulas and whey protein concentrates by ultra-performance liquid chromatography/electrospray ionization tandem mass spectrometry. J. Sep. Sci.2012, 35, 937–946.

115. Martin-Sosa, S.; Martin, M.J.; Castro, M.D.; Cabezas, J.A.; Hueso, P. Lactational changes in the fatty acid composition of human milk gangliosides. Lipids 2004, 39, 111–116.

116. Pan, X.L.; Izumi, T. Variation of the ganglioside compositions of human milk, cow's milk and infant formulas. Early Human Dev. 2000, 57, 25–31.

117. Laegreid, A.; Otnaess, A.B.K.; Fuglesang, J. Human and bovine-milk: Comparison of ganglioside composition and enterotoxin-inhibitory activity. Pediatr. Res. 1986, 20, 416–421.

118. Takamizawa, K.; Iwamori, M.; Mutai, M.; Nagai, Y. Selective changes in gangliosides of human milk during lactation: A molecular indicator for the period of lactation. Biochim. Biophys. Acta 1986, 879, 73–77.

119. Uchiyama, S.-i.; Sekiguchi, K.; Akaishi, M.; Anan, A.; Maeda, T.; Izumi, T. Characterization and chronological changes of preterm human milk gangliosides. Nutrition 2011, 27, 998–1001.

120. Giuffrida, F.; Elmelegy, I.M.; Thakkar, S.K.; Marmet, C.; Destaillats, F. Longitudinal evolution of the conentration of gangliosides GM3 and GD3 in human milk. Lipids 2014, 49, 997–1004.

121. Rueda, R.; Puente, R.; Hueso, P.; Maldonado, J.; Gil, A. New data on content and distribution of gangliosides in human milk. Bioli. Chemi. 1995, 376, 723–727.

122. Park, E.J.; Suh, M.; Clandinin, M.T. Dietary ganglioside and long-chain polyunsaturated fatty acids increase ganglioside GD3 content and alter the phospholipid profile in neonatal rat retina. Invest. Ophthalmol. Vis. Sci. 2005, 46, 2571–2575.

123. Rueda, R.; Garcia Salmeron, J.L.; Maldonado, J.; Gil, A. Changes during lactation in ganglioside distribution in human milk from mothers delivering preterm and term infants. Bioli. Chemi. 1996, 377, 599–601.

124. Puente, R.; Garciapardo, L.A.; Hueso, P. Gangliosides in bovine-milk-changes in content and distribution of individual ganglioside levels during lactation. Bioli. Chemi. 1992, 373, 283–288.

125. Anderson, J.W.; Johnstone, B.M.; Remley, D.T. Breast-feeding and cognitive development: A meta-analysis. Am. J. Clin. Nutr. 1999, 70, 525–535.

126. Horwood, L.J.; Darlow, B.A.; Mogridge, N. Breast milk feeding and cognitive ability at 7–8 years.Arch. Dis. Child. Fetal Neonatal Ed. 2001, 84, F23–F27.

127. Syahrir, L.; Fadlyana, E.; Effendi, S.H. Comparison of language and visual-motor developments between exclusively and non-exclusively breastfed infants through cognitive adaptive test/clinical linguistic and auditory milestone scale. Paediatr. Indones. 2009, 49, 337–341.

128. Lucas, A.; Morley, R.; Cole, T.J. Randomised trial of early diet in preterm babies and later intelligence quotient. Brit. Med. J. 1998, 317, 1481–1487.

129. Isaacs, E.B.; Fischl, B.R.; Quinn, B.T.; Chong, W.K.; Gadian, D.G.; Lucas, A. Impact of breast milk on intelligence quotient, brain size, and white matter development. Pediatr. Res. 2010, 67, 357–362.

130. Willatts, P. Long chain polyunsaturated fatty acids improve cognitive development. J. Fam. Health Care2002, 12, 5.

131. Helland, I.B.; Smith, L.; Saarem, K.; Saugstad, O.D.; Drevon, C.A. Maternal supplementation with very-long-chain n-3 fatty acids during pregnancy and lactation augments children's IQ at 4 years of age. Pediatrics 2003, 111, e39–e44.

132. Lauritzen, L.; Jorgensen, M.H.; Olsen, S.F.; Straarup, E.M.; Michaelsen, K.F. Maternal fish oil supplementation in lactation: Effect on developmental outcome in breast-fed infants. Reprod. Nutr. Devel. 2005, 45, 535–547.

133. Park, E.J.; Suh, M.; Ramanujam, K.; Steiner, K.; Begg, D.; Clandinin, T.T. Diet-induced changes in membrane gangliosides in rat intestinal mucosa, plasma and brain. J. Pediatr. Gastroenterol. Nutr. 2005,40, 487–495.

134. Hungund, B.L.; Morishima, H.O.; Gokhale, V.S.; Cooper, T.B. Placental-transfer of (H-3)-GM1 and its distribution to maternal and fetal tissues of the rat. Life Sci. 1993, 53, 113–119.

135. Watson, M.B.; Clandinin, T. Dietary ganglioside and neurochemistry in the developing rat.Glycobiology 2006, 16, 1148–1149.

136. Mitchell, M.D.; Henare, K.; Balakrishnan, B.; Lowe, E.; Fong, B.Y.; McJarrow, P. Transfer of gangliosides across the human placenta. Placenta 2012, 33, 312–316.

137. Vickers, M.H.; Guan, J.; Gustavsson, M.; Krageloh, C.U.; Breier, B.H.; Davison, M.; Fong, B.; Norris, C.; McJarrow, P.; Hodgkinson, S.C. Supplementation with a mixture of complex lipids derived from milk to growing rats results in improvements in parameters related to growth and cognition. Nutr. Res.2009, 29, 426–435.

138. Gustavsson, M.; Hodgkinson, S.C.; Fong, B.; Norris, C.; Guan, J.A.; Krageloh, C.U.; Breier, B.H.; Davison, M.; McJarrow, P.; Vickers, M.H. Maternal supplementation with a complex milk lipid mixture during pregnancy and lactation alters neonatal brain lipid composition but lacks effect on cognitive function in rats. Nutr. Res. 2010, 30, 279–289.

139. Salvati, S.; Attorri, L.; Avellino, C.; Di Biase, A.; Sanchez, M. Diet, lipids and brain development. Devel. Neurosci. 2000, 22, 481–487.

140. Dobbing, J.; Sands, J. Comparative aspects of the brain growth spurt. Early Human Devel. 1979, 3, 79–83.

141. Liu, H.; Radlowski, E.C.; Conrad, M.S.; Li, Y.; Dilger, R.N.; Johnson, R.W. Early supplementation of phospholipids and gangliosides affects brain and cognitive development in neonatal piglets. J. Nutr.2014, 144, 1903–1909.

142. Xu, X.Z.; Zhu, T.-C. Effect of ganglioside in repairing the neurological function of chinese children with cerebral palsy: Analysis of the

curative efficacy in 2230 cases. Chin. J. Clin. Rehab. 2005, 9, 122–123.

143. Gurnida, D.A.; Rowan, A.M.; Idjradinata, P.; Muchtadi, D.; Sekarwana, N. Association of complex lipids containing gangliosides with cognitive development of 6-month-old infants. Early Human Devel.2012, 88, 595–601.

144. Bouhours, D.; Bouhours, J.F. Developmental changes of hematoside of rat small intestine. Postnatal hydroxylation of fatty acids and sialic acid. J. Biol. Chem. 1983, 258, 299–304.

145. Holgersson, J.; Stromberg, N.; Breimer, M.E. Glycolipids of human large-intestine: Difference in glycolipid expression related to anatomical localization, epithelial non-epithelial tissue and the ABO, LE and SE phenotypes of the donors. Biochimie 1988, 70, 1565–1574.

146. Wu, C.; Bai, L.; Wang, W. Neurobehavior effect of GM-1 on LBW infants. Hei Long Jiang Med. J. 2010,34, 410–412.

147. Hungund, B.L.; Ross, D.C.; Gokhale, V.S. Ganglioside GM1 reduces fetal alcohol effects in rat pups exposed to ethanol in utero. Alcohol. Clin. Exp. Res. 1994, 18, 1248–1251.

148. Laev, H.; Karpiak, S.E.; Gokhale, V.S.; Hungund, B.L. In-utero ethanol exposure retards growth and alters morphology of cortical cultures: GM1 reverses effects. Alcohol. Clin. Exp. Res. 1995, 19, 1226–1233.

149. Saito, M.; Mao, R.F.; Wang, R.; Vadasz, C.; Saito, M. Effects of gangliosides on ethanol-induced neurodegeneration in the developing mouse brain. Alcohol. Clin. Exp. Res. 2007, 31, 665–674.

150. Svennerholm, L.; Brane, G.; Karlsson, I.; Lekman, A.; Ramstrom, I.; Wikkelso, C. Alzheimer disease: Effect of continuous intracerebroventricular treatment with GM1 ganglioside and a systematic activation programme. Dement. Geriatr. Cogn. Disord. 2002, 14, 128–136.

151. Guillermo, R.B.; Yang, P.; Vickers, M.H.; McJarrow, P.; Guan, J. Supplementation with complex milk lipids during brain development promotes neuroplasticity without altering myelination or vascular density. Food Nutr. Res. 2015, 59, 25765.

152. Agnati, L.F.; Fuxe, K.; Calza, L.; Benfenati, F.; Cavicchioli, L.; Toffano, G.; Goldstein, M. Gangliosides increase the survival of lesioned nigral dopamine neurons and favour the recovery of dopaminergic synaptic function in striatum of rats by collateral sprouting. Acta Physiol. Scand. 1983, 119, 347–363.

153. Fong, T.G.; Neff, N.H.; Hadjiconstantinou, M. GM1 ganglioside improves spatial learning and memory of aged rats. Behav. Brain Res. 1997, 85, 203–211.

154. Silva, R.H.; Felicio, L.F.; Nasello, A.G.; Vital, M.; Frussa, R. Effect of ganglioside (GM1) on memory in senescent rats. Neurobiol. Aging 1996, 17, 583–586.

155. Guan, J.; MacGibbon, A.; Zhang, R.; Elliffe, D.M.; Moon, S.; Liu, D.-X. Supplementation of complex milk lipid concentrate (CMLc) improved the memory of aged rats. Nutr. Neurosci. 2015, 18, 22–29.

156. Mattson, M.P. Glutamate and neurotrophic factors in neuronal plasticity and disease. In Neural signaling: Opportunities for novel diagnostic approaches and therapies; Goetzl, E.J., Ed.; Wiley-Blackwell: Boston, MA, USA, 2008; Volume 1144, pp. 97–112.

157. Backman, L.; Lindenberger, U.; Li, S.-C.; Nyberg, L. Linking cognitive aging to alterations in dopamine neurotransmitter functioning: Recent data and future avenues. Neurosci. Biobehav. Rev. 2010,34, 670–677.

158. Nitsche, M.A.; Kuo, M.-F.; Grosch, J.; Bergner, C.; Monte-Silva, K.; Paulus, W. D-1-receptor impact on neuroplasticity in humans. J. Neurosci. 2009, 29, 2648–2653.

159. Ryan, J.M.; Rice, G.E.; Mitchell, M.D. The role of gangliosides in brain development and the potential benefits of perinatal supplementation. Nutr. Res. 2013, 33, 877–887.

160. Poppitt, S.D.; McGregor, R.A.; Wiessing, K.R.; Goyal, V.K.; Chitkara, A.J.; Gupta, S.; Palmano, K.; Kuhn-Sherlock, B.; McConnell, M.A. Bovine complex milk lipid containing gangliosides for prevention of rotavirus infection and diarrhoea in northern indian infants. J. Pediatr. Gastroenterol. Nutr. 2014, 59, 167–171.

4

REVIEW OF INFANT FEEDING: KEY FEATURES OF BREAST MILK AND INFANT FORMULA

Camilia R. Martin[1], Pei-Ra Ling[2] and George L. Blackburn[2]

[1]Department of Neonatology and Division of Translational Research, Beth Israel Deaconess Medical Center, Harvard Medical School, Boston, MA 02215, USA

[2]Department of Surgery, Feihe Nutrition Laboratory, Beth Israel Deaconess Medical Center, Harvard Medical School, Boston, MA 02215, USA

ABSTRACT

Mothers' own milk is the best source of nutrition for nearly all infants. Beyond somatic growth, breast milk as a biologic fluid has a variety of other benefits, including modulation of postnatal intestinal function, immune ontogeny, and brain development. Although breastfeeding is highly recommended, breastfeeding may not always be possible, suitable or solely adequate. Infant formula is an industrially produced substitute for infant consumption. Infant formula attempts to mimic the nutritional composition of breast milk as closely as possible, and is based on cow's milk or soymilk. A number of alternatives to cow's milk-based formula also exist. In this article, we review the nutritional information of breast milk and infant formulas for better understanding of the importance of breastfeeding and the uses of infant formula from birth to 12 months of age when a substitute form of nutrition is required.

Keywords: breast milk; infant formula; cow's milk allergy; cow's milk alternatives

1. INTRODUCTION

Mothers' own milk is considered to be the best source of infant nutrition [1]. Extensive evidence has shown that breast milk contains a variety of bioactive agents that modify the function of the gastrointestinal tract and the immune system, as well as in brain development. Thus, breast milk is widely recognized as a biological fluid required for optimal infant growth and development. Recently, studies have further suggested that breast milk mitigates infant programming of late metabolic diseases, particularly protecting against obesity and type 2 diabetes [2].

The World Health Organization recommends that infants should be exclusively breastfed for the first six month of life [3]. The American Academy of Pediatrics also recommends breastfeeding for at least 12 months [4]. Recently, the Academy of Nutrition and Dietetics reaffirms and updates their mission that exclusive breastfeeding provides optimal nutrition and health protection for the first six months of life, and that breastfeeding with complementary foods from six months until at least 12 months of age is the ideal feeding pattern for infants [1]. In addition to its nutritional advantage, breastfeeding is convenient and inexpensive, and also is a bonding experience for the mother and infant.

The decision to breastfeed is highly personal and is often influenced by many factors [5]. Under certain situations, breastfeeding might not be possible, unsuitable or inadequate, which warrants an interruption or cessation in breastfeeding. Globally, only 38% of infants are exclusively breastfed. In the United States, only 75% of infants initiate breastfeeding from birth; however, by the age of three months, 67%, or 2.7 million, of them rely on infant formula for some portion of their nutrition [6]. Among new mothers, the six-month "any breastfeeding" rate for the total U.S. population is 43%, with only 13% meeting the recommendation to breastfeed exclusively for six months [4].

Infant formula is intended as an effective substitute for infant feeding [7,8]. Although production of an identical product to breast milk is not feasible, every effort has been taken to mimic the nutrition profile of human breast milk for normal infant growth and development. Cow milk or soymilk are most commonly used as the base, with supplemental ingredients added to better approximate the composition to human breast milk and to attain health benefits, including iron, nucleotides and compositions of fat blends. The fatty acids of arachidonic acid (AA) and docosahexenoic acid (DHA) are added. Probiotics and compounds, produced by genetic engineering, are either added or currently being considered for addition to formula.

During the first six months of infant life, providing optimal nutrition is critical as the consequences of inadequate nutrition can be very severe. The purpose of this article is to review nutritional information on breast milk and infant formulas to reinforce the importance of breastfeeding, while also understanding the uses of infant formula.

2. HUMAN BREAST MILK

Human breast milk contains carbohydrates, protein, fat, vitamins, minerals, digestive enzymes and hormones. In addition to these nutrients, it is rich in immune cells, including macrophages, stem cells, and numerous other bioactive molecules. Some of these bioactive molecules are protein-derived and lipid-derived, while others are protein-derived and indigestible, such as oligosaccharides. Human milk oligosaccharides (HMOs) possess anti-infective properties against pathogens in the infant gastrointestinal tract, such as *Salmonella, Listeria,* and *Campylobacter*, by flooding the infant gastrointestinal tract with decoys that bind the pathogens and keep them off the intestinal wall [9]. Oligosaccharides also play a vital role in the development of a diverse and balanced microbiota, essential for appropriate innate and adaptive immune responses, and help colonize up to 90% of the infant biome [10].

2.1. Composition of Human Breast Milk

Human breast milk is a complex matrix with a general composition of 87% water, 3.8% fat, 1.0% protein, and 7% lactose. The fat and lactose, respectively, provide 50% and 40% of the total energy of the milk [11]. However, the composition of human breast milk is dynamic and changes over time, adapting itself to the changing needs of the growing child. For instance, during each nursing session, the milk that is expressed first (foremilk) is thinner with a higher content of lactose, which satisfies a baby's thirst, and following the foremilk, hindmilk, is creamier with a much higher content of fat for the baby's needs. Variations are also present with the stage of nursing (age of infant), maternal diet, maternal health, and environmental exposure. During early lactation, the protein content in human milk ranges from 1.4–1.6 g/100 mL, to 0.8–1.0 g/100 mL after three to four months of lactation, to 0.7–0.8 g/100 mL after six months [11,12]. The fat content varies significantly with maternal diet and is also positively related to weight gain during pregnancy. Remarkably, it has been observed that a mother's breast milk is almost always adequate in essential nutrients for her term infant's growth and development, even when her own nutrition is inadequate. Although the mean concentrations of protein, sodium, chloride and potassium in early preterm milk are adequate to meet the estimated requirements for preterm infants, specific nutritional supplementation is required for mother's milk delivered to preterm infants [13,14].

In contrast to protein and fat, lactose content is fairly constant in mature milk (after 21 days postpartum). The stable concentration of lactose is important in maintaining a constant osmotic pressure in human milk. Lactose also aids the absorption of minerals and calcium. In breast milk, many carbohydrate-based bioactive compounds, such oligosaccharides, are attached to lactose. If the small intestine does not produce enough of an enzyme (lactase) to digest these sugar complexes, lactose malabsorption and intolerance syndromes can be observed. Lactase deficiency malabsorption and disease are extremely rare in the exclusively breastfed infant.

2.2. Protein in Human Breast Milk

There are two classes of protein in breast milk: Casein and whey. Casein becomes clots or curds in the stomach; while whey remains as a liquid and is easier to digest. Depending on the stage of milk, 80% to 50% of protein in breast milk is whey [11]. The whey/casein ratio in human milk fluctuates between 70/30 and 80/20 in early lactation and decreases to 50/50 in late lactation [15]. This proportion is significantly greater compared to the milk of other mammals. In cow's milk, whey proteins represent only 18% of milk protein. Traditionally, infant formulas are high in casein, making them harder to digest compared to human breast milk. Because the amino acid profiles of casein and whey proteins are different, the overall amino acid profile of human milk varies depending on the stage of lactation. Glutamine, the most abundant free amino acid, is nearly 20 times higher in mature milk than its lowest value in colostrum [16]. Glutamine is important for providing ketoglutaric acid for the citric acid cycle, possibly acting as a neurotransmitter in the brain, and serving as a major energy substrate for intestinal cells [17].

The main whey proteins are alpha-lactalbumin, lactaferrin and secretory IgA. Other proteins include lysozyme, folate-binding protein, bifidus factor, casein, lipase and amylase, alpha1-antitrypsin and antichymotrypsin, and haptocorrin [11]. After ingestion, these proteins are broken down rapidly to free amino acids for absorption and utilization. Most of these proteins also have bioactive functions and non-nutritive functions [18]. For instance, alpha-lactalbumin is essential for lactose synthesis and binding of Ca and Zn ions. Casein assists to form masses with calcium and phosphorus. Lactoferrin and lysozyme prevent the spread of potentially pathogenic bacteria, preventing illnesses in infants. The IgA antibody destroys bacteria and protects the mucosal surface of the gut.

2.3. Fats in Human Breast Milk

Fats are the most important composition of breast milk, supplying energy and helping the development of the central nervous system. Moreover, milk fat is a carrier of taste and aroma. In general, human breast milk fat content ranges from 3.5% to 4.5% during lactation. The main lipid fraction are triglycerides, which account for about 95% of total lipids. Near half of milk fatty acids are saturated fatty acids, with 23% palmitic acid (C16:0) in total fatty acids [11]. The monounsaturated fatty acid, oleic acid (18:1w9), is in the highest percentage (36%) in milk. Human breast milk also contains two essential fatty acids, linoleic acid (C18:2w6) at 15% and alpha-linolenic acid (C18:3w3) at 0.35% [11]. These two essential fatty acids are, respectively, converted to arachidonic acid (AA, C20:4w6) and eicosapentaenoic acid (EPA, C20:5w3), the latter of which is further converted to docosahexaenoic acid (DHA, 22:6w3). AA, EPA and DHA are important for regulating growth, inflammatory responses, immune function, vision, cognitive development and motor systems in newborns.

Long chain polyunsaturated fatty acids are transferred from mother to fetus in the third trimester through the placenta, and to infants through breast milk after birth [19]. During the last trimester and neonatal period, brain tissue is rapidly synthesized. Cell differentiation and development of active synapses in the brain need specific requirements of DHA and AA. Eighty percent of brain DHA is acquired from the 26th week of gestation until birth. Notably, the synthesis of AA and DHA from linoleic acid (18:2w6) and alpha-linolenic acid (18:3w3) is limited in the fetus and neonate due to the premature enzyme activity. Thus, the required amounts of AA and DHA must come from the mother during pregnancy, or as breast milk after birth. One study has showed that the fat content and the percentage of all polyunsaturated fatty acids in breast milk increase significantly between the sixth week and sixth month of lactation [20]. There is evidence that slowly turning-over maternal body pools of AA are the major source of milk AA [21]. The AA concentration in breast milk is dose-dependently associated with the consumption of AA-rich foods in lactating mothers [22]. Breast milk EPA and DHA concentrations are also closely linked to maternal dietary EPA and DHA intake [23]. Human milk from lactating women consuming vegan or vegetarian diets has <0.1% DHA, compared to mean levels of 0.2%–0.4% DHA in the United States and ≥0.8% DHA in China, where DHA intakes from fish or other sources are high [24]. It is suggested that intakes of ~300 mg of DHA per day are necessary to achieve human milk levels of 0.3%–0.35% of DHA [25]. However, the effects of human milk fatty acids on neurodevelopment is complex, particularly because neurodevelopment is assessed after the period of the first six month of exclusive human milk feeding.

In premature birth, the transmission of these fatty acids is interrupted from the placenta to the fetus during the critical last trimester. Studies also showed that decreased postnatal docosahexaenoic and arachidonic acid blood levels in premature infants are associated with neonatal morbidities [26]. Thus, after birth, the preterm infant is dependent on an adequate diet for sufficient fatty acid levels. Adding DHA and AA to preterm-infant formulas led to initial beneficial effects on visual acuity, visual attention and cognitive development compared with infant receiving no supplementation [27].

2.4. Vitamins, Minerals and Other Bioactive Components in Breast Milk

Human breast milk contains adequate amounts of most vitamins to support normal infant growth, except for vitamins D and K. Infants who are exclusively breastfeeding receive below the minimum recommended intake of vitamin D, and much lower than the recommended dietary intake. These infants are at the risk for vitamin D deficiency, inadequate bone mineralization and conditions such as rickets. However, the overall risk of vitamin D deficiency in breastfed infants is also correlated with overall sun exposure with increasing risk in climates with a lower sun index. Maternal supplementation with 400–2000 IU (International Unit). of vitamin D/day can increase the levels of vitamin D in breast milk, but only a higher dose (2000 IU) achieves satisfactory levels of 25-OH-D in the infant [11]. Normal vitamin D stores present at birth are depleted within eight weeks. Sunlight exposure and vitamin D supplementation are recommended for breastfed infant. Formula-fed infants often have higher serum concentration of vitamin D metabolites than breastfed-infants. Vitamin K is essential to the protein involved in blood coagulation. However, only limited amounts of vitamin K is transferred from the placenta to fetus. Thus, a newborn infant often has an extremely low concentration of vitamin K, and is at risk of developing hemorrhagic disease. After birth, vitamin K supplementation is recommended.

In human breast milk, minerals contribute to a variety of physiological functions, forming essential parts of many enzymes and are of biological important to molecules and structures. The contents of minerals are comparable between human milk and bovine milk. Over the decades, many other bioactive components have been identified in human milk, including hormones, growth factors and immunological factors.

3. HUMAN MILK OPTIONS—MILK DONORS AND MILK BANKS

The World Health Organization and the American Academy of Pediatrics recommends pasteurized human donor milk for preterm infants when a mother's own milk is unavailable [28]. Donor milk undergoes a pasteurization process, which reduces many of the normal commensal microbes, as well as significantly reducing or obliterating live immune cells, bioactive proteins, and enzymes, collectively limiting some of the health benefit compared to a mother's own milk. Research efforts to optimize donor breast milk are ongoing. Before using donor breast milk, the mother should consult with their baby's health care providers.

4. INFANT FORMULAS

Infant formula is intended as an effective substitute to breast milk and is formulated to mimic the nutritional composition of breast milk. The recently updated FDA (Food and Drug Administration) rule on current Good Manufacturing Practices for infant formula, 21 CFR 106.96 [6], requires, among other things, that formulas satisfy the quality factors of normal physical growth and a sufficient biological quality of protein component (adequate amounts of protein in a form that can be used by infants). Infant formula is only for the health of infants without unusual medical or dietary problems. The manufacturing process is highly regulated and monitored to meet national and international quality criteria [29,30].

4.1. The Infant Formula Market

The United Nations estimates that the world's current population of 7.2 billion will grow by one billion over the next 12 years, reaching 9.6 billion by 2050 [31]. This increase will drive the global demand for infant milk formula, especially innovative products that use ingredients such as prebiotics and specific milk protein fractions [31]. Over the next five years, the $50 billion infant formula market is expected to be the fastest-growing packaged food category, achieving gains in excess of 7% a year [32]. Some industry experts predict an even higher annual growth of 8%–9% [32]. According to analyst Diana Cowland, the rapid expansion of infant formulas is set to continue with a compounded annual growth rate of 11%, with demand driven by Asia, and more particularly, China [33].

Infant formulas are available in three forms: (1) powder: The least expensive form of infant formula that must be mixed with water before feeding; (2) liquid: Concentrated liquid that must be mixed with an equal amount of water; and (3) ready-to-feed: The most expensive form of infant formula that requires no mixing.

4.2. Guidelines for Manufacturing of Infant Formula

Infant formulas must include proper amounts of water, carbohydrate, protein, fat, vitamins and minerals. The composition of infant formula is strictly regulated, and each manufacturer must follow established guidelines set by government agencies. For instance, all the major components added to formula (protein, lipids, carbohydrates) have a range of minimum and maximum values for their effectiveness. These components must have established a history of safe use [34]. The required range of each nutrient must be maintained throughout the shelf life of the product [35]. For amino acids, only L forms of amino acid are allowed to be added, while D forms are not permitted because they may cause D-lactic acidosis [36]. Fructose should be avoided due to fructose intolerance. Hydrogenated fats and oils are also not allowed. Ionizing radiation of the formula product is not permitted because it could cause product deterioration [35]. Infant formula prepared ready for consumption should contain no less than 60 kcal (250 kJ) and no more than 70 kcal (295 kJ) of energy per 100 mL (CAC, 1981) [11]. Furthermore, product reformulation must be based on medical and nutritional findings. The committee of the "Evaluation of the addition of Ingredients New to Infant Formula" has recommended that "manufacturers must demonstrate that the formula containing the new ingredient is capable of sustaining physical growth and development over 120 days when formula is likely to be the sole source of infant nutrition" [7].

In the United States, the Food and Drug Administration (FDA) defines that adding new ingredients to infant formula should have "reasonable certainty of no harm" as the safety standard [7]. The World Health Organization (WHO) has noted that unmodified cow's milk should never be fed to infants, and that unmodified goat's milk is also not recommended for infants. With the WHO guidelines, federal and local agencies of different countries control and monitor infant formula regulations, including requirements for quality and manufacturing practices in their own countries. From a manufacturers perspective, it is in their best interest to continuously improve their products to be as close as possible to human breast milk.

4.3. Classes of Infant Formula Products

There are three major classes of infant formulas: Cow-milk based formula, soy-based formula and specialized formula. They vary in nutrition, calories, taste, digestion, and cost. Specific kinds of formulas are available to meet a variety of needs. Some cow's milk substitutes are amino acid based or contain extensively hydrolyzed whey or casein proteins. Some are rice-based formula.

4.4. Cow Milk-Based Formula

Bovine milk is the basis for most infant formula. However, bovine milk contains higher levels of fat, minerals and protein compared to human breast milk. Therefore, cow milk must be skimmed and diluted to more closely resemble human breast milk composition [34,35]. Cow-milk-based infant formula contains added vegetable oils, vitamins, minerals and iron for consumption by most healthy full term infants.

According to the American Academy of Pediatrics [37], children under one year of age should not be fed raw, unmodified, or unpasteurized cow's milk as a replacement for human milk or infant formula. Additionally, unmodified milk does not provide enough vitamin E, iron or essential fatty acids. Moreover, infants' systems cannot handle the high levels of protein, sodium, and potassium of unmodified cow milk. Formulas with a protein content 2–2.5 g/100 mL and a protein/energy ratio <3 g/100 kcal are used for normal infants, while with higher protein content (2.9 g/100 mL) and higher protein/energy ratio (3.5 g/100 kcal) are for a very low birth weight or preterm infants [38]. Recent studies showed that high protein content in infant formula is associated with excess weight gain in infancy, which can lead to a 20% risk of obesity later in life [39].

Cow's milk is one of the first foods introduced into an infant's diet and one of the most common causes of food allergy [40]. Usually, clinical reactions start very early in life, after breastfeeding has stopped and cow's milk is introduced into the diet; symptoms rarely appear during lactation.

The clinical manifestations of cow milk allergy vary widely in type and severity. It may be defined as a reproducible adverse reaction to one or more milk proteins (usually caseins or whey beta-lactoglobulin) mediated by at least one immune mechanism [41]. The prevalence of cow milk allergy varies across studies, as well as across diagnostic criteria and infant diets. It presents in the first year of life, with an estimated population prevalence of between 2% and 3% [41], or as high as 7% [42]. However, the results from recent cohort studies and from a randomized trial of early introduction of allergenic foods in the diet of breast-fed infant have shown that the incidence for IgE-mediated cow milk allergy could be as low as 0.5% [43,44,45]. Because there is no definition for differentiating IgE-mediated and non-lgE-mediated cow milk allergy, and the clinical symptoms of both overlap significantly, it is possible that, at least, some non-IgE mediated allergy cases may have been included in IgE-mediated cow milk allergy in previous reports.

Symptoms of cow milk allergy may be immediate or delayed. IgE-associated mechanisms are responsible for approximately 60% of cow milk-induced adverse reactions. These typically appear immediately or within 1 to 2 h after ingestion, and tend to affect the skin, respiratory system, and gastrointestinal tract. In severe cases, cow milk allergy can also cause systemic anaphylactic reactions [40].

Non-IgE-associated symptoms are characterized by delayed onset of approximately 2 h to several days after cow milk consumption. The period of 2 h helps to exclude the non-IgE-associated or no-allergic reactions. The non-lgE-associated clinical symptoms mainly affect the gastrointestinal system, and include enterocolitis, proctocolitis, enteropathy, and eosinophilic esophagitis [46]. Non-lgE-mediated gastrointestinal food-induced allergic disorders have a favorable prognosis and majority symptoms dissolve within one to five years [47].

4.5. Soy-Based Formulas

Formulas made from soy proteins are effective options for infants with galactosemia or congenital lactase deficiency. They help with colic and milk allergies, however, rarely, infants who are allergic to cow's milk may also be allergic to soymilk [33]. Soy products should not be used in infants under six months of age with food allergy [46]. Because phytoestrogens are present in soy-based formula, the uses of soy-based formulas are limited by the concern of potential harm for the infant, although this remains controversial [48,49].

4.6. Hypoallergenic Formulas

Protein hydrolysate formulas are meant for infants and babies who are unable to tolerate cow milk or soy-based formulas. They contain protein that has been hydrolyzed—partially or extensively—into smaller sizes than those found in cow or soy-based products. For infants who have a protein allergy, extensively hydrolyzed formulas are a satisfactory alternative.

4.7. Amino Acid Formulas

Amino acid formulas are another option for infants who have severe cow milk allergy with reactions to or refusal to ingest appropriate amounts of extensively hydrolyzed formula. They provide protein in the form of free amino acids with no peptides.

5. NON-BOVINE MILK SOURCES

Elimination of all cow milk products without appropriately modified and fortified substitutions can lead to malnutrition and/or specific nutrient deficiencies at a time when infants and children are growing. Infant milk formulas from different animals (goat, ewe, mare, donkey, or camel), or formulas based on lamb or chicken, have been widely marketed as substitutes for cow milk in the management of cow milk allergy in infants and children. However, other animal-milk-based formulas are currently not acceptable in many places because there are no robust randomized clinical trials.

6. PROBIOTICS AND PREBIOTICS

The high concentration and structural diversity of human milk oligaosaccharides are unique to humans. Without probiotics and prebiotics supplementation, the gut microbiota of formula-fed infants is generally not dominated by the Bifidobacterium species [50,51]. Studies have shown that breastfed newborn carry a more stable and uniform population of oligaosaccharides compared with formula-fed newborns [52]. Adding probiotics to formula represents a key strategy to reduce the incidence and severity of diarrhea in infants [53].

Domestic animal milk contains a large variety of complex oligosaccharides. Sialylated oligosaccharides account for approximately 80%–90% of the total pool from milk of all domestic animals [54]. Milk of grazing cows contains higher concentrations of sialic acid compared to non-grazing cows [55]. Cow's milk might be a useful source of a variety of sialylated oligosaccharides for use as additives in infant formula. It is now also possible to supplement commercial infant formula with synthesized oligosaccharides that are chemically identical to human milk oligosaccharides [56].

Most probiotic strains added to formula have been isolated from food or fecal infant microbiota. Although the use of probiotics is now extending from research to recommendations, rigorous scientific effort is still required to validate specific strains with antiallergenic potential for preventive and therapeutic applications.

Evidence on the use of *Lactobacillus reuteri* for the treatment of infantile colic is mixed. Sung *et al.* [57] found that *L. reuteri* DSM 17938 was of no benefit in a community sample of breast- and formula-fed infants with colic. This outcome differed from those of smaller trials in select populations, but did not support a general recommendation for the use of probiotics to treat colic. Conversely, a prospective, randomized, blinded, controlled trial in 138 infants showed that *L. reuteri* had a protective effect. The treatment group showed a lower number of pediatric consultations related to infant colic than the control group ($p < 0.0001$). It also reduced the use of pain relieving agents and of infant formula [58]. Such new treatments as probiotics have been proposed to treat infantile colic, but only a few strains have been tested. Further investigations are needed to provide evidence-based guidelines.

The impact of early pre- and probiotic intervention on preterm infants' well-being, crying, growth, and microbiological programming were conduced in 94 infants (gestational age 32 to 36 weeks and birth weight >1500 g) who were randomized to receive prebiotics (a mixture of galacto-oligosaccharide and polydextrose 1:1), probiotics (*Lactobacillus rhamnosus* GG), or placebo during the first two months of life and follow-up lasted one year [59]. The results showed that among excessive criers (29% of the infants), there was significantly less crying in the pre- and probiotic groups than in the placebo group (19% *vs.* 19% *vs.* 47%, respectively; $p = 0.02$). The placebo group had a higher percentage of *Clostridium histolyticum* bacteria in their stools than the probiotic group did (13.9% *vs.* 8.9%, respectively; $p = 0.05$). Another study on safe and simple strategies to prevent viral respiratory tract infections between 3 and 60 days of life has demonstrated that prebiotics or probiotics had significantly less ($p < 0.001$ and $p = 0.022$) viral respiratory tract infections compared with those receiving placebo [60]. Additionally, the incidence of rhinovirus-induced episodes, which comprised 80% of all respiratory tract infections, was significantly lower in the prebiotic ($p = 0.003$) and probiotic ($p = 0.051$) groups than in the placebo group.

7. FATTY ACIDS AND MILK FATS FROM DIFFERENT MAMMALIAN SPECIES

The lipid portion of human milk is the major source of energy for growing infants and provides approximately 45% to 55% of total energy. The lipid compositions of mammalian milks (cow, buffalo, donkey, sheep, and camel) were compared with that of human milk on fatty acid profiles and triacylglycerol (TAG), phospholipid, and phospholipid fatty acid compositions, as well as melting and crystallization profiles [59]. The results showed that these milk fats, especially sheep milk fat, had high degrees of similarity to human milk fat in total fatty acid composition. However, other chemical aspects had less similarity. This outcome indicates that these milk fats do not meet the requirements of human milk fat substitutes, but large amounts of these commercialized mammalian milk fats are good raw materials for infant formula production. Milk fat globule membranes are a fraction that has been previously excluded from infant formulas, but its components are active and prevent infection [61]. Milk fat globule membrane supplementation of infant formula also narrows the gap in cognitive development between breastfed and formula fed infants [62].

8. BIOACTIVE PROTEINS

Novel dairy fractions from bovine milk have been isolated and are now commercially available. Many of these components are proteins, such as α-lactalbumin, lactoferrin, osteopontin, and milk fat globule membrane proteins. When adding bioactive proteins to infant formulas, it is important to reduce the total protein content of formula. The amino acid composition of formula is also important; serum concentrations of essential amino acids should not be lower than those in breastfed infants. For example, α-Lactalbumin, often the first limiting amino acid in infant formulas, is digested into smaller peptides with antimicrobial and prebiotic activities and has an immunostimulatory effect. It also enhances mineral absorption. Osteopontin is a heavily phosphorylated and glycosylated protein that modulates immune function and stimulates Th1/Th2 switching. It might also affect bone mineralization and growth, and facilitate the biological function of lactoferrin.

9. CONCLUSIONS

Breast milk is the best nutrition for infant growth and development, and is also rich in antibodies that provide the first source of adaptive immunity in a newborn's intestinal tract. In preterm or low birth weight newborns, a mother's own milk is the first choice for preterm infants; when it is unavailable, donor breast milk is considered as the next best choice. For healthy newborns whose mothers are unable to provide sufficient breast milk, the current option of choice is infant formula.

ACKNOWLEDGMENTS

The authors extend their appreciation to Rita Buckley for research, writing, and editing services provided in the development of this manuscript.

AUTHOR CONTRIBUTIONS

The sole author had responsibility for all parts of the manuscript. All authors have read and approved the manuscript.

CONFLICTS OF INTEREST

The authors declare no conflict of interest.

REFERENCES

1. Lessen, R.; Kavanagh, K. Position of the academy of nutrition and dietetics: Promoting and supporting breastfeeding. J. Acad. Nutr. Diet. 2015, 115, 444–449.
2. Savino, F.; Bebetti, S.; Lignori, S.A.; Sorrenti, M.; Cordero, D.; Montezemolo, L. Advances on human milk hormones and protection against obesity. Cell. Mol. Biol. 2013, 59, 89–98.
3. Infant and Young Child Nutrition. Available online: http://apps.who.int/gb/archive/pdf_files/WHA55/ea5515.pdf (accessed on 14 November 2012).
4. Eidelman, A.I.; Schanler, R.J.; Johnston, M.; Landers, S.; Noble, L.; Szucs, K.; Viehmann, L. Breastfeeding and the use of human milk. Pediatrics 2012, 129, e827–e841.
5. Kozhimannil, K.B.; Jou, J.; Attanasio, L.B.; Joarnt, L.K.; McGovern, P. Medically complex pregnancies and early breastfeeding behaviors: A retrospective analysis. PLoS ONE 2014, 9, e104820.
6. U.S. Food and Drug Administration. Guidance for Industry. Demostration of the Quality Factor Reguirements Under 21 CFR 106.96 (i) for "Eligible Infant Formulas". Available online: http//www.fda.gov/downloads/Food/Guidance Regulation/Guidance Documents Regulatory Information/UCM400136/pdf (accessed on 14 June 2014).

7. Institute of Medicine of the National Academics. Infant Formula: Evaluating The Safety of New Ingredients; The National Academics Press: Washington, DC, USA, 2004.

8. Stevens, E.E.; Patrick, T.E.; Pickler, R. A history of infant feeding. J. Perinat. Educ. 2009, 18, 32–39.

9. Gura, T. Nature's first functional food. Science 2014, 345, 747–749.

10. Walker, W.A. Initial intestinal colonization in the human infant and immune homeostasis. Ann. Nutr. Metab. 2013, 63 (Suppl. 2), 8–15.

11. Guo, M. Human Milk Biochemistry and Infant Formula. In Manufacturing Technology; Elsevier: Cambrdige, UK, 2014.

12. Jackson, J.G.; Janszen, D.B.; Lonnerdal, B.; Lie, E.L.; Pramuk, K.P.; Kuhlman, C.F. A multinational study of α-lactalbumin concentrtions in human milk. J. Nutr. Biochem. 2004, 15, 517–521.

13. Domellof, M. Nutritional Care of Preterm Infants: Scientific Basis and Practical Guidelines; Koletzko, B., Poinderxter, B., Uauy, R., Eds.; Karger Medical and Scientific Publishers: Basel, Switzerland, 2014.

14. Gros, S.J.; Geller, J.; Tomarelli, R.M. Composition of breast milk rom mothers of preterm infants.Pediatrics 1981, 68, 490–493.

15. Lonnerdal, B. Nutritonal and physiologic significance of human milk proteins. Am. J. Clin. Nutr. 2003,77, 1537S–1543S.

16. Zhang, Z.H.; Adelman, A.S.; Rai, D.; Boettcher, J.; Lonnerdal, B. Amino acid profiles in term and preterm human milk through lactation: A systemic Review. Nutrients 2013, 5, 4800–4821.

17. Agostoni, C.; Carratu, B.; Boniglia, C.; Lammardo, A.M.; Riva, E.; Sanzini, E. Free glutamine and glutamic acid in casien in human milk through a three-month lactation period. J. Pediatry. Gastroenterol. Nutr. 2000, 31, 508–512.

18. Bonigla, C.; Carratu, B.; Chiammariol, S.; Sanzini, E. Influence of maternal protein intake on nitrogen fractions of human milk. Int. J. Vitam. Nutr. Res. 2003, 76, 447–457.

19. Herrera, E. Implication of dietary fatty acids during pregnancy on placental, fetal and postnatal development—A review. Placenta 2002, 23, S9–S19.

20. Joardar, A.; Sen, A.K.; Das, S. Docosahexaeoic acid facilitates cell maturation and beta-adrenergic transmission in astrocytes. J. Lipid Res. 2006, 47, 571–581.

21. Szabo, E.; Bowhm, G.; Bwwemn, C.; Weyrman, M.; Brenner, H.; Rothenbacher, D.; Decsi, T. Fatty aicd profile comparisons in human

milk sampled from the same mothers at the sixth week and the sixth month of lactation. J. Pediatr. Gastroenteral. Nutr. 2010, 50, 316–320.

22. Del Prado, M.; Villapando, S.; Elizondo, A.; Rodrignez, M.; Demmelair, H.; Koletzko, B. Contribution of dietary and newly formed arachidonic acid to human milk lipids in women eating a low-fat diet.Am. J. Clin. Nutr. 2001, 74, 242–247.

23. Weseler, A.R.; Dirix, C.E.; Bruins, M.J.; Hornstra, G. Dietary arachidonic acid dose-dependency increases with arachidonic acid concetration in human milk. J. Nutr. 2008, 138, 2190–2197.

24. Makrides, M.; Nenmann, M.A.; Gibson, R.A. Effect of maternal docosahexaenoic acid (DHA) suppplementation on breast milk composition. Eur. J. Clin. Nutr. 1996, 50, 352–357.

25. Fleith, M.; Clandinin, M.T. Dietary PUFA for preterm and term infants. review of clinical studies. Crit. Rev. Food Sci. Nutr. 2005, 45, 205–229.

26. Innis, S.M. Impact of maternal diet on human milk composition and neurological development of infants. Am. J. Clin. Nutr. 2014, 99, 734S–741S.

27. Martin, C.R.; Dasilva, D.A.; Cluette-Brown, J.E.; Dimonda, C.; Hamill, A.; Bhutta, A.Q.; Coronel, E.; Wilschanski, M.; Stephens, A.J.; Driscooo, D.F.; et al. Decreased postnatal docsahexaenoic and arachidonic acid blood levels in premature infants are associated with neonatal morbidities. J. Pediatr.2011, 159, 743–749.

28. Formerly Director of Medical Service Ghana; at present Inter-Regional coordinator of the World Hunger Program. WHO/UNICEF meeting on infant and young child feeding. J. Nurse Medwifery 1980,25, 31–38.

29. Lonnerdal, B. Preclinical assessment of infant formula. Am. Nutr. Metab. 2012, 60, 196–199.

30. Safe Preparation, Storage and Handling of Powdered Infant Formula Guidelines. Available online: http://www.who.int/foodsafety/publication/micro/pit_quideline.pdf (accessed on 13 Feberuary 2006).

31. Michael, M. SPX Benefitting from Increased Global Demand for Powdered Infant Formula. Available online: http://www.spx.com/en/literature/articles/apac-infant-formula (accessed on 2 October 2014).

32. Renfrew, E. 7% Growth for $50 Billion Global Infant Nutrition Market. Available online: http://www.zenithinternational.

com/articles/1355?7%25+growth+for+%2450+Billion+global+infant+nutrition+market (accessed on 1 April 2014).

33. U.S. National Library of Medicine. Infant Formulas—Overview. Available online: https://www.nlm.nih.gov/medlineplus/ency/article/002447.htm (accessed on 15 May 2015).

34. Koletzko, B.; Baker, S.; Cleghorn, G.; Nete, U.F.; Gopalan, F.; Hernall, O.; Hock, Q.S.; Jirapinyo, P.; Lonnerdal, B.; Pencharz, P. Global standard for the composition of infant formula: Recommendations of an ESPGHAN coordinated internation expert group. J. Pediatr. Gastroenteral. Nutr. 2005, 41, 584–599.

35. Cook, D.A. Nutrient levles in infant formulas: Technical considerations. J. Nutr. 1989, 119, 1773–1777.

36. Paragaroufalis, K.; Fotiou, A.; Egli, D.; Tran, L.A.; Steenhout, P. A randomized double blind controlled safety trial evaluating d-lactic acid production in healthy infants fed a lactobacillus reuteri-containing formula. Nutr. Metab. Insights. 2014, 7, 19–27.

37. Koletzko, B.; Beyer, J.; Brands, B.; Demmelmair, H.; Grote, V.; Haile, G.; Gruszfeid, D.; Rzehak, P.; Socha, P.; Weber, M. European Childhood Obesity trial Study. Early influences of nutrition on postnatal growth. Nestle Nutr. Inst. Workshop Ser. 2013, 71, 11–27.

38. Fanaro, S.; Ballardini, E.; Vigi, V. Different pre-term formula for different pre-term infants. Early Hum. Dev. 2010, 86 (Suppl. 1), 27–31.

39. Michaelsen, K.F.; Greer, F.R. Protein needs early in life and long-term health. Am. J. Clin. Nutr. 2014,99, 718S–722S.

40. Hochwallner, H.; Schulmeister, U.; Swoboda, I.; Spitzauer, S.; Valenta, R. Cow's milk allergy: From allergens to new forms of diagnosis, therapy and prevention. Methods 2014, 66, 22–33.

41. Luyt, D.; Ball, H.; Makwana, N.; Green, M.R.; Bravin, K.; Nasser, S.M.; Clark, A.T. Standard of Care Committee (SOCC) of the British Society for Allergy and Clinical Immunology (BSACI). BSACI guideline for the diagnosis and management of cow's milk allergy. Clin. Exp. Allergy 2014, 44, 642–672.

42. Turck, D. Cow's milk and goat's milk. World Rev. Nutr. Diet. 2013, 108, 56–62.

43. Katz, Y.; Rajuan, N.; Golberg, M.R.; Eisenberg, E.; Heyman, E.; Cohen, A.; Leshon, M. Early exposure to cow's milk protein is protective against IgE-mediated cow's milk protien allergy. J. Allergy Clin. Immunol. 2010, 126, 77–81.

44. Schiemare, A.A.; Sprikkelman, A.A.; Grimshaw, K.E.; Roberts, G.; Grabenhenrich, L.; Rosenfeld, L.; Siegert, S.; Dubakiene, R.; Rudzeviciene, O.; Reche, M.; et al. Incidence and natural history of challenge-proven cow's milk allergy in Europea children–EuroPrevall Birth. Allergy 2015, 70, 963–972.

45. Perkin, M.R.; Logan, K.; Tseng, A.; Raji, B.; Ayis, S.; Peacock, J.; Brough, H.; Marrs, T.; Radulovic, S.; Craven, J.; et al. Randomized trial of introduction of allergenic foods in breast-fed infants. N. Engl. J. Med. 2016.

46. Fiocchi, A.; Brozek, J.; Schunemann, H.; Bahna, S.L.; von Borg, A.; Bozzola, M.; Bradsher, J.; Compalati, E.; Ebisawan, M.; Dubakiene, R.; et al. World Allergy Organization (WAO) Diagnosis and Rationale for Action against Cow's Milk Allergy (DRACMA) Guidelines. Pediatr. Allergy Immunol.2010, 21 (Supp. 2l), 1–25.

47. Nowak-Wegrzyn, A.; Kate, Y.; Soneil Mehr, S.; Koletzkon, S. Non-lgE-mediated gastrointestinal food allergy. J. Allergy Clin. Immunol. 2015, 135, 1114–1124.

48. Upson, K.; Sathyanarayana, S.; Scholes, D.; Holt, V.L. Early-life factors and endometriosis risk. Fertil. Steril. 2015, 104, 964–971.

49. Adgent, M.A.; Daniels, J.L.; Pogan, W.J.; Adair, L.; Edwards, L.J.; Westreich, D.; Maisonet, M.; Marcus, M. Early life soy exposure and age at menarahe. Peadiatr. Perinat. Epidemiol. 2012, 26, 163–175.

50. Musilova, S.; Rada, V.; Vlkova, E.; Bunesova, V. Beneficial effects of human milk oligosaccharides on gut microbiota. Benef. Microbes. 2014, 5, 273–283.

51. Oozeer, R.; van Limpt, K.; Ludwig, T.; Ben Amor, K.; Martin, R.; Wind, R.D.; Boehm, G.; Knol, J. Intestinal microbiology in early life: Specific prebiotics can have similar functionalities as human-milk oligosaccharides. Am. J. Clin. Nutr. 2012, 98, 561S–571S.

52. Bezirtzoglou, E.; Tsiotsias, A.; Welling, G.W. Microbiota profile in feces of breast- and formula-fed newborns by using fluorescence in situ hybridization (FISH). Anaerobe 2011, 17, 478–482.

53. Chassard, C.; de Wouters, T.; Lacroix, C. Probiotics tailored to the infant: A window of opportunity.Curr. Opin. Biotechnol. 2014, 26, 141–147.

54. Albrecht, S.; Lane, J.A.; Marino, K.; Al Busadah, K.A.; Carrington, S.D.; Hickey, R.M.; Rudd, P.M. A comparative study of free oligosaccharides in the milk of domestic animals. Br. J. Nutr. 2014, 111, 1313–1328.

55. Asakuma, S.; Ueda, Y.; Akiyama, F.; Uemmnra, Y.; Miyaji, M.; Nakamura, M.; Mural, M.; Urashima, T. Short communication: Effect

of grazing on the concentrations of total sialic acid and hexose in bovine milk. J. Dairy Sci. 2010, 93, 4850–4854.

56. Coulet, M.; Phothirath, P.; Allais, L.; Schilter, B. Pre-clinical safety evaluation of the synthetic human milk, nature-identical, oligosaccharide 2′-O-Fucosyllactose (2′FL). Regul. Toxicol Pharmacol. 2014, 68, 59–69.

57. Sung, V.; Hiscock, H.; Tang, M.L.; Mensah, F.K.; Nation, M.L.; Salzke, C.; Heine, R.G.; Stock, A.; Barr, R.G.; Wake, M. Treating infant colic with the probiotic Lactobacillus reuteri: Double blind, placebo controlled randomised trial. BMJ 2014, 348, g2107.

58. Partty, A.; Luoto, R.; Kalliomaki, M.; Salminen, S.; Isolauri, E. Effects of early prebiotic and probiotic supplementation on development of gut microbiota and fussing and crying in preterm infants: A randomized, double-blind, placebo-controlled trial. J. Pediatr. 2013, 163, 1272–1277.

59. Zou, X.; Huang, J.; Jin, Q.; Guo, Z.; Lin, Y.; Cheong, L.; Xu, X.; Wang, X. Lipid composition analysis of milk fats from different mammalian species: Potential for use as human milk fat substitutes. J. Agric. Food Chem. 2013, 61, 7070–7080.

60. Luoto, R.; Ruuskanen, O.; Waris, M.; Kalliomaki, M.; Salminen, S.; Isolauri, E. Prebiotic and probiotic supplementation prevents rhinovirus infections in preterm infants: A randomized, placebo-controlled trial. J. Allergy Clin. Immunol. 2014, 133, 405–413.

61. Lonnerdal, B. Biological effects of novel bovine milk fractions. Nestle Nutr. Workshop Ser. Pediatr. Program. 2011, 67, 41–54.

62. Timby, N.; Domellof, E.; Hernell, O.; Lonnerdal, B.; Domellof, M. Neurodevelopment, nutrition, and growth until 12 mo of age in infants fed a low-energy, low-protein formula supplemented with bovine milk fat globule membranes: A randomized controlled trial. Am. J. Clin. Nutr. 2014, 99, 860–868.

5

THE ESSENTIALITY OF ARACHIDONIC ACID IN INFANT DEVELOPMENT

Kevin B. Hadley[1], Alan S. Ryan[2,*], Stewart Forsyth[3], Sheila Gautier[1] and Norman Salem Jr.[1]

[1]DSM Nutritional Products, 6480 Dobbin Road, Columbia, MD 21045, USA

[2]Clinical Research Consulting, 9809 Halston Manor, Boynton Beach, FL 33473, USA

[3]School of Medicine, Dentistry & Nursing, University of Dundee, Ninewells Hospital and Medical School, Dundee, UK

ABSTRACT

Arachidonic acid (ARA, 20:4*n*-6) is an *n*-6 polyunsaturated 20-carbon fatty acid formed by the biosynthesis from linoleic acid (LA, 18:2*n*-6). This review considers the essential role that ARA plays in infant development. ARA is always present in human milk at a relatively fixed level and is accumulated in tissues throughout the body where it serves several important functions. Without the provision of preformed ARA in human milk or infant formula the growing infant cannot maintain ARA levels from synthetic pathways alone that are sufficient to meet metabolic demand. During late infancy and early childhood the amount of dietary ARA provided by solid foods is low. ARA serves as a precursor to leukotrienes, prostaglandins, and thromboxanes, collectively known as eicosanoids which are important for immunity and immune response. There is strong evidence based on animal and human studies that ARA is critical for infant growth, brain development, and health. These studies also demonstrate the importance of balancing the amounts of ARA and DHA as too much DHA may suppress the benefits provided by ARA. Both ARA and DHA have been added to infant formulas and follow-on formulas for more than two decades. The amounts and ratios of ARA and DHA needed in infant formula are discussed based on an in depth review of the available scientific evidence.

Keywords: arachidonic acid; docosahexaenoic acid; infant formula; growth; human milk; long-chain polyunsaturated fatty acids

1. INTRODUCTION

During the first year of life, infants have special nutritional requirements to maintain a healthy body and support rapid growth and development. Human milk is typically the sole source of nutrition that must supply the infant with appropriate amounts of energy and nutrients. The long-chain polyunsaturated fatty acids (LCPUFA), docosahexaenoic acid (DHA, 22:6*n*-3) and arachidonic acid (ARA, 20:4*n*-6) are always present in human milk. These fatty acids play key roles in the structure and function of human tissues, immune function, and brain and retinal development during gestation and infancy [1,2]. Although breastfeeding is considered the ideal way to nourish infants, recent nutrition surveys report that the majority of infants in developed countries receive at least some infant formula during the first year of life [3,4].

Both ARA and DHA have been added to infant formulas in the United States since 2001, although supplementation began in Europe much earlier. Most infant formulas contain 0.2% to 0.4% of total fatty acids as DHA and between 0.35% and 0.7% of total fatty acids as ARA based on worldwide averages of DHA and ARA content in human milk [5] and the recommendations from a number of international expert groups [6,7,8,9]. Thus, all commercially available infant formulas contain preformed ARA at levels equal to or higher than the DHA content in order to maintain adequate DHA and ARA status in non-breastfed infants.

Both α-linolenic acid (ALA, 18:3*n*-3) and linoleic acid (LA, 18:2*n*-6) are regarded as nutritionally essential fatty acids [10]. However, as Lauritzen *et al.* [10] point out, all classic signs of essential fatty acid (EFA) deficiency can be completely reversed by the administration of *n*-6 fatty acids alone, particularly ARA. With respect to infants, the presence of a relatively fixed level of preformed ARA in human milk and the active accumulation of ARA by tissues throughout the body support the concept of the essentiality of ARA. Previously, a description of the essentiality of ARA during infancy has not been considered in detail, although a brief outline of the essentiality of *n*-6 and *n*-3 polyunsaturated fatty acids was presented by Lauritzen *et al.* [10] in 2001.

The purpose of this paper is to review the essentiality of ARA for infant growth and development. We consider both animal models and human studies of ARA. We describe: (1) ARA accumulation and function in brain and tissues; (2) ARA content in human milk and in various tissues, including rates of accretion during gestation and early infancy; (3) the structure and biosynthesis of ARA from LA and its role as a precursor to leukotrienes, prostaglandins, and thromboxanes, collectively known as eicosanoids; (4) dietary intakes of ARA during late infancy and early childhood when non-breast milk food items are introduced into the diet; (5) immune system development and the dual role of PGE_2 and its receptors in modulating the inflammatory response during infancy; (6) bone metabolism and growth; (7) regulation of cardiac function; (8) consequences of ARA deficiency; (9) the importance of ARA for optimal brain and central nervous system development; (10) the history, reasons for, and nutritional effects of adding both DHA and ARA to infant formulas, with an emphasis on the effects of ARA; (11) the importance of ARA in infant health; and (12) the regulatory requirements for ARA and DHA in infant formulas. Based on a detailed review of the scientific literature presented herein, recommendations for dietary intakes ARA during infancy are provided.

2. ARA ACCUMULATION AND FUNCTION IN BRAIN AND TISSUE

Over the last decade, there has been increased understanding of the molecular roles that the *n*-3 and *n*-6 PUFA play in brain and cellular function. The variety of functions shown to be related to ARA indicates its importance and essentiality in the metabolic chain of events leading to brain structural lipid development, signaling, and many basic cellular functions.

ARA is indispensable for brain growth where it plays an important role in cell division and signaling [11]. The brain in mammals consists of 60% fat, which requires DHA and ARA for its growth and function [12]. Across different species of mammals there is little variation in DHA and ARA composition of the brain. ARA is one of the most abundant fatty acids in the brain, and compared with DHA, ARA is present in similar quantities [13,14]. The two fatty acids account for approximately ~25% of its total fatty acid content predominately in the form of phospholipids and thus are major structural components of neural cellular membranes.

ARA rapidly accumulates in the brain during development [1,14,15] which takes place from the beginning of the third trimester of gestation up to about 2 years of age [16] (Figure 1). As shown in brain kinetics in fetal baboons, [17] in addition to maternal preformed ARA, LA may be transported across the blood-brain barrier despite its very low content within brain lipids. The brain has an active desaturation/elongation system that converts LA to ARA [17]. ARA activity is higher in brain than in other organs such as the liver. However, the conversion of LA to ARA is low (see below).

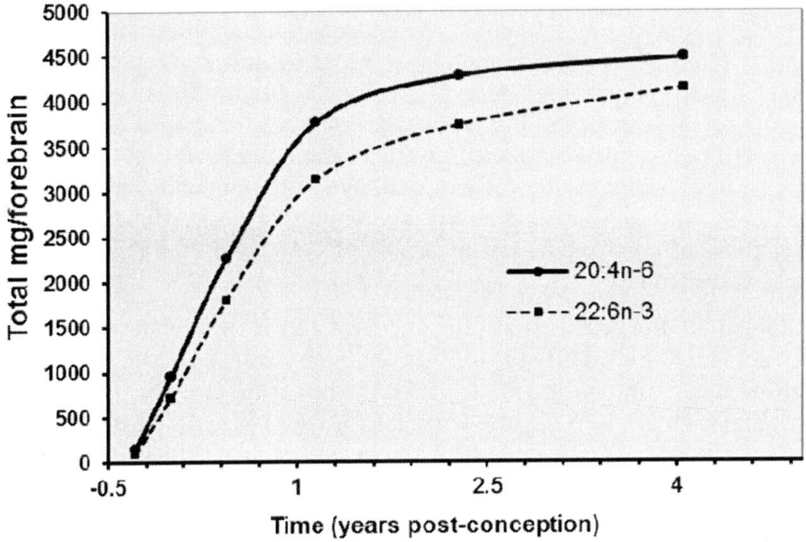

Figure 1. Long-chain polyunsaturated fatty acids (LCPUFA) accretion in the human brain during perinatal development (Data from Martinez [15]).

The maximum rate of brain growth is primarily associated with myelination [14]. In animal models, approximately 50% of the adult amounts of ARA and DHA accumulate in rat brain during the period before myelination and at 15 days after birth when myelination has just started [14]. Diets low in LCPUFA adversely affect the development of the myelin lipids needed early in brain development [14].

ARA has several functions in the brain. ARA mediates neuronal firing [18], signaling [19], and long-term potentiation [20]. ARA also helps maintain membrane order and hippocampal plasticity [21], defends the brain against oxidative stress in the hippocampus by activating the peroxisome proliferator-activated receptor gamma (PPARγ), and aids in the synthesis of new protein in tissue [22].

A potentially important aspect of ARA metabolism *in vivo* is its function as an immediate precursor for adrenic acid (22:4*n*-6) [23]. Adrenic acid is the third most abundant PUFA in the brain that is found in large quantities in myelin lipids, particularly in phosphatidylethanolamine (PE) [1]. Rapid accumulation of adrenic acid, like ARA, occurs during the early post-natal period of the brain growth spurt in infants. The conversion of ARA to adrenic acid may represent an important pathway for ARA utilization in infants in order to meet the rapid increase of adrenic acid needed for neural tissue development.

Using a single dose of U-^{13}C-labeled ARA to investigate preformed ARA utilization in baboon neonates, Wijendran *et al.* [23] reported that a major portion of ARA consumed (79%–93%) was accumulated as ARA in tissue lipids, consistent with its primary function as a principal constituent of membrane lipids. Approximately 5% to 16% of ARA was converted to adrenic acid. Based on tracer data, net accretion of ARA and adrenic acid during the first 4 weeks of age in the neonate baboon brain was 17% and 8%, respectively, corresponding to efficiencies (*i.e.*, percentage of dose recovered in brain) of 0.48% and 0.54% of dietary levels, respectively.

To determine the effects that differing DHA to ARA ratios have on tissue fatty acids, twelve-week-old full term baboons were randomized to one of three diets: control (no DHA or ARA), moderate (0.33% DHA, 0.67% ARA) and high LCPUFA (1.00% DHA, 0.67% ARA) [24]. In all groups, DHA levels increased significantly in liver, heart, plasma and in the central nervous system (CNS) regions (precentral gyrus, frontal cortex, inferior and superior colliculi, globus pallidus, and caudate). The formula with the highest level of DHA significantly reduced ARA levels in two areas of the brain (superior colliculus and globus pallidus), indicating its competition with ARA and the importance of a proper balance of DHA to ARA.

Phosphatidylcholine (PC) is a lipid class that is a major component of most intracellular membranes [25]. Some intracellular lipid bilayers include PC containing ARA (ARA-PC). ARA-PC functions as a retrograde messenger in long-term potentiation of synapses in the hippocampus CA1 region [26] and is involved in migration of neurons in the cerebral cortex [27].

Using imaging mass spectrometry, Yang *et al.* [25] characterized the distribution of ARA-PC within cultured neurons of the superior cervical ganglia and found an increasing gradient of ARA-PC along the proximodistal axonal axis that may provide a source for free ARA release [25]. Released free ARA is known to activate protein kinases and ion channels, inhibit neurotransmitter uptake, and enhance synaptic transmission [11]. Free ARA therefore modulates neuronal excitability. As ARA mediates intracellular signaling the concentration of free ARA must be maintained at precise levels within the cells. A higher concentration of ARA-PC near the axon terminal might provide a timely source of ARA when needed during the activated period [25].

ARA also is responsible for the activation of syntaxin-3 (STX-3), a plasma membrane protein involved in the growth and repair of neurites [28]. Growth of neurite processes from the cell body is a critical step in neuronal development. STX-3 serves as a single effector molecule and direct target for ARA [28]. Neurite growth closely correlates with the ability of ARA to activate STX-3 in membrane expansion at growth cones [28].

ARA also enhances the engagement of STX-3 with the fusogenic soluble N-ethylmaleimide-sensitive factor attachment protein receptors (SNARE complex), proteins that form a ternary complex that drives exocytosis [29]. In the brain, at the neuromuscular junction, and in endocrine organs, a set of three SNARE proteins has a primary role in producing fusion of vesicular and plasma membranes. The formation of this SNARE complex drives membrane fusion which leads to the release of vesicular cargo into the extracellular spaces [29]. Darios and colleagues [29] report that α-synuclein, a synaptic modulatory protein implicated in the development of Parkinson disease, can sequester ARA and thereby block the activation of the SNARE complex. This finding underlines the importance of ARA for the regulation of synaptic transmission and transport.

Detergent resistant microdomains, also referred to as lipid rafts, are specialized regions within plasma membranes [30]. These microdomains serve as platforms for biomechanical interactions between the lipid and protein components of signal transduction pathways [30,31,32]. The outer leaflet of lipid rafts is highly enriched with glycol-sphingolipids and cholesterol [32]. The inner, or cytosol facing leaflet is enriched with alkenyl forms of PE which have been termed plasmenylethanolamine. Electrospray ionization/mass spectrometric analysis has shown that the ARA-containing plasmenylethanolamine represents as much as 50% of the phospholipids of the cytosolic leaflet [31]. This is consistent with a role of PE as an important source of ARA within the cell.

Stearoyl-2-arachidonoyl is a highly abundant species of phosphatidylinositol (PI) found in the phosphorylated forms of PI, the phosphoinositides [33,34,35,36,37,38]. In addition to serving as a substrate for phospholipase C to produce inositol-triphosphate and diacylglycerol, phosphoinositides serve important biochemical functions including lipid signaling, cell signaling and membrane trafficking. Phosphoinositides perform these roles in part by serving as adaptors for protein-protein and protein-membrane interactions in order to facilitate and/or regulate G-receptor protein activity and signal transduction, and trafficking of various metabolites such as cholesterol or calcium, or other ions, between cellular compartments [39,40,41,42,43]. These biochemical functions of ARA demonstrate its importance for cell signaling, trafficking and regulation of spatial-temporal interactions between cellular structures.

3. LEVELS OF ARA IN HUMAN MILK, BRAIN, AND TISSUES

Fat is a critical component of human milk that provides energy and nutrients needed for the development of the CNS [10]. DHA and ARA are the principal LCPUFA found in human milk. The synthesis of DHA and ARA is limited in infants [5] and both DHA and ARA must be obtained from dietary sources. Amounts of DHA and ARA in human milk tend to vary by diet, nutritional status, and other factors [5]. Based on data from 65 studies of human milk from 2474 women, the mean concentration of ARA (by weight) was 0.47% ± 0.13% (range 0.24% to 1.0%) whereas the mean concentration of DHA was 0.32% ± 0.22% (range 0.06% to 1.4%) [5]. The DHA concentration in human milk is lower and more variable than ARA. The level of ARA in human milk is much more stable. The relatively stable content of ARA in human milk is biologically important because it provides preformed ARA consistently at a time when brain growth and development is most critical. The majority of ARA in human milk does not derive from dietary LA but rather from maternal stores of ARA [44]. The correlation between DHA and ARA is low, which may reflect a higher degree of variability in the ratio of DHA to ARA in individual human milk samples [5].

The composition of the brain is dominated by ARA and DHA [45]. During pregnancy, both ARA and DHA are preferentially transferred across the placenta [46] and sequestered in the developing brain from the earliest phases of its growth. After birth, human milk provides both DHA and ARA to the breastfed infant [47] with a rapid rise towards adult levels of DHA and ARA in the brain within the first two years of life [48]. ARA is found at a level comparable to that of DHA in neural membranes, particularly those of the brain [49].

Based on estimated total body content of ARA from fetal organ weights during the last trimester of pregnancy and early infancy the relative amount of brain ARA decreases, but because of brain growth the absolute amount of ARA increases [50]. In fact, the absolute amount of ARA increases in all organs with increasing gestational age while the relative contribution (g per 100 g fatty acids; g %) decreases [50]. At 25 weeks gestation, the whole fetal body contains about 1.1 g ARA which increases to 4.2 g ARA at 35 weeks gestation. A full-term infant (3500 g) has about 7.6 g of ARA. The accretion rate of ARA is estimated to be 6.1 mg/day during the first 25 weeks and increases to 95.2 mg/day by 35–40 weeks gestation. The fetal accretion rate for ARA is 2-fold that of DHA [50]. Most of the bodily ARA at 25 and 40 weeks is located in skeletal muscle, adipose tissue and the brain, in that order [50].

In human infant central nervous tissue (cerebral cortex and retina) ARA comprises approximately 10%–12% of total fatty acids [49]. The amount of ARA in central nervous tissue appears to be influenced to a greater extent by postnatal age than by dietary ARA supply [49]. Samples of frontal cerebral cortex obtained from 58 human autopsies (mean age 40 ± 29 years) indicated that the relative levels of PUFA expressed as a percentage of total fatty acids generally decrease with age with the exception of DHA [51].

The distribution of *n*-6 and *n*-3 PUFA was determined in various viscera and tissues within the whole body of rats fed a diet containing 10 wt % fat (15% linoleate and 3% α-linolenate) until 7 weeks of age when they were sacrificed [52] (Figure 2).

The rat whole body was comprised of primarily saturated fatty acids (48.4% of total fatty acids) while the monounsaturated fatty acids were present in the second greatest amount (34.8%). The total amount of *n*-6 PUFA was 12.0% and was more than 5-fold greater than the total *n*-3 PUFA. The *n*-6 PUFA with the highest content was LA (10.1%) followed by ARA (1.4%). ARA was the major PUFA in nearly every tissue and was the major PUFA in most internal organs. The tissues with the highest content of ARA were plasma (25.3%) followed by kidney, red blood cells (RBC), and spleen, ranging from 18.7% to 23.5%. Brown adipose, white adipose and the eye contained very low amounts of ARA (<1.0%). For each compartment in the rat body, the total ARA/organ was highest for muscle, then liver, adipose, and carcass. In terms of fatty acid composition expressed as a percentage, ARA was highest in the circulation, kidney, and the spleen.

As shown in rat pups, when tissues are deprived of *n*-3 PUFA, the accretion of ARA from LA is increased and ARA is further metabolized to produce docosapentaenoic acid (22:5*n*-6, DPA) which accumulates in tissues [53], particularly in the nervous system. The accumulated DPA in turn reciprocally replaces lost DHA in tissue [53].

Pigs fed varying amounts of ARA and DHA levels after birth and then sacrificed at day 28 showed that dietary ARA had little effect on tissue DHA accretion [54], but heart tissue was particularly sensitive to ARA intake. These observations are particularly notable because the pigs were fed ARA at a level of 0.53% of total fatty acids. This level is slightly above the worldwide ARA mean in human milk and 0.67% of total fatty acids is the level currently added to many infant formulas and is near the high end of human milk ARA levels [5]. Neonatal pigs serve as a practical biomedical model of human infant development due to their similar metabolic responses, genetics of the fatty acid desaturases, and rates of perinatal brain growth [5]. The importance of ARA for the heart is discussed in greater detail in Section 8.

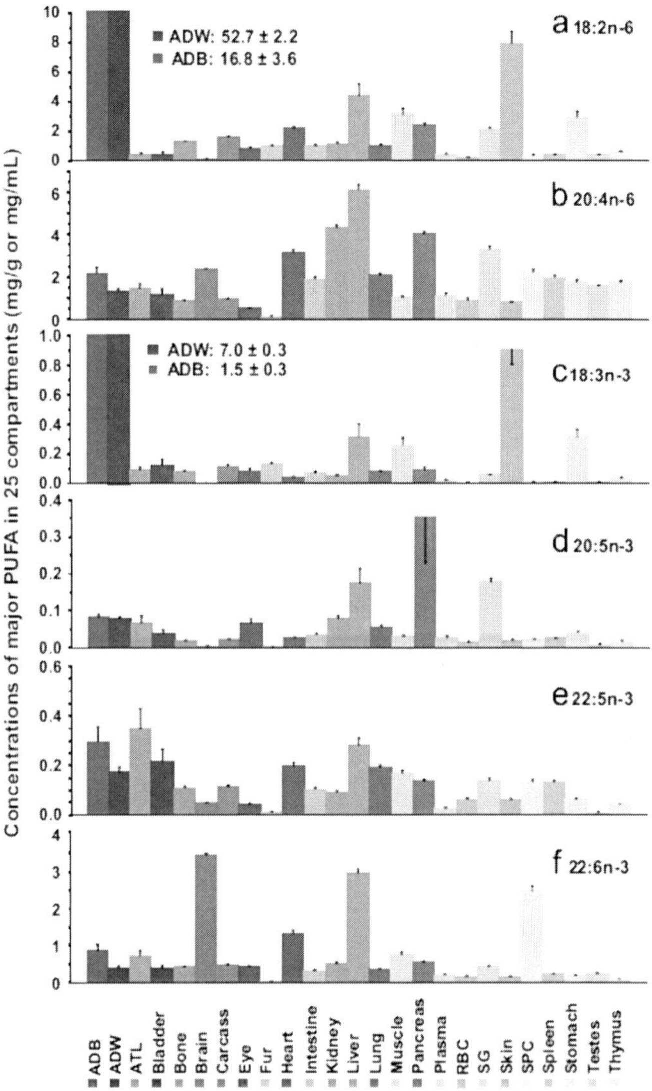

Figure 2. Distribution of fatty acids in 25 different tissue compartments in young male rats. Abbreviations: ATL, adrenal gland, thyroid gland, mandibular gland, and lymph nodes; RBC, red blood cell; SG, salivary gland; ADB, brown adipose tissue; ADW, white adipose tissue (from Salem *et al.* [52]).

4. ARA BIOSYNTHESIS AND METABOLISM

ARA is an *n*-6 polyunsaturated 20-carbon essential fatty acid formed by biosynthesis from LA [10]. ARA is a precursor to leukotrienes, prostaglandins, and thromboxanes, collectively known as eicosanoids [55,56]. ARA is found in membrane phospholipids throughout the body and is particularly abundant in the brain, muscles and liver. The metabolic pathways of the *n*-6 series and *n*-3 series are shown in Figure 3.

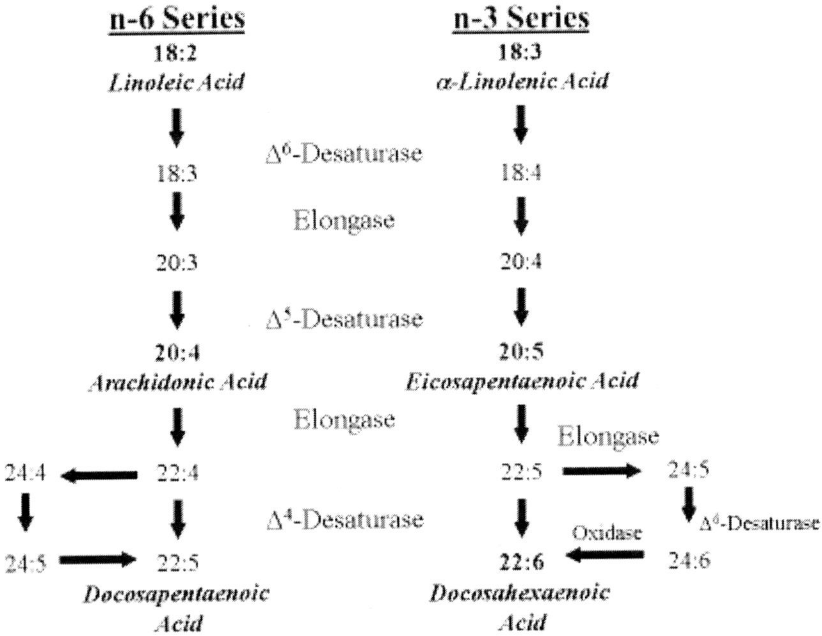

Figure 3. Metabolic pathways of linoleic and α-linolenic acid (Adapted from Lauritzen *et al.* [10]).

The use of stable isotope labelled fatty acids to investigate essential fatty acid metabolism was pioneered in the 1930s with the first identifiable study done by Schoenheimer and Rittenberg [57]. Several decades later, Nichaman *et al.* [58] gave four adult subjects ^{14}C-labeled LA and found a very small but significant incorporation into plasma phospholipid ARA acid based on responses in gas-radiochromatography. Similarly, ^{14}C-labeled LA was shown to be converted to ^{14}C-ARA in human fetal liver microsomes, *in vitro*[59]. El Boustani *et al.* [60] studied the conversion of deuterated dihommo-gamma linoleic acid (20:3*n*-6, DGLA) into ARA in plasma phospholipid and triglyceride fractions *in vivo* in diabetic patients. After a 2 g isotope ingestion, a maximum of 5 mg of labelled

ARA/L was observed in plasma. The authors stated that "this was consistent with the very low Δ5 desaturase activity observed *in vitro* in the human liver".

In the 1980s, Emken and colleagues [61] developed stable isotope technology in adult humans. In an early study of deuterated-LA metabolism *in vivo*, when a large dose of over 14 g of isotope was given, the authors concluded that "interconversion products such as deuterium-labeled ...20:3 and 20:4 were not detected in any of the lipid classes" [62]. They calculated that a conversion of as little as 0.00012% would have been detectable. The absence of any LA metabolism to ARA was confirmed in a subsequent study even where labelled-ALA was clearly incorporated into EPA and DHA [61].

In 1995, Demmelmair *et al.* [63] used natural abundance ^{13}C measurements in corn oil fed infants to demonstrate LA conversion to ARA. They observed conversion but concluded that "the activity of the enzyme system seems to be limited". Shortly afterwards, the conversion of LA to ARA was conclusively and directly demonstrated in newborn infants using stable isotope technology by Salem *et al.* [64]. The D5-LA was used together with a highly sensitive NCI GC/MS method after PFB derivatization [65]. With this new methodology, the deuterated fatty acid could be chromatographically separated from its corresponding endogenous analogue and so the signal of the stable isotope labeled metabolite would not be obscured by the much larger signal from the endogenous fatty acid. A crude estimate was made of the net accretion of ARA over the six day period of the study which was 53 mg, or about 9 mg ARA/day. Such estimates of "net synthesis" treated the organism as if the synthesis was occurring within the bloodstream and this is clearly not the case. In addition, what was being measured is the synthesis, minus the catabolism, plus the transport/release into a compartment such as the bloodstream where it can be sampled. Carnielli *et al.* [66] used a similar methodology to confirm the conversion of LA to ARA in very low birth-weight infants using ^{13}C-labeled LA.

Pawlosky *et al.* [67] studied stable isotope labeled LA and DGLA metabolism to ARA in 10 newborn human infants within the first week of life *in vivo* and performed compartmental modeling to provide an estimate of the synthetic rates. Formula and breast milk intakes were considered so that ARA and other PUFA intake could be estimated; LA and ARA intake was estimated at 3 g/kg/day and 2.8 mg/kg/day, respectively. They concluded that "the mean daily rate of synthesis and turnover of 20:4*n*-6 in plasma of infants were estimated to be from 0.06 to 2.1 mg/day...and from 0 to 51 mg/day (mean 10.2)" [67]. They went on to say that "such rates of synthesis are incapable of sustaining plasma 20:4*n*-6 concentrations in nearly all of these subjects necessitating an intake of ~4 mg/kg/day from either human milk or a supplement". The fractional rate of conversion (FRC) observed in this study was 2.7%

which is even more than that observed by Sauerwald *et al.* [68] who calculated an FSR of 0.4% to 1.1% depending upon the ALA content of the formula.

Carnielli and colleagues in 2007 [69] studied LA conversion to ARA using natural abundance ^{13}C measurements in preterm infants *in vivo* in those fed LCPUFA or no LCPUFA-containing formulas at 1, 3 and 7 months of age. These authors show that ARA synthesis is decreasing with age as it fell from 26.7 mg/kg/day to 14.4 mg/kg/day and then to 11.6 mg/kg/day from 1 to 3 to 7 months of age, respectively. It seems that the endogenous synthesis rate in these infants was inadequate as the ARA plasma phospholipid level fell from 5.6 mol% in the ARA fed group to 1.9 mol% in the no ARA group, a 66% drop. This underlines the inadequacy of LA alone as a source of ARA and the requirement for preformed ARA in the infant diet if blood levels of ARA are to be maintained similar to those in breastfed infants.

5. GLOBAL INTAKE OF ARA IN EARLY LIFE

In contrast to *n*-3 LCPUFA, there are few data relating to dietary intakes of *n*-6 LCPUFA in early life. In relation to dietary ARA, many regulatory agencies have tended to assume that beyond the age of 6 months, the endogenous synthesis of ARA will meet the needs of infants and young children during this period of rapid growth and development [70,71]. However, studies have shown that the endogenous synthesis of both DHA and ARA may be insufficient with evidence of blood and tissue concentrations decreasing after birth if there is not an exogenous supply [49,72].

The World Health Organization (WHO) [73,74] and the American Academy of Pediatrics [75] recommend that infants should be exclusively breastfed for the first six months of life to achieve optimal growth, development and health. Thereafter, to meet their dietary requirements during growth, infants should receive nutritionally adequate and safe complementary foods while breastfeeding continues for up to two years of age or beyond [73,74]. However, there is widespread variation in compliance with this recommendation in both developed and developing countries. In an evaluation of 33 developing countries, where the health benefits of this policy could have the greatest impact, exclusive breastfeeding occurred in 46% of countries, the median duration of breastfeeding was 18.6 months and over 30% received complementary foods before 6 months of age [76]. The extent to which variation in feeding practices may influence global intakes of ARA and infant growth and development in early life needs to be further evaluated.

5.1. ARA Intake from Human Milk

In exclusively breastfed infants, the mean human milk intake at 6 months has been measured to be 854 g/day [73,74]. Based on those data and an estimation that 4.2% of human milk is composed of fatty acids [77] the average ARA and DHA intakes in exclusively breastfed infants at 6 months of age are about 169 mg/day and 115 mg/day, respectively. Moreover, many infants continue to receive human milk throughout the first year of life and longer. It is estimated that at 12 months of age the intake of human milk is in the range of 600–900 g/day [73,74]. This amount provides infants with an ARA intake from human milk in the range of 118–178 mg/day. The mean estimated ARA intake is approximately 12–18 mg/kg/day when adjusted for body weight using weight-for-age percentiles [78].

5.2. ARA Intake from Infant Formula

Infant formulas typically contain levels of ARA and DHA at 140 mg/day and 100 mg/day, respectfully, based on worldwide averages of ARA and DHA content in human milk [5]. Therefore, intakes of ARA and DHA from infant formula are similar to those provided from human milk.

5.3. ARA Intake from Weaning Foods

In both developed and developing countries weaning foods contain low amounts of fat, which results in a sharp transition from adequate fat intake during breastfeeding to significantly lower fat intake when children are weaned from the breast [79,80]. The main food sources of ARA are beef, poultry, eggs and seafood. Complementary foods in low-income countries are typically cereal-based and therefore LCPUFA dietary intake from these weaning foods may be minimal [79]. Countries with the lowest gross national product (GNP) (e.g., Malawi, Ethiopia, Bangladesh, Burkina Faso, Ghana and India) had a mean percentage of total PUFA from animal source foods of 4.9% *vs.* countries with a higher GNP (Vietnam, Bolivia, Indonesia, Guatemala, China, South Africa, Mexico) where the mean percentage of total PUFA from animal food sources was 18.1% [79].

Intakes of ARA (mean mg/day and estimated mean mg/kg/day) from several developing and developed countries are presented in Table 1. In the village of Keneba, Gambia, estimated mean intake of ARA during the period of 0–6 months when infants are predominantly breastfed was 90 mg/day and as complementary foods were introduced the ARA intake fell steadily to 10 mg/day at 24 months [81]. In Heqing County, Yunnan Province China, the mean intake of ARA was 55 mg/day at 1 to 3 years of age and 50 mg/day at 4 to 5 years of age [82].

Table 1. ARA intakes in developed and developing countries during the first 2 years of life.

Country	Age	Method	Mean ARA Intake (mg/Day) (mg/kg/Day) [1]
Australia [91]	2–3 years	1-day weighed food record	16 (1.3)
		1-day weighed food record	22 (1.8)
Belgium [90]	2–5 years	3 days food record	17 (1.4)
	4–6.5 years	3 days food record	18 (1.0)
Canada [92,93]	1.5–2 years	1 day food frequency	133 (11.0)
	2.1–3 years	I day food frequency	260 (22.0)
	3.1–5 years	1 day food frequency	226 (15.0)
	4–7 years	3-days food records	57 (2.9)
China [82]	1–3 years	3 days 24 h recall	55 (4.6)
	4–5 years	3 days 24 h recall	50 (2.5)
Gambia [82]	0–6 months	1 day weighed food monthly	90 (15.0)
	7–12 months	1 day weighed food monthly	70 (7.8)
	13–17 months	1 day weighed food monthly	60 (6.7)
	24 months	1 day weighed food monthly	10 (0.8)
Germany [87,88]	6 months	3 days weighed food record	72 (12.0)
	9 months	3 days weighed food record	24 (2.7)
Italy [89]	1 month	Human milk composition	95.6 (29.0)
	2 months	Human milk composition	109.6 (33.0)
	3 months	Human milk composition	101.1 (16.9)
	6 months	Human milk composition	58.7 (9.8)
U.S. 2003–2008 [94]	1–4 years	1 day weighed food record	60 (5.0)
U.S. 2015 [95]	2–5 years	1 day weighed food record	80 (6.7)

Notes: [1] Estimated mean intake for ARA (9 mg/kg/day) for ages 0 month to 3 years was calculated using median weight-for-age percentiles for boys, birth to 36 months, and from median body mass index for ages 4 through 19 years; from the Centers for Disease Control and Prevention-Growth Charts (CDC, [78]).

Vulnerable infants and young children need energy- and nutrient-dense foods to grow and develop both physically and mentally [83]. For these reasons, dietary diversity is now included as a specific recommendation in the guidance for complementary feeding of the breastfed child aged 6 to 23 months [83]. Many factors contribute to limited dietary diversity including economic limitations, religious beliefs, and a concern that infants under 1 year of age cannot digest animal sourced foods [84,85]. There is also a widely held perception by parents that fish may be associated with allergic reactions [85,86].

Even in developed countries where dietary diversity is higher and meat and eggs contribute more to the complementary diet, the detrimental impact of the introduction of complementary feeding with low amounts of ARA and DHA content is evident. For example, mean ARA intake in German infants/toddlers decreased from 72 mg/day at 6 months of age to 24 mg/day at 9 months of age [87] (Table 1). In a separate study, these authors reported that predominately breastfed German infants had an ARA intake of 103 mg/day at 3 months of age and this amount declined to 24 mg/day at 9 months when human milk represented only 20% of the diet [88].

One hundred-seventy-four Italian breastfed children were followed from birth to 12 months of age [89]. Human milk samples were analyzed. The mean ARA intake from human milk was 95.6 mg/day at 1 month, 109.6 mg/day at 2 months, and 101.1 mg/day at 3 months. However, at 6 months of age, ARA intake sharply declined to 58.7 mg/day.

In Belgium, mean intakes of ARA were very low at 2.5 to 3 years of age and at 4 to 6.5 years of age (17 and 18 mg/day, respectively) [90] (Table 1). In Australia, national intake data indicated that 2 to 3 year-old and 4 to 7 year-old children consumed 16 mg/day and 22 mg/day of ARA, respectively [91]. Much higher mean ARA intakes were reported for Canadian children living in Vancouver where intakes ranged from 133 to 260 mg/day among children 1.5 to 5 years of age [92]. However, in children aged 4 to 7 years of age from Ontario, Canada, mean intake of ARA was lower (57 mg/day) [93].

Based on food records from the National Health and Nutrition Examination Survey (NHANES 2003–2008), the mean intake of ARA in American children at 1 to 4 years of age was 60 mg/day [94]. Most of the ARA was obtained from poultry (32.5%), eggs (27.5%), and meat dishes (20.9%). The latest NHANES data from 2015 indicate that the mean ARA intake of American children at 2 to 5 years of age increased to 80 mg/day [95].

Based on these dietary intakes from local and national surveys, it is clear that the diets of young children contain low levels of ARA. Reported mean intakes of ARA at 10 to 18 mg/day in developing and developed countries are only about 10% of the amount of ARA available to infants fed human milk or infant formulas containing DHA and ARA.

Birch *et al.* [96] reported that despite the introduction of a variety of solid foods at 17 weeks of age, infants who did not receive an infant formula supplemented with ARA and DHA throughout the first year of life had significantly lower levels of both of these fatty acids in plasma. The clinical consequences of low intake of ARA have not been adequately investigated.

6. ARA AND ITS ROLE IN IMMUNE SYSTEM DEVELOPMENT AND FUNCTION

There is growing evidence from preclinical and clinical studies that ARA plays an important role in maintaining infant health through its effects on the immune system and through the modulation of the inflammatory response [97]. The eicosanoids that ARA produces serve as both mediators and regulators of inflammation [98] (Table 2). These immunomodulatory effects have generated much interest in the potential roles that LCPUFA in general and ARA in particular have in common inflammatory conditions in childhood such as asthma, eczema, atopic dermatitis, and food allergies [97].

Table 2. Pro- and anti-inflammatory effects of prostaglandin E_2 (PGE_2) and leukotriene B_4 (LTB_4) [1].

Eicosanoid	Effects
PGE_2	**Proinflammatory**
	Induces fever Increases vascular permeability Increases vasodilatation Causes pain Enhances pain caused by other agents
	Anti-inflammatory
	Inhibits production of TNF and IL-1 Inhibits 5-LOX (decreases 4-series LT production Induces 15-LOX (increases lipoxin production)
LTB_4	**Proinflammatory**
	Increases vascular permeability Enhances local blood flow Chemotactic agent for leukocytes Induces release of lysomal enzymes Induces release of oxygen species by granulocytes Increases production of TNF, IL-8, and IL-6

Notes: [1] IL, interleukin; LOX, lipoxygenase; TNF, tumor necrosis factor. From Calder [98].

In cell membranes, ARA contributes to membrane order, has roles in signal transduction, and gene expression, and provides substrate for production of important chemical mediators [99]. Although ARA has been widely viewed as a pro-inflammatory agent, the eicosanoids that ARA produces serve as both mediators and regulators of inflammation [98]. ARA-derived eicosanoids, and other oxidized derivatives [98] are generated by the metabolic processes as shown in Figure 4.

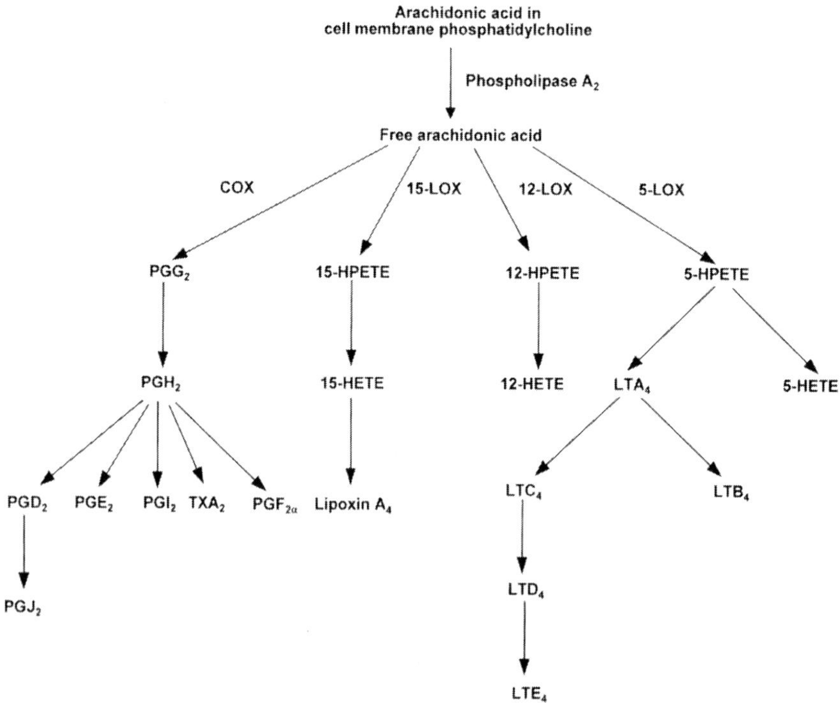

Figure 4. Generalized pathway for the conversion of ARA to eicosanoids. COX, cyclooxygenase; HETE, hydroxyeicosatetraenoic acid; HPETE, hydroperoxyeicosatetraenoic acid; LOX, lipoxygenase; LT, leukotriene; PG, prostaglandin; TX, thromboxane (from Calder [98]).

Another example of the dual role of PGE_2 and its receptors in modulating the inflammatory response has been described by Riccioti and FitzGerald [100]. During neuro-inflammation, the LPS-induced PGE_2synthesis causes adverse effects in neurons resulting in lesions and enhanced pain [101]. However, PGE_2also mediates bradykinin-induced neuroprotection by blocking LPS and ATP-induced cytokine synthesis in microglia and in neuron-glia co-cultures [102]. The anti-inflammatory and neuro-protective effects of $PGE_2$are mediated by microglial EP2- and EP4-receptors [100].

ARA is the substrate for the biosynthesis of prostaglandins. Prostaglandins and thromboxane A_2 are collectively called prostanoids. Prostanoids are formed when ARA is released from the plasma membrane by phospholipase and metabolized by the sequential actions of prostaglandin G/H synthase, or cyclooxygenase (COX), and by respective synthesis [100]. Prostanoids serve a variety of functions. The adhesion-type prostaglandins as well as prostacyclin are important in vasodilation and in anti-thrombus formation [103]. The E series prostaglandins act to dilate arterioles and capillaries to bring about a drop in blood pressure, relax vascular smooth muscle, open the bronchi of the lungs, and enhance blood flow through the kidney [104]. Prostaglandins are also involved in sleep regulation [105], febrile response [106], and in pain perception [107].

As discussed by Calder [98], the overall physiologic (or pathophysiologic) outcome associated with the production of eicosanoids depends on the cells present, the nature of the stimulus, the timing of eicosanoid production, and the sensitivity of the target cells and tissues to the type of eicosanoids that are produced [98,108]. For example, studies have shown that prostaglandin PGE_2, acting as a pro-inflammatory agent, induces cyclooxygenase 2 (COX-2) in fibroblast cells and by doing so up-regulates its own production [109] which in turn stimulates the production of IL-6 by macrophages (see Astudillo *et al.* [108] for a review of ARA metabolism by inflammatory cells). As an anti-inflammatory agent, PGE_2 inhibits 5-lipoxygenase (5-LOX) thereby decreasing production of the 4-series leukotrienes [110].These 4-series leukotrienes induce 15-LOX which in turn promotes the formation of lipoxins that aid the resolution phase of inflammation [110]. Thus, the ARA-derived PGE_2 has both pro- and anti-inflammatory effects (Table 2).

Leukotrienes, eoxins, lipoxins, and hydroperoxyeicosatetranoic acids (HPETEs) are synthesized from ARA by lipoxgenase enzymes and metabolized to LTA_4 [108]. LTA_4 is unstable and can be rapidly converted into LTB_4 or LTC_4. These three leukotrienes constitute the slow-reacting substances involved in anaphylaxis that act in allergic response [111]. They contract smooth muscle and affect vascular permeability. Eoxins are generally proinflammatory and are produced in the same manner as leukotrienes, but by the action of 15-lipoxgenase [108]. Lipoxins are produced by transcellular biosynthesis and have anti-inflammatory properties and are involved in the resolution of inflammation [112].

In vitro and animal studies suggest that ARA has a critical role in immune cell growth in the thymus, and in differentiation, migration, and proliferation of immune cells [98]. During early growth, there is substantial accretion of ARA in the mouse thymus which corresponds to the enrichment of the placental ARA for the fetus [98].

The immune system is composed of an integrated network of organs, tissues, cells and molecules that work together to resist infection, but maintain tolerance to harmless factors such as "self", antigens, and allergens [113]. When a challenge is detected (e.g., an allergen or pathogen), cell signaling between immune cells produces a coordinated immune response involving the release of cytokines and eicosanoids, which under normal circumstances allows cells to communicate with each other to neutralize and eliminate the challenge [114,115].

ARA is highly abundant in platelet membranes and is closely linked to many platelet functions [116]. Due to their high numbers (*i.e.*, normal platelet count of 1.50–4.00 × 10^{11} platelets per liter of blood [117] and their ability to release inflammatory mediators, platelets perform several sentinel tasks and can quickly communicate with the cells of the immune system [118]. For example, in inflammatory skin disorders, platelets recognize bacterial pathogens through interactions with Toll-like receptors leading to the elimination of bacteria by release of antimicrobial peptides or by aggregation of platelets around the bacteria [119]. An array of receptors present on platelet membranes facilitate transduction of signals and coordinate release of chemokines, cytokines, and other inflammatory mediators to regulate inflammation and respond to invading pathogens [118,119].

Inflammation is the immune's systems response to infection and injury [98]. Inflammation disorders are observed in infants, particularly those born prematurely [120]. Inflammation is characterized by the production of inflammatory cytokines, inflammatory agents such as reactive oxygen species, adhesion molecules, and the ARA-derived eicosanoids, and other oxidized derivatives [98]. *N*-3 LCPUFA decrease the production of the inflammatory mediators (eicosanoids, cytokines, and reactive oxygen species) and expression of adhesion molecules [98] by replacing ARA as an eicosanoid substrate and inhibiting ARA metabolism [98,121]. Aspirin and nonsteroidal anti-inflammatory drugs (NSAIDs) also inhibit the cyclooxygenase-catalyzed conversion of ARA to prostaglandins [122].

Although inflammation is perceived to be a serious health problem, the inflammation process is in fact an intrinsically beneficial event. Offending factors are removed or destroyed and as a result the affected tissues and physiological functions are restored. During the acute phase of inflammation, there is a rapid influx of blood granulocytes, typically neutrophils, followed by monocytes that mature into inflammatory macrophages [100]. The macrophages proliferate and affect the functions of resident tissue macrophages. This initial acute phase causes the usual signs of inflammation: redness, heat, swelling, and pain [100]. Once the initial adverse stimulus is removed via phagocytosis, the inflammation reaction typically decreases and ultimately resolves. During the resolution phase of inflammation, granulocytes are eliminated and macrophages and lymphocytes return to their normal pre-inflammatory levels [100]. The usual outcome of the acute inflammatory process is successful resolution and repair of tissue damage.

Eicosanoids and Their Effects on Hormones and Bone Formation

The typical definition of a hormone is a chemical substance produced in the body that controls and regulates the activity of certain cells or organs [123]. Many hormones are secreted by special glands, such as thyroid hormone produced by the thyroid gland. Eicosanoids are recognized as different from hormones because they are not synthesized or stored in select tissues or endocrine organs. Eicosanoids are synthesized in almost all tissues and exert their biological effect near the site of their synthesis rather at a distance such as other hormones [124]. Despite these differences, eicosanoids are generally classified as hormones [125]. Eicosanoids directly affect other hormones including glycoprotein hormones. The glycoprotein hormones include luteinizing hormone, somatostatin, and glucagon. Somatostatin is an important growth hormone that controls and regulates growth and cell division [126]. It is the main hormone that stimulates cell proliferation and growth, and this hormone must be regulated so that growth is controlled [126]. Insulin and glucagon release are also affected by the eicosanoid derivatives, epoxy-eicosatrienoic acids [124].

ARA also plays an important role in the hormonal regulation of normal bone formation and whole body mineral metabolism during infant and childhood growth (Table 3). During skeletal development, the eicosanoids relay cellular, organ, and systemic signals to balance the calcium and phosphate needs for bone formation and other metabolic activities [127,128]. During long bone growth, when bone tissue is created, [127,128,129,130] ARA mediates vitamin D_3-regulated chondrocyte maturation [131] and proliferation [127,132,133,134] for the mineralization of skeletal growth plates (Figure 5).

Table 3. Roles of ARA in bone formation, metabolism, and mineral balance.

Metabolic Effector	Physiological Roles of ARA
ARA [149]	Maintain normal balance between bone mineral accrual and bone resorption during infant development
ARA, growth hormones [150,151]	Increase insulin-like growth factor gene expression and induction of osteoblast-dependent bone formation
Vitamin D_3 [128,133]	Mediate vitamin D_3 coordination of chondrocyte proliferation in the epiphyseal growth plates of long bones. Parathyroid hormone secretion
Calcium and phosphorous [152]	Regulation of parathyroid hormone secretion in response to blood mineral concentrations
Parathyroid hormone [149,150]	ARA mediated/activated pathway involved in the activation and secretion of vitamin D_3 by kidneys
Physical activity [153]	ARA mediates bone adaptation to changes in physical stress through mechanisms which mediate resorption and remodeling

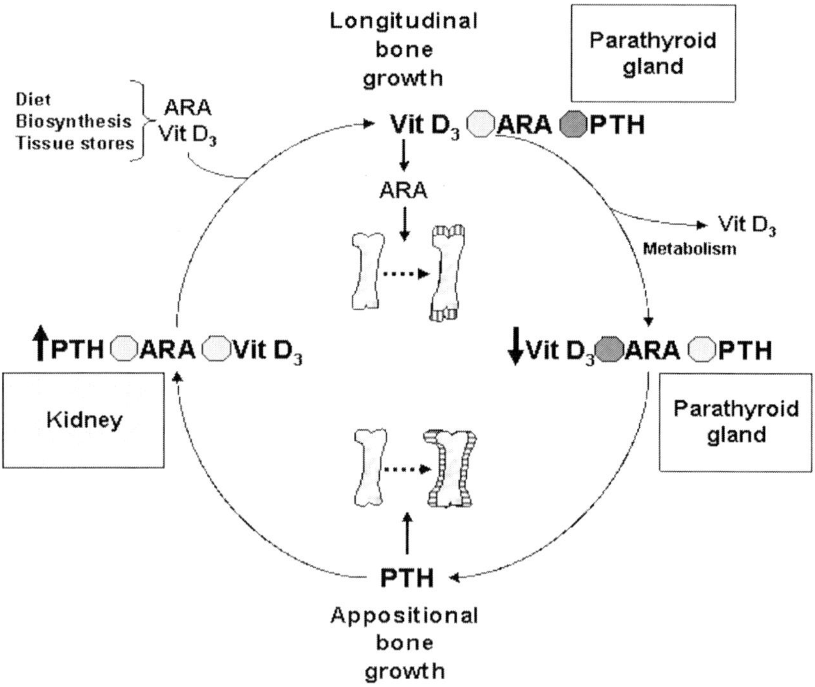

Figure 5. The role of ARA in bone development and homeostatic regulation of vitamin D_3 and parathyroid hormone (PTH) levels along the parathyroid gland-kidney axis during growth. ARA and vitamin $D_3$are acquired from the diet and/or from endogenous sources. ARA mediates vitamin D_3 regulation of chondrocyte proliferation and growth plate mineralization during bone elongation. As vitamin D_3 is metabolized and levels subside, ARA-dependent PTH suppression is diminished and PTH production by the parathyroid gland is upregulated. This results in increased periosteal bone mineral content (appositional bone growth). In kidney, PTH induces the ARA-mediated increase in vitamin D_3 activation and secretion, elevating the amount of vitamin D_3 in circulation. The cycle continues as the restoration of vitamin D_3 results in the ARA-dependent suppression of PTH and stimulates longitudinal bone growth.

A product of ARA, prostaglandin PGE_2 is a potent regulator of cartilage formation or chondrogenesis and resorption [135,136,137,138,139]. At low levels, PGE_2 stimulates bone formation by increasing the production of insulin-like growth factor, a powerful growth stimulator for bone, cartilage, and muscle [140]. At high levels, PGE_2 has the opposite effect: bone formation is reduced and resorption is increased [140].

These differential effects of osteoclast formation and resorption are mediated through multiple subtypes of G-protein coupled PGE_2 cell surface receptors (EP1, EP2, and EP4) [141]. Activation of the EP2 and EP4 receptor subtypes are linked to an elevated level of cyclic adenosine monophosphate (cAMP) and bone formation. EP2 also acts as a selective agonist which has the ability to heal long bone fractures as demonstrated in animal models [141,142].

PGE_2 is also critically important for bone strength [141]. When different doses (3 or 6 mg PGE2/kg/day) of prostaglandin PGE_2 were given to Sprague-Dawley rats for 3 weeks an increase in bone and hard tissue mass, calcified cartilage cores, and a decrease in osteoclasts were observed [143]. PGE_2 increased metaphyseal calcified tissue mass by depressing hard tissue resorption and stimulating the replication and differentiation of osteoblast precursors to form new bone [143].

Other prostaglandins play multiple roles for bone metabolism and remodeling by regulating various signaling pathways [144]. For example, PGF2α, through the activation of protein kinase C (PKC), stimulates the Na-dependent inorganic phosphate transport in osteoblasts [144]. PGF2α also up-regulates interleukin (IL-6) to stimulate osteoclast formation and increases vascular endothelial growth factor (VEGF) associated with the growth of blood vessels from pre-existing vasculature [144]. PKCα, in particular, appears to play a critical role in the regulation of osteoblastic function under load-bearing conditions [145]. During exposure to mechanical strain, PKCα is activated in osteoblast-like cells [146] while PKC signaling has been implicated in the regulation of various mechanically response genes including the osteoblast differentiation marker osteocalcin [147,148].

In studies of piglets fed formulas with differing levels of ARA (0.30%, 0.45%, 0.60% or 0.75% of fat) plus the same level of DHA (0.1% of fat), proportions of ARA in plasma, liver and adipose were dose dependent but bone modeling was not [150]. Whole-body bone mineral content was elevated in the piglets fed the highest levels of ARA (0.60% and 0.75%) and was best predicted by dietary ARA [150,154]. In addition, the 0.60% and 0.75% ARA groups had bone mineral content values closest to that of a reference group of suckled piglets [150,155].

Overall, dietary provision of ARA serves a number of important roles in skeletal metabolism. ARA functions as an important modulator of vitamin D regulation of chondrocyte proliferation and growth plate mineralization. ARA derived metabolites are important inducers of osteoclast [156,157] and osteoblast differentiation [158,159], and in modulating resorption of bone [139,149] by increasing IGF-1 gene expression [104,151] and circulating levels of IGF-1 [150]. ARA also responds to changes in stress and mechanical loading [153,160], and accelerates bone repair and healing [142,161]. Additionally, in term infants, cord blood ARA levels correlate positively with bone mineral density [155]. Thus, ARA represents an important nutrient for infant and childhood bone development and metabolism.

7. ARA IN SKELETAL AND CARDIAC MUSCLE

Several animal studies have shown that the concentration of ARA in the heart is highly sensitive to levels and ratios of ARA and DHA of the diet [24,54,162,163,164]. The amount of ARA in cardiac tissue muscle is at concentrations 2 to 3 times greater than observed in skeletal muscle [52]. Analysis of the phospholipid composition of skeletal muscle biopsies collected from 56 children <2 years of age indicated that ARA represented 16.5% of the total percentage of LCPUFA in muscle phospholipids [165].

Repetitive force loading and unloading during ATP-dependent contraction of actin filaments are major mechanical functions of heart muscle, and to a lesser extent, skeletal muscle [166]. ARA is critical for muscle contraction [166,167,168]. In skeletal muscle, excitation–contraction coupling is the process by which muscles contract [166] when a muscle action potential in the muscle fiber causes the myofibrils to contract [169]. Excitation–contraction coupling relies on a direct coupling between key proteins, the sarcoplasmic reticulum calcium release channel (the release of Ca^{2+} ions), and the voltage-gated L-type calcium channels [170]. The release of Ca^{2+} ions from the sarcoplasmic reticulum causes binding between actin and myosin to induce muscle contraction. This cycle is reset as calcium declines back to resting levels [166]. Cardiac and skeletal muscle require tight regulation of voltage-gated calcium channels and calcium homeostasis to coordinate the excitation-contraction coupling process [170].

Phosphatidylinositol (4,5) bisphosphate (PIP_2), a phospholipid component of cell membranes, serves as an important regulator for Ca^{2+} release from the sarcoplasmic reticulum and assists in the maintenance of normal calcium signaling to control contractile forces [167,168,171]. The fatty acids of PIP_2 are variable in different species and tissues, but studies show the most common fatty acids are stearic in position 1 and ARA in position 2 [37].

Calcium homeostasis, regulation and maintenance are critical elements for normal muscle function. Wolf*et al.* [34] have shown that the endoplasmic reticulum is directly responsible for the regulation of intracellular Ca^{2+} concentrations. ARA plays an important cooperative role with myo-inositol 1,4,5-triphosphate (IP_3) in glucose-induced calcium mobilization and insulin secretion by pancreatic islets.

PIP_2 and phosphatidylinositol 3,4,5-triphosphate (PIP_3) are also critical for cardiac function [172]. In the heart, PIP_2, as a key second messenger, controls the activity of ion channels involved with the modulation of heart rhythm. PIP_3, on the other hand, is primarily involved in the control of cardiomyocyte apoptosis, hypertrophy, and contractility [172]. In adults, deregulation of the phosphoinositide metabolism is associated with the onset and progression of several cardiovascular pathologies including atherosclerosis and heart failure [172].

Muscle growth and atrophy depend on the balance between the rates of protein synthesis and degradation [173]. *In vitro* experiments with animal and human skeletal and cardiac muscle tissue indicate that prostaglandins are involved in the regulation of protein synthesis and degradation in various types of striated muscle [173,174]. While PGE_2 increases degradation of muscle in young rats and causes net protein balance to become more negative, $PGF_{2\alpha}$ causes a large stimulation of protein synthesis in muscle tissue [173]. These findings are consistent with the many important roles played by prostaglandins PGE_2 and $PGF_{2\alpha}$ in muscle protein balance and indicate that overall, ARA serves multiple functions in cardiac and skeletal muscle function and physiology.

8. BIOMAGNIFICATION AND ACCRETION OF ARA IN INFANTS

Biomagnification is when infants have higher levels of LCPUFA in plasma lipid fractions and erythrocytes as compared with their mothers [45]. Biomagnification can be especially marked for ARA with levels more than 2-fold of that from the maternal side and independent from the amount of its precursor LA available maternally. The stability of LA content implies that any conversion to ARA is not keeping up with the fetal demand for ARA [45]. Biomagnification by the placenta serves to preferentially obtain preformed ARA and DHA from the mother in order to deliver it to and nourish the fetus [45].

As shown by Kuipers *et al.* [175], biomagnification is independent of maternal ARA status at both delivery and at 3 months of age and is found to be similar across different populations with differing diets. These findings indicate that biomagnification as a biological process seeks to achieve a uniform ARA status in infants at the expense of their mothers. The process of biomagnification suggests that a certain level of infant prenatal ARA status must be maintained for optimal infant growth.

Infants with the lowest birthweights have the lowest levels of ARA, and those born earliest have the lowest levels of DHA [45]. The process of biomagnification initially protects vascular growth which is a requirement for brain growth. Vascular growth must precede brain growth to meet the brain's demand for energy, which can be as high as 70% of the total fetal demand for energy in the last trimester of pregnancy [45].

At delivery, as shown by Luxwolda *et al.* [176], the maternal RBC-ARA content is consistently higher than that at 3 months postpartum. At delivery, infant RBC-ARA content is similar or higher than their mother's RBC-ARA contents. From delivery to 3 months postpartum, maternal RBC-ARA increases while infant RBC-ARA decreases. The decrease in RBC-ARA content may be due to a lower conversion of LA to ARA since the infant's capacity to synthesize LCPUFA decreases dramatically after delivery [69] and has been shown to decrease with gestational age at birth [64].

There appears to be a tightly regulated synergism between DHA and ARA at low DHA status and an antagonism at high DHA status [50]. Intrauterine DHA biomagnification in mothers with low fish intakes aims at a synergistic increase of fetal DHA to maintain a balance with ARA. Bioattenuation at higher DHA status may in turn prevent abundant passage of DHA across the placenta that leads to antagonism with ARA. Since ARA is important for fetal growth [177] and is rapidly accreted in the fetal brain [178,179] any competition from gestational DHA must be tightly regulated and balanced for optimal neurodevelopment after birth [50]. Dietary depletion of ARA in early infancy may have adverse consequences for brain development [178,179].

9. CONSEQUENCES OF ARA DEFICIENCY

Essential fatty acid (EFA) deficiency impairs lipid and energy metabolism, cell membrane structures, lipid signaling pathways, and ultimately leads to death [180,181]. Mammals are dependent on a dietary supply of LA and ALA which are converted into *n*-6 and *n*-3 PUFA, respectively. Δ6-fatty acid desaturase (FADS2) converts LA to γ-linolenic acid (C18:3*n*-6) and Δ5-fatty acid desaturase (FADS1) converts dihommo-γ-linolenic acid (C20:3*n*-6) to ARA [182].

Early studies of EFA deficiency considered the effect that various dose levels of intake of LA, ARA and ALA esters (0% to 10% of total calories for 100 days) had on the fatty acid composition in the liver of rats [183]. Fat deficiency symptoms (necrotic tail and scaly feet) appeared in all animals fed LA at less than 0.6% of calories and ARA at less than 0.25% of calories. ARA was 3-fold more effective than LA in liver incorporation and mitigating deficiency. Fat deficiency symptoms affecting the skin were not surprising. In the skin, as in all organs, EFA are principally found in glycolipids and phospholipids. Most of the epidermal fatty acid

PUFA is ARA [184]. EFA deficiency causes skin flaking in humans, dogs, and mice, symptoms that can be restored with LA dietary therapy [184].

Since LA deficiency results in disruption of the skin's water barrier function [185] and heat loss from skin [186] the side effects make it difficult to distinguish the specific effects of ARA deficiency independent from those related to LA deficiency. The fads2$^{-/-}$ mouse allows for the specific investigation of ARA deficiency without the underlying complications of LA deficiency [181]. The mutation eliminates Δ-6 desaturase activity leading to a dramatic decrease in the accumulation of ARA in tissues and subsequently, ARA conversion to PGE, TXB, prostacyclin and leukotrienes. Platelet aggregation and thrombosis are therefore also limited.

When fads2$^{-/-}$ mice were followed for several weeks and fed a diet lacking Δ6-fatty acid desaturase products but containing ample amounts of LA, the lack of PUFA and eicosanoids did not impair lifespan but all the mice were sterile, developed ulcerative dermatitis, splenomegaly, and ulceration in the duodenum and ileocecal junction [182]. Liver levels of ARA and DHA declined by 95% and somewhat smaller decreases were observed in the brain and testes (~50%). The absence of γ-linolenic conversion in the fads2$^{-/-}$ mouse deprived the cyclooxygenase and lipoxygenase pathways of their substrates, including the elimination of PGE synthesis, the failure of synthesis of TXBs by thrombocytes, and the failure to produce leukotrienes [181]. PUFA supplementation completely restored the adverse symptoms observed in fads2$^{-/-}$ mice. The mechanism by which ARA prevented dermatitis may be due, at least in part, to lower levels of prostaglandin D_2 (PGD_2) when skin ARA is decreased [182].

The ulceration of the small intestine in the fads2$^{-/-}$ mouse may have been associated with the decline of prostaglandin synthesis, similar to the effect often seen with the long-term use of NSAIDs [187]. NSAIDs block prostaglandin synthesis by inhibiting cyclooxygenases, leading to an erosion and then ulceration of the mucosal layer of the stomach and small intestine. However, loss of organized stratification of proliferating cells into defined zones is a common feature of EFA deficiency [188].

The Δ6-fatty acid desaturase gene FADS2 was cloned in 1999 [189]. An adult human case of Δ6-fatty acid desaturase deficiency was identified and described in the literature [190]. The patient exhibited growth retardation accompanied by skin abnormalities, corneal ulceration, and feeding intolerance. Treatment with dietary DHA and ARA restored normal growth and eliminated most of the symptoms [190].

A novel genetic model, the FADS1 (Δ5 desaturase) knockout mouse was used to determine the role that the ARA-derived 2-series eicosanoids had in mucosal physiology and inflammation [191]. Fads1$^{-/-}$ mice have very low levels of ARA in tissues (colon mucosa, liver, spleen, serum and fatty acid profiles). The deficiency in ARA resulted in a massive enhancement of dihomo-y-linolenic acid, the 1-series prostaglandin substrate in tissues and a decrease in 2-series-derived prostaglandins or PGE_2. Fads1$^{-/-}$ mice failed to thrive, gradually dying at 5 to 6 weeks of age with no survivors past 12 weeks of age [191]. The lack of PGE_2 was associated with disturbed intestinal crypt proliferation, altered immune cell homeostasis, and a heightened sensitivity to acute inflammatory challenge [191]. Dietary supplementation with ARA extended the longevity of fads1$^{-/-}$ mice to levels comparable with normal wild-type mice (Figure 6).

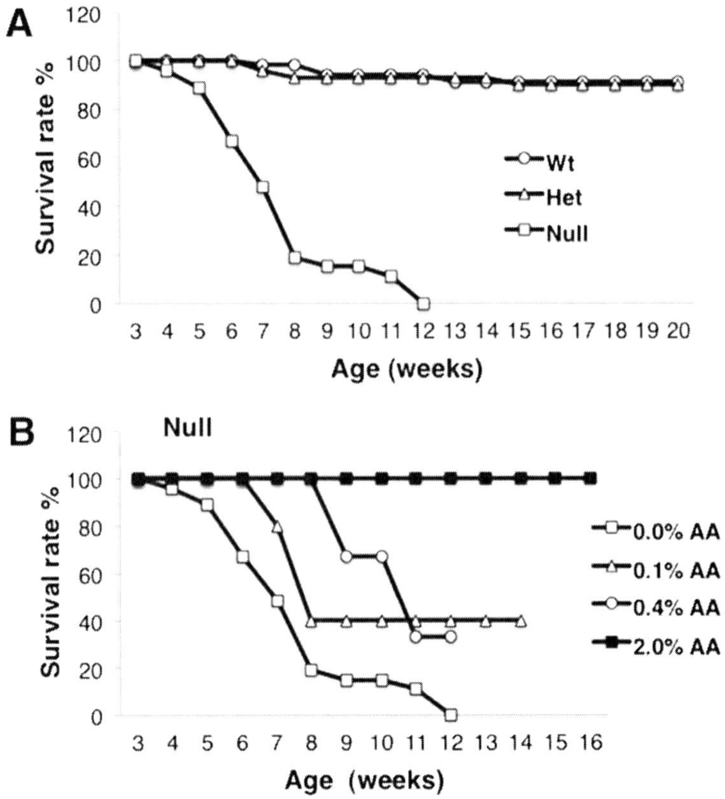

Figure 6. Kaplan-Meier survival curves of *Fads1* mice, AA = ARA. (**A**)*Fads1* null mice exhibited low viability when fed a standard AA-free diet; *n* = 37 for wild-type, *n* = 44 for heterozygous, *n* = 11 for Null; (**B**) Dietary supplementation with AA (0.1% and 0.4%, *w/w*) partially reversed the *Fads1* null mouse phenotype; *n* = 5 for Null + 0.1% AA, *n* = 3 for Null + 0.4% AA. Supplementation with 2.0% AA completely reverse the Null phenotype; *n* = 4 for Null ± 2% AA (from Fan *et al.*[191]).

Although fads1$^{-/-}$ and fad2$^{-/-}$ mice are useful to examine the function of ARA *in vivo* PUFA are transferred through the placenta from the heterozygous mother into the homozygous fetus. Additionally, the amount of DHA and ARA in the brain tends to remain tightly controlled even under conditions of PUFA deprivation, but at the expense of other tissues to protect the brain [182]. Lpiat1$^{-/-}$ mice have a mutation that affects the synthesis of ARA-containing PI. PI is unique in its fatty acid composition, *i.e.*, most of the fatty acid that is attached to the sn-2 position of PI is ARA [33]. Other membrane phospholipids such as PC and PE contain other PUFA including DHA.

Lee *et al.* [33] showed that Lpiat1$^{-/-}$ mice had a reduced content of ARA in PI and had deficits in cortical lamination during brain development, delayed neuronal processes in the cortex, and reduced neurite outgrowth *in vitro*. Lpiat1$^{-/-}$ mice died within a month and showed atrophy of the cerebral cortex and hippocampus. These results demonstrate the importance of ARA-containing PI in normal cortical development in mice. By eliminating LPIAT1 in Lpiat1$^{-/-}$ mice, the enzyme responsible for the incorporation of ARA into PI, it was shown that the ARA-containing PI is essential for brain development in mammals [33].

Newborn pups of Δ6-fatty acid desaturase knockout mice were administered artificial milks that contained 3.7% ALA and 16% LA with or without 1.2% ARA and/or 1.2% DHA for 18 days immediately after birth [192]. Compared with wild-type mice, the body weight of the mice fed the control diet was significantly lower, particularly after 6 weeks of age. However, body weights of knockout mice fed milks with DHA and ARA+DHA were similar to that of the wild-type mice. Motor activities of the knockout mice fed ARA were elevated compared with the wild-type mice and those fed the control diet. Better motor performance was also observed in knockout mice fed the ARA + DHA diet. The authors concluded that ARA corrected the decrease in body weight and the combination of ARA and DHA improved the motor dysfunction caused by the deficit of Δ6-fatty acid desaturase.

Taken together, results from these investigations indicate the importance of ARA and its derivatives for the coordination of cellular differentiation, organogenesis, and function during early growth and development.

10. ANIMAL STUDIES OF ARA SUPPLEMENTATION

10.1. Immunomodulatory Effects of ARA and DHA Supplementation

The activation of peroxisomal proliferator-activated receptors (PPARs) has been shown to be protective in brain ischemic and oxidative injury and in many neurological diseases that may affect infants [193]

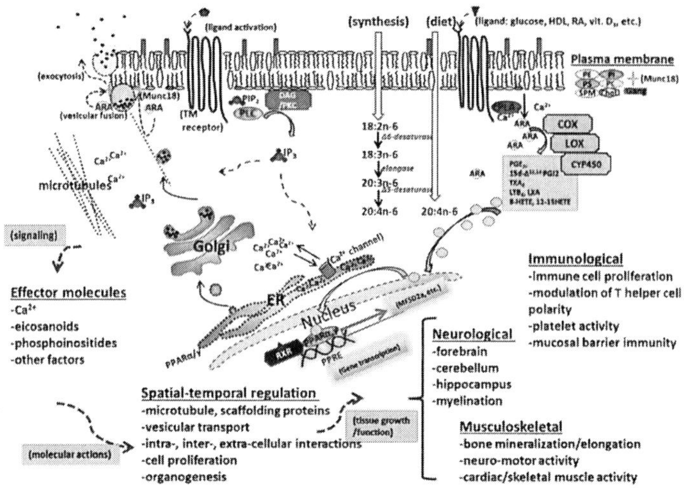

Figure 7. Schematic summary of molecular events and functional outcomes involved in metabolism of ARA. ARA is derived from endogenous synthesis or directly from the diet and is incorporated into cellular membrane complex lipids. Within the lipid bilayer, ARA is enriched in PE and PI in the inner membrane. Coordination of spatial-temporal interactions between molecular and cellular components and activities are mediated by metabolites of, or molecules associated with metabolism of ARA. Metabolism of ARA is triggered by activation of transmembrane receptors as a result of binding a ligand. A few examples of receptor-mediated activation of ARA metabolism include glucose, vitamin D_3, Ca^{2+}, or antigen presentation or detection by immune cells. ARA released from the membrane by the actions of PLA_2or metabolized by enzymes such as COX, CYP450, and/or LOX can act directly or serve as a substrate for various enzymes to produce second-messengers. ARA-derived eicosanoids, including prostaglandins, leukotrienes, lipoxins, and HETEs regulate numerous activities including passage of ions between subcellular compartments, interactions between various structures or cells, and nuclear regulation of gene transcription by PPARs activators. Within the inner leaflet of cell membranes, ARA is enriched in micro-domains and is involved in regulation of receptor mediated activities. In addition, micro-domains serve as foundations for biophysical interactions between subcellular structures such as microtubules and other cytoskeletal activities including vesicular transport. The consequences of temporal-spatial regulation include coordinated release of hormones, expression of various cell functions, and/or alterations in phenotypes, and cellular motility. Examples of PPAR-regulated gene products involved tissue uptake of LCPUFA and oxidation of stored lipids: MFSD2A, major facilitator of superfamily domain-containing protein 2A. Membrane components: Chol, cholesterol; Gang, gangliosides; PC, phosphatidylcholine; PE, phosphatidylethanolamine; PI, phosphatidylinositol; PS, phosphatidylserine; SPM, Sphingomyelin. Nuclear transcription factors: PPAR, peroxisome-proliferator activator receptors; RXR, retinoid X receptors. TM, transmembrane receptors. Enzymes: COX, cyclooxygenase; CYP450, cytochrome P450; LOX, lipoxygenase; PLA_2, Phospholipase A2; PLC, Phospholipase C. Signaling molecules: PG, prostaglandin (Adapted from Pike [32]).

Figure 7). In addition, transcription of the gene for the Major Facilitator Superfamily of the domain-containing protein 2a (MFSD2A) has been identified as being an important transporter for the uptake of DHA across the blood brain barrier [194] and is under the control of PPAR [195]. Studies indicate that LCPUFA and their metabolites are ligands to PPARs. Diets containing an *n*-6:*n*-3 ratio of about 1-2:1 supplied during pregnancy and lactation appear to be optimal for the expression of neuron-specific enolase, glial fibrillary acidic protein and myelin basic protein, markers related to the growth and maturation of neurons, astrocytes and myelin [193].

To investigate the immunomodulatory effects of different PUFA, weanling rats were fed a high-fat diet (178 g/kg) that contained 4.4 g of ALA, γ-linolenic, ARA, EPA, or DHA/100 g total diet [196] for 6 weeks. The proportion of total PUFA content (~35 wt %) was held constant and the *n*-6 to *n*-3 ratio was maintained at 5.8 to 7.0. PGE_2 production was enhanced in leukocytes from rats fed the ARA-rich diet and was decreased from leukocytes in rats fed the EPA or DHA diets. ARA did not affect lymphocyte proliferation, NK cell activity, or the cell-mediated immune response. Lack of an effect on T-lymphocyte proliferation and Con A in splenocyte cultures was also observed in mice fed a safflower oil ethyl ester diet +1% ARA for 10 days [197]. The lack of an immunological effect of ARA agrees with findings from a human study that considered 1.5 g of ARA/day for 50 days on the proliferation response of peripheral blood mononuclear cells to Con A, phytohemagglutinin, or poke-weed mitogen [198]. Human peripheral blood NK activity was also unaffected by the consumption of ARA.

Prostaglandins which are involved in inflammatory processes also play a major role in the recovery of intestinal barrier function in ischemia-injured porcine ileum by converting ARA to PGH_2 [199]. The importance of ARA and ARA-derived eicosanoids in the intestinal epithelium was reviewed by Ferrer and Moreno [200]. In a study that considered the effect of supplemental ARA on intestinal barrier repair in ischemia-injured porcine ileum pigs were fed a formula containing no LCPUFA (0% ARA), 0.5% ARA, 5% ARA, or 5% EPA for 10 days. Piglets that were fed 5% ARA exhibited enhanced recovery compared with piglets fed 0% ARA or 0.5% ARA [201]. The EPA-fed piglets had enhanced recovery comparable with piglets fed 0% ARA. The enhanced recovery response observed with 5% ARA was supported by reduction in the mucosal-to-serosal flux of ^{3}H-mannitol and ^{14}C-inulin compared with the other dietary groups. Jacobi *et al.*[201] concluded that piglets fed a high-ARA diet are less susceptible to ischemia-induced epithelial cell sloughing and that feeding elevated levels of LCPUFA, including ARA, enhances acute recovery of ischemia-injured porcine ileum. For infants affected by necrotizing enterocolitis (NEC) where physiological repair of the intestines is necessary elevated LCPUFA intake including ARA enhances recovery of damaged tissues [201].

EFA deficiency also leads to hepatic steatosis. When rats were administered varying amounts of DHA and ARA to determine whether exclusive supplementation with DHA or ARA could prevent EFA deficiency and inhibit the development of hepatic steatosis mice fed at least 2% of their calories from DHA and 1% of the calories from ARA did not develop clinical or biochemical evidence of EFA deficiency disease or hepatic steatosis [202]. Although hepatic steatosis is an adult disease, the fact that mice fed at least 2% of their calories from DHA and 1% of the calories from ARA prevent the development of EFA deficiency suggests the importance of ARA throughout the lifespan.

To investigate the ability of ARA- and ARA + DHA-enriched formula to modulate immune response in neonatal piglets to an inactivated influenza virus vaccine Bassaganya-Riera *et al.* [203] considered a diet with ARA + DHA in sow milk fed at birth. The diet modulated antigen-specific T-cell responses to an inactivated influenza virus and up-regulated IL-10 expression [203]. Although ARA and DHA have been suggested to elicit opposing immunomodulatory actions, the immunologic outcome in the study was beneficial [203]. The authors concluded that ARA + DHA enriched formulas, with the approximate 2:1 ratio fed during the neonatal period, may prevent or manage autoimmune and allergic reactions in infants by down-modulating T-cell reactions.

10.2. Retinal and Neurodevelopmental Effects of ARA and DHA Supplementation

The retinal DHA content in guinea pigs was considered in relation to diets containing different n-6:n-3 ratios (from 72.0 to 2.5) [204]. Not surprisingly, diets with the highest n-6:n-3 PUFA ratios had the highest n-6 retinal fatty acid profiles. Weisinger *et al.* [204] reported that retinal function was altered by tissue DHA levels and responded according to an inverted "U-shaped" function. As DHA levels increased past an optimal amount found to be 19%, the result was poorer electroretinographic scores. However, there was no mention that as DHA increased there was a corresponding decrease in ARA levels due to ARA antagonism. The marked decrease in ARA levels may have been the variable of interest that was not fully considered and responsible for the electroretinographic changes at high DHA intakes.

Champoux *et al.* [205] used a neurodevelopmental battery to test the neurological behavior in rhesus macaques neonates fed a control formula without LCPUFA or a LCPUFA-supplemented formula with 1 wt % each of DHA and ARA. Macaque neonates fed the supplemented formula obtained higher scores on motor maturity and orientation than those fed a control formula. Champoux *et al.* [205] concluded that the results supported the view that preformed DHA and ARA in infant formulas are required for optimal neurological development.

Learning behavior in rats fed a diet supplemented with 3% safflower oil (Safflower, *n*-3 fatty acid deficient, high LA acid) was compared to those fed 3% perilla oil (Perilla, high ALA) [206]. Through two generations, the *n*-3 fatty acid deficient group exhibited decreased correct response ratios in a brightness-discrimination behavior test. The altered learning ability in the brightness-discrimination test was restored with supplementation of DHA after weaning, only after levels of ARA in the brain lipids were normalized. The authors concluded that *n*-3 fatty acid is essential for the maintenance of learning performance and that *n*-3 deficiency in the presence of *n*-6 fatty acid during gestation did not lead to irreversible damage to the brain [206]. Thus, both DHA and ARA affected learning performance and a balance of ARA and DHA levels must be maintained.

To investigate the effects that varying dietary levels of LCPUFA have on growth, brain fatty acid composition and behavior in mice, 5 groups of pregnant and lactating mice were fed diets with either very high *n*-6 to *n*-3 ratio of 49 (*n*-3 deficient), a more usual ratio of 4.0, or a low ratio of 0.32 for 15 weeks [207]. There was no effect of diet on birth weights of pups, but on days 15 and 22 the pups in the low *n*-6 to *n*-3 groups weighed less than those in the other treatment groups. Increasing levels of DHA in the diet increased brain DHA and decreased brain ARA. The differing ratios of *n*-6 to *n*-3 had no effect on the ability of mice to learn the place test or perform in the Morris water maze. However, the mice fed the low *n*-6 to *n*-3 ratio swam more slowly, unless ARA was substituted for LA as the source of the *n*-6. The lower body weight in the high *n*-6 to *n*-3 fed mice was not attributed to simply *n*-6 deficiency. The high *n*-3 to *n*-6 ratio led to the inhibition of Δ6-desaturase [207]. Thus, the conversion of LA to ARA was impeded and ARA became unavailable for growth. Mice fed high levels of DHA also had high levels of EPA showing a considerable amount of retroconversion. The findings showed the importance of balancing the amounts of ARA to DHA and that some deficits can be overcome if LA is replaced by ARA as the source of *n*-6 fatty acid [207]. Wainright *et al.* [208] also reported that ARA supplementation increased ARA levels and decreased DHA levels in forebrain membrane phospholipids in Long-Evans rats, whereas DHA supplementation increased DHA levels and decreased ARA levels. Correlational analyses did not show a relationship between DHA and ARA levels in the forebrain and working memory performance [208,209].

Newborn infants of diabetic mothers have lower ARA and DHA levels in cord blood than newborns of normal mothers [210]. The lower levels of the LCPUFA in the newborns of diabetic mothers were associated with impaired and altered sensory-cognitive and psychomotor functions at birth and reduced visual and memory performance at 8 and 12 months [211,212]. Compared with normal controls, most rat models of diabetes are characterized by a lower level of brain ARA only and not a lower level of DHA [213]. Even though both ARA and DHA are important for neurodevelopment, brain accretion of ARA exceeds that of DHA during gestation [214], especially in the first two trimesters during the period of rapid proliferation of neuronal and glial elements [179,215]. When Sprague-Dawley diabetic, pregnant rats were fed either a control diet or an ARA (0.5%) supplemented diet throughout reproduction, the weaned offspring in the ARA group performed significantly better in the water maze and rotarod tests and showed greater exploratory behavior than control-diet offspring [216]. The results indicated that maternal hyperglycemia has long-term consequences during the initial stages of learning and that maternal supplementation with ARA positively influences learning outcomes.

Amusquivar *et al.* [217] reported that rat pups of dams fed diets with *n*-3 fatty acids from fish oil compared with those fed *n*-6 fatty acids from olive oil during pregnancy and lactation had smaller increases in postnatal body weight and length, and delayed body and psychomotor maturation indices. Slower growth and brain development occurred when both dams and fetuses were fed a moderate amount of fish oil (10%) as the only fat source [217]. In the study, the ARA level was lower than the DHA level in brain tissue of the offspring of dams fed high *n*-3 fatty acid diets. The differences in postnatal development disappeared when the fish oil was supplemented with γ-linolenic acid, a precursor of ARA. The growth deficits were also eliminated by the inclusion of ARA in the diet [207]. The studies demonstrate the importance of maintaining adequate levels of ARA during development, and suggest that diets too high in *n*-3 fatty acids during development may have negative effects on development by reducing tissue levels of ARA [218].

To investigate the effects that a DHA-rich maternal diet compared with an ARA-only diet have on brain fatty acid composition of Sprague-Dawley rats, Elsherbiny *et al.* [219] considered a control diet containing ARA (0.4 g/100 g of total fatty acid) *vs.* a DHA + ARA diet (0.9 g/100 of DHA and 0.4 g/100 g of ARA of total fatty acid). The results indicated that at three weeks postnatally the DHA-rich diet increased levels of DHA in the brain and decreased ARA by 12.8%. The brain of a three-week-old rat is at a comparable stage as that of a human toddler at 2–3 years of age [219]. At six weeks (comparable to a 12–18 years old human), the DHA-induced decreases in ARA were reversed and disappeared when DHA was continued (*i.e.*, DHA/control group). Thus, elevated dietary levels of DHA decrease the amount of ARA in brain without an adequate supply of dietary ARA.

Prepulse inhibition (PPI) is a normal suppression of a startle response when a low intensity stimulus that elicits little or no behavioral response immediately precedes an unexpected stronger startling response [220]. Deficits in PPI have been reported in individuals that have mental disorders including schizophrenia [220]. Various brain regions including the hippocampus have been associated with PPI problems. To determine whether dietary administration of LCPUFA enhances neurogenesis in the rat hippocampus and improves PPI response in wild-type mice and $Pax6^{+/-}$ mice (that exhibit PPI deficits) a control diet or diets supplemented with ARA (4%), DHA (4%) or ARA (4%) + DHA (4%) were administered [220]. Compared with the other diets, the administration of the ARA diet successfully increased neurogenesis not only in the $Pax6^{+/-}$ mice but also in the wild-type mice. Treating the $Pax6^{+/-}$ mice with ARA also resulted in alleviating their PPI deficits. The authors suggested that the ARA diet as compared with the DHA or ARA + DHA diets positively affected postnatal neurogenesis in several regions of the brain including the hippocampus by influencing the fluidity of neuronal membranes and by regulating neuronal transmission [220].

11. INTRODUCTION OF DHA AND ARA IN INFANT FORMULAS

Both DHA and ARA have been added to infant formulas in the United States since 2001. In Europe, the addition of DHA and ARA in infant formulas began much earlier in 1994. Most infant formulas contain 0.2% to 0.4% total fatty acids of DHA and between 0.35% and 0.7% total fatty acids of ARA based on worldwide averages of DHA and ARA content in human milk [5]. Several international expert groups [6,7,8,9] support the addition of these levels of DHA and ARA in infant formulas to ensure optimal infant growth and development. Thus, all commercially available infant formulas contain preformed ARA at levels equal to or higher than the DHA content in formulas where these LCPUFA are added.

To determine the necessity of adding LCPUFA to infant formula, several studies were performed in the 1990s with preterm and term infants fed formulas containing DHA or EPA with and without ARA (see Fleith and Clandinin [221] for a review). No studies have examined supplementation of infant formula with ARA alone. In most studies, a control group without LCPUFA and/or a breastfed group were included. Studies also investigated the effect of adding ALA to ensure a sufficient endogenous synthesis of DHA [222,223], but not surprisingly, due to the limited conversion of ALA to DHA, the added ALA was not effective in raising DHA plasma status to the same level as that observed in breastfed infants ([223]. Some studies also considered experimental formulas containing added γ-linolenic acid from black current-seed oil, borage oil, or evening primrose oil, DHA and EPA from marine oil, and ARA from egg phospholipids [223,224,225,226,227,228,229]. The effects of feeding formula supplemented with soy oil and marine oils containing DHA/EPA showed no abnormalities on growth, tolerance, clotting function, erythrocyte membrane fluidity and vitamin A or E levels in low-birth-weight-term infants [230].

Infant formulas containing DHA and ARA from single cell oils (DHASCO® and ARASCO®, DSM, Columbia, MD, USA), respectively, were evaluated and found to maintain both DHA and ARA status in infants [178,231,232,233]. After 2001, DHASCO and ARASCO (DSM, Columbia, MD, USA) became predominant as the sources of DHA and ARA added to infant formulas in the United States. Both DHASCO and ARASCO are general recognized as safe (GRAS) for use in infant formulas in the United States and approved as novel foods in Canada [234,235]. DHASCO and ARASCO have an established history of use in Europe, Australia and New Zealand and are not considered novel foods and can be added to infant formulas (see Ryan *et al.* [236] for a review of the safety of single-cell oils).

The clinical studies used to evaluate the effects of DHA and ARA added to infant formulas measured infant growth, body and fatty acid composition, behavioral and sensory functions (retinal function, visual acuity and auditory function). Many of the early studies focused on preterm infants because they provided an opportunity to evaluate the effects of DHA- and ARA-enriched formulas in infants who may be deficient in these LCPUFA. Transfer across the placenta and accretion of ARA and DHA in the developing human brain and retina occurs mainly during the last trimester of pregnancy [179,215] when the rate of brain growth is most rapid [237]. Infants born prematurely thus may have an increased need for dietary ARA due to the interrupted supply during the last trimester [220,238,239].

Four studies with preterm infants considered formulas without added preformed ARA [240,241,242,243]. Each was influential in recognizing the importance of ARA for optimal growth. Two of the studies reported an increase in visual acuity at 2 or 4 months postmenstrual age (PMA) in preterm infants fed formulas supplemented with DHA and EPA from fish oil with a low or high ratio of DHA to EPA (2:1, 5:1, respectively) and no ARA [241,244,245,246].

In one study [240], infants fed formula with a low ratio of DHA to EPA until 79 weeks PMA, compared with controls, had significantly lower z-score values for weight, length and head circumference beginning at 40 weeks PMA. Poorer growth was also associated with lower scores of psychomotor development [244]. The supplemented group also had lower blood levels of ARA [177] suggesting that the effects of growth may have been related to the reduced availability of ARA as a result of the competitive inhibition by the high levels of EPA in marine oil. This finding was supported in a second study [241] in which preterm infants were fed formula with a high DHA to EPA ratio until 48 weeks PMA. Compared with controls, preterm infants fed a high ratio of DHA to EPA and with no ARA consistently had lower mean weight-for-length values at 2, 6, 9, and 12 months PMA, weighed less at 6 and 9 months PMA, and had smaller head circumferences at 9 months PMA [246].

In a third study, three premature infant formulas were compared in a double-blind parallel-group study of the growth of healthy, very-low-birth-weight infants (846–1560 g at birth) [242]. The DHA formula contained 0.34% of fat as DHA and the DHA +ARA formula contained 0.33% as DHA and 0.60% as ARA. The control formula contained no DHA or ARA. A reference group consisted of term infants who were predominately breastfed. Results indicated that infants fed formula with DHA + ARA gained significantly more weight than infants fed formula without DHA+ARA. At 48 and 57 weeks, weight of infants in the DHA + ARA group did not differ from the reference group of term infants. Length of infants in the DHA + ARA group was significantly greater than that of infants fed DHA alone at 40 and 48 weeks, but not at 57 weeks. The authors concluded that supplementation with ARA in addition to DHA supported growth of preterm infants [242].

In a fourth study [243], male but not female preterm infants fed formula with DHA and EPA from fish oil (0.2 wt %, 5:1 ratio) and no ARA, or a control formula to 59 weeks PMA, had significantly smaller gains in weight, length, and head circumference and lower fat-free mass as determined from total body electrical conductivity (TOBEC).

In the Carlson *et al.* [177,240] studies that reported slower growth in preterm infants a positive association between plasma ARA concentration and measures of growth (weight, length, z-score, weight-to-length, and head circumference) was observed. In the 1990s, when the studies were conducted, it was unknown whether the negative effects of DHA and/or EPA supplementation on growth could be overcome by adding ARA to infant formulas [177]. However, it was known that the bioactive metabolites of ARA mediate the secretion of several hormones associated with growth and basic metabolic functions [247]. These include luteinizing hormone, prolactin, adrenocorticotropic hormone (ATCH), and corticotropin-releasing hormone (CRH) [241] which could influence growth. ARA and its second messengers also appear to be involved in bone formation and resorption [248].

To maintain ARA status in preterm infants, additional studies were performed with infant formulas containing both ARA and DHA [221]. The addition of ARA at levels found in breast milk produced growth comparable to that observed in breastfed infants [221]. Adding both ARA and DHA to preterm infant formulas also resulted in beneficial effects on visual acuity as compared with infants fed a control formula [221].

Concern with a high level of DHA without a concomitant increase of ARA was raised in a randomized, controlled clinical trial of term infants administered formula with no LCPUFA, or differing levels of DHA intakes of 0.32%, 0.64% and 0.96% at the same ARA level of 0.64% [249,250]. There were no formula effects on tests of behavioral and psychophysiological indices of attention at 4, 6, and 9 months of age. However, infants supplemented at the two lower doses of DHA spent proportionately more time engaged in active stimulating processes (increased attention) than infants in the unsupplemented group [249]. Positive results were also observed on vocabulary (Peabody Picture Vocabulary), a card-shorting task, and an intelligence test (Wechler Primary Preschool Scales of Intelligence) at 3 to 6 years of age in the two lower doses of DHA (0.32% and 0.64%) [250]. However, performance of infants and children who were administered the highest dose of DHA (0.96%) but with a reduced ratio of ARA to DHA was attenuated [250]. The results demonstrated that a proper balance of DHA and ARA is needed for optimal cognitive performance as too much DHA may suppress the benefits provided by ARA.

The effects that different ratios of *n*-6 to *n*-3 had on preterm infant neurodevelopment were recently considered by Alshweki *et al.* [251]. Preterm infants (<1500 g and/or <32 weeks gestational age) were given infant formula with an *n*-6 to *n*-3 ratio of 2:1 or 1:1. The infants were followed for up to 2 years. Preterm infants fed formula with a 2:1 ratio of ARA to DHA had higher ARA blood levels during the first year of life and better psychomotor development compared with those fed a 1:1 ratio of ARA to DHA. However, despite the fact that one group received twice the amount of ARA than DHA (66% *vs.* 33%), there was almost no difference between the two groups in plasma *n*-6 to *n*-3 ratio. The balance between *n*-6 and *n*-3 is very complex, but appears to be maintained at a steady level when adequate supplementation of ARA and DHA are available [251].

To date, several systematic reviews of the literature and meta-analyses have been published to evaluate the effects of LCPUFA in preterm and term infants [221,229,252] on various outcomes including growth, cognition and vision. The reviews considered both the earlier and more recent studies on LCPUFA. Fleith and Clandinin [221], in one of the earliest reviews, reported that collectively the body of literature supported the view that LCPUFA are important for growth and development of preterm and term infants. Formula levels of ARA and DHA should be in the same range as those found in human milk and with the same ratio [221]. There needs to be a dietary supply of ARA and DHA to achieve similar accretion levels in plasma and in RBC as compared with breastfed infants [221,253,254].

A Cochrane review of LCPUFA supplementation in term infants reported that there was little evidence that supplementation conferred a benefit on visual or cognitive development [229]. In a recent meta-analysis of four clinical trials including data for preterm and term infants, LCPUFA supplementation was also shown to have no effect on Bailey Developmental scores at 18 months of age [252]. However, as pointed out by Colombo *et al.* [250], the lack of an observed effect at 18 months of age is consistent with the view that the Bailey Scales of Infant Development are not very sensitive to the effects of LCPUFA supplementation. The Bailey Scales of Infant Development yield a composite score obtained from the infant's attainment of normal developmental milestones and may not be able to provide detailed assays of specific cognitive mechanisms that are measured using more sophisticated laboratory tasks [250]. This raises the question of whether the Bailey Scales are appropriate for measuring the effects of LCPUFA on cognitive development in older infants/toddlers.

A systematic review of 20 randomized, controlled trials of term infants who received DHA and ARA supplemented formula or a control formula indicated that infants given formulas containing DHA levels close to the worldwide human milk mean of 0.32% of total fatty acids were more likely to yield positive results on cognitive and visual tests [255]. There was also clinical evidence to suggest than an ARA:DHA ratio greater than 1:1 was associated with improved cognitive outcomes [255].

Since the publication of the Hoffman *et al.* [255] review, several epidemiological and interventional studies of LCPUFA supplementation during infancy have appeared in the literature. These recent publications have been reviewed by Ryan *et al.* [256]. The most recent data indicate that maternal supplementation during pregnancy and/or lactation support the role for LCPUFA in the neurodevelopment of infants [256]. Supplementation with LCPUFA-containing infant formula for more than 6 months increased the likelihood of observing improved cognitive function during infancy [256,257].

The reasons that some studies failed to show a statistically significant association between LCPUFA intake and better neurological performance may have been related to limitations of study design and the use of varying amounts and sources of LCPUFA. For example, in the United States, most of the recent studies that have considered both neurocognitive function and growth have used infant formulas for preterm and term infants that contain DHA and ARA from single cell oils (DHASCO® and ARASCO®). Outside the United States, DHA may be obtained from fish oil [258,259] and ARA may be obtained from eggs [260,261]. As mentioned above, the early studies of the 1990s used a variety of experimental formulas with different sources and amounts of LCPUFA (many of these studies are cited by EFSA in their review, [71]). This is noteworthy because only DHA and ARA from single cell oils have been recently shown to enhance neurodevelopment in both preterm [262,263,264] and term infants [257,265]. The source of the oil is important because it significantly affects growth. Clandinin *et al.* [264] have shown that body weight at 118 days and length at 79 and 92 weeks of age in preterm infants fed formula containing ARA and DHA from single-cell oils were greater than in those than those fed a formula containing DHA from tuna oil. Additionally, for both weight and length, there were no differences between breastfed infants and those fed ARA and DHA from single-cell oils [264]. The possibility that ARA and DHA derived from single-cell oils or fish oil have differing effects on growth and neurodevelopment may also be due to the EPA at too high a level in fish oil and its propensity to antagonize ARA [257]. An ARA:DHA ratio greater than 1:1 with up to 0.65% of ARA of total fatty acids is associated with improved cognitive performance [255] and balances the potential competition caused by high levels of DHA.

12. THE BENEFITS OF ARA FOR INFANT HEALTH

LCPUFA are not only important for growth and neurodevelopment, but recent studies have shown that LCPUFA are also critical for infant health. In an early study, infants fed formula with ARA and DHA developed significantly less stage II and III NEC than those fed a control formula but had similar rates of bronchopulmonary dysplasia, septicemia and retinopathy of prematurity (ROP) [266]. In a recent retrospective cohort study of premature infants (<30 weeks gestation), the relationship between fatty acid profiles during the first postnatal month and infant morbidity due to chronic lung disease, ROP, and late-onset sepsis was analyzed [120]. Results indicated that fatty acid levels of DHA and ARA declined rapidly with a concomitant increase in LA. While the decreased DHA level was associated with increased risk of chronic lung disease, decreased ARA was associated with increased risk of late-onset sepsis. The authors noted that in premature infants, low levels of DHA and ARA contribute to dysregulation of immune and inflammatory responses, predisposing these infants to chronic lung disease and late-onset sepsis [120]. The DHA-derived resolvins decrease neutrophil infiltration and enhance macrophage phagocytosis [267]. DHA also downregulates nuclear factor kB activity in cells either directly or by stimulating the activation of peroxisome proliferator-activated receptors thereby limiting the pro-inflammatory signaling mediated by ARA [267]. Thus, a low level of DHA would predispose these infants to an increased inflammatory response as seen in chronic lung disease. In late-onset sepsis, a decreased ARA level increases the risk of inhibiting the innate immune response through decreased production of eicosanoids, particularly leukotrienes, which enhance chemotaxis of leukocytes, neutrophil activation, and activity of natural killer cells [97,120]. As a consequence, in premature infants, a balance of DHA and ARA levels must be maintained to help reduce the development of morbidity due to prematurity.

Studies have demonstrated that LCPUFA in human milk can modulate immunological responses and affect the T-helper cell (Th) type-1 (Th1)/Th2 balance [268,269]. Th1 cell effectors produce interferon-γ (IFN-γ) and TNF-α which regulate cellular immunity against infection whereas Th2 cells produce interleukin (IL)-4, IL-5, and IL-13 which help mediate immunity against parasitic infections [270,271]. For example, Barakat *et al.* [272] reported that supplementation for 14 days with 10 mg/kg of ARA in *Schistosoma mansoni*-infected schoolchildren induced moderate cure rates (50%) in children with light infection and modest cure rates (21%) in those with high infection. The cure rates associated with ARA were comparable to those produced by 40 mg/kg of praziquantel. The combination of ARA and praziquantel elicited 83% and 78% cure rates in children with light and heavy infections, respectively [272].

The relationship between maternal fatty acid desaturase (FADS) genotype and LCPUFA levels in human milk on infant blood T-cell profiles and cytokine production in 6-month old infants was recently considered [271]. LCPUFA levels in human milk were measured at 4 weeks of age and the FADS genotype was determined in both mothers and infants. Results indicated that ARA levels in human milk were inversely correlated with the production of the cytokines IL-10, IL-17, IL-5 and IL-13 and EPA levels were positively associated with counts of regulatory T-cells and cytotoxic T-cells and decreased T-helper cell counts. The minor FADS alleles were associated with lower ARA and EPA levels in human milk and a higher production of IL-10, Il-17, and IL-5. The major FADS alleles were associated with an increase in the level of ARA in human milk (19%–22%) compared with the minor alleles. There were no association between T-cell distribution and maternal or infant gene variants. Also, there was no relationship between cytokine levels in plasma and levels of LA, EPA, or DHA in human milk.

It has been shown that the FADs gene polymorphism may influence the risk of developing allergies in children [273]. In the study by Muc *et al.* [271] the strongest association between LCPUFA levels and cytokines was observed among those related to the activity of type-2 and type-17 *vs.* those from type-1 responses. The expression of type-2 and type-17 cells have been linked to increased airway inflammation in severe asthma [274]. By reducing type-2 and type-17 activity, LCPUFA including ARA found in human milk may help reduce the risk of childhood asthma and allergies [271].

Notably, Th17 cytokines were initially identified as key factors in the induction of inflammation and tissue destruction associated with a variety of autoimmune response such as multiple sclerosis, arthritis, colitis, celiac disease, and gluten sensitivity [275,276]. However, it is becoming apparent that T17 cells also provide protective immunity against various pathogens at different mucosal sites [270]. Thus, there is a fine balance between protection and pathological manifestation of Th17 responses. As a consequence, a balance of LCPUFA levels similar to that found in human milk is needed to help reduce the risk of developing autoimmune diseases in childhood.

Two studies have compared the frequency of common illnesses in infants fed formula with and without DHA and ARA [277,278]. Both studies used the same LCPUFA-supplemented formula that contained 0.32% DHA and 0.64% ARA of total fatty acids (17 mg of DHA/100 kcal and 34 mg of ARA/100 kcal). In the first study, infants fed the LCPUFA-supplemented formula experienced a lower incidence of bronchiolitis/bronchitis compared with infants fed formula without DHA and ARA [277]. The results from the second study were similar to those of the first. Infants who consumed formula with DHA and ARA had a lower incidence of bronchiolitis/bronchitis, nasal congestion, cough and diarrhea requiring medical attention than infants fed formula without DHA and ARA. The authors indicated that DHA and ARA at present levels in infant formula and follow-on formulas may have a positive effect on moderate to severe common infant illnesses, including diarrhea [277,278].

Two cohorts of children who had previously completed randomized, double-blind trials (one published [72], one unpublished) in which they received a LCPUFA-supplemented formula that contained 0.32%–0.36% DHA and 0.64%–0.72% ARA of total fatty acids or an unsupplemented formula (control) fed during the first year of life were followed up to 3 years of age to determine the incidence of allergies and common respiratory illnesses [279]. The LCPUFA-supplemented group had a significantly lower risk for developing upper respiratory infections, wheezing/asthma, atopic dermatitis, any allergy, and a longer time to first diagnosis than those given an unsupplemented formula.

A subset of children from the Kansas City cohort of the DIAMOND (DHA Intake and Measurement of Neural Development) study [96] were followed to 4 years of age to determine the incidence of childhood allergies [280]. As infants, they were fed either a control unsupplemented formula or one of three formulas with either 0.32%, 0.64% or 0.96% of total fatty acids as DHA with the same amount of ARA (0.64% of total fatty acids). All the different DHA dose and ARA supplemented subjects were pooled into a single supplemented cohort. Results indicated that the incidence of allergic illnesses in the first year of life was lower in the combined LCPUFA group compared with the control. By 4 years of age, LCPUFA supplementation significantly delayed time to first allergic illness and skin allergic illness. LCPUFA supplementation also reduced the risk of any allergic diseases and skin allergic diseases. If the mother had allergies, LCPUFA supplementation reduced the incidence of wheezing/asthma in her offspring. The results of these allergy studies add to the evidence that supplementation of infant formula with both ARA and DHA in the first year of life delays the onset of allergy and may have a protective effect against allergy in early childhood.

Crawford *et al.* [45] reviewed the potential role that ARA and DHA play in protecting against some central nervous system injuries in preterm infants. Deficits of ARA and DHA may contribute to the complications related to prematurity [45]. The mechanism of action responsible for central nervous system injury is reduced vascular or endothelial integrity leading to hemorrhage or ischemia. ARA acts as an endothelial relaxation factor and plays a dominant role in endothelial membrane lipids. The inner cell membrane lipid of the endothelium is especially rich in ARA which provides for membrane properties, signaling and protein kinase C activation [281]. ARA is also a precursor for a range of small molecules that play a key role in cell trafficking, communication and vaso-regulation. As a result, any ARA or DHA deficiency in very preterm infants will be exacerbated after birth during a period of rapid growth. This deficiency may then lead to fragile, leaking vessels and rupture as seen in ROP [45].

The proportion of ARA the placenta delivers to the fetus ranges from 14% to 20% [45]. Infant formula for preterm infants only delivers ~0.4%–0.6% of ARA. For DHA, the placenta delivery to the fetus is ~6% whereas infant formula delivers ~0.3%. That is a 50-fold reduction in ARA and a >10-fold reduction in DHA compared with the infant's apparent physiological need. When plasma levels of ARA and DHA are followed from birth, they continue to decrease to about one third of that of the fetus. Crawford *et al.* [45] argue that higher intakes of ARA and DHA are needed to correct for deficiency during the first year of life.

To explore the impact that deficiency of ARA and DHA may have on immune cell function Moodley *et al.*[282] examined the fatty acid profile and main phosphoglyceride content of cord blood mononuclear cell (CBMC) membranes in healthy preterm infants (30 to 35 weeks) and term infants (37 to 40+ weeks). Results indicated that ARA was the dominant LCPUFA present in both PC and PE membrane fractions of CBMCs in both preterm and term infants. The proportions of ARA, DHA and other LCPUFA were significantly lower in PE and PC of preterm infants compared with those in term infants. The dominance of ARA was consistent with the process of biomagnification that preferentially selects ARA rather than other LCPUFA for transfer to the fetus. Preterm infants also had significantly lower absolute numbers of CD4+ leukocytes and CD4+ and CD8+ naïve T-cells. At birth, there is a period of transition from a sterile environment to one of higher infectious risk. The elevated levels of ARA in CBMCs concomitant with lower levels of other LCPUFA suggest that the acquisition of ARA is needed in preparation for a responsive immune system after birth. These findings indicate that in preterm infants a deficiency in the supply of ARA exists which may compromise their immune system [282].

Infants with mildly abnormal physical movements at 12 weeks of age are reported to have lower ARA content in erythrocyte membranes [283]. This abnormality occurred with maternal supplementation of DHA alone but was not seen when DHA was combined with ARA during pregnancy and lactation [283]. Mildly abnormal movements have been also observed during infancy and linked to increased prevalence of minor neurologic dysfunction and attention deficits at school age [284]. These findings imply that during early brain development of neonates, a supply of ARA is critical.

The effects that feeding preterm infants human milk (HM), infant formula without DHA and ARA (F) or formula with DHA (0.35%) and ARA (0.49%) have on isolated peripheral blood lymphocytes and lipid composition was evaluated by Field *et al.* [285]. Adding DHA and ARA to a preterm infant formula resulted in lymphocyte and cytokine production, phospholipid composition, and antigen maturity similar to those observed in infants fed human milk. These findings suggest that the addition of both DHA and ARA may improve the ability of the infant to respond to immune challenges in a manner similar to breastfed infants [285].

Several epidemiological studies have shown that individuals with learning disorders including attention deficit hyperactivity disorder (ADHD), dyslexia, and autism have signs of EFA deficiency or have lower than normal blood levels of DHA and ARA [286,287,288,289]. A meta-analysis of pooled data from RBC and plasma/serum samples indicated that ARA and DHA concentrations were significantly lower than normal in individuals with learning/developmental disorders [290]. In absolute amounts, the level of ARA was as severely depressed as DHA within RBC (both ~0.58 mg/100 mg of fatty acid below normal) but much lower than DHA within plasma/serum (−0.71 *vs.* −0.34). The reason for lower than normal blood levels of ARA in children with learning disorders is unknown but could be related to a low dietary intake of ARA relative to metabolic requirements or that ARA is not synthesized efficiently from precursor fatty acids, or not delivered or properly incorporated into the brain [290].

The effects of subnormal ARA on brain function seems to be independent of those associated with *n*-3 deficiency. In Japan where intakes of *n*-3 fatty acids are relatively high, the incidence of dyslexia in children is similar to that observed in Westernized countries (~6%) [291]. Therefore, although the Japanese population consumes sufficient amounts of *n*-3 fatty acids, there still may be insufficient intake of ARA to meet the needs for normal brain function during childhood [290]. The important role of ARA in normal growth and development requires as much research emphasis as DHA has received. ARA should be the focus of preclinical and clinical research for a detailed assessment of dietary requirements.

13. THE REGULATORY REQUIREMENTS FOR ARA AND DHA IN INFANT FORMULAS

A joint International Expert Consultation of the Food and Agricultural Organization of the United Nations (FAO) and the WHO was assembled in 1976 to review the literature on "The Role of Dietary Fats and Oils in Human Nutrition" [292]. The section dealing with infant growth and development indicated that the ideal recommendation for infant formulas would be to match the essential fatty acids found in human milk with respect to LCPUFA content [292]. The FAO further stated that LCPUFA were particularly important during fetal and infant growth when there is a high demand for the synthesis of cell structured lipid [292]. FAO issued a follow-up report in 1994 which provided additional supportive recommendations for adding both ARA and DHA to infant formulas. The FAO stated that "the *n*-6 and *n*-3 fatty acids have critical roles in the membrane structure and as precursors of eicosanoids, which are potent and highly reactive compounds. Various eicosanoids have widely divergent, and often opposing effects on, for example, smooth muscle cells, platelet aggregation, vascular parameters (permeability, contractility), and on the inflammatory processes and the immune system. Since they compete for the same enzymes and have different biological roles, the balance between the *n*-6 and the *n*-3 fatty acids in the diet can be of considerable importance" [293]. In a follow-up report in 2008–2010, the FAO/WHO Expert Consultancy on Fats and Fatty Acids further concluded that "There can be little doubt about the essentiality of DHA and ARA for the brain" [294].

A global standard for infant formula was established by the Codex Alimentarius Commission in 1981, and revised over the years [295]. The latest revision was issued in 2007 and the latest amendment was added in 2015 [296]. The standard includes details on essential composition of nutrients and a list of food additives that are allowed to be added. Quality control measures such as labeling, packaging, contaminants and hygiene are also specified. In the United States, standards for infant formula are the responsibility of the U.S. Food and Drug Administration (FDA). The U.S. Code of Federal Regulations Title 21, Part 106 specifies infant formula quality control procedures and Part 107 lists the nutrient requirements and other rules concerning labeling for infant formulas. Not surprisingly, the quality and safety standards for infant formulas are extremely high, exceeding most requirements for other food products [297].

In the European Union (EU), the legislation on infant formula and follow-on formulas was adopted in 2006 [298] and at the time of this writing is being revised. Before revising the legislation the European Commission requested the European Food Safety Authority (EFSA) Panel on Dietetic Products, Nutrition and Allergies to provide their scientific advice on the essential nutrient composition of infant and follow-on formulas [70,71]. In the first report of 2013, dedicated to nutrient requirements and dietary intakes of infants and young children in the EU, the EFSA Panel reviewed a variety of nutrients, including the levels of DHA and ARA. In the 2013 EFSA report, adequate intakes were defined as 100 mg/day of DHA and 140 mg/day of ARA from birth to six months of age. From 6 to 24 months of age, 100 mg/day of DHA were considered adequate. These recommendations were also supported by a global expert panel, based on a systematic review of the available scientific literature [299]. However, in the subsequent report of 2014, dedicated to essential composition of infant and follow-on formulae, the EFSA Panel advised that infant and follow-on formulas should contain relatively higher amounts of DHA (20–50 mg/100 kcal). Mandatory addition of ARA was not supported in this EFSA report. Still, the Panel noted that feeding an infant formula containing DHA alone resulted in lower concentration of ARA in erythrocytes compared with a control formula without DHA.

At an assumed mean fat content of 5.2 g 100 kcal in a typical infant formula, this means that the recommendation of higher levels of DHA would result in a DHA content of 0.38% to 0.96% of fatty acids, higher than the 0.2% to 0.3% of DHA currently found in most infant formulas available in the marketplace [2]. Notably, PUFA-supplemented commercially available infant formulas contain preformed ARA at levels equal to or higher than the DHA content. The ESFA Panel's advice of providing up to 1% DHA and no ARA is a unique approach and directly opposite to a consensus reached by international expert groups who have recommended that infant formulas for term infants should contain ARA at levels that range from 0.4% to 0.7% fatty acids (at a 1:1–2:1 ratio to DHA) based on the median worldwide range of ARA and DHA concentrations in breast milk [6,7,8,9].

14. DISCUSSION

ARA is the principle LCPUFA in the inner cell membrane lipid of muscle, heart, vascular endothelium, adrenals, kidneys, liver, the placenta, and in almost all other organs [300]. ARA is essential for cell integrity. The cell membrane separates the interior structures of cells from the outside environment. It also controls the movement of substances in and out of the cell [300]. These membranes contain signalers, receptors, ion channels, antioxidant defense enzymes, and rafts. Changing any aspect of the composition of the cell membrane may alter its function [300]).

ARA has very different biological functions than DHA [300]. While DHA controls signaling membranes in the photoreceptor, brain and nervous system, ARA is indispensable in the vasculature and in specific aspects of immunity. ARA is important for brain growth during gestation and early infancy where it plays a critical role in cell division and signaling [11]. A potentially important aspect of ARA metabolism is its function as a precursor for leukotrienes, prostaglandins, and thromboxanes, collectively known as eicosanoids. Eicosanoids have numerous critical and specific functions occurring in almost every tissue of the body. Eicosanoids function to modulate the release of somatostatin, the principal hormone that stimulates cell proliferation and growth. Eicosanoids also have important roles in immunity and inflammation.

This review focused on the essentiality of ARA for infant growth and development. Animal studies demonstrated the importance of ARA for growth and maturation of neurons and myelin and the resolution of inflammation in models of NEC, influenza and EFA deficiency. These studies also provided compelling evidence that both preformed DHA and ARA are required for optimal cognitive and neurological development. The ratio of ARA and DHA added as supplements really matters. Brain tissue analysis of neonatal baboons fed formula with a high level of 0.96% DHA significantly reduced ARA levels in two regions of the brain indicating the importance of a proper balance of DHA and ARA [24].

For over 10 years, both DHA and ARA have been added to infant formulas worldwide in an attempt to match the nutrient supply and functional benefits achieved with human milk. The combination of ARA and DHA in infant formulas has been shown to be safe in many millions of infants globally. The DHA concentration in human milk is lower and more variable than for ARA and the level of ARA in human milk is more stable [5]. The relatively stability of the ARA level in human milk is biologically important because it provides preformed ARA consistently at a time when brain growth and development is most critical [300]. Although DHA is more variable than ARA it is always present in human milk and the balance between ARA and DHA can be as much as 2 to 1.

The biosynthetic capability for providing ARA and DHA for brain growth is low and preformed ARA and DHA are preferentially incorporated into the brain during gestation and early infancy [2,46]. Infant formulas devoid of ARA results in a dramatic decrease of up to 40% of ARA in plasma shortly after birth [223,230,238,239,300,301], especially in preterm infants who do not receive the third trimester's maternal supply of ARA and DHA. The finding that there is a decrease in ARA shortly after birth shows that biosynthetic capability is insufficient to meet the infant's demand [300]. The process of biomagnification and its resulting fatty acid profile further highlights the importance of ARA in infant growth [45]. In several clinical studies, the provision of high amounts of DHA/EPA without a concomitant supply of ARA has been associated with adverse effects on growth in premature infants [240,241,242,243].

EFSA [71] recently concluded that "there is no necessity to add ARA to infant formula even in the presence of DHA". This recommendation needs further explanation. One of the possible reasons for this recommendation is that it is generally believed that LA is converted into ARA in sufficient quantities, even though EFSA noted that feeding an infant formula containing DHA alone resulted in lower concentration of ARA in erythrocytes compared with a control formula without DHA. From the limited dietary intake data presented here for non-breastfed infants and young children (Table 1) living in developing and developed countries there is evidence that intakes of ARA from dietary sources are very low, much lower than the average amount of ARA available in human milk or infant formulas containing ARA and DHA. The composition of infant formulas and follow-on formulas should therefore not only be based on human milk composition but also on food/nutrient intake data to address the assumption that complementary foods fill nutrient gaps.

The EFSA Panel's advice of providing higher amounts of DHA (20–50 mg/100 kcal, 0.38% to 0.96% of fatty acids) without a concomitant supply of ARA is also questionable. As discussed here, DHA suppresses ARA concentration in membranes and its function. As a result, an infant formula with DHA and no ARA may result in a potential higher risk of morbidity due to the suppression of favorable eicosanoids that play a key role in cell trafficking, communication and vaso-regulation [45].

EFSA did not take into account the original FAO/WHO publications as well as the earlier 2008–2010 publication which strongly concluded "There can be little doubt about the essentiality of DHA and ARA for the brain" [294]. When infants are exclusively breastfed during the first 6 months of life, "there is evidence of a requirement for preformed ARA and DHA after 6 months of life" [294]. According to the 2010 FAO/WHO statement, DHA and ARA should be included in infant formula with DHA (from 0.2% to 0.5% of total fatty acids) and added ARA should be at least equal to the amount of DHA [294].

The clinical trials considered by EFSA were not designed to consider the specific physiological outcomes related to ARA. Most studies included both ARA and DHA. There were no clinical trials that evaluated the effects of ARA in the absence of DHA. The benefits of ARA + DHA supplementation cannot be ascribed to DHA alone but logically must be ascribed to the variables used in most of these studies, the combination of the two. The combination of ARA and DHA has shown benefits for cognitive development, visual function, and blood pressure well beyond the period of supplementation and into early childhood [249,302].

EFSA's lack of support of ARA in DHA-containing formulas relied almost exclusively on the meta-analysis of infant growth conducted by Makrides *et al.* [303]. However, the meta-analysis of Makrides *et al.*[303] was not comprehensive. The majority of subjects (n = 1050) included in the meta-analysis participated in 8 clinical trials using formula containing both DHA and ARA. Only 341 subjects in the 8 trials were provided infant formula with DHA/EPA without ARA. At 12 months of age, only 99 subjects were supplemented with DHA in the absence of ARA. Additionally, the analysis excluded studies that included preterm infants who are most vulnerable to growth faltering due to nutrition, and excluded studies in which DHA plus ARA supplementation was less than 3 months during which growth velocity is particularly sensitive to nutritional inadequacy. The reason for excluding these studies was not reported.

In the meta-analysis, Makrides *et al.* [303] noted that the results were inconclusive, *i.e.*, LCPUFA supplementation had no detrimental effect on growth. The importance of ARA as a structural and metabolically active lipid, was not addressed. Better visual and mental performance was attributed to the contribution of DHA. However, in one of the studies considered, DHA plus ARA improved mental function [265] compared with an unsupplemented control. In fact, in one study, DHA alone did not perform better than did the unsupplemented controls [72].

Clinical evidence to support the safe removal of ARA from infant formula and follow-on formula containing DHA is lacking. The human and nonhuman primate studies described herein question the EFSA recommendation to provide infant formula from birth with up to 1% of DHA without a proportional amount of ARA [2]. Any major change in infant formula composition should be subjected to a full preclinical and clinical evaluation of safety and nutritional adequacy before its introduction into the marketplace [2]. Without such an assessment, and in light of the universal presence of ARA in human milk and the numerous essential ARA functions for cell structure and function, the most judicious approach is to include ARA in DHA-containing infant formulas to promote optimal infant growth and development.

ACKNOWLEDGMENTS

The corresponding author, Alan S. Ryan, has received funds from DSM Nutritional Products to cover the costs to publish this article.

AUTHOR CONTRIBUTIONS

Kevin B. Hadley, Sheila Gautier and Norman Salem, Jr. conceived the structure of the review. Alan S. Ryan, Kevin B. Hadley, Stewart Forsyth, Sheila Gautier, and Norman Salem, Jr. were involved in drafting and writing of the manuscript, reviewing of scientific literature, and manuscript revision. All authors agreed on the final form of the manuscript.

CONFLICTS OF INTEREST

The corresponding author, Alan S. Ryan received funds to help draft the manuscript. Kevin B. Hadley, Sheila Gautier and Norman Salem Jr. are employees of DSM Nutrition Products that manufactures and sells polyunsaturated fatty acids.

REFERENCES

1. Martinez, M. Tissue levels of polyunsaturated fatty acids during early human development. J. Pediatr.1992, 120, S129–S138.
2. Koletzko, B.; Carlson, S.E.; van Goudoever, J.B. Should infant formula provide both omega-3 DHA and omega-6 arachidonic acid? Ann. Nutr. Metab. 2015, 66, 137–138.
3. Bolling, K. Infant Feeding Survey, 2005. Available online: http://www.hscic.gov.uk/pubs/ifs2005(accessed on 17 August 2015).
4. Centers for Disease Control and Prevention, Division of Nutrition, Physical Activity, and Obesity. Breastfeeding Report Card. 2014. Available online: http://www.cdc.gov/breastfeeding/data/reportcard.htm (accessed on 18 August 2015).
5. Brenna, J.T.; Varamini, B.; Jensen, R.G.; Diersen-Schade, D.A.; Boettcher, J.A.; Arterburn, L.M. Docosahexaenoic and arachidonic acid concentrations in human milk worldwide. Am. J. Clin. Nutr.2007, 85, 1457–1464.
6. British Nutrition Foundation. Unsaturated Fatty Acids: Nutritional and Physiological Significance; Chapman & Hall: London, UK, 1992; pp. 152–163.
7. Food and Agricultural Organization of the United Nations/World Health Organization Joint Expert Consultation. Lipids in early

development. In Fats and Oils in Human Nutrition; FAO Food and Nutrition Papers; FAO: Rome, Italy, 1994; Volume 57, pp. 49–55.

8. Simopoulos, A.P.; Leaf, A.; Salem, N., Jr. Workshop on the essentiality of and recommended dietary intakes for omega-6 and omega-3 fatty acids. J. Am. Coll. Nutr. 1999, 18, 487–489.

9. Koletzko, B.; Baker, S.; Cleghorn, G.; Neto, U.F.; Gropalan, S.; Hernell, O.; Hock, Q.S.; Jirapinyo, P.; Lonnerdal, B.; Pencharz, P.; et al. Global standard for the composition of infant formula: Recommendations of an ESPAGHAN coordinated international expert group. J. Pediatr. Gastroenterol. Nutr. 2005, 41, 584–599.

10. Lauritzen, L.; Hensen, H.S.; Jorgensen, M.H.; Michaelsen, K.F. The essentiality of long chain n-3 fatty acids in relation to development and function of the brain and retina. Prog. Lipid Res. 2001, 40, 1–94.

11. Katsuki, H.; Okuda, S. Arachidonic acid as a neurotoxic and neurotrophic substance. Prog. Neurobiol.1995, 46, 607–636.

12. Crawford, M.A.; Broadhurst, C.L. The role of docosahexaenoic and the marine food web as determinants of evolution and hominid brain development: The challenges for human sustainability.Nutr. Health 2012, 21, 17–39.

13. Bazan, N.G.; Reddy, T.S.; Bazan, H.E.P.; Birkle, D.L. Metabolism of arachidonic acid and docosahexaenoic acid in the retina. Prog. Lipid Res. 1886, 25, 595–606.

14. Crawford, M.A.; Sinclair, A.J. Nutritional influences in the evolution of mammalian brain. In Lipids, Malnutrition & the Developing Brain; A Ciba Foundation Symposium: Amsterdam, The Netherlands, 1971; pp. 267–292.

15. Martinez, M. Polyunsaturated fatty acids in the developing human brain, red cells and plasma: Influence of nutrition and peroxisomal disease. In Fatty Acids and Lipids: Biological Aspects; Galli, C., Simopoulos, A.P., Tremoli, E., Eds.; Karger: Basel, Switzerland, 1994; Volume 75, pp. 70–78.

16. Dobbing, J.; Sands, J. Comparative aspects of the brain growth spurt. Early Hum. Dev. 1979, 3, 79–83.

17. Su, H.-M.; Corso, T.N.; Nathanielsz, P.W.; Brenna, J.T. Linoleic acid kinetics and conversion to arachidonic acid in the pregnant and fetal baboon. J. Lipid Res. 1999, 40, 1304–1311.

18. Sanchez-Mejia, R.O.; Newman, J.W.; Toh, S.; Yu, G.; Zhou, G.Q.; Halabisky, B.; Cissé, M.; Scearce-Levie, K.; Cheng, I.H.; Gan, L.; et al. Phospholipase A2 reduction ameliorates cognitive deficits in a

mouse model of Alzheimer's disease. Nat. Neurosci. 2008, 11, 1311–1318.

19. Vijayaraghaven, S.; Huang, B.; Blumenthal, E.M.; Berg, D.L. Arachidonic acid as a possible negative feedback inhibitor of nicotinic acetylcholine receptors on neurons. J. Neurosci. 1995, 15, 3679–3687.

20. Williams, J.H.; Errington, M.L.; Lynch, M.A.; Bliss, T.V. Arachidonic acid induces a long term activity-dependent enhancement of synaptic transmission in the hippocampus. Nature 1989, 341, 739–742.

21. Fukaya, T.; Gondaira, T.; Kashiyae, Y.; Kotani, S.; Ishikura, Y.; Fujikawa, S.; Kiso, Y.; Sakakibara, M. Arachidonic acid preserves hippocampal neuron membrane fluidity in senescent rats. Neurobiol. Aging2007, 28, 1179–1186.

22. Wang, Z.-J.; Liang, C.-L.; Li, G.-M.; Yu, C.-Y.; Yin, M. Neuroprotective effects of arachidonic acid against oxidative stress on rat hippocampal slices. Chem. Biol. Interact. 2006, 163, 207–217.

23. Wijendran, V.; Lawrence, P.; Diau, G.-Y.; Boehm, G.; Nathanielsz, P.W.; Brenna, J.T. Significant utilization of dietary arachidonic acid is for brain adrenic acid in baboon neonates. J. Lipid Res. 2002,43, 762–767.

24. Hsieh, A.T.; Anthony, J.C.; Diersen-Schade, D.A.; Rumsey, S.C.; Lawrence, P.; Li, C.; Nathanielsz, P.W.; Brenna, J.T. The influence of moderate and high dietary long chain polyunsaturated fatty acids (LCPUFA) on baboon neonate tissue fatty acids. Pediatr. Res. 2007, 61, 537–545.

25. Yang, H.-J.; Sugiura, Y.; Ikegami, K.; Konishi, Y.; Setou, M. Axonal gradient of arachidonic acid-containing phosphatidylcholine and its dependence on actin dynamics. J. Biol. Chem. 2012, 287, 5290–5300.

26. Bazan, N.G. The neuromessenger platelet-activation factor in plasticity and neurodegeneration. Prog. Brain Res. 1998, 118, 281–291.

27. Hattori, M.H.; Adachi, H.; Tsujimoto, M.; Arai, H.; Inoue, K. Miller-Dieker lissencephaly gene encodes a subunit of brain platelet-activation factor acetylhydrolase. Nature 1994, 370, 216–218.

28. Darios, F.; Davletov, B. Omega-3 and omega-6 fatty acids stimulate cell membrane expansion by acting on syntaxin 3. Nature 2006, 440, 813–817.

29. Darios, F.; Ruiperez, V.; Lopez, I.; Villanueva, J.; Gutierrez, L.M.; Davletov, B. α-synuclein sequesters arachidonic acid to modulate SNARE-mediated exocytosis. EMBO Rep. 2010, 11, 528–533.

30. Smart, E.J.; Graf, G.A.; McNiven, M.A.; Sessa, W.C.; Engelman, J.A.; Scherer, P.E.; Okamoto, T.; Lisanti, M.P. Caveolins, liquid-ordered domains, and signal transduction. Mol. Cell. Biol. 1999, 19, 7289–7304.

31. Pike, L.J.; Han, X.; Chung, K.-N.; Gross, R.W. Lipid rafts are enriched in arachidonic acid and plasmmenylethanolamine and their composition is independent of caveolin-1 expression: A quantitative electrospray ionization/mass spectrometric analysis. Biochemistry 2002, 41, 2075–2088.

32. Pike, L. Lipid rafts: Bringing order to chaos. J. Lipid Res. 2003, 44, 655–667.

33. Lee, H.-C.; Sasaki, J.; Kubo, T.; Matsuda, S.; Nakasaki, Y.; Hattori, M.; Tanaka, F.; Udagawa, O.; Kono, N.; Itoh, T.; et al. LPIAT1 regulates arachidonic acid content in phosphatidylinositol and is required for cortical lamination in mice. Mol. Biol. Cell 2012, 23, 4689–4697.

34. Wolf, B.A.; Turk, J.; Sherman, W.R.; McDaniel, M.L. Intracellular Ca2+ mobilization of arachidonic acid. J. Biol. Chem. 1986, 261, 3501–3511.

35. Cao, Y.; Pearman, A.T.; Zimmermann, G.A.; McIntyre, T.M.; Prescott, S.M. Intracellular unesterified arachidonic acid signals apoptosis. PNAS 2000, 97, 11280–11285.

36. Hicks, A.M.; DeLong, C.J.; Thomas, M.J.; Samuel, M.; Cui, Z. Unique molecular signatures of glycerophospholipid species in different rat tissues analyzed by tandem mass spectrometry. Biochim. Biophys. Acta 2006, 71, 1022–1029.

37. Tanaka, T.; Iwawaki, D.; Sakamoto, M.; Takai, Y.; Morishige, J.-I.; Murakami, K.; Satouchi, K. Mechanisms of accumulation of aracidonate in phoshatidylinositol in yellowtail. Eur. J. Biochem. 2003,270, 1466–1473.

38. Jungalwala, F.B.; Evans, J.E.; McCluer, R.H. Compositional and molecular species analysis of phospholipids by high performance liquid chromatography couples with chemical ionization mass spectrometry. J. Lipid Res. 1984, 25, 738–749.

39. Szentpetery, Z.; Varnai, P.; Balla, T. Acute manipulation of Golgi phosphoinositides to assess their importance in cellular trafficking and signaling. PNAS 2010, 107, 8225–8230.

40. Di Paolo, G.; De Camilli, P. Phosphoinositides in cell regulation and membrane dynamics. Nature2006, 44, 12.

41. Malaiyandi, L.M.; Honick, A.S.; Rintoul, G.L.; Wang, Q.L.; Reynolds, I.J. Zn2+ inhibits mitochondrial movement in neurons by phosphatidylinositol 3-kinase activation. J. Neurosci. 2005, 25, 9507–9514.

42. De Vos, K.J.; Sable, J.; Miller, K.E.; Sheetz, M.P. Expression of phosphatidylinositol (4,5) bisphosphate-specific pleckstrin homology domains alters direction but not the level of axonal transport of mitochondria. Mol. Biol. Cell 2002, 14, 3636–3649.

43. Caroni, P. New EMBO members' review: Actin cytoskeleton regulation through modulation of PI(4,5)P(2) rafts. EMBO J. 2001, 20, 4332–4336.

44. Del Prado, M.; Villalpando, S.; Elizondo, A.; Rodriguez, M.; Demmelmair, H.; Koletzko, B. Contribution of dietary and newly formed arachidonic acid to human milk lipids in women eating a low-fat diet. Am. J. Clin. Nutr. 2001, 74, 242–247.

45. Crawford, M.A.; Golfetto, I.; Ghebremeskel, K.; Min, Y.; Moodley, T.; Poston, L.; Phylactos, A.; Cunnane, S.; Schmidt, W. The potential role for arachidonic and docosahexaenoic acids in protection against some central nervous system injuries in preterm infants. Lipids 2003, 38, 303–315.

46. Larque, E.; Ruiz-Palacios, M.; Koletzko, B. Placental regulation of fetal nutrient supply. Curr. Opin. Clin. Nutr. Metab. Care 2013, 16, 292–297.

47. Koletzko, B.; Agostini, C.; Bergmann, R.; Ritzenthaler, K.; Shamir, R. Physiological aspects of human milk lipids and implications for infant feeding: A workshop report. Acta Paediatr. 2011, 100, 1405–1415.

48. Sinclair, A.J.; Crawford, M.A. The accumulation of arachidonate and docosahexaenoate in the developing rat brain. J. Neurochem. 1972, 19, 1753–1758.

49. Makrides, M.; Neumann, M.; Byard, R.W.; Simmer, K.; Gibson, R.A. Fatty acid composition of brain, retina, and erythrocytes in breast- and formula-fed infants. Am. J. Clin. Nutr. 1994, 60, 189–194.

50. Kuipers, R.S.; Luxwolda, M.F.; Offringa, P.J.; Boersma, E.R.; Dijck-Brouwer, D.A.; Muskiet, F.A.J. Fetal intrauterine whole body linoleic, arachidonic, and docosahexaenoic acid contents and accretion rates.Prostaglandins Leukot. Essent. Fat. Acids 2011, 86, 13–20.

51. Carver, J.D.; Benford, V.J.; Han, B.; Cantor, A.B. The relationship between age and the fatty acid composition of cerebral cortex and erythrocytes in human subjects. Brain Res. Bull. 2001, 56, 79–85.

52. Salem, N.M.; Lin, Y.H.; Moriguchi, T.; Lim, S.Y.; Salem, N., Jr.; Hibbelin, J.R. Distribution of omega-6 and omega-3 polyunsaturated fatty acids in the whole rat body and 25 compartments. Prostaglandins Leukot. Essent. Fat. Acids 2015, 100, 13–20.

53. DeMar, J.C., Jr.; DiMartino, C.; Baca, A.W.; Lefkowitz, W.; Salem, N., Jr. Effect of dietary docosahexaenoic acid on biosynthesis of docosahexaenoic acid from alpha-linolenic acid in young rats. J. Lipid Res. 2008, 49, 1963–1980.

54. Tyburczy, C.; Kothapalli, K.S.D.; Park, W.J.; Blank, B.S.; Bradford, K.L.; Zimmer, J.P.; Butt, C.M.; Salem, N., Jr.; Brenna, J.T. Heart arachidonic acid is uniquely sensitive to dietary arachidonic acid and docosahexaenoic acid content in domestic piglets. Prostaglandins Leukot. Essent. Fat. Acids 2011, 85, 335–343.

55. Axelrod, J. Receptor-mediated activation of phospholipase A2 and arachidonic acid release in signal transduction. Biochem. Soc. Trans. 1990, 18, 503–507.

56. Piomelli, D. Eicosanoids in synaptic transmissions. Crit. Rev. Neurobiol. 1994, 11, 367–373.

57. Schoenheimer, R.; Rittenberg, D. Deuterium as an indicator in the study of intermediary metabolism V. The desaturation of fatty acids in the organism. J. Biol. Chem. 1936, 113, 505–510.

58. Nichaman, M.Z.; Olson, R.E.; Sweeley, C.C. Metabolism of linoleic acid-1-14C in normolipidemic and hyperlipidemic humans fed linoleate diets. Am. J. Clin. Nutr. 1967, 20, 1070–1083.

59. Chambaz, J.; Ravel, D.; Manier, M.-C.; Pepin, D.; Mulliez, N.; Bereziat, G. Essential fatty acids interconversion in the human fetal liver. Neonatology 1985, 47, 136–140.

60. El Boustani, S.; Descomps, B.; Monnier, L.; Warnant, J. In vivo conversion of dihommogamma linolenic acid into arachidonic acid in man. Prog. Lipid Res. 1986, 25, 67–71.

61. Emken, E.A.; Adlof, R.O.; Rakoff, H.; Rohwedder, W.K. Metabolism of deuterium-labeled linolenic, linoleic, oleic, stearic and palmitic acid in human subjects. In Synthesis and Applications of Isotopically Labelled Compounds, Proceedings of the Third International Symposium, Innsbruck, Austria, 17–21 July 1988; Baillie, T.A., Jones, J.R., Eds.; Elsevier Science Publishers: Amsterdam, The Netherlands, 1988; pp. 713–716.

62. Emken, E.A.; Rohwedder, W.K.; Adlof, R.O.; Gulley, R.M. Metabolism in humans of cis-12, trans-15-octadecadienoic acid relative to palmitic, stearic, oleic and linoleic acids. Lipids 1987, 22, 495–504.

63. Demmelmair, H.; von Schenck, U.; Behrendt, E.; Sauerwald, T.; Koletzko, B. Estimation of arachidonic acid synthesis in full-term neonates using natural variation of 13C content. J. Pediatr. Gastroenterol. Nutr. 1995, 21, 31–36.

64. Salem, N., Jr.; Wegher, B.; Mena, P.; Uauy, R. Arachidonic and docosahexaenoic acids are biosynthesized from their 18-carbon precursors in human infants. Proc. Natl. Acad. Sci. USA 1996, 93, 49–54.

65. Pawlosky, R.J.; Sprecher, H.W.; Salem, N., Jr. High sensitivity negative ion GC/MS method for detection of desaturated and chain-elongated products of deuterated linoleic and linolenic acids. J. Lipid Res. 1992, 33, 1711–1717.

66. Carnielli, V.P.; Wattimea, D.J.L.; Luijendijk, I.H.T.; Boerlage, A.; Degenhart, H.J.; Sauer, P.J.J. The very low weight premature infant is capable of synthesizing arachidonic and docosahexaenoic acids from linoleic and linolenic acids. Pediatr. Res. 1996, 40, 169–174.

67. Pawlosky, R.J.; Lin, Y.H.; Llanos, A.; Mena, P.; Uauy, R.; Salem, N., Jr. Compartmental analysis of plasma 13C- and 2H-labelled n-6 fatty acids arising from oral administrations of 13C-U-18:2n-6 and2H5-20:3n-6 in newborn infants. Pediatr. Res. 2006, 60, 327–333.

68. Sauerwald, T.U.; Hachey, D.L.; Jensen, C.L.; Chen, H.; Andersen, R.E.; Heird, W.C. Effect of dietary α-linolenic intake on incorporation of docosahexaenoic and arachidonic acids into plasma phospholipids of term infants. Lipids 1996, 31, S131–S135.

69. Carnielli, V.P.; Simonato, M.; Verlato, G.; Luijendijk, I.; De Curtis, M.; Sauer, P.J.J.; Cogo, P.E. Synthesis of long-chain polyunsaturated fatty acids in preterm newborns fed formula with long-chain polyunsaturated fatty acids. Am. J. Clin. Nutr. 2007, 86, 1323–1330.

70. EFSA Panel on Dietetic Products. Scientific opinion on nutrient requirements and dietary intakes on infants and young children in the European Union. EFSA J. 2013, 11, 3408.

71. EFSA Panel on Dietetic Products, Nutrition and Allergies (NDA). Scientific opinion on the essential composition of infant and follow-on formulae. EFSA J. 2014, 12, 3760.

72. Birch, E.E.; Castañeda, Y.S.; Wheaton, D.H.; Birch, D.G.; Uauy, R.D.; Hoffman, D.R. Visual maturation of term infants fed long-chain polyunsaturated fatty acid-supplemented or control formula for 12 mo.Am. J. Clin. Nutr. 2005, 81, 871–879.

73. World Health Organization (WHO). Global Strategy on Infant and Young Child Feeding, 2002. Available online:

http://www.who.int/nutrition/topics/infantfeeding_recommendation/en/ (accessed on 15 October 2015).

74. World Health Organization (WHO). Nutrient Adequacy of Exclusive Breastfeeding for the Term Infant during the First Six Months of Life, 2002. Available online: http://www.who.int/nutrition/publications/infantfeeding/9241562110/en/ (accessed on 15 October 2015).

75. American Academy of Pediatrics. Breastfeeding and the use of human milk. Pediatrics 2012, 129, e827.

76. International Baby Food Action (IBFAN). The State of Breastfeeding in 33 Countries, 2010. Available online: https://www.google.co.uk/?gws_rd=ssl#q=IBFAN+33+countries (accessed on 15 October 2015).

77. U.S. Department of Agriculture, Agricultural Research Service. USDA National Nutrient Database for Standard Reference, Release 27. Nutrient Data Laboratory Home Page. 2014. Available online: http://www.ars.usda.gov/nutrientdata (accessed on 15 October 2015).

78. Centers for Disease Control and Prevention. Clinical Growth Charts, 2015. Available online: http://www.cdc.gov/growthcharts/cdccharts.htm (accessed on 19 November 2015).

79. Michaelsen, K.F.; Dewey, K.G.; Perez-Exposito, A.B. Food sources and intake of n-6 and n-3 fatty acids in low-income countries with emphasis on infants, young children (6–24 months), and pregnant and lactating women. Mater. Child Nutr. 2011, 7 (Suppl. S2), 124–140.

80. Agostoni, C. Docosahexaenoic acid (DHA): From the maternal-foetal dyad to the complementary feeding period. Early Hum. Dev. 2010, 86 (Suppl. S1), 3–6.

81. Prentice, A.M.; Paul, A.A. Fat and energy needs of children in developing countries. Am. J. Clin. Nutr.2000, 72, 1253S–1265S.

82. Barbarich, B.N.; Willows, N.D.; Wang, L.; Clandinin, M.T. Polyunsaturated fatty acids and anthropometric indices of children in rural China. Eur. J. Clin. Nutr. 2006, 60, 1100–1107.

83. PAHO/WHO. Guiding Principles for Complementary Feeding of the Breastfed Child; PAHO/WHO: Washington, DC, USA, 2003.

84. Joshi, N.; Agho, K.E.; Dibley, M.J.; Senarath, U.; Tiwari, K. Determinants of inappropriate complementary feeding practices in young children in Nepal: Secondary data analysis of Demographic and Health Survey 2006. Mater. Child Nutr. 2012, 1, 45–59.

85. Dutta, T.; Sywulka, S.M.; Frongillo, E.A.; Lutter, C.K. Characteristics attributed to complementary foods by caregivers in four countries of Latin America and the Caribbean. Food Nutr. Bull. 2006, 27, 316–326.

86. Agostoni, C.; Decsi, T.; Fewtrell, M.; Goulet, O.; Kolacek, S.; Koletzko, B.; Michaelsen, K.F.; Moreno, L.; Puntis, J.; Rigo, J.; et al. Complementary feeding: A commentary by the ESPGHAN Committee on Nutrition. J. Pediatr. Gastroenterol. Nutr. 2008, 46, 99–110.

87. Schwartz, J.; Dube, K.; Alexy, U.; Kalhoff, H.; Kersting, M. PUFA and LC-PUFA intake during the first year of life: Can dietary practice achieve a guideline diet? Eur. J. Clin. Nutr. 2010, 64, 124–130.

88. Schwartz, J.; Dube, K.; Sichert-Hellert, W.; Kannenberg, F.; Kunz, C.; Kalhoff, H.; Kersting, M. Modification of dietary polyunsaturated fatty acids via complementary food enhances n-3 long-chain polyunsaturated fatty acid synthesis in healthy infants: A double blinded randomised controlled trial.Arch. Dis. Child. 2009, 94, 876–882.

89. Grote, V.; Verduci, E.; Scalioni, S.; Vecchi, F.; Contarini, G.; Giovannini, M.; Koletzko, B.; Agostoni, C. Breast milk composition and infant nutrient intakes during the first 12 months of life. Eur. J. Clin. Nutr. 2015, 70, 250–256.

90. Sioen, I.; Matthys, C.; De Backer, G.; Van Camp, J.; De Henauw, S. Importance of seafood as nutrient source in the diet of Belgian adolescents. J. Hum. Nutr. Diet. 2007, 20, 580–589.

91. Meyer, B.J.; Mann, N.J.; Lewis, J.L.; Milligan, G.C.; Sinclair, A.J.; Howe, P.R. Dietary intakes and food sources of omega-6 and omega-3 polyunsaturated fatty acids. Lipids 2003, 38, 391–398.

92. Innis, S.M.; Vaghri, Z.; King, D.J. n-6 docosapentaenoic acid is not a predictor of low docosahexaenoic acid status in Canadian preschool children. Am. J. Clin. Nutr. 2004, 80, 768–773.

93. Lien, V.W.; Clamdinin, M.T. Dietary assessment of arachidonic acid and docosahexaenoic acid intake in 4–7 year-old children. J. Am. Coll. Nutr. 2009, 28, 7–15.

94. Keim, S.A.; Branum, A.M. Dietary intake of polyunsaturated fatty acids and fish among US children 12–60 months of age. Mater. Child Health Nutr. 2015, 11, 987–998.

95. U.S. Department of Agriculture, Agricultural Research Service. What We Eat in America. 2015. Available online: http://www.ars.usda.gov/Services/docs.htm?docid=13793# (accessed on 15 October 2015).

96. Birch, E.E.; Khoury, J.C.; Berseth, C.L.; Castaneda, Y.S.; Couch, J.M.; Bean, J.; Tamer, R.; Harris, C.L.; Mitmesser, S.H.; Scalabrin, D.M. The impact of early nutrition on incidence of allergic manifestations and common respiratory illnesses in children. J. Pediatr. 2010, 156, 902–906.

97. Calder, P.C. Polyunsaturated fatty acids and inflammation: From molecular biology to the clinic.Lipids 2003, 38, 343–352.

98. Calder, P.C. n-3 polyunsaturated fatty acids, inflammation, and inflammatory diseases. Am. J. Clin. Nutr. 2006, 83, 1505S–1519S.

99. Jones, P.J.H.; Kubow, S. Lipids, sterols, and their metabolites. In Modern Nutrition in Health and Disease; Shils, M.E., Shike, M., Ross, A.C., Caballero, B., Cousins, B., Eds.; Lippincott Williams & Wilkins: Philadelphia, PA, USA, 2006; pp. 92–122.

100. Ricciotti, E.; FitzGerald, G.A. Prostaglandins and inflammation. Arterioscler. Thromb. Vasc. Biol. 2011,31, 986–1000.

101. Reinhold, H.; Ahmadi, S.; Depner, U.B.; Layh, B.; Heindl, C.; Hamza, M.; Pahl, A.; Brune, K.; Narumiya, S.; Müller, U.; et al. Spinal inflammatory hyperalgesia is mediated by prostaglandin E receptors of the EP2 subtype. J. Clin. Investig. 2005, 115, 673–679.

102. Noda, M.; Kariura, Y.; Pannasch, U.; Nishikawa, K.; Wang, L.; Seike, T.; Ifuku, M.; Kosai, Y.; Wang, B.; Nolte, C.; et al. Neuroprotective role of bradykinin because of the attenuation of pro-inflammatory cytokine release from activated microglia. J. Neurochem. 2007, 101, 397–410.

103. Min, Y.; Crawford, M.A. Essential fatty acids. In The Eicosanoids; Curtis, P., Ed.; John Wiley & Sons, Ltd.: West Sussex, UK, 2004; pp. 257–276.

104. McCarthy, T.L.; Casinghino, S.; Mittanck, D.W.; Ji, C.H.; Centrella, M.; Rotwein, P. Promoter-dependent and -independent activation of insulin-like growth factor binding protein-5 gene expression by prostaglandin E2 in primary rat osteoblasts. J. Biol. Chem. 1996, 271, 6666–6671.

105. Urade, Y.; Hayaishi, O. Prostaglandin D2 and sleep regulation. Biochem. Biophys. Acta 1999, 1436, 606–615.

106. Ushikubi, F.; Sergi, E.; Sugimoto, Y.; Murata, T.; Matsuoka, T.; Kobayashi, T.; Hizaki, H.; Tuboi, K.; Katsuyama, M.; Ichikawa, A.; et al. Impaired febrile response in mice lacking the prostaglandin E receptor subtype EP3. Nature 1998, 395, 281–284.

107. Murata, T.; Ushikubi, F.; Matsuoka, T.; Hirata, M.; Yamasaki, A.; Sugimoto, Y.; Ichikawa, A.; Aze, Y.; Tanaka, T.; Yoshida, N.; et al.

Altered pain perception and inflammatory response in mice lacking prostacyclin receptor. Nature 1997, 388, 678–682.

108. Astudillo, A.M.; Balgoma, D.; Balboa, M.A.; Balsinde, J. Dynamics of arachidonic acid mobilization by inflammatory cells. Biochim. Biophys. Acta 2012, 1821, 249–256.

109. Bagga, D.; Wang, L.; Faris-Eisner, R.; Glaspy, J.A.; Reddy, S.T. Differential effects of prostaglandin derived from ω-6 and ω-3 polyunsaturated fatty acids on COX-2 expression and IL-6 secretion. Proc. Natl. Acad. Sci. USA 2003, 100, 1751–1756.

110. Levy, B.D.; Clish, C.B.; Schmidt, B.; Gronert, K.; Serhan, C.N. Lipid mediator class switching during acute inflammation signals in resolution. Nat. Immunol. 2001, 2, 612–619.

111. Samuelsson, B.; Dahlen, S.E.; Lindgren, J.A.; Rouzer, C.A.; Serhan, C.N. Leukotrienes and lipoxins—Structures, biosynthesis, and biological effects. Science 1987, 237, 1171–1176.

112. Fredman, G.; Serhan, C.N. Specialized proresolving mediator targets for RvE1 and RvD1 in peripheral blood and mechanisms of resolution. Biochem. J. 2011, 437, 185–197.

113. Kelley, D.S. Modulation of human and inflammatory responses by dietary fatty acids. Nutrition 2001,17, 669–673.

114. Calder, P.C.; Kew, S. The immune system: A target for functional foods? Br. J. Nutr. 2002, 88 (Suppl. S2), S165–S176.

115. Lentz, A.K.; Feezor, R.J. Principles of immunology. Nutr. Clin. Pract. 2003, 18, 451–460.

116. Mahadevappa, V.G.; Holub, B.J. The molecular species composition of individual diacyl phospholipids in human platelets. Biochim. Biophys. Acta 1982, 713, 73–79.

117. Kaushansky, K. Lineage-specific hematopoietic growth factors. N. Engl. J. Med. 2006, 354, 2034–2045.

118. Semple, J.W.; Italiano, J.E., Jr.; Freedman, J. Platelets and the immune system. Nature 2011, 11, 264–274.

119. Tamagawa-Mineoka, R. Important roles of platelets as immune cells in the skin. J. Dermatol. Sci. 2015,77, 93–101.

120. Martin, C.R.; DaSilva, D.A.; Cluette-Brown, J.E.; DiMonda, C.; Hamill, A.; Bhutta, A.Q.; Coronel, E.; Wilschanski, M.; Stephens, A.J.; Driscoll, D.F.; et al. Decreased postnatal docosahexaenoic and arachidonic acid blood levels in premature infants are associated with neonatal morbidities. J. Pediatr.2011, 159, 743–749.

121. Lands, W.F.; LeTellier, P.R.; Rome, L.H.; Vanderhoek, J.Y. Inhibition of prostaglandin biosynthesis.Adv. Biosci. 1973, 9, 15–227.

122. Vane, J.R. Inhibition of prostaglandin synthesis as a mechanism of action for aspirin-like drugs. Nature1971, 231, 232–235.

123. Hormones. Available online: http://www.medicinenet.com/script/main/art.asp?articlekey=3783(accessed on 6 February 2016).

124. McMurray, W.C. A Synopsis of Human Biochemistry; Harper and Row Publishers: New York, NY, USA, 1982; pp. 193–199.

125. Bowen, R.A. Hormone Chemistry, Synthesis and Elimination. Available online: http://www.vivo.colostate.edu/hbooks/pathphys/endocrine/basics/chem.html (accessed on 15 October 2015).

126. Thorner, M.O.; Vance, M.L.; Hartman, M.L.; Holl, R.W.; Evans, W.S.; Veldhuis, J.D.; Van Cauter, E.; Copinschi, G.; Bowers, C.Y. Physiological role of somatostatin on growth hormone regulation in humans. Metabolism 1990, 39, 40–42.

127. Brochhausen, C.; Neuland, P.; Kirkpatrick, C.J.; Nusing, R.M.; Klaus, G. Cyclooxygenases and prostaglandin E2 receptors in growth plate chondrocytes in vitro and in situ-prostaglandin E2dependent proliferation of growth plate chondrocytes. Arthritis Res. Ther. 2006, 8, R78.

128. Boyan, B.D.; Sylvia, V.L.; Dean, D.D.; Del Toro, F.; Schwartz, Z. Differential regulation of growth plate chondrocytes by 1alpha,25-(OH)2D3 and 24R,25-(OH)2D3 involves cell-maturation-specific membrane-receptor-activated phospholipid metabolism. Crit. Rev. Oral Biol. Med. 2002, 13, 143–154.

129. Specker, B.; Binkley, T. Randomized trial of physical activity and calcium supplementation on bone mineral content in 3- to 5-year-old children. J. Bone Miner. Res. 2003, 418, 885–892.

130. Specker, B.L.; Mulligan, L.; Ho, M. Longitudinal study of calcium intake, physical activity, and bone mineral content in infants 6–18 months of age. J. Bone Miner. Res. 1999, 14, 569–576.

131. Del Toro, F., Jr.; Sylvia, V.L.; Schubkegel, S.R.; Campos, R.; Dean, D.D.; Boyan, B.D.; Schwartz, Z. Characterization of prostaglandin E(2) receptors and their role in 24,25-(OH)(2)D(3)-mediated effects on resting zone chondrocytes. J. Cell Physiol. 2000, 182, 196–208.

132. Boyan, B.D.; Sylvia, V.L.; Dean, D.D.; Pedrozo, H.; Del Toro, F.; Nemere, I.; Posner, G.H.; Schwartz, Z. 1,25-(OH)2D3 modulates growth plate chondrocytes via membrane receptor-mediated protein kinase C by a mechanism that involves changes in

phospholipid metabolism and the action of arachidonic acid and PGE2. Steroids 1999, 64, 129–136.

133. Schwartz, Z.; Sylvia, V.L.; Curry, D.; Luna, M.H.; Dean, D.D.; Boyan, B.D. Arachidonic acid directly mediates the rapid effects of 24,25-dihydroxyvitamin D3 via protein kinase C and indirectly through prostaglandin production in resting zone chondrocytes. Endocrinology 1999, 140, 2991–3002.

134. Sylvia, V.L.; Schwartz, Z.; Curry, D.B.; Chang, Z.; Dean, D.D.; Boyan, B.D. 1,25(OH)2D3 regulates protein kinase C activity through two phospholipid-dependent pathways involving phospholipase A2 and phospholipase C in growth zone chondrocytes. J. Bone Miner. Res. 1998, 13, 559–569.

135. Kosher, R.A.; Walker, K.H. The effect of prostaglandins on in vitro limb cartilage differentiation. Exp. Cell Res. 1983, 145, 145–153.

136. Copray, J.C.; Jansen, H.W. Cyclic nucleotides and growth regulation of the mandibular condylar cartilage of the rat in vitro. Arch. Oral Biol. 1985, 30, 749–752.

137. Li, T.F.; Zuscik, M.J.; Ionescu, A.M.; Zhang, X.; Rosier, R.N.; Schwarz, E.M.; Drissi, H.; O'Keefe, R.J. PGE2 inhibits chondrocyte differentiation through PKA and PKC signaling. Exp. Cell Res. 2004, 300, 159–169.

138. Raisz, L.G. Prostaglandins and bone: Physiology and pathophysiology. Osteoarthr. Cartil. 1999, 7, 419–421.

139. Suzawa, T.; Miyaura, C.; Inada, M.; Maruyama, T.; Sugimoto, Y.; Ushikubi, F.; Ichikawa, A.; Narumiya, S.; Suda, T. The role of prostaglandin E receptor subtypes (EP1, EP2, EP3, and EP4) in bone resorption: An analysis using specific agonists for the respective EPs. Endocrinology 2000, 141, 1554–1559.

140. Baylink, D.J.; Finkelman, R.D.; Mohan, S. Growth factors to stimulate bone formation. J. Bone Min. Res.1993, 8, S565–S572.

141. Paralkar, V.M.; Borovecki, F.; Ke, H.Z.; Cameron, K.O.; Lefker, B.; Grasser, W.A.; Owen, T.A.; Li, M.; DaSilva-Jardine, P.; Zhou, M.; et al. An EP2 receptor-selective prostaglandin E2 agonist induces bone healing. PNAS 2003, 100, 6736–6740.

142. Tanaka, M.; Sakai, A.; Uchida, S.; Tanaka, S.; Nagashima, M.; Katayama, T.; Yamaguchi, K.; Nakamura, T. Prostaglandin E2 receptor (EP4) selective agonist (ONO-4819.CD) accelerates bone repair of femoral cortex after drill-hole injury associated with local upregulation of bone turnover in mature rats. Bone2004, 34, 940–943.

143. Jee, W.S.; Ueno, K.; Kimmel, D.B.; Woodbury, D.M.; Price, P.; Woodbury, L.A. The role of bone cells in increasing metaphyseal hard tissue in rapidly growing rats treated with prostaglandin E2. Bone 1987,8, 171–178.

144. Agas, D.; Marchetti, L.; Hurley, M.M.; Sabbieti, D. Prostaglandin F2α: A bone remodeling mediator. J. Cell. Physiol. 2013, 228, 25–29.

145. Galea, G.L.; Meakin, L.B.; Williams, C.M.; Hulin-Curtis, S.L.; Lanyon, L.E.; Poole, A.W.; Price, J.S. Protein kinase Cα (PKCα) regulates bone architecture and osteoblast activity. J. Biol. Chem. 2014, 289, 25509–25522.

146. Zaman, G.; Sunters, A.; Galea, G.L.; Javaheri, B.; Saxon, L.K.; Moustafa, A.; Armstrong, V.J.; Price, J.S.; Lanyon, L.E. Loading-related regulation of transcription factor EGR2/Krox-20 in bone cells is ERK1/2 protein-mediated and prostaglandin, Wnt-signaling pathway-, and insulin-like growth factor-I axis-dependent. J. Biol. Chem. 2012, 287, 3946–3962.

147. Kido, S.; Kuriwaka-Kido, R.; Umino-Miyatani, Y.; Endo, I.; Inoue, D.; Taniguchi, H.; Inoue, Y.; Imamura, T.; Matsumoto, T. Mechanical stress activates Smad pathway through PKCδ to enhance interleukin-11 gene transcription in osteoblasts. PLoS ONE 2010, 5, e13090.

148. Nakura, A.; Higuchi, C.; Yoshida, K.; Yoshikawa, H. PkCα suppresses osteoblastic differentiation. Bone2011, 48, 476–484.

149. Weiler, H.A.; Fitzpatrick-Wong, S. Dietary long-chain polyunsaturated fatty acids minimize dexamethasone-induced reductions in arachidonic acid status but not bone mineral content in piglets.Pediatr. Res. 2002, 51, 282–289.

150. Blanaru, J.L.; Kohut, J.R.; Fitzpatrick-Wong, S.C.; Weiler, H.A. Dose response of bone mass to dietary arachidonic acid in piglets fed cow milk-based formula. Am. J. Clin. Nutr. 2004, 79, 139–147.

151. Pash, J.M.; Canalis, E. Transcriptional regulation of insulin-like growth factor-binding protein-5 by prostaglandin E2 in osteoblast cells. Endocrinology 1996, 137, 2375–2382.

152. Almaden, Y.; Canalejo, A.; Ballesteros, E.; Anon, G.; Rodriguez, M. Effect of high extracellular phosphate concentration on arachidonic acid production by parathyroid tissue in vitro. J. Am. Soc. Nephrol. 2000, 11, 1712–1718.

153. Klein-Nulend, J.; Burger, E.H.; Semeins, C.M.; Raisz, L.G.; Pilbeam, C.C. Pulsating fluid flow stimulates prostaglandin release and inducible prostaglandin G/H synthase mRNA expression in primary mouse bone cells. J. Bone Miner. Res. 1997, 12, 45–51.

154. Weiler, H.A. Dietary supplementation of arachidonic acid is associated with higher whole body weight and bone mineral density in growing pigs. Pediatr. Res. 2000, 47, 692–697.

155. Weiler, H.; Fitzpatrick-Wong, S.; Schellenberg, J.; McCloy, U.; Veitch, R.; Kovacs, H.; Kohut, J.; Kin Yuen, C. Maternal and cord blood long-chain polyunsaturated fatty acids are predictive of bone mass at birth in healthy term-born infants. Pediatr. Res. 2005, 58, 1254–1258.

156. Akatsu, T.; Takahashi, N.; Udagawa, N.; Imamura, K.; Yamaguchi, A.; Sato, K.; Nagata, N.; Suda, T. Role of prostaglandins in interleukin-1-induced bone resorption in mice in vitro. J. Bone Miner. Res.1991, 6, 183–189.

157. Liu, X.H.; Kirschenbaum, A.; Yao, S.; Levine, A.C. Interactive effect of interleukin-6 and prostaglandin E2 on osteoclastogenesis via the OPG/RANKL/RANK system. Ann. N. Y. Acad. Sci. 2006, 1068, 225–233.

158. Zhang, X.; Schwarz, E.M.; Young, D.A.; Puzas, J.E.; Rosier, R.N.; O'Keefe, R.J. Cyclooxygenase-2 regulates mesenchymal cell differentiation into the osteoblast lineage and is critically involved in bone repair. J. Clin. Investig. 2002, 109, 1405–1415.

159. Samoto, H.; Shimizu, E.; Matsuda-Honjyo, Y.; Saito, R.; Nakao, S.; Yamazaki, M.; Furuyama, S.; Sugiya, H.; Sodek, J.; Ogata, Y. Prostaglandin E2 stimulates bone sialoprotein (BSP) expression through cAMP and fibroblast growth factor 2 response elements in the proximal promoter of the rat BSP gene. J. Biol. Chem. 2003, 278, 28659–28667.

160. Cherian, P.P.; Cheng, B.; Gu, S.; Sprague, E.; Bonewald, L.F.; Jiang, J.X. Effects of mechanical strain on the function of Gap junctions in osteocytes are mediated through the prostaglandin EP2 receptor. J. Biol. Chem. 2003, 278, 43146–43156.

161. Miyaura, C.; Inada, M.; Matsumoto, C.; Ohshiba, T.; Uozumi, N.; Shimizu, T.; Ito, A. An essential role of cytosolic phospholipase A2alpha in prostaglandin E2-mediated bone resorption associated with inflammation. J. Exp. Med. 2003, 197, 1303–1310.

162. Boswell, K.; Koskelo, E.-K.; Carl, L.; Glaze, S.; Hensen, D.J.; Williams, K.D.; Kyle, D.J. Preclinical evaluation of single-cell oils that are highly enriched with arachidonic acid and docosahexaenoic acid.Food Chem. Toxicol. 1996, 34, 585–593.

163. Suarez, A.; del Carmen Ramirez, M.; Faus, M.J.; Gil, A. Dietary long-chain polyunsaturated fatty acids influence tissue fatty acid composition in rats at weaning. J. Nutr. 1996, 126, 887–897.

164. De la Presa-Owens, S.; Innis, S.M.; Rioux, F.M. Addition of triglycerides with arachidonic acid or docosahexaenoic acid to infant formula has tissue- and lipid class-specific effects on fatty acids and hepatic desaturase activities in formula-fed piglets. J. Nutr. 1998, 128, 1376–1384.

165. Baur, L.A.; O'Connor, J.; Pan, D.A.; Kriketos, A.D.; Storlien, L.H. The fatty acid composition of skeletal muscle membrane phospholipid: Its relationship with the type of feeding and plasma levels in young children. Metabolism 1998, 47, 106–112.

166. Blaauw, B.; Del Piccolo, P.; Rodriguez, L.; Hernandez Gonzales, V.H.; Agata, L.; Solagna, F.; Mammano, F.; Pozzan, T.; Schiaffino, S. No evidence for inositol 1,4,5-triphosphate-dependent Ca2+release in isolated fibers of adult mouse skeletal muscle. J. Gen. Physiol. 2012, 140, 235–241.

167. Berthier, C.; Kutchukian, C.; Bouvard, C.; Okamura, Y.; Jacquemond, V. Depression of voltage-activated Ca2+ release in skeletal muscle by activation of a voltage-sensing phosphatase. J. Gen. Physiol. 2015, 145, 315–330.

168. Ohizumi, Y.; Hirata, Y.; Suzuki, A.; Kobayashi, M. Two novel types of calcium release from skeletal sarcoplasmic reticulum by phosphatidylinositol 4,5 biphosphate. Can. J. Physiol. Pharmacol. 1999, 77, 276–285.

169. Sandow, A. Excitation-contraction coupling in muscular response. Yale J. Biol. Med. 1952, 25, 176–201.

170. Melzer, W.; Herrmann-Frank, A.; Luttgau, H.C. The role of Ca2+ ions in excitation-contraction coupling of skeletal muscle fibres. Biochim. Biophys. Acta 1995, 1241, 59–116.

171. Kobayashi, M.; Muroyama, A.; Ohizumi, Y. Phosphatidylinositol 4,5-bisphosphate enhances calcium release from sarcoplasmic reticulum of skeletal muscle. Biochem. Biophys. Res. Commun. 1989, 29, 1487–1491.

172. Ghigo, A.; Perino, A.; Hirsch, E. Phosphoinositides and cardiac function. In Phosphoinositides and Disease, Current Topics in Microbiology and Immunology; Falasca, M., Ed.; Springer Science + Business Media: Dordrecht, The Netherlands, 2012; pp. 43–60.

173. Rodemann, H.P.; Goldberg, A.L. Arachidonic acid, prostaglandin E2 and F2α influence rates of protein turnover in skeletal and cardiac muscle. J. Biol. Chem. 1982, 25, 1632–1638.

174. Standley, R.A.; Liu, S.; Jemiolo, B.; Trappe, S.W.; Trappe, T.A. Prostaglandin E2 induces transcription of skeletal mass regulators

interleukin-6 and muscle RING finger-1 in humans. Prostaglandins Leukot. Essent. Fat. Acids 2013, 88, 361–364.

175. Kuipers, R.S.; Luxwolda, M.F.; Dijck-Brouwer, J.; Muskiet, F.A.J. Intrauterine, postpartum and adult relationships between arachidonic acid (AA) and docosahexaenoic acid (DHA). Prostaglandins Leukot. Essent. Fat. Acids 2011, 85, 245–252.

176. Luxwolda, M.F.; Kuipers, R.S.; Sango, W.S.; Kwesigabo, G.; Dijck-Brouwer, D.A.J.; Muskiet, F.A.J. A maternal erythrocyte DHA content of approximately 6 g% is the DHA status at which intrauterine DHA biomagnifications turns into bioattenuation and postnatal infant DHA equilibrium is reached.Eur. J. Nutr. 2012, 51, 665–675.

177. Carlson, S.E.; Werkman, S.H.; Peoples, J.M.; Cooke, R.J.; Tolley, E.A. Arachidonic acid status correlates with first year of growth in preterm infants. Proc. Natl. Acad. Sci. USA 1993, 90, 1073–1077.

178. Clandinin, M.T. Brain development and assessing the supply of polyunsaturated fatty acids. Lipids1999, 34, 131–137.

179. Clandinin, M.T.; Chappell, J.E.; Leong, S.; Heim, T.; Sayer, P.R.; Chance, G.W. Intrauterine fatty acid accretion in infant brain: Implications for fatty acid requirements. Early Hum. Dev. 1980, 4, 121–129.

180. Cunnane, S. Problems with essential fatty acids: Time for a new paradigm? Prog. Lipid Res. 2003, 42, 544–568.

181. Stoffel, W.; Holz, B.; Jenke, B.; Binczek, E.; Günter, R.H.; Kiss, C.; Karakesisoglou, I.; Thevis, M.; Weber, A.A.; Arnhold, S.; et al. Δ6-desaturase (FADS2) deficiency unveils role of ω3- and ω6-polyunsaturated fatty acids. EMBO J. 2008, 27, 2281–2292.

182. Stroud, C.K.; Nara, T.Y.; Roqueta-Rivera, M.; Radlowski, E.C.; Lawrence, P.; Zhang, Y.; Cho, B.H.; Segre, M.; Hess, R.A.; Brenna, J.T.; et al. Disruption of FADS2 gene in mice impairs male reproduction and causes dermal and intestinal ulceration. J. Lipid Res. 2009, 50, 1870–1880.

183. Mohrhauer, H.; Holman, R.T. The effect of dose level of essential fatty acids upon fatty acid composition of the rat liver. J. Lipid Res. 1963, 4, 151–159.

184. Prottey, C. Essential fatty acids and the skin. Br. J. Dermatol. 1976, 94, 579–587.

185. Hansen, H.S.; Jensen, B. Essential function of linoleic acid esterified in acylglucosylceramide and acylceramide in maintaining the epiderma; water permeability barrier. Evidence from feeding studies with oleate, linoleate, arachidonate, columbinate and alpha-linolenate. Biochem. Biophys. Acta 1985,834, 357–363.

186. Hartop, P.J.; Prottey, C. Changes in transepidermal water loss and the composition of epidermal lecithin after application of pure fatty acid triglycerides to the skin of essential fatty acid-deficient rats.Br. J. Dermatol. 1976, 95, 255–264.

187. Bertram, T.A. Gastrointestinal tract. In Handbook of Toxicologic Pathology; Haschek, W.M., Rousseaux, C.G., Wallig, M., Eds.; Academic Press: San Diego, CA, USA, 2002; pp. 121–186.

188. Miller, C.C.; Ziboh, V.A. Induction of epidermal hyperproliferation by topical n-3 polyunsaturated fatty acids on guinea pig skin linked to decreased levels of 13-hydroxyoctadecadienoic acid (13-Hode). J. Investig. Dermatol. 1990, 94, 353–358.

189. Cho, H.P.; Nakamura, T.; Clarke, S.D. Cloning, expression, and nutritional regulation of the mammalian delta-6 desaturate. J. Biol. Chem. 1999, 274, 471–477.

190. Williard, D.E.; Nwankwo, J.O.; Kaduce, T.L.; Harmon, S.D.; Irons, M.; Moser, H.W.; Raymond, G.V.; Spector, A.A. Identification of a fatty acid delta6-desaturase deficiency in human skin fibroblasts. J. Lipid Res. 2001, 42, 501–508.

191. Fan, Y-Y.; Monk, J.M.; Hou, T.Y.; Callway, E.; Vincent, L.; Weeks, B.; Yang, P.; Chapkin, R.S. Characterization of an arachidonic acid-deficient (Fads1 knockout) mouse model. J. Lipid Res. 2012, 53, 1287–1295.

192. Hatanaka, E.; Yasuda, H.; Harauma, A.; Watanabe, J.; Konishi, Y.; Nakamura, M.; Salem, N., Jr.; Moriguchi, T. The Effects of Arachidonic Acid and/or Docosahexaenoic Acid on the Brain Development Using Artificial Rearing of Delta-6-Desaturase Knockout Mice. In Proceedings of the Asian Conference of Nutrition, Yokohama, Japan, 14–18 May 2015.

193. Tian, C.; Fan, C.; Liu, X.; Xu, F.; Qi, K. Brain histological changes in young mice submitted to diets with different ratios of n-6/n-3 polyunsaturated fatty acids during maternal pregnancy and lactation.Clin. Nutr. 2011, 30, 659–667.

194. Nguyen, L.N.; Ma, D.; Shui, G.; Wong, P.; Cazenave-Gassiot, A.; Zhang, X.; Wenk, M.R.; Goh, E.L.K.; Silver, D.L. Mfsd2a is a transporter for the essential omega-3 fatty acid docosahexaenoic acid. Nature2014, 509, 503–506.

195. Berger, J.H.; Charron, M.J.; Silver, D.L. Major Facilitator superfamily domain-containing protein 2a (MFSD2A) has roles in body growth, motor function, and lipid metabolism. PLoS ONE 2012, 7, e50629.

196. Peterson, L.D.; Jeffrey, N.M.; Sanderson, P.; Newholme, E.A.; Calder, P.C. Eicosapentaenoic and docosahexaenoic acids alter rat spleen

leukocyte fatty acid composition and prostaglandin E2production but have different effects on lymphocyte functions and cell-mediated immunity. Lipids1998, 33, 171–180.

197. Jolly, C.A.; Jiang, Y-H.; Chapkin, R.S.; McMurray, D.N. Dietary (n-3) polyunsaturated fatty acids suppress murine lymphoproliferation, interleukin-2 secretion, and the formation of diacylglycerol and cereamide. J. Nutr. 1997, 127, 37–43.

198. Kelley, D.S.; Taylor, P.C.; Nelson, G.J.; Schmidt, P.C.; Mackey, B.E.; Kyle, D. Effects of dietary arachidonic acid on human immune response. Lipids 1997, 32, 449–456.

199. Blikslager, A.T.; Moeser, A.J.; Gookin, J.L.; Jones, S.L.; Odle, J. Restoration of barrier function in injured intestinal mucosa. Physiol. Rev. 2007, 87, 545–564.

200. Ferrer, R.; Moreno, J.J. Role of eicosanoids on intestinal epithelial homeostatis. Biochem. Pharmacol.2010, 80, 431–438.

201. Jacobi, S.K.; Moeser, A.J.; Corl, B.A.; Harrell, R.J.; Bilksager, A.T. Dietary long-chain PUFA enhances acute repair of ischemia-injured intestine of suckling pigs. J. Nutr. 2012, 142, 1266–1271.

202. Le, H.D.; Meisel, J.A.; de Meijer, V.E.; Fallon, E.M.; Gura, K.M.; Nose, V.; Bistrian, B.R.; Puder, M. Docosahexaenoic acid and arachidonic acid prevent essential fatty acid deficiency and hepatic steatosis. J. Parenter. Enter. Nutr. 2012, 36, 431–441.

203. Bassaganya-Riera, J.; Guri, A.J.; Noble, A.M.; Reynolds, K.A.; King, J.; Wood, C.M.; Ashby, M.; Rai, D.; Hontecillas, R. Arachidonic acid- and docosahexaenoic acid-enriched formulas modulate antigen-specific T cell responses to influenza virus in neonatal piglets. Am. J. Clin. Nutr. 2007, 85, 824–836.

204. Weisinger, H.S.; Vingrys, A.J.; Sinclair, A.J. The effect of docosahexaenoic acid on the electroretinogram of the guinea pig. Lipids 1996, 31, 65–70.

205. Champoux, M.; Hibbeln, J.R.; Shannon, C.; Majchrzak, S.; Suomi, S.J.; Salem, N., Jr.; Higley, J.D. Fatty acid formula supplementation and neuromotor development in rhesus monkey neonates. Pediatr. Res.2002, 51, 273–281.

206. Ikemoto, A.; Ohishi, M.; Sato, Y.; Hata, N.; Misawa, Y.; Fujii, Y.; Okuyama, H. Reversibility of n-3 fatty acid deficiency-induced alterations of learning behavior in the rat: Level of n-6 fatty acids as another factor. J. Lipid Res. 2001, 42, 1655–1663.

207. Wainwright, P.E.; Xing, H.-C.; Mutsaers, L.; McCutcheon, D.; Kyle, D. Arachidonic acid offsets the effects on mouse brain and behavior of

a diet with a low (n-6):(n-3) ratio and very high levels of docosahexaenoic acid. J. Nutr. 1997, 127, 184–193.

208. Wainwright, P.E.; Xing, H.-C.; Ward, G.R.; Huang, Y.-S.; Bobik, E.; Auestad, N.; Montalto, M. Water maze performance is unaffected in artificially reared rats fed diets supplemented with arachidonic acid and docosahexaenoic acid. J. Nutr. 1999, 129, 1079–1089.

209. Wainwright, P.E.; Huang, Y.S.; Bulman-Fleming, B.; Dalby, B.; Mills, D.E.; Redden, P.; McCutcheon, D. The effect of dietary n-3/n-6 ration on brain development in the mouse: A dose response study with long-chain n-3 fatty acids. Lipids 1992, 27, 98–103.

210. Min, Y.; Lowry, C.; Ghebremeskel, K.; Thomas, B.; Offley-Shore, B.; Crawford, M. Unfavorable effect of type 1 and type 2 diabetes on maternal and fetal essential fatty acid status: A potential marker of fetal insulin resistance. Am. J. Clin. Nutr. 2005, 82, 1162–1168.

211. Siddappa, A.M.; Georgieff, M.K.; Wewerka, S.; Worwa, C.; Nelson, C.A.; Deregnier, R.A. Iron deficiency alters auditory recognition memory in newborn infants of diabetic mothers. Pediatr. Res.2004, 55, 1034–1041.

212. DeBoer, T.; Wewerka, S.; Bauer, P.J.; Geogieff, M.K.; Nelson, C.A. Explicit memory performance in infants of diabetic mothers at 1 year of age. Dev. Med. Child Neurol. 2005, 47, 525–531.

213. Holman, R.T.; Johnson, S.B.; Gerrand, J.M.; Mauer, S.M.; Kupcho-Sandberg, S.; Brown, D.M. Arachidonic acid deficiency in streptozotocin-induced diabetes. Proc. Natl. Acad. Sci. USA 1983, 80, 1375–2379.

214. Hadders-Algra, M. Prenatal long-chain polyunsaturated fatty acid status: The importance of a balanced intake of docosahexaenoic acid and arachidonic acid. J. Perinat. Med. 2008, 36, 101–109.

215. Clandinin, M.T.; Chappell, J.E.; Leong, S.; Heim, T.; Sayer, P.R.; Chance, G.W. Extrauterine fatty acid accretion in infant brain: Implications for fatty acid requirements. Early Hum. Dev. 1980, 4, 131–138.

216. Zhao, J.; Bigio, M.R.; Weiler, H.A. Maternal arachidonic acid supplementation improves neurodevelopment in young adult offspring from rat dams with and without diabetes. Prostaglandins Leukot Essent. Fat. Acids 2011, 84, 63–70.

217. Amusquivar, E.; Ruperez, F.J.; Barbas, C.; Herrera, E. Low arachidonic acid rather than α-tocopherol is responsible for the delayed postnatal development of offspring of rats fed fish oil instead of olive oil during pregnancy and lactation. J. Nutr. 2000, 13, 2855–2865.

218. Haubner, L.; Sullivan, J.; Ashmeade, T.; Saste, M.; Wiener, D.; Carver, J. The effects of maternal dietary docosahexaenoic acid intake on rat pup myelin and the auditory startle response. Dev. Neurosci. 2007,29, 460–467.

219. Elsherbiny, M.E.; Goruk, S.; Monckton, E.A.; Richard, C.; Brun, M.; Emara, M.; Field, C.J.; Godbout, R. Long-term effect of docosahexaenoic acid feeding on lipid composition and brain fatty acid-binding protein expression in rats. Nutrients 2015, 7, 8802–8817.

220. Maekawa, M.; Takashima, N.; Matsumata, M.; Ikegami, S.; Kontani, M.; Hara, Y.; Kawashima, H.; Owada, Y.; Kiso, Y.; Yoshikawa, T.; et al. Arachidonic acid drives postnatal neurogenesis and elicits a beneficial effect on prepulse inhibition, a biological trait of psychiatric illnesses. PLoS ONE 2009, 4, e5085.

221. Fleith, M.; Clandinin, T. Dietary PUFA for preterm and term infants: Review of clinical studies. Crit. Rev. Food Sci. Nutr. 2005, 3, 205–229.

222. Jensen, C.L.; Prager, T.C.; Fraley, J.K.; Chen, H.; Anderson, R.E.; Heird, W.C. Effect of dietary linoleic/alpha-linolenic acid ratio on growth and visual function of term infants. J. Pediatr. 1997, 131, 200–209.

223. Makrides, M.; Neumann, M.A.; Jeffrey, B.; Lien, E.L.; Gibson, R.A. A randomized trial of different ratios of linoleic to alpha-linolenic acid in the diet of term infants: Effect on visual function and growth. Am. J. Clin. Nutr. 2000, 71, 120–129.

224. Makrides, M.; Neumann, M.A.; Simmer, K.; Gibson, R.A. Erythrocyte fatty acids of term infants fed either breast milk, standard formula, or formula supplemented with long-chain polyunsaturates.Lipids 1995, 30, 941–948.

225. Makrides, M.; Neumann, M.A.; Simmer, K.; Pater, J.; Gibson, R. Are long-chain polyunsaturated fatty acids essential nutrients in infancy? Lancet 1995, 345, 1463–1468.

226. Makrides, M.; Neumann, M.A.; Simmer, K.; Gibson, R.A. Dietary long-chain polyunsaturated fatty acids do not influence growth in term infants: A randomized clinical trial. Pediatrics 1999, 104, 468–475.

227. Scott, D.T.; Janowsky, J.S.; Carroll, R.E.; Taylor, J.A.; Auestad, N.; Montalto, M.B. Formula supplementation with long-chain polyunsaturated fatty acids: Are there developmental benefits?Pediatrics 1998, 102, e59.

228. Gibson, R.A.; Neumann, M.; Makrides, M. The effects of diets rich in docosahexaenoic acid and/or gamma-linolenic acid on plasma fatty

acid profiles in term infants. In Lipids in Infant Nutrition; Huang, Y.S., Sinclair, A., Eds.; AOCS Press: Champaign, IL, USA, 1998; pp. 19–28.

229. Simmer, K. Longchain polyunsaturated fatty acid supplementation in infants born at term. Cochrane Database Syst. Rev. 2001, 4.

230. Uauy, R.D.; Hoffman, D.R.; Birch, E.E.; Birch, D.G.; Jameson, D.M.; Tyson, J.E. Safety and efficacy of omega-3 fatty acids in the nutrition of very low birth weight infants: Soy oil and marine oil supplementation of formula. J. Pediatr. 1994, 124, 612–620.

231. Clandinin, M.T.; Van Aerde, J.E.; Parrott, A.; Field, C.J.; Euler, A.R.; Lien, E.L. Assessment of the efficacious dose of arachidonic and docosahexaenoic acids in preterm infant formula: Fatty acid composition of erythrocyte membrane lipids. Pediatr. Res. 1997, 42, 819–825.

232. Vanderhoof, J.; Gross, S.; Hegyi, T.; Clandinin, T.; Porcelli, P.; DeCristofaro, J.; Rhodes, T.; Tsang, R.; Shattuck, K.; Cowett, R.; et al. Evaluation of a long-chain polyunsaturated fatty acid supplemented formula on growth, tolerance, and plasma lipids in preterm infants up to 48 weeks postconceptional age. J. Pediatr. Gastr. Nutr. 1999, 29, 318–326.

233. Vanderhoof, J.; Gross, S.; Hegyi, T. A multicenter long-term safety and efficacy trial of preterm formula with long-chain polyunsaturated fatty acids. J. Pediatr. Gastr. Nutr. 2000, 31, 121–127.

234. Food and Drug Administration. Agency Response Letter. GRAS Notice No. GRN 000041; U.S. Food and Drug Administration, Department of Health and Human Services: Washington, DC, USA, 2001.

235. Health Canada. Novel Food Decision, DHASCO and ARASCO Oils as Sources of Docosahexaenoic (DHA) and Arachidonic Acid (ARA) in Human Milk Substitutes, 2002. Available online: http://www.novelfoods.gc.ca (accessed on 19 October 2015).

236. Ryan, A.S.; Zeller, S.; Nelson, E.B. Safety evaluation of single cell oils and the regulatory requirements for use as food ingredients. In Single Cell Oils. Microbial and Algal Oil, 2nd ed.; Cohen, Z., Ratledge, C., Eds.; AOCS Press: Urbana, IL, USA, 2010; pp. 317–350.

237. Dobbing, J. Vulnerable periods in developing brain. In Brain, Behavior, and Iron in the Infant Diet; Dobbing, J., Ed.; Springer-Verlag: London, UK, 1990; pp. 1–25.

238. Uauy, R.D.; Birch, D.G.; Birch, E.E.; Hoffman, D.R.; Tyson, J.E. Effect of dietary essential ω-3 fatty acids on retinal and brain

development in premature infants. In Essential Fatty Acids and Eicosanoids; Sinclair, A., Gibson, E., Eds.; American Oil Chemists' Society: Champaign, IL, USA, 1992; pp. 197–202.

239. Uauy, R.D.; Treen, M.; Hoffman, D.R. Essential fatty acid metabolism and requirements during development. Semin. Perinatol. 1989, 13, 118–130.

240. Carlson, S.E.; Crooke, R.J.; Werkman, S.H.; Tolley, E.A. First year growth of preterm infants fed standard compared to marine oil n-3 supplemented formula. Lipids 1992, 27, 901–907.

241. Carlson, S.E.; Werkman, S.H. A randomized trial of visual attention of preterm infants fed docosahexaenoic acid until two months. Lipids 1996, 31, 85–90.

242. Diersen-Schade, D.A.; Hansen, J.W.; Harris, C.L.; Merkel, K.L.; Wisont, K.D.; Boettcher, J.A. Docosahexaenoic acid plus arachidonic acid enhance preterm infant growth. In Essential Fatty Acids and Eicosanoids: Invited Papers from the Fourth International Congress; Riemersma, R.A., Armstrona, R., Kelly, W., Wilson, R., Eds.; AOCS Press: Champaign, IL, USA, 1998; pp. 123–127.

243. Ryan, A.S.; Montalto, M.B.; Groh-Wargo, S.; Mimouni, F.; Sentipal-Walerius, J.; Doyle, J.; Siegman, J.; Thomas, A.J. Effect of DHA-containing formula on growth of preterm infants to 59 weeks postmenstrual age. Am. J. Hum. Biol. 1999, 11, 457–467.

244. Carlson, S.E.; Werkman, S.H.; Peeples, J.M.; Wilson, W.M. Long-chain fatty acids and early visual and cognitive development of preterm infants. Eur. J. Clin. Nutr. 1994, 48, S27–S30.

245. Werkman, S.H.; Carlson, S.E. A randomized trial of visual attention of preterm infants fed docosahexaenoic acid until nine months. Lipids 1996, 31, 91–97.

246. Carlson, S.E.; Werkman, S.H.; Tolley, E.A. Effect of long-chain n-3 fatt acid supplementation on visual acuity and growth of preterm infants with and without bronchopulmonary dysplasia. Am. J. Clin. Nutr. 1996, 63, 687–697.

247. Root, A.W. Mechanisms of hormone action: General Principals. In Clinical Pediatric Endocrinology; Hung, W., Ed.; Mosby-Year Book: St. Louis, MO, USA, 1992; pp. 1–12.

248. Watkins, B.A.; Shen, C.-L.; Allen, K.G.D.; Siefert, M.F. Dietary (n-3) and (n-6) polyunsaturates and acetylsalicylic acid alter ex vivo PGE2 biosynthesis, tissue IGF-I levels, and bone morphometry in chicks. J. Bone Miner. Res. 1996, 11, 1321–1332.

249. Colombo, J.; Carlson, S.E.; Cheatham, C.L.; Fitzgerald-Gust Afson, K.M.; Kepler, A.; Doty, T. Long-chain polyunsaturated fatty acid

supplementation in infancy reduces heart rate and positively affects distribution of attention. Pediatr. Res. 2011, 70, 406–410.

250. Colombo, J.; Carlson, S.E.; Cheatham, C.L.; Shaddy, D.J.; Kerling, E.H.; Thodosoff, J.M.; Gustafson, K.M.; Brez, C. Long-term effects of LCPUFA supplementation on childhood cognitive outcomes. Am. J. Clin. Nutr. 2013, 98, 403–412.

251. Alshweki, A.; Munuzuri, A.P.; Bana, A.M.; de Castro, J.; Andrade, F.; Aldamiz-Echevarria, L.; de Pipaon, M.S.; Fraga, J.M.; Couce, M.L. Effects of different arachidonic acid supplementation on psychomotor development in very preterm infants; a randomized trial. Nutr. J. 2015, 14, 101.

252. Beyerlein, A.; Hadders-Algra, M.; Kennedy, K.; Fewtrell, M.; Singhal, A.; Rosenfeld, E.; Lucas, A.; Bouwstra, H.; Koletzko, B.; von Kries, R. Infant formula supplementation with long-chain polyunsaturated fatty acids has no effect on Bayley developmental scores at 18 months of age-IPD meta-analysis of 4 large clinical trials. J. Pediatr. Gastroenterol. Nutr. 2010, 50, 79–84.

253. Innis, S.M.; Auestad, N.; Siegman, J.S. Blood lipid docosahexaenoic and arachidonic acid in term gestation infants fed formula with high docosahexaenoic, low eicosapentaenoic acid fish oil. Lipids1996, 31, 617–625.

254. Desci, T.; Keleman, B.; Minda, H.; Burus, I.; Kohn, G. Effect of type of early infant feeding on fatty acid composition of plasma lipid classes in full-term infants during the second 6 months of life. J. Pediatr. Gastroenterol. Nutr. 2000, 30, 547–551.

255. Hoffman, D.R.; Boettcher, J.A.; Diersen-Schade, D.A. Toward optimizing vision and cognition in term infants by dietary docosahexaenoic and arachidonic acid supplementation: A review of randomized clinical trials. Prostaglandins Leukot. Essent. Fat. Acids 2009, 81, 151–158.

256. Ryan, A.S.; Entin, E.K.; Hoffman, J.P.; Kuratko, C.N.; Nelson, E.B. Role of fatty acids in the neurological development of infants. In Nutrition in Infancy, Volume 2, Nutrition and Health; Watson, R.R., Ed.; Springer Science + Business Media: New York, NY, USA, 2013; pp. 331–346.

257. Drover, J.R.; Hoffman, D.R.; Casteneda, Y.S.; Morale, S.E.; Garfield, S.; Wheaton, D.H.; Birch, E.E. Cognitive function in 18-month-old term infants of the DIAMOND study: A randomized, controlled trial with multiple dietary levels of docosahexaenoic acid. Early Hum. Dev. 2011, 87, 223–230.

258. Makrides, M.; Gibson, R.A.; McPhee, A.J.; Collins, C.T.; Davis, P.G.; Doyle, L.W.; Simmer, K.; Colditz, P.B.; Morris, S.; Smithers, L.G.; et al.

Neurodevelopment outcomes of preterm infants fed high-dose docosahexaenoic acid: A randomized controlled trial. JAMA 2009, 301, 175–182.

259. Makrides, M.; Gibson, R.A.; McPhee, A.J.; Yelland, L.; Quinlivan, J.; Ryan, P. Effect of DHA supplementation during pregnancy on maternal depression and neurodevelopment of young children: A randomized controlled trial. JAMA 2010, 304, 1675–1683.

260. Lucas, A.; Stafford, M.; Morley, R.; Abbott, R.; Stephenson, T.; MacFadyen, U.; Elias-Jones, A.; Clements, H. Efficacy and safety of long-chain polyunsaturated fatty acid supplementation of infant formula milk: A randomized trial. Lancet 1999, 354, 1948–1954.

261. Fewtrell, M.S.; Morley, R.; Abbott, R.A.; Singhal, A.; Isaacs, E.B.; Stephenson, T.; MacFadyen, U.; Lucas, A. Double-blind, randomized trial of long-chain polyunsaturated fatty acid supplementation in formula fed to preterm infants. Pediatrics 2002, 110, 73–82.

262. Henriksen, C.; Haugholt, K.; Lindgren, M.; Aurvåg, A.K.; Rønnestad, A.; Grønn, M.; Solberg, R.; Moen, A.; Nakstad, B.; Berge, R.K.; et al. Improved cognitive development among preterm infants attributable to early supplementation of human milk with docosahexaenoic acid and arachidonic acid.Pediatrics 2008, 121, 1137–1145.

263. Westerberg, A.C.; Schei, R.; Henriksen, C.; Smith, L.; Veierød, M.B.; Drevon, C.A.; Iversen, P.O. Attention among very low birth weight infants following early supplementation with docosahexaenoic acid and arachidonic acid. Acta Pediatr. 2011, 100, 47–52.

264. Clandinin, M.T.; Van Aerde, J.E.; Merkel, K.L.; Harris, C.L.; Springer, M.A.; Hansen, J.W.; Diersen-Schade, D.A. Growth and development of preterm infants fed infant formulas containing docosahexaenoic acid and arachidonic acid. J. Pediatr. 2005, 146, 461–468.

265. Birch, E.E.; Garfield, S.; Hoffman, D.R.; Uauy, R.; Birch, D.G. A randomized controlled trial of early dietary supply of long-chain polyunsaturated fatty acids and mental development in term infants.Dev. Med. Child Neurol. 2000, 42, 174–181.

266. Carlson, S.E.; Montalto, M.B.; Ponder, D.L.; Werkman, S.H.; Korones, S.B. Lower incidence of necrotizing enterocolitis in infants fed a preterm formula with egg phospholipids. Pediatr. Res. 1998,44, 491–498.

267. Seki, H.; Sasaki, T.; Ueda, T.; Arita, M. Resolvins as regulators of the immune system. Sci. World J.2010, 10, 18–31.

268. Das, U.N. Perinatal supplementation of long-chain polyunsaturated fatty acids, immune response and adult diseases. Med. Sci. Monit. 2004, 10, HY19–HY25.

269. D'Vas, N.; Meldrum, S.J.; Dunstan, J.A.; Lee-Pullen, T.F.; Metcalfe, J.; Holt, B.J.; Serralha, M.; Tulic, M.K.; Mori, T.A.; Prescott, S.L. Fish oil supplementation in early infancy modulates developing infant immune responses. Clin. Exp. Allergy 2012, 42, 1206–1216.

270. Khader, S.A.; Gaffen, S.L.; Kolls, J.K. Th17 cells at the crossroads of innate and adaptive immunity against infectious diseases at the mucosa. Mucosal Immunol. 2009, 2, 403–411.

271. Muc, M.; Kreiner-Moller, E.; Larsen, J.M.; Birch, S.; Brix, S.; Bisgaard, H.; Lauritzen, L. Maternal fatty acid desaturase genotype correlates with infant immune responses at 6 months. Br. J. Nutr. 2015, 114, 891–898.

272. Barakat, R.; Abou El-Ela, N.E.; Sharaf, S.; El Sagheer, O.; Selim, S.; Tallima, H.; Bruins, M.J.; Hadley, K.B.; El Ridi, R. Efficacy and safety of arachidonic acid for treatment of school-aged children inSchistosoma mansoni high-endemicity regions. Am. J. Trop. Med. Hyg. 2015, 92, 797–804.

273. Standl, M.; Lattka, E.; Stach, B.; Koletzko, S.; Bauer, C.P.; von Berg, A.; Berdel, D.; Kramer, U.; Schaaf, B.; Roder, S.; et al. FADS1 FADS2 gene cluster, PUFA intake and blood lipids in children: Results from the GINIplus and LISAplus Study Group. PLoS ONE 2012, 7, e37780.

274. Makajima, H.; Hirose, K. Role of IL-23 and Th17 cells in airway inflammation in asthma. Immune Netw.2010, 10, 1–4.

275. Dong, C. Regulation and pro-inflammatory function of interleukin-17 family cytokines. Immunol. Rev.2008, 226, 80–86.

276. Sapone, A.; Lammers, K.M.; Casolaro, V.; CAmmarota, M.; Giulano, M.T.; De Rosa, M.; Stefanile, R.; Mazzarella, G.; Tolone, C.; Russo, M.I.; et al. Divergence of gut permeability and mucosal immune gene expression in two gluten-associated conditions: Celiac disease and gluten sensitivity. BMC Med.2011, 9, 23.

277. Pastor, N.; Soler, B.; Mitmesser, S.H.; Ferguson, P.; Lifschitz, C. Infants fed docosahexaenoic acid- and arachidonic acid-supplemented formula have decreased incidence of bronchiolitis/bronchitis the first year of life. Clin. Pediatr. 2006, 45, 850–855.

278. Lapillone, A.; Pastor, N.; Zhuang, W.; Scalabrin, D.M.F. Infants fed formula with added long chain polyunsaturated fatty acids have reduced incidence of respiratory illnesses and diarrhea during the first year of life. BMC Pediatr. 2014, 14, 168.

279. Birch, E.E.; Carlson, S.E.; Hoffman, D.R.; Fitzgerald-Gustafson, K.M.; Fu, V.L.; Drover, J.R.; Castañeda, Y.S.; Minns, L.; Wheaton, D.K.; Mundy, D.; et al. The DIAMOND (DHA Intake and Measurement of Neural Development) Study: A double-masked, randomized controlled clinical trial of the maturation of infant visual acuity as a function of the dietary level of docosahexaenoic acid. Am. J. Clin. Nutr.2010, 91, 848–859.

280. Foiles, A.M.; Kerling, E.H.; Wick, J.A.; Scalabrin, D.M.; Colombo, J.; Carlson, S.E. Formula with long chain polyunsaturated fatty acids reduces incidence of allergy in early childhood. Pediatr. Allergy Immunol. 2015.

281. Hindenes, J.O.; Nerdal, W.; Guo, W.; Di, L.; Small, D.M.; Holmsen, H. Physical properties of the transmembrane signal molecule, sn-1-stearoyl-2-arachidonylglycerol. Acyl chain segregation and its biochemical implications. J. Biol. Chem. 2000, 275, 6857–6867.

282. Moodley, T.; Vella, C.; Djahanbakhch, O.; Branford-White, C.J.; Crawford, M.A. Arachidonic and docosahexaenoic acid deficits in preterm neonatal mononuclear cell membranes. Implications for the immune response at birth. Nutr. Health 2009, 20, 167–185.

283. Van Goor, S.A.; Schaafsma, A.; Erwich, J.J.; Dijck-Brouwer, D.A.; Muskiet, F.A. Mildly abnormal general movement quality in infants is associated with high Mead acid and lower arachidonic acid shows a U-shaped relation with the DHA/AA ratio. Prostaglandins Leukot. Essent. Fat. Acids 2010, 82, 15–20.

284. Groen, S.E.; de Blecourt, A.C.; Postema, K.; Hadders-Algra, M. General movements in early infancy predict neuromotor development at 9 to 12 years of age. Dev. Med. Child Neurol. 2005, 47, 731–738.

285. Field, C.J.; Thompson, C.A.; Van Aerde, J.E.; Parrott, A.; Euler, A.; Lien, E.; Clandinin, M.T. Lower proportion of CD45R0+ cells and deficient interleukin-10 production by formula-fed infants, compared with human-fed, is correlated with supplementation of long-chain polyunsaturated fatty acids. J. Pediatr. Gastroenterol. Nutr. 2000, 31, 291–299.

286. Richardson, A.J.; Calvin, C.M.; Clisby, C.; Schoenheimer, D.R.; Montgomery, P.; Hall, J.A. Fatty acid deficiency signs predict the severity of reading and related difficulties in dyslexic children.Prostaglandins Leukot. Essent. Fat. Acids 2000, 63, 69–74.

287. Burgess, J.R.; Stevens, L.; Zhang, W.; Peck, L. Long-chain polyunsaturated fatty acids in children with attention-deficit hyperactivity disorder. Am. J. Clin. Nutr. 2000, 71, 327S–330S.

288. Chen, J.R.; Hsu, S.F.; Hsu, C.D.; Hwang, L.H.; Yang, S.C. Dietary patterns and blood fatty acid composition in children with attention-deficit hyperactivity disorder in Taiwan. J. Nutr. Biochem. 2004,15, 467–472.

289. Young, G.S.; Maharaj, N.J.; Conquer, J.A. Blood phospholipid fatty acid analysis of adults with and without attention deficit/hyperactivity disorder. Lipids 2004, 39, 117–123.

290. Morse, N.L. A meta-analysis of blood fatty acids in people with learning disorders with particular interest in arachidonic acid. Prostaglandins Leukot. Essent. Fat. Acids 2009, 81, 373–389.

291. Yamada, J.; Banks, A. Evidence for and characteristics of dyslexia among Japanese children. In Annals of Dyslexia; Schatschneider, C., Compton, D., Eds.; Springer: New York, NY, USA, 1994; pp. 103–119.

292. FAO. Dietary Fats and Oils in Human Nutrition; FAO/WHO: Rome, Italy, 1978.

293. FAO/WHO. Fats and Oils in Human Nutrition; Report of a Joint FAO/WHO Expert Consultation, 19 to 26 October 1993; FAO/WHO: Rome, Italy, 1994.

294. FAO. Fats and Fatty Acids in Human Nutrition; FAO/WHO: Rome, Italy, 2010.

295. Codex Alimentarius Commission. Standards for Infant Formula and Formulas for Special Medical Purposes Intended for Infants; CODEX STAN 72-1981, Last Revised 2007; Codex Alimentarius Commission: Rome, Italy, 2007.

296. Codex Alimentarius Commission. Amendments, 2015. Available online: http://www.fao.org/fao-who-codexalimentarius/en (accessed on 26 January 2016).

297. Infant Formula Act. H.R.6940—An Act to Amend the Federal Food, Drug, and Cosmetic Act to Strengthen the Authority under that Act to Assure the Safety and Nutrition of Infant Formulas, and for Other Purposes. Fed. Regist. 1980, 50, 45106–45108.

298. European Commission. Commission Directive 2006/141/EC of 22 December 2006 on Infant Formulae and Follow-on Formulae and Amending Directive 1999/21/EC; L.401/1; Official Journal of the European Union: Brussels, Belgium, 2008.

299. Koletzko, B.; Boey, C.C.; Campoy, C.; Carlson, S.E.; Chang, N.; Guillermo-Tuazon, M.A.; Joshi, S.; Prell, C.; Quak, S.H.; Sjarif, D.R.; et al. Current information and Asian perspectives on long-chain polyunsaturated fatty acids in pregnancy, lactation, and infancy:

Systematic review and practice recommendations from an early nutrition academy workshop. Ann. Nutr. Metab. 2014, 65, 49–80.

300. Crawford, M.A.; Wang, Y.; Forsyth, S.; Brenna, J.T. The European Food Safety Authority recommendation for polyunsaturated fatty acid composition of infant formula overrules breast milk, puts infants at risk, and should be revised. Prostaglandins Leukot. Essent. Fat. Acids 2015, 102–103, 1–3.

301. Leaf, A.A.; Leighfield, M.J.; Costeloe, K.L.; Crawford, M.A. Long chain polyunsaturated fatty acids and fetal growth. Early Hum. Dev. 1992, 30, 183–191.

302. Forsyth, J.S.; Willatts, P.; Agostoni, C.; Bissenden, J.; Casaer, P.; Boehm, G. Long chain polyunsaturated fatty acid supplementation in infant formula and blood pressure in later childhood: Follow up of a randomized controlled trial. Br. Med. J. 2003, 326, 953.

303. Makrides, M.; Gibson, R.A.; Udell, T.; Ried, K. Supplementation of infant formula with long-chain polyunsaturated fatty acids does not influence the growth of term infants. Am. J. Clin. Nutr. 2005, 81, 1094–1101.

6

THE RELATIONSHIP OF DOCOSAHEXAENOIC ACID (DHA) WITH LEARNING AND BEHAVIOR IN HEALTHY CHILDREN: A REVIEW

Connye N. Kuratko [1,*], Erin Cernkovich Barrett [1], Edward B. Nelson [1] and Salem Norman Jr. [1]

DSM Nutritional Products, 6480 Dobbin Road Columbia, MD 21045, USA

ABSTRACT

Childhood is a period of brain growth and maturation. The long chain omega-3 fatty acid, docosahexaenoic acid (DHA), is a major lipid in the brain recognized as essential for normal brain function. In animals, low brain DHA results in impaired learning and behavior. In infants, DHA is important for optimal visual and cognitive development. The usual intake of DHA among toddlers and children is low and some studies show improvements in cognition and behavior as the result of supplementation with polyunsaturated fatty acids including DHA. The purpose of this review was to identify and evaluate current knowledge regarding the relationship of DHA with measures of learning and behavior in healthy school-age children. A systematic search of the literature identified 15 relevant publications for review. The search found studies which were diverse in purpose and design and without consistent conclusions regarding the treatment effect of DHA intake or biomarker status on specific cognitive tests. However, studies of brain activity reported benefits of DHA supplementation and over half of the studies reported a favorable role for DHA or long chain omega-3 fatty acids in at least one area of cognition or behavior. Studies also suggested an important role for DHA in school performance.

Keywords: docosahexaenoic acid; children; learning; behavior; school performance

1. INTRODUCTION

The period from birth to 2 years of age is considered the primary growth phase for the human brain when measured in terms of brain weight. However, certain areas of the brain are not fully developed by the age of two, and development as well as growth continues throughout childhood and adolescence [1]. Myelination of brain frontal lobes begins as early as 6 months of age and continues throughout childhood and adolescence with spurts of development identified at 2 years of age, 7–9 years of age, and during mid-adolescence [1]. Tissue content of the long chain, omega-3 fatty acid (*n*-3 LC-PUFA) docosahexaenoic acid (DHA, 22:6*n*-3) is important for this development. DHA-rich frontal lobes are thought to be responsible for executive and higher-order cognitive activities such as planning, problem solving, and focused attention [2]. Investigators report an association of these prefrontal cortex structures with the limbic system, where the development of high-level cognitive function also correspond to a child's social, emotional and behavioral development [3,4].

Many components of the diet are known to affect cognition and influence learning. DHA in particular is recognized as essential for normal brain function. DHA is the principal omega-3 fatty acid in brain gray matter representing about 15% of all fatty acids in the human frontal cortex [5] and is known to affect neurotransmitter pathways, synaptic transmission, and signal transduction. Its multiple double bonds and unique structure allow DHA to impart a disorder to membranes which allows for effective cell signaling [6,7]. Studies in animals and humans show that adequate levels of DHA in neural membranes are important for cortical astrocyte maturation and vascular coupling, and for cortical glucose uptake and metabolism [8,9,10,11,12,13]. In addition, certain metabolites of DHA are bioactive molecules which protect tissues from oxidative injury and stress [14,15,16,17]. In animals, low brain DHA results in changes in behavior and is associated with learning problems and memory deficits [18]. In humans, studies at various life stages indicate that DHA supports normal IQ [19] and preserves visuo-spatial learning and memory [20]. For the brain and retina, therefore, it is clear that adequate DHA composition allows for optimal function [12]. Low blood levels of *n*-3 LC-PUFA, as well as high *n*-6 to *n*-3 ratios are reported in children with certain developmental and behavioral disorders such as attention deficit hyperactivity disorder (ADHD), dyslexia, or dyspraxia. Although inconsistent, various fatty acid supplementation strategies, some of which included DHA, proved to benefit measures of learning and behavior in these children. Some investigators even report clinical signs of fatty acid deficiency in these children which respond to *n*-3 LC-PUFA supplementation [21,22]. Reasons for this low omega-3 status are not fully understood, but could include disorders of fatty acid metabolizing enzymes and pathways which are unique to the disease. However, the explanation could also include a childhood history of inadequate *n*-3 fatty acid intake [23,24,25]. This brings to question whether poor *n*-3 LC-PUFA status in otherwise healthy children might also impact learning, memory, and behavior. If so, could improvement in *n*-3 LC-PUFA status, and DHA status in particular, improve learning and performance in healthy children? The current review was conducted to assess current knowledge regarding the relationship of DHA with learning (including such measures as reading and spelling), memory, and behavior in healthy children.

2. EXPERIMENTAL SECTION

All observational studies evaluating DHA intake or status as well as randomized controlled trials (RCT) evaluating DHA supplementation (with or without the addition of other fatty acids and nutrients) in healthy children, age 4 to 18 years, were considered for review. Eligible outcomes included all measures of cognitive function, learning, and behavior, with special interest in school performance or assessments such as those for reading, spelling, or listening comprehension. For inclusion, the publication had to quantify DHA in the diet or supplement, or include DHA as a biomarker measured in tissue or blood. Analyses had to report neurocognitive outcomes in relation to a concurrent indicator of DHA intake or status. Studies were included if subjects were generally described as healthy, without diagnosis of a neuropsychological condition, and participating in the available mainstream school setting. Studies in children with known deficiencies in nutrients other than DHA were included if no psychological or learning disabilities were apparent. Studies in children diagnosed with a learning disability or with special educational needs were excluded. Many of the authors of the studies included in this review performed multiple analyses due to an interest in several outcomes and various subgroups. While it is true that the more analyses that are done, the more likely that some of them will be statistically significant by chance, the correction for multiple comparisons applied by the study authors was not consistent. In order to provide a comprehensive review of the literature and avoid reinterpretation of the reported outcomes, no studies were excluded due to method of statistical assessment.

Studies which included α-linolenic acid (ALA, 18:3 *n*-3) or sources of ALA as the only treatment intervention were excluded. However, studies which used micronutrient or fatty acid mixtures were included if the formulation included DHA. Whenever multiple publications reported unique analyses pertaining to the same subjects, all publications were retained. The database of PubMed was searched in November, 2012 using a predefined algorithm. References from pertinent systematic reviews were screened for potential articles that might have been missed in the search.

Search results were sorted by an investigator at the title/abstract level, eliminating inappropriate articles. Two investigators independently reviewed the full papers of the remaining studies to determine the final accepted list. The search criteria were built on the following PubMed Mesh terms:

((“Fatty Acids, Omega-3” [Mesh] OR “Fish Oils” [Mesh]) OR “Docosahexaenoic Acids” [Mesh]) OR “Eicosapentaenoic Acid” [Mesh]

And

((((("Learning” [Mesh] OR “Memory” [Mesh]) OR “Reading” [Mesh]) OR “Students” [Mesh]) OR “Schools” [Mesh]) OR “Education” [Mesh]) OR “Educational Status” [Mesh]

And

Filters activated: English, Preschool Child: 2–5 years, Child: 6–12 years, Adolescent: 13–18 years.

Search terms to select the populations of interest included: All infant birth-23 months; Preschool Child 2–5 years, Child 6–12 years.

Exclusion criteria included subject populations of children diagnosed with or a history of ADHD, dyslexia, autism, dyspraxia, or learning disabilities requiring special education, or medication use for any of these conditions. Other exclusions included fish-only interventions without calculation or confirmation of the fatty acids provided; analyses reporting only fish-related outcomes, with no report of fatty acid contribution; reviews or mechanistic papers; outcomes from *in vitro* or animal models; and follow-up studies of prenatal or infant interventions failing to report intake or biomarker data corresponding with outcome assessments in the older child.

3. RESULTS

3.1. Search Results

Using the above criteria, the PubMed search resulted in 38 citations. Level 1 Screening, based on the abstract excluded 24 articles. Full papers were retrieved for the 14 remaining studies, two of which were excluded. Three additional studies were located from reviewing the references in the retrieved papers. Fifteen articles conducted in healthy children ages 4–14 years met the full criteria for inclusion in this review. Table 1, Table 2 include a list of the final accepted studies with reported outcomes. Table 3 is a summary of DHA biomarker data reported by study.

Table 1. Observational studies reporting docosahexaenoic acid (DHA)-related outcomes of cognition and behavior in children.

Reference	Study Design	Participants	Outcome Measures	Results	
				Cognition	Behavior
Boucher *et al.*, 2011 [27]	Prospective	*n* = 154 School-aged children (10–13 years) with available cord blood samples	**Cognition** Digit span forward subtest of the Wechsler Intelligence Scales for Children (WISC-IV) The California Verbal Learning Test Children's Version (CVLT) Continuous Recognition Task (CRT) Event-related potentials (ERPs) were acquired via electroencephalogram recordings while the CRT was performed **Biological Samples** Fatty acid composition of umbilical cord blood Fatty acid composition of child blood	Neuropsychological Assessments Cord blood DHA levels were positively associated with performance on the digit span forward test Cord blood DHA levels were positively associated with performance on the CVLT Electroencephalogram Recordings Higher cord DHA levels were associated with a shorter latency of the FN400 component and a larger amplitude of the LPC Higher current DHA levels were associated with a greater FN400 amplitude	
Kirby *et al.*, 2010 [28]	Cross-Sectional	*n* = 411 School-aged children (8–10 years)	**Cognition** Kaufman Brief Intelligence Test (KBIT-2) Working Memory Test Battery for Children (WMTB-C) Wechsler Individual Achievement Test (WIAT-II) Test of Everyday Attention for Children (TEA-Ch) Matching Familiar Figures Task (MFFT) **Behavior** Swanson, Nolan, and Pelham (SNAP-IV; parents/guardians and teachers) Strengths and Difficulties Questionnaire (SDQ: parents/guardians and teachers) Developmental Coordination Disorder Questionnaire (DCDQ: parents/guardians) **Biological Samples** Cheek cell samples for measurement of fatty acid composition	Cheek cell DHA levels were positively associated with non-verbal IQ (KBIT-2)	Cheek cell DHA levels were negatively associated with teacher rated hyperactivity and total difficulties (SDQ)

Table 1. *Cont.*

Bakker *et al.*, 2003 [26]	Prospective	*n* = 306 School-aged children (7 years of age) with available cord blood samples	**Cognition** Kaufman Assessment Battery for Children (K-ABC) **Biological Samples** Fatty acid composition of umbilical cord blood Fatty acid composition of child blood at age 7	There was no association between cord blood DHA levels and cognitive performance (K-ABC) There was no association between plasma DHA at age 7 and cognitive performance (K-ABC)

Abbreviations: California Verbal Learning Test (CVLT). Continuous Recognition Task (CRT). Developmental Coordination Disorder Questionnaire (DCDQ). Event-Related Potentials (ERP). Kaufman Assessment Battery for Children (K-ABC). Kaufman Brief Intelligence Test (KBIT). Matching Familiar Figures Task (MFFT). Test of Everyday Attention for Children (TEA-Ch). Strengths Difficulties Questionnaire (SDQ). Swanson, Nolan, and Pelham (SNAP). Wechsler Individual Achievement Test (WIAT). Wechsler Intelligence Scales for Children (WISC). Working Memory Test Battery for Children (WMTB-C)

Table 2. Clinical trials of DHA supplementation on outcomes of cognition and behavior in children.

Reference	Study Design	Participants	Intervention	Time (week)	Outcome Measures	Results: Cognition	Results: Behavior
Baumgartner *et al.*, 2012 [29]	RCT	**Randomized,** *n* = 321 School-aged children with iron deficiency (6–11 years) Completed cognitive follow up, *n* = 288 Placebo, *n* = 73 Iron, *n* = 70 DHA/EPA, *n* = 72 Iron + DHA/EPA, *n* = 73	1. Placebo 2. Iron: 50 mg/day 3. DHA/EPA: 420 mg/80 mg/day 4. Iron + DHA/EPA: 420 mg/80 mg/day NOTE: Supplements were provided 4 days/week on school days only. Overall 4.8 g iron (45.7 mg/day), 41.2 g DHA (392 mg/day), and 7.8 g EPA (74 mg/day) were provided	38	**Cognition** Kaufman Assessment Battery (K-ABC-II) Hopkins Verbal Learning Test (HVLT)	**K-ABC-II** In all groups, tests of learning abilities (Atlantis, Atlantis Delayed) and test of simultaneous processing (Triangles) improved significantly (time-effect) Scores for sequential processing (Hand movement) decreased in all but the placebo plus DHA/EPA group Among all children, there were no significant intervention effects There was a significant effect of DHA/EPA for lower Atlantis test scores in children with anemia There was a significant negative effect of DHA/EPA supplementation on Atlantis Delayed test performance in girls. **HVLT** Among all children, DHA/EPA supplementation had no effect on HVLT scores Girls supplemented with DHA/EPA alone recalled significantly more words than girls given the combined treatments or placebo	
Dalton *et al.*, 2009 [30]	RCT	**Randomized,** *n* = 183 Low-income, marginally nourished school aged-children (7–9 years) **Per-Protocol,** *n* = 155 Control bread spread, *n* = 78 Fish-flour bread spread, *n* = 77	1. Control spread 2. Fish flour spread: ~892 mg DHA/week NOTE: The spread was provided on school days only and not on weekends or during school holidays	24	**Cognition** Hopkins Verbal Learning Test (HVLT) Reading Spelling	**Verbal Learning** Fish flour spread improved verbal learning ability including identification of true positive ($p = 0.0191$), and false negatives ($p = 0.0075$) **Spelling** Fish flour spread protected against declines in spelling	
Hamazaki *et al.*, 2008 [31]	RCT	**Randomized,** *n* = 233 Healthy, school-aged children (9–14 years) **Per-Protocol,** *n* = 189 Placebo tablets, *n* = 96 Fish oil, *n* = 93	1. Placebo capsules 2. Fish oil: 650 mg DHA/day + 100 mg EPA/day	12	**Behavior** Hostility-Aggression Questionnaire for Children (HAQ-C) Barratt Impulsiveness Scale (BIS-11) School attendance		**Aggression/ Impulsiveness** Fish oil had no effect on aggression or impulsiveness **School Attendance** Fish oil improved school attendance rate ($p = 0.003$)

Table 2. *Cont.*

Dalton *et al.*, 2009 [30]	RCT	**Randomized**, $n = 183$ Low-income, marginally nourished school aged-children (7–9 years) **Per-Protocol**, $n = 155$ Control bread spread, $n = 78$ Fish-flour bread spread, $n = 77$	1. Control spread 2. Fish flour spread: ~892 mg DHA/week NOTE: The spread was provided on school days only and not on weekends or during school holidays	24	**Cognition** Hopkins Verbal Learning Test (HVLT) Reading Spelling	**Verbal Learning** Fish flour spread improved verbal learning ability including identification of true positive ($p = 0.0191$), and false negatives ($p = 0.0075$) **Spelling** Fish flour spread protected against declines in spelling	
Hamazaki *et al.*, 2008 [31]	RCT	**Randomized**, $n = 233$ Healthy, school-aged children (9–14 years) **Per-Protocol**, $n = 189$ Placebo tablets, $n = 96$ Fish oil, $n = 93$	1. Placebo capsules 2. Fish oil: 650 mg DHA/day + 100 mg EPA/day	12	**Behavior** Hostility-Aggression Questionnaire for Children (HAQ-C) Barratt Impulsiveness Scale (BIS-11) School attendance		**Aggression/ Impulsiveness** Fish oil had no effect on aggression or impulsiveness **School Attendance** Fish oil improved school attendance rate ($p = 0.003$)
Itomura *et al.*, 2005 [32]	RCT	**Randomized, $n = 179$** Healthy, school-aged children (9–12 years) **Per-Protocol, $n = 166$** Control foods, $n = 83$ Fish oil fortified foods, $n = 83$	1. Control foods 2. Fortified foods: ~514 mg DHA/day ~120 mg EPA/day	12	**Behavior** Hostility-Aggression Questionnaire for Children (HAQ-C) Picture Frustration (PF) Study Diagnostic questionnaires for ADHD		**Aggression** Among females, fish oil protected against increases in aggression as assessed by the HAQ-C ($p = 0.008$) Among males, aggression against others as assessed by the PF study, increased in the fish oil group but not in the control group **Impulsivity (DSM-IV)** Among females, fish-oil reduced impulsivity ($p = 0.008$)
Kennedy *et al.*, 2009 [33]	RCT	**Randomized** $n = 90$ Healthy, school-aged children (10–12 years) **Per-Protocol** $n = 88$ Placebo tablets, $n = 30$ Low-dose algal DHA, $n = 28$ High-dose algal DHA, $n = 30$	Placebo capsules Algal DHA: 400 mg/day Algal DHA: 1000 mg/day	8	**Cognition** Cognitive Drug Research (CDR) battery Internet Battery	**CDR** Except for the word recognition task, DHA had no effect on cognition For the word recognition task, children given low-dose DHA performed significantly faster ($p < 0.05$) while children given high-dose DHA performed significantly slower ($p < 0.05$)	**Internet battery** Except for the visual analog ratings of "relaxed", DHA had no effect on cognition For ratings of "relaxed", children given low and high dose DHA rated themselves as "more relaxed"
Kirby *et al.*, 2010 [34]	RCT	**Randomized**, $n = 450$ Healthy, school-aged children (8–10 years) **Per-Protocol**, $n = 348$ Placebo tablets, $n = 177$ Fish oil + micronutrients, $n = 171$ NOTE: After 16 weeks of supplementation, all children received the active supplement for an additional 8 weeks (one-way cross-over)	Placebo capsules Fish oil: 400 mg DHA/day 56 mg EPA/day 800 vitamin A 60 g vitamin C 5.0 µg vitamin D 3.0 mg vitamin E	16	**Cognition** Kaufman Brief Intelligence Test (KBIT-II) Wechsler Individual Achievement Test (WIAT-II) Working Memory Test Battery for Children (WMTB-C) Test of Everyday Attention for Children (TEA-Ch) Matching Familiar Figures Task (MFFT) Computerized Penmanship Evaluation Tool (ComPET) **Behavior** Swanson, Nolan, and Pelham (SNAP-IV) Strengths and Difficulties Questionnaires (SDQ)	In the ITT analysis, fish oil had no effect on IQ, reading and spelling, working memory, attention, visual attention, impulsivity, or handwriting process Among compliant children (≥80% of dosing requirement), fish oil improved visual attention and impulsivity as assessed by the MFFT ($p < 0.001$)	**SDQ** Among all children, teacher rated difficulties score improved in the control group but not in the fish oil group ($p < 0.001$) Among compliant children (≥80% of dosing requirement), fish oil protected against declines in parent rated prosocial behavior ($p = 0.001$)

McNamara *et al.*, 2010 [12]	RCT	**Randomized** $n = 38$ Healthy, school-aged boys (8–10 years) **Per-Protocol** $n = 33$ Placebo tablets, $n = 10$ Low-dose algal DHA, $n = 10$ High-dose algal DHA, $n = 13$	Placebo capsules Algal DHA: 400 mg/day Algal DHA: 1200 mg/day	8	**Cognition** Functional magnetic resonance imaging (fMRI) Identical-Pairs version of the Continuous Performance Task (CPT-IP)	**Whole-brain activation patterns during performance of sustained attention tasks** Low-dose DHA improved brain activation (increased activation in the dorsolateral prefrontal cortex, decreased activation in the occipital cortex) High-dose DHA improved brain activation (increased activation in the dorsolateral prefrontal cortex, decreased activation in the cerebellar cortex) Erythrocyte DHA positively correlated with dorsolateral prefrontal cortex activation **Sustained attention** Low- and high-dose DHA had no effect on sustained attention performance Erythrocyte DHA inversely correlated with reaction time on the sustained attention test ($p = 0.02$)	
Muthayya *et al.*, 2009 [35]	RCT	**Randomized,** $n = 598$ Low income, marginally nourished school-aged children (6–10 years) **Per-Protocol,** $n = 550$ Low micronutrient/ low omega-3, $n = 140$ Low micronutrient/ high omega-3, $n = 139$ High micronutrient/ low omega-3, $n = 133$ High micronutrient/ high omega-3, $n = 138$	**Low micronutrient** Vitamin A: 75 µg RE/day Riboflavin: 0.14 mg/day Vitamin B6: 0.15 mg/day Vitamin B12: 0.27 µg/day Folate: 45 µg/day Vitamin C: 5.25 mg/day Calcium: 105 mg/day Iodine: 15 µg/day Iron: 2.7 mg/day Zinc: 1.7 mg/day **High micronutrient** Vitamin A: 500 µg RE/day Riboflavin: 0.9 mg/day Vitamin B6: 1 mg/day Vitamin B12: 1.8 µg/day Folate: 300 µg/day Vitamin C: 227.1 mg/day Calcium: 231 mg/day Iodine: 100 µg/day Iron: 18 mg/day Zinc: 10.5 mg/day **Low omega-3** Total omega-3: 0.14 g/day ALA: 0.14 g/day DHA: 0 g/day **High omega-3** Total omega-3: 1.03 g/day ALA: 0.93 g/day DHA: 0.10 g/ day	52	**Cognition** Kaufman Assessment Battery for Children (K-ABC) Wechsler Intelligence Scales for Children (WISC-R and WISC-4) Rey Auditory Verbal Learning Test (RAVLT) Neuropsychological Assessment tool (NEPSY) Number cancellation	All four groups improved in short-term memory, retrieval ability, fluid reasoning, cognitive speediness, and the Mental Processing Index compared with baseline There was no significant differences between the high and the low omega-3 groups on any cognitive measure	
Richardson *et al.*, 2012 [36]	RCT	**Randomized** $n = 362$ Healthy, school-aged children (7–9 years), who were underperforming in reading (<33rd% tile) **Per-Protocol** $n = 359$ Placebo tablets, $n = 180$ Algal DHA, $n = 179$	Placebo capsules Algal DHA: 600 mg/day	16	**Cognition** British Ability Scale (BAS-II) **Behavior** Conners' Parent Rating Scales (CPRS) Conners' Teacher Rating Scales (CTRS)	**Reading** For children with baseline reading scores ≤33rd% tile, DHA had no effect on reading For children with baseline reading scores ≤20th ($p = 0.041$) and ≤10th ($p = 0.011$) % tiles, DHA improved reading For children with baseline reading scores ≤10th% tile, DHA led to a 1.9 month gain in reading age ($p = 0.032$) **Working memory** DHA had no effect on working memory	For children with baseline reading scores ≤33rd% tile, DHA improved parent-rated behavior including oppositional ($p = 0.004$), hyperactivity ($p = 0.007$), ADHD index (0.042), Global Restless-Impulsive ($p = 0.001$), Global Emotional Liability ($p = 0.001$), Global Index Total ($p = 0.001$), DSM-IV Hyperactive Impulsive ($p = 0.021$), and DSM-IV Total ADHD ($p = 0.031$) scores For children with baseline reading scores ≤33rd% tile, DHA had no effect on teacher rated behavior

Ryan *et al.*, 2008 [37]	RCT	**Randomized** $n = 175$ Healthy 4 year olds **Per-Protocol** $n = 175$ Placebo tablets, $n = 90$ Algal DHA, $n = 85$	Placebo capsules Algal DHA: 400 mg/day	16	**Cognition** Leiter-R test of Sustained Attention Peabody Picture Vocabulary Test Day-Night Stroop Test Conners' Kiddie Continuous Performance Test	In the ITT analysis, DHA had no effect on cognition Among children who provided a blood sample, DHA in whole blood was positively associated with scores on the Peabody Picture Vocabulary Test ($p = 0.018$)	
Sinn *et al.* 2011 [38]	Open label	$n = 47$ School-aged children (3–14 years) in a remote, primarily Indigenous, Northern Territory school **Per-Protocol** $n = 37$ Fish-oil	Fish oil: 558 mg EPA 174 mg DHA 60 mg GLA 10.8 mg vitamin E NOTE: The capsules were consumed on school days with a 1 week break after 5 weeks	12	**Cognition** Wide Range Achievement Test (WRAT) Raven's Colored Matrices **Behavior** Conners' Behavior Ratings Scales (CBRS)	Fish oil improved reading ($p = 0.01$) and spelling ($p = 0.01$) as assessed by the WRAT Fish oil improved non-verbal intelligence as assessed by the Raven's Colored Matrices ($p < 0.01$)	Behavior was not assessed at week 12 (teachers who completed the CBRS at baseline left the school)
The NEMO Study Group 2007 [39]	RCT	**Randomized,** $n = 396$ Well-nourished, school-aged children (6–10 years) from South Australia **Per-Protocol,** $n = 276$ Placebo drink, $n = 71$ Flavored drink + micronutrients, $n = 67$ Flavored drink + omega-3s, $n = 67$ Flavored drink + micronutrients and omega-3s, $n = 71$ **Randomized,** $n = 384$ Marginally nourished, school-aged children (6–10 years) from Indonesia **Per-Protocol,** $n = 367$ Placebo drink, n = 88 Flavored drink + micronutrients, $n = 91$ Flavored drink + omega-3s, $n = 94$ Flavored drink + micronutrients and omega-3s, $n = 94$	**Micronutrient mx** Iron: 10 mg Zinc: 5 mg Vitamin A: 400 μg Folate: 150 μg Vitamin B6: 1 mg Vitamin B12: 1.5 μg Vitamin C: 45 mg **Omega-3 mix** DHA: 8 mg EPA: 22 mg NOTE: The fruit flavored drink was consumed 6 days/week	52	**Cognition** *Both Countries* Wechsler Intelligence Scales for Children (WISC-III: digits backwards, coding, block design, vocabulary) Neuropsychological Assessment tool (NEPSY: visual attention, fluency structured and random) Rey Auditory Verbal Learning Test (RAVLT) Wechsler Individual Achievement Test (WIAT-II: mathematical reasoning) *Australia* WIAT-II (reading, spelling) *Indonesia* Neale Analysis of Reading ability	**Australia** Among well-nourished school-aged children, omega-3 fatty acids had no effect on any measure of cognitive function or school performance **Indonesia** Among marginally nourished school-aged children, omega-3 fatty acids had no effect on any measure of cognitive function or school performance	

Abbreviations: Barratt Impulsiveness Scale (BIS-II), British Ability Scale (BAS), Cognition Drug Research (CDR), Computerized Penmanship Evaluation Tool (ComPET), Conners' Behavior Ratings Scales (CBRS), Conners' Parent Rating Scales (CPRS), Conners' Teacher Rating Scales (CTRS), Functional Magnetic Resonance Imaging (fMRI), Hostility-Aggression Questionnaire for Children (HAQ-C), Hopkins Verbal Learning Test (HVLT), Identical-Pairs verson of the Continuous Performance Task (CPT-IP), Intention-to-treat (ITT), Kaufman Assessment Battery (K-ABC), Kaufman Brief Intelligence Test (KBIT), Matching Familiar Figures Task (MFFT), Neuropsychological Assessment tool (NEPSY), Picture Frustration (PF), Ray Auditory Verbal Learning Test (RAVLT), Strengths and Difficulties Questionnaire (SDQ), Swanson, Nolan, and Pelham (SNAP), Test of Everyday Attention for Children (TEA-Ch), Wechsler Individual Achievement Test (WIAT), Wechsler Intelligence Scales for Children (WISC), Wide Range Achievement Test (WRAT), Working Memory Test Battery for Children (WMTB-C).

Three studies were observational in design [26,27,28]. Bakker *et al.* [26] and Boucher *et al.* [27] provided a prospective analysis of cohorts identified in infancy and followed through age 7 and through 10–13 years, respectively. The cross-sectional study by Kirby *et al.* [28] reported correlations between cheek cell fatty acids, including DHA, with relevant cognitive and behavioral outcomes.

The remaining 12 studies were randomized, controlled trials of varied purpose and design. The sources of DHA administered as the intervention included triglycerides from fish oil and algal oil. The oils were given as supplements in capsules or chewable form, or were incorporated into fortified foods. Experimental designs often included additional nutrients, such as vitamins, minerals, or other fatty acids. In most of the studies, the placebo consisted of corn, soy, or olive oil, or a combination of those vegetable oils. Daily doses of DHA ranged from 88 to 1200 mg per day.

Table 3. DHA status reported in clinical trials of DHA supplementation in children.

Reference	Time Point	Placebo		Treatment		Biomarker
Dalton *et al.*, 2009 [30]	Baseline					% of total fatty acids (PC in RBCs)
	Endpoint	2.81 ± 0.70		3.71 ± 0.66		
	Baseline					% of total fatty acids (PE in RBCs)
	Endpoint	7.77 ± 2.16		8.98 ± 2.61		
	Baseline					% of total fatty acids (PC + PE in RBCs)
	Endpoint	10.58		12.69		
Hamazaki *et al.*, 2008 [31]	Baseline	4.4 ± 0.9		4.4 ± 1.1		% of total fatty acids (phospholipid fraction in RBCs)
	Endpoint	4.9 ± 1.2		7.8 ± 1.1		
Itomura *et al.*, 2005 [32]	Baseline	6.4 ± 0.9		6.1 ± 0.9		% of total fatty acids (phospholipid fraction in RBCs)
	Endpoint	6.6 ± 0.9		7.1 ± 1.8		
Kennedy *et al.*, 2009 [33]				NR		
Kirby *et al.*, 2010 [34]	Baseline	0.07 ± 0.007		0.08 ± 0.009		% of total fatty acid methyl esters (buccal cell sample)
	Endpoint	0.20 ± 0.16		0.37 ± 0.23		
Richardson *et al.*, 2012 [36]				NR		
Ryan *et al.*, 2008 [37]	Baseline	1.0 ± 0.34		1.0 ± 0.34		% of total fatty acids (capillary whole blood)
	Endpoint	1.1 ± 0.40		3.3 ± 1.54		
Sinn *et al.*, 2011 [38]				NR		
		Placebo	**Iron**	**DHA/EPA**	**Iron + DHA/EPA**	
Baumgartner *et al.*, 2012 [29]	Baseline	3.07 ± 0.69	3.07 ± 0.64	3.05 ± 0.63	3.01 ± 0.58	% of total fatty acids (phospholipid fraction in RBCs)
	Endpoint	3.79 ± 0.96	3.83 ± 1.06	5.94 ± 1.71	5.90 ± 2.11	
		Placebo		**Low Dose Treatment**	**High Dose Treatment**	
McNamara *et al.*, 2010 [12]	Baseline	3.3 ± 1.3		3.3 ± 1.3	3.3 ± 1.3	% of total fatty acids (RBC)
	Endpoint	2.5 ± 1.6		7.5 ± 1.3	10.3 ± 1.5	
		Low Micro/ Low Omega-3	Low Micro/ High Omega-3	High Micro/ Low Omega-3	High Micro/ High Omega-3	
Muthayya *et al.*, 2009 [35]	Baseline	3.3 ± 0.9	3.2 ± 0.9	3.2 ± 0.8	3.2 ± 0.7	% of total fatty acids (phospholipid fraction in RBCs)
	Endpoint	3.6 ± 1.0	5.2 ± 1.2	3.6 ± 0.8	5.2 ± 1.2	
		Placebo	**Micronutrients**	**DHA/EPA**	**Micronutrients + DHA/EPA**	
The NEMO Study Group 2007 [39]	Baseline	33.2 ± 11.0	35.2 ± 10.4	31.1 ± 10.6	32.7 ± 10.3	Plasma (μg/mL)
	Endpoint	38.4 ± 14.0	36.8 ± 8.5	47.4 ± 12.7	47.4 ± 10.7	

Abbreviations: Red blood cells (RBCs), Phosphatidylcholine (PC), Phosphatidylethanolamine (PE), Not reported (NR).

Five of the included studies, or study arms, were conducted in populations of children with multiple nutrient deficiencies [29,30,35,38,39]. It is commonly known that many nutrients other than DHA affect neurological development and cognition in children. Although the question has been raised, it is unknown to what extent cognitive function can be affected based on replenishment of *n*-3 LC-PUFA alone in the case of generalized malnutrition [40]. Further investigation of this question should be considered in future studies. However, since the studies met the pre-defined inclusion criteria for this review by enrolling children without a neuropsychological diagnosis, and analyzing data in a *n*-3 LC-PUFA specific manner, those studies conducted in populations with an increased incidence of malnutrition were included in this review.

The number of subjects varied greatly among the studies, ranging from 33 in the study by McNamara *et al.* [12] to 598 in the study by Muthayya *et al.* [35]. Study duration was also variable and ranged from 2

[33] to 12 months [35,39]. Many of the studies reported results as analyzed both in the "intention to treat" as well as the "per protocol" subject groups, and most reported results as a degree of change from baseline. Compliance was monitored and assessed in all studies and many authors acknowledged the difficulty associated with supplement protocols that often involved subjects, parents, caregivers, and teachers.

The accepted studies included a wide number of assessments of cognition and behavior. All measures were included in this review in order to represent the broadest possible assessment from the limited number of studies. Outcome measures included objective and subjective methodologies including: electrical potential readings and functional magnetic resonance imaging; standardized tests of cognition, including reading, spelling, listening comprehension, memory, attention, and IQ; as well as self- ,parent-, and teacher-assessed behavioral questionnaires. While outcomes were often reported in relation to total omega-3 fatty acids, ALA, or ratios of *n*-3 and *n*-6 fatty acids, the focus of this review was to determine the effect of DHA, specifically, on cognitive and behavioral outcomes.

Perhaps most interesting were the results of DHA supplementation on school performance as reported in seven of the studies [12,27,30,34,36,37,39]. Of the seven studies, five reported that DHA status or supplementation improved measures of school performance including learning ability, reading, and spelling as assessed by sub-tests of cognitive batteries [12,27,30,36,37] while two, in contrast, found no effect [34,39]. Two studies directly assessed brain activity and both reported benefit as the result of DHA intake or status [12,27]. Behavioral outcomes were measured in six studies [28,31,32,33,34,36], with five reporting at least one significant improvement related to DHA supplementation or status [28,32,33,34,36].

3.2. Review of Studies

Below are brief summaries of the included studies. Three of the studies were observational in design, seven were clinical studies that provided a lipid intervention, and five were clinical studies that provided other nutrients in addition to a lipid intervention.

3.2.1. Observational Studies

In 2003, Bakker *et al.* [26] reported results of a prospective, longitudinal study investigating the relationship of cognitive performance at 7 years of age with current fatty acid status and fatty acid status at birth. A total of 306 children with available cord blood samples were followed up at age 7 for assessments of cognitive function and achievement using the Kaufman Assessment Battery for Children (K-ABC). The investigators found no association between cord or plasma DHA and cognitive performance in the 7 year olds.

In 2011, Boucher *et al.* [27] reported results from a prospective, longitudinal study examining the effect of prenatal and childhood intake of DHA on memory function in school-aged children from Arctic Quebec. A total of 154 children aged 10–13 years, who had available umbilical cord blood samples, constituted the study population. Visual recognition memory was assessed by event-related potentials (ERPs) recordings during performance of a continuous recognition memory task. Memory was also assessed via a battery of neuropsychological tests including the "Digit span forward" subtest of the Wechsler Intelligence Scales for Children, 4th edition (WISC-IV) and The California Verbal Learning Test—Children's Version (CVLT). Current blood samples, along with cord blood samples, were analyzed for DHA. The pattern of ERP recordings showed that higher prenatal DHA status positively affected visual information processing and brain activity in later childhood; and, that current DHA intake was positively associated with brain activity during performance of a specific task requiring familiarity processing. Higher cord blood DHA was associated with better performance on both the 'Digit span forward' test and the CVLT suggesting better memory development. Neither cord-nor current-DHA blood levels were associated with any of the tasks of continuous recognition memory.

Kirby *et al.* [28] reported results of a cross-sectional study examining the relationship between cheek cell fatty acids and multiple assessments of learning, and behavior in 411 children aged 8–10 years. Assessments included: the Kaufman Brief Intelligence Test (KBIT-2) to assess verbal and non-verbal IQ; the Working Memory Test Battery for Children (WMTB-C) to assess working memory; and, the Wechsler Individual Achievement Test (WIAT-II) to assess reading and spelling ability. Test of Everyday Attention for Children (TEA-Ch) assessed attention capabilities; and, visual attention and impulsivity were assessed using the Matching Familiar Figures Task (MFFT). Aspects of behavior were assessed by parents and teachers using the Swanson, Nolan, and Pelham (SNAP-IV) and the Strengths and Difficulties Questionnaire (SDQ) scales; parents also completed the Developmental Coordination Disorder Questionnaire (DCDQ). Results revealed a positive association of cheek cell DHA with non-verbal IQ and a negative association with teacher rating of hyperactivity and total difficulties.

3.2.2. Clinical Studies (Lipid-Only Intervention)

Seven of the studies which met inclusionary criteria were randomized, clinical trials that included supplementation with either fish oil or algal oil as a source of DHA. Ryan and Nelson [37], reported results of a multi-center RCT designed to determine the effects of DHA supplementation on cognition in children. A total of 175 healthy, 4 year olds were supplemented with 400 mg DHA/day or placebo for 4 months. Cognitive function was measured using the Leiter-R test of Sustained Attention (Leiter-R SA), the Peabody Picture Vocabulary Test (PPVT), the Day-Night Stroop Test (DNST), and the Conners' Kiddie Continuous Performance Test (Conners' KCPT). Results showed no effect of DHA supplementation on any of the tests of cognitive function. There was a reported ceiling effect on the Leiter-R SA and the DNST in the children tested, (*i.e.*, a significant number of children had high level scores at baseline leaving little ability to show improvement on the tests) which likely influenced the results. Interestingly, a pre-planned analysis of 93 children showed a significant and positive association of whole-blood DHA and PPVT scores.

An RCT by McNamara *et al.* [12] was designed to determine the effects of DHA supplementation on functional cortical activity during sustained attention in healthy boys. A total of 33 healthy boys aged 8–10 years were supplemented with 400 or 1200 mg DHA/day, or a matching placebo for 8 weeks. The primary outcome included change from baseline in cortical activation patterns, as measured by functional magnetic resonance imaging (fMRI). fMRI scans were recorded at baseline and after 8 weeks of supplementation and whole-brain activation patterns were determined during a task of sustained attention. Results of the study confirmed that 8 weeks of supplementation with either low- or high-dose DHA significantly increased functional activation in the dorsolateral prefrontal cortex (DLPFC) during performance of the sustained-attention task (CPT-IP) as compared to placebo. The authors reported a ceiling effect for the CPT-IP task with an 80%–90% accuracy rate exhibited by all groups and, therefore, performance of the DHA-supplemented children on the CPT-IP did not differ from controls. However, the children's erythrocyte DHA was positively correlated with DLPFC activation and inversely correlated with reaction time during the attention task.

In an RCT by Kennedy *et al.* [33], 90 healthy children aged 10–12 years were supplemented with 400 or 1000 mg DHA/day, or a matching placebo for 8 weeks. Cognition and mood were assessed using two batteries of computer-based cognitive tests, the Cognitive Drug Research Battery (CDR) and the Internet Battery. The authors reported no consistent cognitive benefit as assessed by the CDR as a result of DHA supplementation. After 8 weeks, children supplemented with 400 mg DHA/day performed faster on word recognition tasks than children given placebo while those supplemented with 1000 mg DHA/day performed slower than children given placebo. In addition, performance on the Internet Battery did not differ by treatment group except for the visual analogue measure of "relaxed", where children supplemented with both 400 mg and 1000 mg DHA/day showed greater improvement from baseline than controls. The validity of this difference is questionable since the groups differed significantly on this measure at baseline.

The study by Dalton *et al.* [30] was a randomized, single-blind, placebo-controlled clinical study investigating the effect of a spread, rich in omega-3 fatty acids, on cognition in children. A total of 183 low-income, marginally nourished schoolchildren aged 7–9 years were given a bread spread containing marine fish flour or an analogous spread without fish flour. Children were given the spread on two slices of bread per day, on school days, for 6 months. The fish flour spread provided ~892 mg of DHA/week. The primary outcomes were change from baseline on scores from the Hopkins Verbal Learning Test battery (HVLT), a reading test, and a spelling test. After 6 months, children supplemented with fish flour spread performed better on the HVLT compared to unsupplemented children, scoring higher on measures of recognition and discrimination. Children consuming the *n*-3 spread did not experience the decline in spelling observed in the unsupplemented children over the 6 months, and there was a marginally significant benefit of supplementation on reading.

Two studies investigated the effect of omega-3 supplementation on behavior, exclusively [31,32]. Hamazaki *et al.* [31] reported results from an RCT in healthy school children from Lampung Province, Indonesia. A total of 233 children aged 9–14 years were supplemented with DHA-rich fish oil containing 650 mg DHA and 100 mg EPA/day or placebo for 3 months. Aggression and impulsivity were assessed using the Hostility-Aggression Questionnaire for Children (HAQ-C) and the Barratt Impulsiveness Scale, version 11 (BIS-11). School attendance was recorded as well. While school attendance was higher in the DHA group, no behavioral benefit as the result of supplementation was found.

Itomura *et al.* [32] reported findings from an RCT investigating the effect of omega-3 fatty acid supplementation on behavior (aggression) in healthy school children from Japan. A total of 179 healthy children aged 9–12 years were given fish oil-fortified foods (rolls, bread, sausage, and spaghetti) providing approximately 3600 mg DHA/week (~514 mg/day) plus 840 mg EPA/week (~120 mg/day) or similar unfortified foods for 3 months. Aggression was assessed using the HAQ-C and the Picture Frustration (PF) Study. Attention deficit, hyperactivity and impulsivity were assessed by parents/guardians using diagnostic questionnaires of DSM-IV for ADHD. After 3 months of supplementation, DHA-rich fish oil-fortified foods protected against increases in aggression (as assessed by the HAQ-C). Specifically, in a per-protocol analysis, HAQ-C or physical aggression, scores increased significantly between baseline and week 12 in the female placebo group, but remained unchanged in the female active treatment group. Supplementation was also associated with reduced impulsivity among females as assessed by parents/guardians. No significant changes in assessment of aggression or impulsivity were observed in males, however.

Richardson *et al.* [36] reported the results of an RCT designed to determine the effects of dietary supplementation with DHA on reading, working memory, and behavior in healthy schoolchildren. A total of 362 healthy children aged 7–9 years who were underperforming in reading (below the 33rd percentile/equivalent to ~18 months behind chronological age) were enrolled in the study. Subjects were supplemented with 600 mg DHA/day or a matching placebo for 16 weeks. The primary outcome measures were reported as change from baseline in reading, working memory, and behavior. Reading was assessed using the Word Reading Achievement sub-test of the British Ability Scale (BAS II). Working memory was assessed using two sub-tests of the BAS II (Recall of Digits Forward and Recall of Digits Backward), and behavior was assessed by parents or guardians and teachers using the Conners' Ratings Scales (CPRS and CTRS respectively). After 16 weeks of supplementation, performance on the reading test did not differ by treatment. However, in a pre-planned analysis of 224 children with baseline reading scores ≤20th percentile, and in the subgroup of 105 children with baseline reading scores ≤10th percentile, DHA supplementation significantly improved reading. DHA supplementation also improved reading age, above the expected 4 month gain over the 16 week study period. For children with baseline readings scores ≤20th percentile, DHA supplementation led to an additional 0.8 month gain in reading age (approximately 20% greater than expected). For children with baseline reading scores ≤10th percentile, supplementation resulted in an additional 1.9 month gain in reading age (approximately 50% greater than expected. DHA also significantly improved behavior as rated by parents. Compared to children given placebo, children supplemented with DHA experienced significant improvements in eight of fourteen scales of the CPRS. No effects were seen on behavior as rated by the teachers and there were no significant effects of supplementation on memory.

3.2.3. Clinical Studies (Multiple Nutrient Interventions)

Five studies included supplementation with either fish oil or algal oil as a source of DHA in addition to other fatty acids and/or micronutrients [29,34,35,38,39]. The RCT by Muthayya *et al.* [35] compared the effects of a combination of micronutrients and omega-3 fatty acids on indicators of cognitive performance in low-income, marginally nourished schoolchildren in Bangalore, India. A total of 598 children aged 6–10 years were allocated to 1 of 4 intervention groups to receive foods fortified with micronutrients at either a high (100% RDA) or low (15%

RDA) level in combination with either 930 mg ALA/day plus 100 mg DHA/day (high omega-3) or 140 mg ALA/day (low omega-3) for 12 months. The four groups were defined as: high micronutrients, high omega-3; low micronutrients, high omega-3; high micronutrients, low omega-3; or, low micronutrient, low omega-3. The cognitive test battery consisted of 11 sub-tests to evaluate: fluid reasoning; short-term memory; retrieval ability; and, cognitive speediness. Outcomes were measured at baseline and after 6 and 12 months of supplementation. After 12 months, all 4 groups had significant improvements in short-term memory, retrieval ability, fluid reasoning, cognitive speediness, and overall cognitive performance. Analyses showed differences in cognitive scores according to micronutrient supplementation, but no differences were detected related to DHA or other omega-3 fatty acid supplementation.

Kirby *et al.* [34] reported the results of a one-way crossover design RCT investigating the effect of omega-3 fatty acid supplementation on cognition and behavior in healthy schoolchildren from a mainstream school population. A total of 450 healthy children aged 8–10 years were supplemented with 400 mg DHA + 56 mg EPA/day plus other micronutrients or placebo for 16 weeks. Following 16 weeks of supplementation, both groups received the active supplement for an additional 8 weeks. Outcomes included verbal and non-verbal IQ as measured by the KBIT-2, reading and spelling ability as measured by the WIAT-II, working memory as measured by the WMTB-C, attention as measured by the TEA-Ch, and visual attention and impulsivity as measured by the MFFT. In addition, handwriting process was measured using the Computerized Penmanship Evaluation Tool (ComPET), and parent- and teacher-rated behavior was assessed using the SNAP-IV and SDQ. Results showed that performance on the cognitive tests did not differ by treatment group. However, in a per-protocol analysis of the 235 children who reached compliance criteria, DHA supplementation significantly improved visual attention and impulsivity as assessed by the MFFT. The number of first correct responses was significantly higher in the active treatment group compared to the placebo group. With regard to behavior in the per-protocol analysis, supplementation protected against declines in parent-rated pro-social behavior. In the intention-to-treat analysis, however, teachers saw improvement in the "difficulties" score in unsupplemented children only.

Sinn *et al.* [38] reported results from an open-label pilot study designed to investigate the feasibility of providing fish oil supplements to children in a remote, Northern Territory school in Australia. A total of 47 children, aged 3–14 years, were given a supplement during school days providing 558 mg EPA + 174 mg DHA, plus 60 mg GLA, plus 10.8 mg vitamin E/day for 12 weeks. Reading and spelling were assessed using the Wide Range Achievement Test (WRAT); non-verbal intelligence was assessed using the Raven's Colored Matrices; and behavior was assessed by teachers using the Conners' Behavior Ratings Scales (CBRS). Outcomes were measured at baseline and after 12 weeks. After 12 weeks of supplementation, reading and spelling scores and performance on the Raven's Coloured Matrices improved significantly. It is important to note, however, that this pilot study was not placebo-controlled and the observed changes may not be due to supplementation.

The Nutrition Enhancement for Mental Optimization (NEMO) study group [39] reported results of a RCT which assessed the effect of micronutrients, long-chain omega-3 fatty acids, or both on cognitive performance in well-nourished and marginally nourished children age 6–10 years. A total of 396 well-nourished children from Adelaide, South Australia, and a total of 384 marginally nourished children from Jakarta, Indonesia, were enrolled in the study. Children received a fruit-flavored drink containing: a micronutrient mix alone; an omega-3 fatty acid mix alone; a combination of the micronutrient and omega-3 fatty acid mixes; or a placebo mix. The fruit-flavored drink was consumed 6 days/week for 12 months. The micronutrient mix provided 100% of the RDA for iron, folate, vitamin B6, vitamin B12, vitamin A, and vitamin C and 50% of the RDA for zinc. The omega-3 fatty acid mix provided 88 mg DHA/day and 22 mg EPA/day. Cognitive function and school performance were assessed after 6 and 12 months of supplementation using a series of standardized neuropsychological tests. Subjects from the two test sites were different with regard to several characteristics including nutritional status; therefore data for Indonesia and Australia were analyzed separately. From the Australian site, results showed that in well-nourished, school-aged children, 12 months of supplementation with multiple micronutrients improved verbal learning and memory, but had no effect on tests of general intelligence or attention. Supplementation with omega-3 fatty acids did not affect cognitive function or school performance. In the Indonesian children who were marginally nourished, results showed that fortification with multiple micronutrients improved verbal learning and memory in girls but not in boys. As in the Australian cohort, omega-3 fatty acids had no effect on any measure of cognitive function or school performance.

Baumgartner *et al.* [29] reported the findings of an RCT investigating the effects of iron and DHA/EPA supplementation alone and in combination, in children with poor iron and omega-3 fatty acid status. A total of 321 children, aged 6–11 years, with iron deficiency were allocated to receive iron, DHA/EPA, iron plus DHA/EPA, or a matching placebo 4 days/week for 34 weeks. Iron was provided at a dose of 50 mg/day and fish oil was provided at a dose of 420 mg DHA/day and 80 mg EPA/day. Because supplementation was interrupted by holidays and teacher strikes, supplements were provided for 105 days. Mean total iron intake over the 105 day supplementation period was 4.8 g (45.7 mg/day). Mean total

DHA intake was 41.2 g (392 mg/day), and mean total EPA intake was 7.8 g (74 mg/day). The primary outcome measures were change from baseline in tests of cognition including 4 subtests of the Kaufman Assessment Battery for Children (KABC-II) and the Hopkins Verbal Learning Test (HVLT). Results of the study indicated that while children supplemented with iron performed better than un-supplemented children on the HVLT test of learning and memory, DHA/EPA supplementation had no effect on learning or memory. Results of the study also found that in a sub-group analysis stratified by anemia status, children with iron deficiency anemia supplemented with DHA/EPA performed significantly worse than children given placebo on the KABC-II test of long-term memory. In a second sub-group analysis stratified by sex, there was a significant negative effect of DHA/EPA supplementation on the Atlantis Delayed test performance (long term memory and retrieval) in girls, while boys benefitted from supplementation on this measure. On the HVLT, girls receiving DHA/EPA supplementation alone had better recall than placebo or those receiving the combination with iron.

4. DISCUSSION

Infancy, childhood, and adolescence are times of rapid neuronal maturation, synaptogenesis, and gray matter expansion, all of which are associated with brain DHA accumulation [5,13]. DHA synthesis from its fatty acid precursors is known to be inefficient, making dietary sources of preformed DHA important [41]. However, DHA is found at high levels in only a very limited number of foods and the typical intake by children worldwide is surprisingly low.

Studies in animals show that omega-3 deficiency causes structural and functional changes in the hippocampus, hypothalamus, and cortex areas of the brain [18,42,43]. In animals, reduced levels of brain DHA are associated with problems of spatial and serial learning and memory [42,44] as well as increases in depressive symptoms and aggressive behavior [45,46]. In humans, DHA is particularly important during gestation and infancy for early brain and visual development [47,48]. There is also growing understanding of the role that *n*-3 LC-PUFA may play in the prevention and treatment of several neuropsychiatric conditions common to children and adults [49,50]. Recently, polyunsaturated fatty acids, including DHA, have been studied in children with certain developmental disorders characterized by learning and behavior difficulties. Many of these conditions are associated with abnormal fatty acid status, and whether due to altered fatty acid metabolism or poor intake, some studies show an improvement in symptoms following supplementation with polyunsaturated fatty acids [22,51,52]. Healthy children without these neuropsychiatric conditions may also have poor DHA status, due to poor intake. It is therefore plausible that improvements in DHA status could benefit learning and behavior in healthy children as well. The purpose of this review was to determine whether current literature defines a relationship of DHA with learning and behavior in healthy children.

Although there appears to be interest in investigating the effects of DHA and other fatty acids on cognition, learning and behavior in healthy children, the results of our search revealed studies which were very diverse in both research design and focus. For assessment of cognition, the investigators of studies in this review used many and varied age-standardized tests, most of which contained multiple-subtests each measuring a different aspect of cognitive function. There were no apparent consistent results among the studies with regard to treatment effects or DHA biomarker status on specific cognitive tests. In spite of this heterogeneity, however, the studies as a group provide a global view of DHA impact on brain activity and cognition, with over half of the studies reporting significant results favoring a role of DHA or *n*-3 LC-PUFA in at least one area of cognition, learning, or behavior.

Two studies included in this review, demonstrated a benefit of DHA by direct measurement of brain activity using neurophysiological measures of cognitive function during performance of a standardized test. First, the study by McNamara *et al.* [12] showed the effects of DHA supplementation on brain activity in children as measured by fMRI during performance of a task of sustained attention. Even though outcomes on the standardized task showed no difference as a result of supplementation, probably due to a ceiling effect for the CPT-IP in the children, fMRI data confirmed enhanced activation of the dorsolateral prefrontal cortex (DLPFC) in the DHA-supplemented children. There is a normal age-related increase in DHA content in the prefrontal cortex which occurs during adolescence and young adulthood in healthy well-nourished subjects [13] and the study by McNamara *et al.* [12] provides a better understanding of the extent to which apparently well-nourished youth may be deficient in DHA and how this deficiency can affect brain activity. Second, a study by Boucher *et al.* used electroencephalographic data to directly assess brain activity in healthy children. In that study, ERP results showed a significant positive association of DHA status with brain activity during performance of a standardized test [27]. While the association with ERP outcomes was stronger for cord blood DHA status than for the child's current plasma DHA, it was still apparent that DHA status in the 10–13 year old children was not optimal and that brain activity was, to some degree, correctable or modifiable through supplementation. There was no effect of DHA status on the outcome of the standardized test.

The findings of McNamara *et al.* [12] and Boucher *et al.* [27] are supported by a similar study in an older population. Jackson *et al.* used near-infrared spectroscopy to assess the relative changes in concentration of oxyhemoglobin and deoxyhemoglobin in the prefrontal cortex of young adults during performance of standardized cognitive tasks [53]. In this study, supplementation with 1- or 2-g DHA-rich fish oil increased concentrations of oxyhemoglobin and total hemoglobin in participants during completion of a standardized test. This pattern of oxygenation is indicative of increased blood flow. As in the studies by Boucher *et al.* [27] and McNamara *et al.* [12], there were no consistent changes in performance on the cognitive test in spite of the changes in the hemodynamic response.

Studies such as those by Boucher *et al.* [27], McNamara *et al.* [12], and Jackson *et al.* [52] which demonstrate a DHA relationship with measurable brain activity, but without a concurrent measurable effect on test performance, bring into question the appropriateness or sensitivity of many commonly administered cognitive tests used for assessing DHA adequacy. It is possible that certain tests do not assess the particular cognitive activity affected by DHA. In addition, it is more likely that larger group sizes are required in order achieve the statistical power needed for valid comparisons. From our review, it is interesting that DHA-related improvements in school performance were, however, detectable in some of the studies. In particular, reading performance improved in DHA-supplemented children with poor reading skills in the study by Richardson *et al.* [36]; and in the study by Dalton *et al.* [30], spelling performance was maintained in a DHA-supplemented group while the control group experienced a loss of skills. Functions of the brain are often described in modular terms and the type of learning detected in these two studies may involve a coordination of several functions, including memory and attention. Providing adequate DHA may provide small improvements in multiple areas of cognitive function which are ultimately sufficient to affect reading or spelling.

The implication of improving school performance in reading or spelling for struggling students is significant. Reading is described as the major foundational skill for all school-based learning. Students who do not achieve proficiency in reading during the early years of school are known to have difficulty comprehending subject matter in the grades that follow. Educational experts suggest that when students have difficulty learning to read, their love of learning and motivation diminishes [54]. In addition, experts define reading as a critical skill needed not only for academic success, but also for social and economic success [55]. Children with poor reading skills have been shown to experience decreases in self-esteem, self-concept, and further motivation for learning [56] and are more likely to become frustrated, overwhelmed, or disinterested [57]. Given the importance of academic success to children's well-being and ultimate success in life, all educational and environmental factors, including improved nutrition, are worthy of consideration for optimization of learning.

Since changes in normal, healthy children may be difficult to detect, the study design, subject selection, and outcome measure selection are increasingly important. In the study by Richardson *et al.* [36], the authors state that the original trial design required children to rank below the 20th percentile (2 years behind actual age) in reading. However, recruitment concerns led the investigators to broaden the inclusion criteria to those scoring below the 33rd percentile (18 months behind actual age) in reading with a pre-planned analysis of children below the 20th and 10th percentile on reading. While no treatment effect was detectable in the whole population, a significant effect on the change from baseline reading scores was detected in children scoring below the 20th percentile and an apparent greater and significant effect was observed for those who scored below the 10th percentile at baseline. Similar findings were reported in a smaller study conducted in children with ADHD which included a subgroup of children who were also behind in their age level in reading and spelling by approximately 2 years. Following 4 months of supplementation with either high EPA or high DHA fish oil, no group differences in behavior or cognition were detected in the group as a whole. However, higher erythrocyte DHA levels were associated with improved word reading, spelling, attention, and oppositional behavior, particularly in the subgroup with reading and spelling difficulties [58].

The doses of DHA supplements used in the studies included in this review varied widely. The 600 mg daily dose of DHA used by Richardson *et al.* [36] is similar to that used in studies of children with learning disorders; it establishes an effective dose for use in healthy children as well. Whether 600 mg DHA/day is the optimal dose for healthy children of this age remains to be tested. Future studies are needed to define blood DHA levels, as a surrogate for brain DHA level, which are associated with both the low and improved scores on tests of reading and spelling. Due to differences in study design, methods of measuring and reporting blood DHA, level of DHA supplementation, and compliance, the current literature is still incomplete in this regard. In the study by McNamara *et al.* [12], mean erythrocyte DHA composition at baseline was 3.3% of total fatty acids. Following supplementation with 400 mg DHA/day or 1200 mg DHA/day for 8 weeks, erythrocyte DHA composition increased to 7.5% and 10.3%, respectively. Since these levels were associated with activation of the dorsolateral prefrontal cortex, they could be considered as appropriate goals for future RCTs. Only four other RCTs reported erythrocyte DHA composition as seen inTable 3 [29,31,32,35]. Of these, the studies by Itomura *et al.* [32] and Hamazaki *et al.* [31] achieved erythrocyte DHA levels of 7.1% and 7.5% total fatty acids, respectively. Baumgartner *et al.* [29] and Muthayya *et al.*

[35] reported erythrocyte DHA levels of less than 6% following the supplementation period. The remaining studies either did not report DHA blood status or reported DHA status in a different blood compartment or tissue, making comparisons questionable.

The oils used as placebo in the included studies varied greatly with regard to fatty acid composition. Some, such as medium chain triglycerides provided no additional *n*-6 or *n*-3 PUFA while others provided additional linoleic acid from corn or sunflower oil, ALA from soybean and rapeseed oil, or monounsaturated fatty acids and polyphenols from olive oil. The search for the most appropriate placebo for both human and animal studies has been a topic of debate, with virtually all substitute oils providing fatty acids or other nutrients which on their own can be considered bioactive [59]. For studies specifically designed to increase blood and tissue levels of DHA, oils which supply additional *n*-3 fatty acids in the form of ALA may have minimal effect due to poor conversion of ALA to DHA in humans [41]. However, oils that contribute significant linoleic acid to the diet may reduce DHA accrual in tissues. Studies in animals show decreased incorporation of DHA in retinal tissue in diets high in linoleic acid, and that diets low in linoleic acid support higher brain DHA [60,61]. None of the studies in this review reported fatty acid composition of the background diet; therefore it is not clear whether the small amount of oil used in supplementation significantly impacted fatty acid intake of ALA or LA.

In addition to the dose and duration of DHA supplementation, many studies suggest that blood and tissue biomarkers of *n*-3 PUFAs are associated with several common single nucleotide polymorphisms (SNPs) in the genes responsible for encoding the delta 5, and delta 6 desaturases (FADS1 and FADS2), as well as ELOVL2 elongase. Because there is a distinct association of the haplotypes with both *n*-3 and *n*-6 PUFA biomarker status, individuals with variations of these mutations respond differently to *n*-3 supplementation as compared to the larger population. The fatty acid profile associated with SNPs in the FADS1-FADS2 gene cluster includes increased proportions of precursors such as ALA and decreased proportions of desaturation products such as EPA and DHA. However, the biosynthesis of DHA is very limited in the general population and direct consumption of DHA is the primary way to increase DHA status making distinction between the various haplotypes of less significance for this fatty acid. Large observational studies as well as metabolic tracer studies of ALA metabolism in adults indicate that polymorphisms in the FADS1-FADS2 gene cluster do not result in changes in plasma DHA composition under conditions of a controlled diet. Since this review focused on studies which specifically modified dietary DHA, and not its precursors, such genetic variations would be expected to play only a minor role on their outcomes [62,63,64,65,66].

The criteria for inclusion in this review specified that studies report a biomarker of the children's DHA status from the same time-period as the cognitive or behavioral testing. Studies in animals, however, suggest that earlier periods of development may represent a more critical window of time when adequate DHA is particularly essential. Such findings indicate that *n*-3 deficient diets limit DHA accrual during the perinatal period, and if not corrected early, are associated with deficits in neuronal arborization and synaptic formation. Deficient accrual of brain DHA is also associated with deficits in dopamine and serotonin neurotransmission. Development of these two systems of neurotransmission is of particular interest due to their well-known involvement in the control of learning and memory. In a series of studies using a rodent model, Chalon *et al.* [67] demonstrated that brain DHA decreased in those fed an *n*-3 deficient diet and that this was accompanied by a corresponding decrease in the amount of dopamine in the frontal cortex. Animals fed an *n*-3 repletion diet at birth recovered to normal levels of both brain DHA and dopamine. However, a delay in feeding the *n*-3 repletion diet to the period following weaning resulted in only partial recovery of brain DHA and no recovery of the neurochemical factors. The implication is that early and prolonged insults from a severe *n*-3 deficiency may lead to irreversible damage to specific brain functions. While no critical window of development has been defined in humans, it is possible that differences in early DHA status affect the outcomes of these studies in later childhood [67,68].

5. CONCLUSIONS

In summary, the studies included in this review generally indicate that improvements in DHA status may initiate brain changes which are observable in activities of learning and behavior. Results of neurophysiologic measures that directly assessed brain activity indicate that brain changes occur in healthy children as the result of DHA supplementation. However, standardized tests of cognition do not show consistent changes. The improvement seen in reading and spelling skills following DHA supplementation may represent an accumulation of many subtle changes made over multiple domains that are not readily detectable on other types of tests. These changes may be particularly subtle in healthy children. The reviewed studies conducted in healthy children build on animal data that demonstrate DHA is an essential brain component affecting learning and behavior. While the number of studies is limited, and the design features are diverse, the studies implicate problems in learning and behavior as detrimental effects of DHA deficiency in otherwise healthy children.

CONFLICT OF INTEREST

Kuratko, Barrett, Nelson, and Salem, Jr. are employees of DSM Nutritional Products, a producer and seller of polyunsaturated fatty acids, including DHA.

REFERENCES

1. Brown, T.T.; Jernigan, T.L. Brain development during the preschool years. Neuropsychol. Rev. 2012, 22, 313–333.
2. Anderson, V.; Fenwick, T.; Manly, T.; Robertson, I. Attentional skills following traumatic brain injury in childhood: A componential analysis. Brain Inj. 1998, 12, 937–949.
3. Barkley, R.A. The executive functions and self-regulation: An evolutionary neuropsychological perspective. Neuropsychol. Rev. 2001, 11, 1–29.
4. Steinberg, L. Psychological control: Style or substance? New Dir. Child Adolesc. Dev. 2005, 2005, 71–78.
5. Carver, J.D.; Benford, V.J.; Han, B.; Cantor, A.B. The relationship between age and the fatty acid composition of cerebral cortex and erythrocytes in human subjects. Brain Res. Bull. 2001, 56, 79–85.
6. Feller, S.E.; Gawrisch, K.; MacKerell, A.D., Jr. Polyunsaturated fatty acids in lipid bilayers: Intrinsic and environmental contributions to their unique physical properties. J. Am. Chem. Soc. 2002, 124, 318–326.
7. Niu, S.L.; Mitchell, D.C.; Lim, S.Y.; Wen, Z.M.; Kim, H.Y.; Salem, N., Jr.; Litman, B.J. Reduced G protein-coupled signaling efficiency in retinal rod outer segments in response to n-3 fatty acid deficiency. J. Biol. Chem. 2004, 279, 31098–31104.
8. Champeil-Potokar, G.; Chaumontet, C.; Guesnet, P.; Lavialle, M.; Denis, I. Docosahexaenoic acid (22:6n-3) enrichment of membrane phospholipids increases gap junction coupling capacity in cultured astrocytes. Eur. J. Neurosci. 2006, 24, 3084–3090.
9. Joardar, A.; Sen, A.K.; Das, S. Docosahexaenoic acid facilitates cell maturation and β-adrenergic transmission in astrocytes. J. Lipid Res. 2006, 47, 571–581.
10. Pifferi, F.; Roux, F.; Langelier, B.; Alessandri, J.M.; Vancassel, S.; Jouin, M.; Lavialle, M.; Guesnet, P. (n-3) Polyunsaturated fatty acid deficiency reduces the expression of both isoforms of the brain glucose transporter GLUT1 in rats. J. Nutr. 2005, 135, 2241–2246.

11. Ximenes da Silva, A.; Lavialle, F.; Gendrot, G.; Guesnet, P.; Alessandri, J.M.; Lavialle, M. Glucose transport and utilization are altered in the brain of rats deficient in n-3 polyunsaturated fatty acids. J. Neurochem. 2002, 81, 1328–1337.

12. McNamara, R.K.; Able, J.; Jandacek, R.; Rider, T.; Tso, P.; Eliassen, J.C.; Alfieri, D.; Weber, W.; Jarvis, K.; DelBello, M.P.; et al. Docosahexaenoic acid supplementation increases prefrontal cortex activation during sustained attention in healthy boys: A placebo-controlled, dose-ranging, functional magnetic resonance imaging study. Am. J. Clin. Nutr. 2010, 91, 1060–1067.

13. Giedd, J.N.; Blumenthal, J.; Jeffries, N.O.; Castellanos, F.X.; Liu, H.; Zijdenbos, A.; Paus, T.; Evans, A.C.; Rapoport, J.L. Brain development during childhood and adolescence: A longitudinal MRI study.Nat. Neurosci. 1999, 2, 861–863.

14. Bazan, N.G. Cellular and molecular events mediated by docosahexaenoic acid-derived neuroprotectin D1 signaling in photoreceptor cell survival and brain protection. Prostaglandins Leukot. Essent. Fatty Acids 2009, 81, 205–211.

15. Belayev, L.; Khoutorova, L.; Atkins, K.D.; Bazan, N.G. Robust docosahexaenoic acid-mediated neuroprotection in a rat model of transient, focal cerebral ischemia. Stroke 2009, 40, 3121–3126.

16. Blondeau, N.; Widmann, C.; Lazdunski, M.; Heurteaux, C. Polyunsaturated fatty acids induce ischemic and epileptic tolerance. Neuroscience 2002, 109, 231–241.

17. Ozyurt, B.; Sarsilmaz, M.; Akpolat, N.; Ozyurt, H.; Akyol, O.; Herken, H.; Kus, I. The protective effects of omega-3 fatty acids against MK-801-induced neurotoxicity in prefrontal cortex of rat. Neurochem. Int. 2007, 50, 196–202.

18. Salem, N., Jr.; Litman, B.; Kim, H.Y.; Gawrisch, K. Mechanisms of action of docosahexaenoic acid in the nervous system. Lipids 2001, 36, 945–959.

19. Cohen, J.T.; Bellinger, D.C.; Connor, W.E.; Shaywitz, B.A. A quantitative analysis of prenatal intake of n-3 polyunsaturated fatty acids and cognitive development. Am. J. Prev. Med. 2005, 29, 366–374.

20. Yurko-Mauro, K.; McCarthy, D.; Rom, D.; Nelson, E.B.; Ryan, A.S.; Blackwell, A.; Salem, N., Jr.; Stedman, M. Beneficial effects of docosahexaenoic acid on cognition in age-related cognitive decline.Alzheimers Dement. 2010, 6, 456–464.

21. Richardson, A.J.; Montgomery, P. The Oxford-Durham study: A randomized, controlled trial of dietary supplementation with fatty

acids in children with developmental coordination disorder.Pediatrics 2005, 115, 1360–1366.

22. Richardson, A.J. Clinical trials of fatty acid treatment in ADHD, dyslexia, dyspraxia and the autistic spectrum. Prostaglandins Leukot. Essent. Fatty Acids 2004, 70, 383–390.

23. Stevens, L.J.; Zentall, S.S.; Deck, J.L.; Abate, M.L.; Watkins, B.A.; Lipp, S.R.; Burgess, J.R. Essential fatty acid metabolism in boys with attention-deficit hyperactivity disorder. Am. J. Clin. Nutr. 1995, 62, 761–768.

24. Mitchell, E.A.; Aman, M.G.; Turbott, S.H.; Manku, M. Clinical characteristics and serum essential fatty acid levels in hyperactive children. Clin. Pediatr. (Phila.) 1987, 26, 406–411.

25. Richardson, A.J.; Ross, M.A. Fatty acid metabolism in neurodevelopmental disorder: A new perspective on associations between attention-deficit/hyperactivity disorder, dyslexia, dyspraxia and the autistic spectrum. Prostaglandins Leukot. Essen. Fatty Acids 2000, 63, 1–9.

26. Bakker, E.C.; Ghys, A.J.; Kester, A.D.; Vles, J.S.; Dubas, J.S.; Blanco, C.E.; Hornstra, G. Long-chain polyunsaturated fatty acids at birth and cognitive function at 7 years of age. Eur. J. Clin. Nutr. 2003,57, 89–95.

27. Boucher, O.; Burden, M.J.; Muckle, G.; Saint-Amour, D.; Ayotte, P.; Dewailly, E.; Nelson, C.A.; Jacobson, S.W.; Jacobson, J.L. Neurophysiologic and neurobehavioral evidence of beneficial effects of prenatal omega-3 fatty acid intake on memory function at school age. Am. J. Clin. Nutr. 2011, 93, 1025–1037.

28. Kirby, A.; Woodward, A.; Jackson, S.; Wang, Y.; Crawford, M.A. Childrens' learning and behaviour and the association with cheek cell polyunsaturated fatty acid levels. Res. Dev. Disabil. 2010, 31, 731–742.

29. Baumgartner, J.; Smuts, C.M.; Malan, L.; Kvalsvig, J.; van Stuijvenberg, M.E.; Hurrell, R.F.; Zimmermann, M.B. Effects of iron and n-3 fatty acid supplementation, alone and in combination, on cognition in school children: A randomized, double-blind, placebo-controlled intervention in South Africa. Am. J. Clin. Nutr. 2012, 96, 1327–1338.

30. Dalton, A.; Wolmarans, P.; Witthuhn, R.C.; van Stuijvenberg, M.E.; Swanevelder, S.A.; Smuts, C.M. A randomised control trial in schoolchildren showed improvement in cognitive function after consuming a bread spread, containing fish flour from a marine source. Prostaglandins Leukot. Essent. Fatty Acids 2009, 80, 143–149.

31. Hamazaki, K.; Syafruddin, D.; Tunru, I.S.; Azwir, M.F.; Asih, P.B.; Sawazaki, S.; Hamazaki, T. The effects of docosahexaenoic acid-rich fish oil on behavior, school attendance rate and malaria infection in school children—A double-blind, randomized, placebo-controlled trial in Lampung, Indonesia.Asia Pac. J. Clin. Nutr. 2008, 17, 258–263.

32. Itomura, M.; Hamazaki, K.; Sawazaki, S.; Kobayashi, M.; Terasawa, K.; Watanabe, S.; Hamazaki, T. The effect of fish oil on physical aggression in schoolchildren—A randomized, double-blind, placebo-controlled trial. J. Nutr. Biochem. 2005, 16, 163–171.

33. Kennedy, D.O.; Jackson, P.A.; Elliott, J.M.; Scholey, A.B.; Robertson, B.C.; Greer, J.; Tiplady, B.; Buchanan, T.; Haskell, C.F. Cognitive and mood effects of 8 weeks' supplementation with 400 mg or 1000 mg of the omega-3 essential fatty acid docosahexaenoic acid (DHA) in healthy children aged 10–12 years. Nutr. Neurosci. 2009, 12, 48–56.

34. Kirby, A.; Woodward, A.; Jackson, S.; Wang, Y.; Crawford, M.A. A double-blind, placebo-controlled study investigating the effects of omega-3 supplementation in children aged 8–10 years from a mainstream school population. Res. Dev. Disabil. 2010, 31, 718–730.

35. Muthayya, S.; Eilander, A.; Transler, C.; Thomas, T.; van der Knaap, H.C.; Srinivasan, K.; van Klinken, B.J.; Osendarp, S.J.; Kurpad, A.V. Effect of fortification with multiple micronutrients and n-3 fatty acids on growth and cognitive performance in Indian schoolchildren: The CHAMPION (Children's Health and Mental Performance Influenced by Optimal Nutrition) Study. Am. J. Clin. Nutr. 2009, 89, 1766–1775.

36. Richardson, A.J.; Burton, J.R.; Sewell, R.P.; Spreckelsen, T.F.; Montgomery, P. Docosahexaenoic acid for reading, cognition and behavior in children aged 7–9 years: A randomized, controlled trial (the DOLAB Study). PLoS One 2012, 7, e43909.

37. Ryan, A.S.; Nelson, E.B. Assessing the effect of docosahexaenoic acid on cognitive functions in healthy, preschool children: A randomized, placebo-controlled, double-blind study. Clin. Pediatr. (Phila.) 2008, 47, 355–362.

38. Sinn, N.; Cooper, P.; O'Dea, K. Fish oil supplementation, learning and behaviour in Indigenous Australian children from a remote community school: A pilot feasibility study. Aust. N. Z. J. Public Health 2011, 35, 493–494.

39. Osendarp, S.J.; Baghurst, K.I.; Bryan, J.; Calvaresi, E.; Hughes, D.; Hussaini, M.; Karyadi, S.J.; van Klinken, B.J.; van der Knaap, H.C.;

Lukito, W.; et al. Effect of a 12-mo micronutrient intervention on learning and memory in well-nourished and marginally nourished school-aged children: 2 Parallel, randomized, placebo-controlled studies in Australia and Indonesia. Am. J. Clin. Nutr. 2007, 86, 1082–1093.

40. Bryan, J.; Osendarp, S.; Hughes, D.; Calvaresi, E.; Baghurst, K.; van Klinken, J.W. Nutrients for cognitive development in school-aged children. Nutr. Rev. 2004, 62, 295–306.

41. Brenna, J.T.; Salem, N., Jr.; Sinclair, A.J.; Cunnane, S.C. α-Linolenic acid supplementation and conversion to n-3 long-chain polyunsaturated fatty acids in humans. Prostaglandins Leukot. Essent. Fatty Acids 2009, 80, 85–91.

42. Moriguchi, T.; Salem, N., Jr. Recovery of brain docosahexaenoate leads to recovery of spatial task performance. J. Neurochem. 2003, 87, 297–309.

43. Lim, S.Y.; Hoshiba, J.; Salem, N., Jr. An extraordinary degree of structural specificity is required in neural phospholipids for optimal brain function: n-6 Docosapentaenoic acid substitution for docosahexaenoic acid leads to a loss in spatial task performance. J. Neurochem. 2005, 95, 848–857.

44. Fedorova, I.; Hussein, N.; Baumann, M.H.; di Martino, C.; Salem, N., Jr. An n-3 fatty acid deficiency impairs rat spatial learning in the Barnes maze. Behav. Neurosci. 2009, 123, 196–205.

45. DeMar, J.C., Jr.; Ma, K.; Bell, J.M.; Igarashi, M.; Greenstein, D.; Rapoport, S.I. One generation of n-3 polyunsaturated fatty acid deprivation increases depression and aggression test scores in rats. J. Lipid Res. 2006, 47, 172–180.

46. Fedorova, I.; Salem, N., Jr. Omega-3 fatty acids and rodent behavior. Prostaglandins Leukot. Essent. Fatty Acids 2006, 75, 271–289.

47. Jensen, C.L.; Voigt, R.G.; Prager, T.C.; Zou, Y.L.; Fraley, J.K.; Rozelle, J.C.; Turcich, M.R.; Llorente, A.M.; Anderson, R.E.; Heird, W.C. Effects of maternal docosahexaenoic acid intake on visual function and neurodevelopment in breastfed term infants. Am. J. Clin. Nutr. 2005, 82, 125–132.

48. Birch, E.E.; Garfield, S.; Castaneda, Y.; Hughbanks-Wheaton, D.; Uauy, R.; Hoffman, D. Visual acuity and cognitive outcomes at 4 years of age in a double-blind, randomized trial of long-chain polyunsaturated fatty acid-supplemented infant formula. Early Hum. Dev. 2007, 83, 279–284.

49. Bloch, M.H.; Hannestad, J. Omega-3 fatty acids for the treatment of depression: Systematic review and meta-analysis. Mol. Psychiatry 2012, 17, 1272–1282.

50. Freeman, M.P.; Hibbeln, J.R.; Wisner, K.L.; Davis, J.M.; Mischoulon, D.; Peet, M.; Keck, P.E., Jr.; Marangell, L.B.; Richardson, A.J.; Lake, J.; et al. Omega-3 fatty acids: Evidence basis for treatment and future research in psychiatry. J. Clin. Psychiatry 2006, 67, 1954–1967.

51. Bloch, M.H.; Qawasmi, A. Omega-3 fatty acid supplementation for the treatment of children with attention-deficit/hyperactivity disorder symptomatology: Systematic review and meta-analysis. J. Am. Acad. Child Adolesc. Psychiatry 2011, 50, 991–1000.

52. Milte, C.M.; Sinn, N.; Buckley, J.D.; Coates, A.M.; Young, R.M.; Howe, P.R. Polyunsaturated fatty acids, cognition and literacy in children with ADHD with and without learning difficulties. J. Child Health Care 2011, 15, 299–311.

53. Jackson, P.A.; Reay, J.L.; Scholey, A.B.; Kennedy, D.O. Docosahexaenoic acid-rich fish oil modulates the cerebral hemodynamic response to cognitive tasks in healthy young adults. Biol. Psychol. 2012, 89, 183–190.

54. Stanovich, K. Matthew effects in reading: Some consequences of individual differences in the acquisition of literacy. Read. Res. Q. 1986, 22, 360–406.

55. Snow, C.; Burns, S.; Griffin, P. Preventing Reading Difficulties in Young Children; National Academies Press: Washington, DC, USA, 1998.

56. Armbruster, B.; Lehr, F.; Osborn, J. Putting Reading First. In The Research Building Blocks for Teaching Children to Read: Kindergarten through Grade 3, 3rd ed.; Adler, R., Ed.; Diane Publishing Co.: Darby, PA, USA, 2010.

57. Morrison, F.; Cosden, M. Risk, resilience, and adjustment of individuals with learning disabilities.Learn. Disabil. Q. 1997, 20, 43–60.

58. Milte, C.M.; Parletta, N.; Buckley, J.D.; Coates, A.M.; Young, R.M.; Howe, P.R. Eicosapentaenoic and docosahexaenoic acids, cognition, and behavior in children with attention-deficit/hyperactivity disorder: A randomized controlled trial. Nutrition 2012, 28, 670–677.

59. Dyall, S.C. Methodological issues and inconsistencies in the field of omega-3 fatty acids research.Prostaglandins Leukot. Essent. Fatty Acids 2011, 85, 281–285.

60. Weisinger, H.S.; Vingrys, A.J.; Sinclair, A.J. The effect of docosahexaenoic acid on the electroretinogram of the guinea pig. Lipids 1996, 31, 65–70.

61. Novak, E.M.; Dyer, R.A.; Innis, S.M. High dietary omega-6 fatty acids contribute to reduced docosahexaenoic acid in the developing brain and inhibit secondary neurite growth. Brain Res. 2008,1237, 136–145.

62. Gillingham, L.G.; Harding, S.V.; Rideout, T.C.; Yurkova, N.; Cunnane, S.C.; Eck, P.K.; Jones, P.J. Dietary oils and FADS1-FADS2 genetic variants modulate [13C] alpha-linolenic acid metabolism and plasma fatty acid composition. Am. J. Clin. Nutr. 2013, 97, 195–207.

63. Malerba, G.; Schaeffer, L.; Xumerle, L.; Klopp, N.; Trabetti, E.; Biscuola, M.; Cavallari, U.; Galavotti, R.; Martinelli, N.; Guarini, P.; et al. SNPs of the FADS gene cluster are associated with polyunsaturated fatty acids in a cohort of patients with cardiovascular disease. Lipids 2008, 43, 289–299.

64. Schaeffer, L.; Gohlke, H.; Muller, M.; Heid, I.M.; Palmer, L.J.; Kompauer, I.; Demmelmair, H.; Illig, T.; Koletzko, B.; Heinrich, J. Common genetic variants of the FADS1 FADS2 gene cluster and their reconstructed haplotypes are associated with the fatty acid composition in phospholipids. Hum. Mol. Genet. 2006, 15, 1745–1756.

65. Rzehak, P.; Heinrich, J.; Klopp, N.; Schaeffer, L.; Hoff, S.; Wolfram, G.; Illig, T.; Linseisen, J. Evidence for an association between genetic variants of the fatty acid desaturase 1 fatty acid desaturase 2 (FADS1 FADS2) gene cluster and the fatty acid composition of erythrocyte membranes. Br. J. Nutr.2009, 101, 20–26.

66. Tanaka, T.; Shen, J.; Abecasis, G.R.; Kisialiou, A.; Ordovas, J.M.; Guralnik, J.M.; Singleton, A.; Bandinelli, S.; Cherubini, A.; Arnett, D.; et al. Genome-wide association study of plasma polyunsaturated fatty acids in the InCHIANTI Study. PLoS Genet. 2009, 5, e1000338.

67. Chalon, S. Omega-3 fatty acids and monoamine neurotransmission. Prostaglandins Leukot. Essent. Fatty Acids 2006, 75, 259–269.

68. Kodas, E.; Vancassel, S.; Lejeune, B.; Guilloteau, D.; Chalon, S. Reversibility of n-3 fatty acid deficiency-induced changes in dopaminergic neurotransmission in rats: Critical role of developmental stage. J. Lipid Res. 2002, 43, 1209–1219.

7

NEUROENHANCEMENT WITH VITAMIN B12—UNDERESTIMATED NEUROLOGICAL SIGNIFICANCE

Uwe Gröber [1,*], Klaus Kisters [1,2] and Joachim Schmidt [1]

[1]Academy of Micronutrient Medicine, Zweigertstr. 55, Essen 45130, Germany

[2]St. Anna Hospital, Hospitalstr. 19, Herne 44649, Germany

ABSTRACT

Vitamin B12 is a cofactor of methionine synthase in the synthesis of methionine, the precursor of the universal methyl donor S-Adenosylmethionine (SAMe), which is involved in different epigenomic regulatory mechanisms and especially in brain development. A Vitamin B12 deficiency expresses itself by a wide variety of neurological manifestations such as paraesthesias, skin numbness, coordination disorders and reduced nerve conduction velocity. In elderly people, a latent Vitamin B12 deficiency can be associated with a progressive brain atrophy. Moderately elevated concentrations of homocysteine (>10 µmol/L) have been associated with an increased risk of dementia, notably Alzheimer's disease, in many cross-sectional and prospective studies. Raised plasma concentrations of homocysteine is also associated with both regional and whole brain atrophy, not only in Alzheimer's disease but also in healthy elderly people. Clinician awareness should be raised to accurately diagnose and treat early Vitamin B12 deficiency to prevent irreversible structural brain damage.

Keywords: Vitamin B12; elderly; Vitamin B12 deficiency; diagnostic of Vitamin B12 deficiency; brain atrophy; neuroenhancement

1. INTRODUCTION

Vitamin B12 is an essential water-soluble vitamin that is vitally important in haematopoiesis, nervous system functions, maintenance of intact gastrointestinal mucosa and regulation of numerous other B12-dependent metabolic processes. The structural framework of Vitamin B12 is based on the corrin ring system comprising four reduced pyrrole rings and a central cobalt atom (Figure 1).

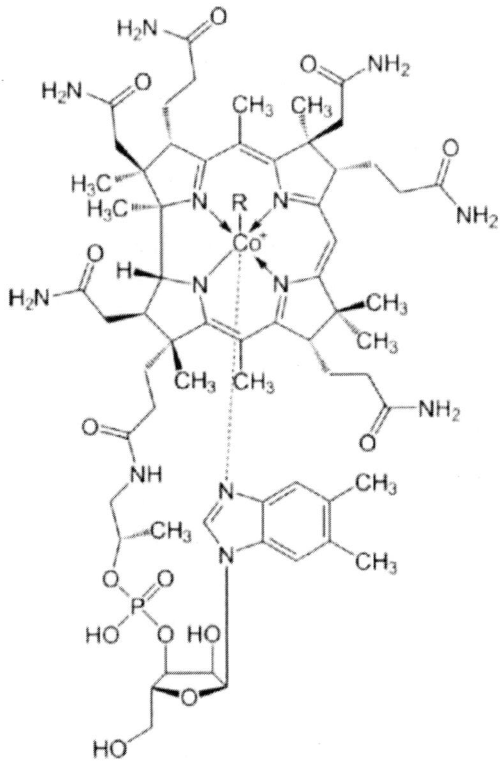

Figure 1. Chemical structure of cobalamin.

Substitution at the sixth ligand of the cobalt atom

- -with CN results in cyanocobalamin,
- -with OH- results in hydroxycobalamin,
- -with H_2O results in aquocobalamin,
- -with NO results in nitrocobalamin,
- -with CH_3 results in methylcobalamin,
- -with 5-Desoxyadenosyl results in adenosylcobalamin.

Vitamin B12 is a collective term for these variously substituted corrinoids. The principal biochemical participants are two coenzyme forms of Vitamin B12 that are produced and activated in two separate cellular compartments: methylcobalamin in the cytosol and adenosylcobalamin in the mitochondria.

Methylcobalamin functions as a cofactor to methionine synthase, which catalyses the remethylation of homocysteine to methionine, whereby the methyl group comes from the 5-methyl-THF. As the most important coenzyme in methylgroup transfer, it plays a significant role in transfer of C1 bodies, e.g., for choline synthesis and regeneration of methionine from homocysteine (methionine synthase) with participation of 5-methyltetrahydrofolic acid for the formation of the C1 body THF. Methylation of the cobalamin takes place in the cytosol.

Adencosylcobalamin is an important cofactor in rearrangement reactions, for example, of d-methylmalonyl-CoA-mutase in degradation of odd-numbered carbon skeleton of amino acids (threonine, valine, methionine) and fatty acids (propionyl-CoA). If this reaction cannot take place due to a lack of cobalamin, methylmalonyl-CoA is hydrolized to methylmalonic acid and eliminated in the urine. The formation of adenosylcobalamin takes place in the mitochondria. This requires that the central atom Co3+ is reduced to Co1+ in a reaction dependent on NADH/H+ and FADH2. The remaining adenosyl is supplied by ATP.

Cobalamin contributes significantly to haematopoiesis, myelin synthesis and synthesis of epithelial tissue. As a coenzyme, it is also a principal component of fatty acid, carbohydrate and nucleic acid metabolism [1].

2. VITAMIN B12 DEFICIENCY SYMPTOMS

A Vitamin B12 deficiency manifests in humans mainly as

-haematopoietic disorders, affecting in particular the formation of erythrocytes,

-neurological/psychiatric disorders and

-epithelial changes in the mucosa of the digestive tract [2].

The haematopoietic disorders and neurological sequelae have a special clinical relevance since they may be linked to severe or even life-threatening diseases. Whereas the haematopoietic changes are highly characteristic and are therefore primary indicators in diagnosis of a Vitamin B12 deficiency, the neurological disorders show a much greater range of variation and do not receive due recognition as effects of a B12 deficiency in clinical practice. Neurological disorders, however, are often the earliest and, in some cases, the only clinical symptoms of a functional Vitamin B12 deficiency. The incidence data vary. According to the IOM (Institute of Medicine, Washington, DC, USA), 75%–90% of persons with a clinically relevant B12 deficiency have neurological disorders, and in about 25% of cases these are the only clinical manifestations of the B12 deficiency [1,3,4]. The neurological disorders may occur together with the haematological changes or independently of them. On the whole, it can be assumed that about 60% of patients with pernicious anaemia will also manifest symptoms of a funicular myelosis. About one-quarter of the patients with confirmed Vitamin B12 deficiency and neurological disorders showed no haematological changes. Interestingly enough, there is also an inverse correlation between the severity of the haematological and neurological disorders [5]. The more severe the neurological disorders, the less significant the haematological changes, and vice versa. The causes of this are unknown.

Vitamin B12 Deficiency: Wide Variety of Neurological Manifestations

The most prominent neurological symptoms are paraesthesias or skin numbness, hands or feet that have "gone to sleep", unsteady gait and coordination disorders up to and including paralyses. These symptoms are expressions of funicular myelosis (funicular spinal cord disease, subacute combined degeneration (SCD) of the spinal cord) [6,7]. This condition develops as a result of combined degeneration of the lateral and posterior funiculi of the spinal cord due to defective myelin sheaths and is a demyelinating disease. Affected structures:

- -posterior funicular tracts (which convey tactile perception and proprioception),
- -cerebellar lateral funicular tracts (which also contribute to proprioceptive perception) and
- -the medullary pyramidal tract (which contributes to motor control).

The resulting neuropathy is symmetrical and is more pronounced in the legs than in the arms. In most cases, this neuropathy takes the form of a sensorimotor peripheral polyneuropathy, although mononeuropathies (optical or olfactory), autonomous neuropathies (impotence, incontinence), and combined forms (myelopathy and neuropathy) are also possible [3,7]. Patients initially notice a paraesthesia, at first in the feet and sometimes the hands, which then spreads over the limbs in the course of the disease. The sensory disturbances are followed by motor disorders (muscle weakness, paralysis symptoms, and motor coordination disorders). The nerve conduction velocity is reduced in motor and sensory nerves. Left untreated, the worst-case outcome of this disease would be paraplegia. Also possible are damage to central nerve tracts and cerebral disorders resulting in psychiatric symptoms. Remarkably there have been some cases of funicular myelosis with normal Vitamin B12 serum levels but with metabolical Vitamin B12 deficiency [8].

The symptoms of the cerebral disorders vary and may include confusion, stupor, apathy, memory and judgment disorders or even psychoses, depressions and dementia. The cerebral disorders, albeit less frequent than the peripheral symptoms, are underestimated in practice. Catatonia is also described as a psychiatric form of Vitamin B12 deficiency. Age-related limitations of mental performance, progredient cerebral atrophy, and Vitamin B12 status show clinically relevant interrelationships. The clinical manifestations of B12 deficiency in infants and small children whose mothers suffered from Vitamin B12 deficiency require special attention. Such infants may develop severe haematological and neurological disorders with lasting harmful effects on child development [9].

The diseases and disorders mentioned above may be caused by a variety of factors (e.g., drugs), among which B12 deficiency is often recognized too late. For this reason, the possibility of a Vitamin B12 deficiency should always be considered when such neurological disorders occur. In these cases, treatment with Vitamin B12 leads to rapid symptom alleviation.

3. DIAGNOSTICS OF B12 DEFICIENCY

There is no "gold standard" for laboratory chemistry confirmation of a clinically relevant Vitamin B12 deficiency. In practice, diagnosis of a Vitamin B12 deficiency is primarily done by determining the serum Vitamin B12 level (serum cobalamin level). This is a low-cost test with limited specificity and sensitivity, particularly in persons with Vitamin B12 concentrations <400 ng/L. The normal serum levels based on modern laboratory chemistry methods are 200–1000 ng/L.

Levels <200 ng/L (<150 pmol/L) are sure signs of a B12 deficiency; but a functional B12 deficiency may also be present at levels under 450 ng/L. Persons with B12 concentrations within the reference range may already manifest clinical signs of a Vitamin B12 deficiency [10]. Characteristic examples are cases of clinically and MRI-confirmed severe funicular myelosis despite normal Vitamin B12 levels and prompt response to B12 substitution [11].

Further investigative work is therefore required in this borderline area. This includes:

3.1. Determination of Holotranscobalamin

In the blood, Vitamin B12 is bound mainly to the proteins transcobalamin and haptocorrin, whereby only the smaller portion bound to transcobalamin (about 10%–25%) is bioavailable and active [12]. This complex of transcobalamin and Vitamin B12 is called holotranscobalamin (holo-TC) and its level is reduced in Vitamin B12 deficiency. Levels <35 pmol/L are indicative of a deficiency of Vitamin B12, and levels from 35 to 50 pmol/L represent a "grey area". A vitamin B12 deficiency is unlikely at levels >50 pmol/L. A reduced holo-TC level is considered the earliest marker for a B12 deficiency. Holo-TC is therefore considered a better marker than serum B12 for determination of B12 status. Theoretically HoloTC should be a better marker than serum B12 but the fact that its confounders are not well investigated, its short half-life in plasma (90 min) and the poor specificity of the HoloTC test should be acknowledged [13].

3.2. Determination of Methylmalonic Acid

Methylmalonic acid (MMA) is considered a highly sensitive functional indicator for a Vitamin B12 deficiency [14,15]. MMA is a metabolic product, the breakdown of which requires Vitamin B12. A lack of Vitamin B12 allows MMA levels to increase significantly. This also makes it possible to detect a functional B12 deficiency in which raised MMA levels and clinical symptoms of a B12 deficiency in the form of neurological and/or haematological disorders may occur despite normal serum cobalamin levels [14]. MMA can be measured in both serum and urine. Cases of impaired renal function, however, may also manifest raised MMA, for which reason the creatinine level should also be determined to confirm sufficient renal concentration capacity. The normal range for serum MMA is 50–300 nmol/L. MMA determination is a highly sensitive test that can confirm a diagnosis even in the early stages of a B12 deficiency and is also a suitable indicator for treatment response (B12 substitution). In contrast to determination of the serum cobalamin level, which is only sufficiently sensitive in cases of alimentary B12 deficiency and disturbed enteral absorption of Vitamin B12, determination of MMA also facilitates detection of less frequent dysfunctions in the Vitamin B12 transport path and intracellular synthesis of active adenosylcobalamin. A round-table discussion in 2010 on selection of the biomarkers for determination of B12 status in continued NHANES studies concluded that determination of serum B12 and methylmalonic acid (MMA) should be recommended for further studies [16].

Another indirect functional parameter of Vitamin B12 status is homocysteine. This potentially toxic amino acid results from demethylation of the essential amino acid methionine, which reaction requires Vitamin B12. A deficiency of Vitamin B12 leads to accumulation and thus raised blood levels of homocysteine. A serum homocysteine level >10 μmol/L thus indicates a possible deficiency of Vitamin B12. However, homocysteine is not a specific B12 deficiency marker, since deficiencies of Vitamin B6 and folic acid can also raise the homocysteine level.

Determination of methylmalonic acid and homocysteine are particularly recommended in cases of diagnostically unclarified B12 deficiency. A Vitamin B12 deficiency can be excluded with something approaching 100% certainty if levels of these metabolites are within the normal ranges [16].

4. INCIDENCE OF B12 DEFICIENCY

The incidence of Vitamin B12 deficiency is underestimated in practice. This is due mainly to statements issued repeatedly regarding provision of a sufficient supply of Vitamin B12 to the populace. Representative data on current nutritional behavior in Germany is provided by the National Nutrition Survey II (2008) [17]. This survey also included vitamin consumption in Germany. The investigations revealed a deficient supply of Vitamin B12 in a considerable number of persons of different ages and sexes (Table 1).

Table 1. Number of persons with dietary Vitamin B12 intake below the reference ranges, expressed as percentages.

Sex	Age in years				
	19–24	**25–34**	**35–50**	**51–64**	**65–80**
Men	7.4	6.8	8.4	7.9	9.8
Women	32.7	26.4	24.5	23.0	26.3

Thus, the daily dietary intake levels alone reveal a risk of pathological B12 deficiency status in a considerable number of persons due to malnutrition.

However, not every case of insufficient vitamin intake leads to a vitamin deficiency. Insufficient vitamin intake is defined as intake of an amount below the reference daily vitamin intake amounts, but vitamin deficiency means development of clinically relevant, measurable disorders or characteristic deficiency symptoms due to insufficient intake. Nonetheless, any person with insufficient vitamin intake is also at risk for development of a functional vitamin deficiency. The incidence of Vitamin B12 deficiency confirmed by laboratory chemical analysis is reported as 5%–7% for younger persons [8]. Vitamin B12 deficiency is widespread in the elderly in particular and has been diagnosed in 10%–30% of "healthy" persons above the age of 65 [3,18]. Despite intake of Vitamin B12 in amounts in excess of the reference levels, a person may suffer from a functional B12 deficiency [10]. Vitamin B12 deficiency is considered to be a worldwide problem [3].

5. CAUSES OF B12 DEFICIENCY

The causes of Vitamin B12 deficiency are varied and can be classified in four groups:

1. Dietary deficiency
   - Due to insufficient dietary intake of Vitamin B12; risk groups include alcoholics, vegetarians/vegans and older persons.
2. Malabsorption (disturbed uptake of Vitamin B12)
   - due to lack of intrinsic factor or parietal cells, e.g., in cases of pernicious anaemia, atrophic gastritis, postgastrectomy syndrome
   - disturbed uptake of Vitamin B12 from food due to gastric acid deficiency (e.g., in cases of long-term intake of acid secretion inhibiting substances (proton pump inhibitors, H-2 blockers)
   - drug interactions (e.g., antibiotics, anticonvulsants, colchicine, metformin, N_2O)
3. Intestinal diseases
   - intestinal resection, tropical sprue, Crohn's disease, Zollinger-Ellinger syndrome, intestinal bypass, Imerslund-Grasbeck syndrome
   - bacterial overgrowth (e.g., helicobacter)

- Fish tapeworm infestation
- drug interactions (e.g., metformin, chronic exposure to N_2O)

4. Disturbed utilization of Vitamin B12 and inborn errors of Vitamin B12 and metabolism [19] Malabsorption is the underlying cause in most cases.

5.1. Proton Pump Inhibitors (e.g., Omeprazole) and Vitamin B12

A number of prescription drugs can deplete vitamin B12 stores, particularly in older patients. Over 60% of sales of proton pump inhibitors are accounted for by gastrointestinal drugs. In patients whose stomach acid content has been reduced by long-term treatment with a proton pump inhibitor (e.g., omeprazole) due to peptic disease, special attention should be directed to Vitamin B12 intake. Proton pump inhibitors reduce gastric acid secretion and thus the intestinal release of Vitamin B12 from foods. Current research suggests that up to 40% of persons over 65 have an insufficient intake of Vitamin B12 (Vitamin B12 serum levels < 450 ng/L) [20]. Particularly problematical is the fact that a Vitamin B12 deficiency is often associated with a folic acid deficiency and significantly increases the risk for a hyperhomocysteinaemia (>10 µmol/L). Folate and Vitamin B12 play an important role in the function of the CNS. Both vitamins are involved in methionine-homocysteine metabolism. Methionine becomes activated to S-adenosylmethionine (SAM), which is indispensable for numerous methylation reactions. Disturbances of these processes are involved in neurodegenerative diseases.

In older individuals, Vitamin B12 deficiency is due in most cases to insufficient production of gastric juice (achlorhydria). The main cause is inflammatory processes in the gastric mucosa, which develop primarily as a result of type B atrophic gastritis type B with subacidity and anacidity, as well as reduced intrinsic factor production. Protein-bound Vitamin B12 is therefore insufficiently released and absorbed (Figure 2). Cobalamins from food bound to protein are released in the stomach by hydrochloric acid and pepsin and bound to intrinsic factor (IF-B12 complex) in a pH-dependent reaction. Cellular cobalamin uptake in the mucosal epithelium takes place in the terminal ileum with the help of specific receptors of the brush-border membrane. The IF-B 12 complex binds to cubilin, which initiates calcium-dependent, receptor-mediated endocytosis together with another protein, megalin. Vitamin B12 uptake can also take place independently of the intrinsic factor by means of passive diffusion in the small intestine. This mechanism, however, is not very efficient, since only about 1% of the ingested Vitamin B12 dose is absorbed.

Reduced gastric acid production in the stomachs of older persons leads to alkalizing of the small intestine milieu, which eliminates the physiological barrier to microorganisms. Bacteria from lower intestinal segments can then increasingly enter the jejunum and ileum. Bacterial overgrowth with Clostridia and Campylobacter is accompanied by increased consumption of Vitamin B12 (conversion into inactive cobalamide) and by bacterial synthesis of substances that compete with the vitamin for receptors in the ileum mucosa. This further reduces the availability of Vitamin B12. In older persons (>65 years), atrophy of the gastric mucosa is frequently due to an infection with Helicobacter pylori. As many as 60% of elderly persons exhibit an increased infestation of the gastric mucosa with Helicobacter pylori and thus a high risk of developing chronic atrophic gastritis [2,21,22]. A regular supplementation of at least 100 µg Vitamin B12 daily in combination with folic acid and vitamin B6 is recommended for persons who regularly take proton pump inhibitors to reduce gastric acid secretion. Since the amount of actively absorbed Vitamin B12 depends on the availability of intrinsic factor (IF), some authors consider a maximum drug dosage (e.g., 1000 µg B12/day p.o.), thus allowing uptake of even small amounts of the vitamin through passive absorption [2].

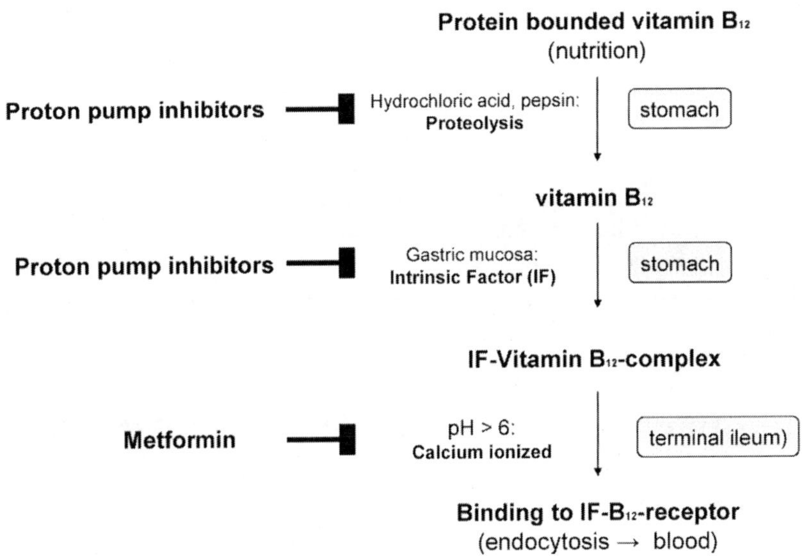

Figure 2. Inhibition of the active absorption of Vitamin B12 by proton pump inhibitors and metformin (inhibition).

5.2. Metformin and Vitamin B12

Metformin is used mainly in treatment of overweight type 2 diabetics since, in contrast to the insulinotrophic antidiabetic drugs, it does not generally initiate weight gain and/or hypoglycaemic attacks. In earlier clinical studies, Vitamin B12 malabsorption and a drop in Vitamin B12 serum levels by up to 30% were repeatedly observed under long-term treatment with metformin (Figure 2) [23]. Metformin leads to a disruption of calcium-dependent and receptor-mediated endocytosis of the intrinsic factor-Vitamin B12 complex in the terminal ileum [2].

These data are confirmed by the findings of current studies. In a case-control study, 155 diabetics who had developed a Vitamin B12 deficiency under treatment with metformin (average Vitamin B12 serum levels 148.6 ± 40.4 pg/mL (110 ± 30 pmol/L)) were compared with 310 control subjects who showed no deficiency of Vitamin B12 under the same medication. After adjustment for potential influencing factors, this resulted in a statistically significant correlation between the duration and dosage of metformin treatment and Vitamin B12 deficiency. Each dose increase by 1 g/day increased the risk of a Vitamin B12 deficiency by more than double (odds ratio 2.88; 95% confidence interval, 2.15–3.87; $p < 0.001$). In comparison with a treatment duration of less than 3 years, the odds ratio for a treatment duration of 3 years and longer was 2.39 (95% confidence interval, 1.46–3.91; $p = 0.001$) [23]. In another study of 165 type 2 diabetics, the influence of metformin and rosiglitazone on Vitamin B12 and folic acid status and homocysteine levels was observed. In this study, under the 6-week therapy with metformin, homocysteine levels rose by 2.36 μmol/L, and blood levels of folic acid and Vitamin B12 dropped. Rosiglitazone, on the other hand, showed no impact on Vitamin B12 and folic acid status [24]. Similar results were observed in controlled interventional studies [25]. In a recent study with 126 patients with diabetes metformin treatment was associated with impaired cognitive function. Vitamin B12 and calcium supplements may alleviate metformin-induced vitamin B12 deficiency and were associated with better cognitive outcomes [26].

6. GROUPS AT RISK FOR VITAMIN B12 DEFICIENCY

The groups at risk for a Vitamin B12 deficiency include mainly

- older persons
- vegetarians/vegans
- persons with gastrointestinal diseases
- persons with raised Vitamin B12 requirements (pregnant women, breastfeeding women, patients with autoimmune diseases or an HIV infection)

- persons under long-term treatment with proton pump inhibitors, metformin or H-2 blockers
- patients with renal diseases.

Deficiency of Vitamin B12 in elderly persons (>65 years) is due mainly to malfunction of the uptake of Vitamin B12 in the gastrointestinal tract (malabsorption). In studies of older patients with Vitamin B12 deficiency, 53% suffered from malabsorption and 33% from pernicious anaemia, and in only 2% of cases was the condition ascribed to a dietary cause. The aetiology of the Vitamin B12 deficiency remained unclear in 11% [27]. In addition, older people often have atrophic gastritis or a lack of stomach acid from other causes. Vegetarians, and vegans in particular, have an increased risk of developing a B12 deficiency in view of the fact that foods from animal sources are the main sources of Vitamin B12 [28]. In a study of lacto-vegetarians and lacto-ovo-vegetarians, 63% of the subjects showed raised methylmalonic acid levels (>271 nmol/L) and 73% reduced holotranscobalamin levels (<35 pmol/L). Vegans had raised methylmalonic acid levels in 86% of the cases and reduced holotranscobalamin levels in 90% [29]. Also particularly at risk are infants from mothers with a Vitamin B12 deficiency. These infants develop growth disorders, severe inhibition of the psychomotor development, muscular hypotonia, and brain atrophy [29]. For this reason it is essential to ensure pregnant women and nursing mothers have a sufficient intake of Vitamin B12. The importance of infestation with Helicobacter pylori or treatment with drugs such as metformin or proton pump inhibitors resulting in development of a B12 deficiency is also underestimated [2,21]. Another clinically relevant factor is "cobalamin resistance", in which a functional Vitamin B12 deficiency can develop despite normal serum B12 values and adequate intake of Vitamin B12 with the diet [30]. This aspect must be considered in particular in older persons and patients with renal diseases and diabetes mellitus.

7. PREVENTION AND TREATMENT OF B12 DEFICIENCY

The cause of B12 deficiency-induced haematological and neuropsychiatric diseases is the functional deficiency of Vitamin B12. This represents confirmed, internationally recognized state-of-the-art science. Therefore, diseases of this kind must be treated by administration of Vitamin B12. It has been satisfactorily demonstrated that these conditions can be effectively treated with Vitamin B12 substitution. Substitution must begin as early as possible to avoid irreversible damage. Both hydroxycobalamin and methylcobalamin and cyanocobalamin are suitable treatment. The most comprehensive catalogue of experiential data available is on cyanocobalamin.

The recommended intake of Vitamin B12 (according to the German, Austrian and Swiss Nutrition Societies, known as DACH) is 3 μg/day for adults (healthy) and 3.5–4.0 μg/day for pregnant and lactating women. In deficiencies due to alimentary problems, dosages of 10–100 μg/day lead to normalization of the levels. Higher dosages, however, are needed in cases of malabsorption, intestinal disease, or disorders affecting the utilization of Vitamin B12.

Absorption of physiological doses takes place via an active absorption mechanism. Protein-bound dietary cobalamin is released in the stomach by pepsin and hydrochloric acid and bound to an r-protein in the gastric juice produced by the salivary glands. The cobalamin is released in the duodenum and bound to intrinsic factor (IF). The IF-cobalamin complex is then absorbed in the ileum with the help of specific receptors. However, the amount absorbed through this active mechanism is limited, due to the limited intake capacity of the ileum mucosa, to 1.5 μg. This maximum amount is achieved by oral administration of 10 μg Vitamin B12. Vitamin B12 uptake is also possible by means of passive diffusion. This, however, requires higher dosages because this absorption pathway is less efficient. In cases of severe disease, especially those involving limitations on enteral intake, larger amounts of B12 are administered in initial parenteral doses to achieve a rapid onset surge [31].

Oral or Parenteral Administration?

Parenteral administration is still the dominant method in clinical practice. The reason for this is the assumption that oral Vitamin B12 administration is ineffective. This is not in agreement with the current state of knowledge. Experimental and clinical studies have shown that treatment is also perfectly feasible with oral administration of Vitamin B12. The oral dosage required for normalization of Vitamin B12 status depends on the causes of the deficiency and the severity of the disease. In cases of deficiency due to alimentary causes, dosage 10–100 μg/day leads to normalization of the levels. However, higher dosages are needed in cases of malabsorption, intestinal disease, or disorders affecting the utilization of Vitamin B12. Randomized double-blind dosage determination studies with daily administration of 2.5, 100, 250, 500 and 1000 μg of cyanocobalamin have shown that to normalize the levels (cobalamin and holo-TC as well as MMA and homocysteine), doses of >500–1000 μg/day are required [32]. Hence, the lowest dose of oral cyanocobalamin required to normalize mild Vitamin B12 deficiency is more than 200 times the recommended dietary allowance of approximately 3 μg daily (*i.e.*, >500 μg per day). This is also confirmed by clinical findings on factors influencing the haematological and/or neurological Vitamin B12 deficiency symptoms.

The clinical studies confirm the efficacy of orally administered cyanocobalamin in treatment of B12 deficiency-related diseases. In summing up all of the available studies, it can be assumed that in milder forms of Vitamin B12 deficiency, e.g., due to alimentary causes, doses of >100 µg/day cyanocobalamin would be recommended for prophylaxis and treatment of deficiencies and for forms resulting from absorption disorders, doses of 500–2000 µg/day would be required. According to a review by the *Cochrane Group*, the efficacy of orally administered cyanocobalamin as a curative treatment with initial doses of 1000–2000 µg daily, then weekly, is confirmed and is just as effective as parenteral administration. Initial parenteral administration is recommended only in severe neurological disorders in order to achieve a rapid onset surge of Vitamin B12 [33].

8. COGNITIVE IMPROVEMENT

Among the most important parameters influencing quality of life in the elderly is their nutritional and micronutrient status. For a variety of reasons, older persons are at increased risk for micronutrient deficiencies. In addition to age-related changes in organs and chronic diseases, regular intake of various medicinal drugs (such as metformin, proton pump inhibitors) also plays a role [2]. A micronutrient critical in this regard is Vitamin B12, the absorption and utilization of which is often limited in elderly subjects (>60 years). As shown by current results of an Oxford University study, improved intake of Vitamin B12 in the elderly may significantly reduce the risk of age-related brain atrophy [34].

Vitamin B12-dependent methylation reactions play a central role in intermediary metabolism. Vitamin B12 regulates, together with 5-methyl-tetrahydrofolic acid, the remethylation of homocysteine to L-methionine and the subsequent ATP-dependent formation of S-adenosylmethionine (SAM). SAM is essential to most biological methylation reactions including the methylation of myelin, neurotransmitters, and phospholipids (e.g., phosphatidylcholine). A dietary Vitamin B12 and/or folic acid deficiency is one of the most common causes of hyperhomocysteinaemia.

An increase in plasma homocysteine levels is an indication of a disorder of the methyl group metabolism, which is associated with an increased risk for neuronal damage, impairment of cell proliferation, and an increased tendency to chromosome strand breaks. According to studies, a latent inadequate supply of Vitamin B12 and elevated homocysteine plasma levels (>10 µmol/L) in older persons are associated with a decline in cognitive performance [35,36,37]. The variance in cognitive performance tests in elderly subjects is attributed, regardless of the intelligence quotient, to the homocysteine levels in approximately 11% of cases [38]. The occurrence of depression in the elderly is also frequently associated with a low Vitamin B12 status and increased homocysteine levels. Hyperhomocysteinaemia, and the associated hypomethylation of the CNS, is also considered to be an independent risk factor for Alzheimer's-type dementia (DAT) (Figure 3). In the major Framingham study, the risk of Alzheimer's disease was nearly doubled at homocysteine levels >14 μmol/L [39]. Raised homocysteine levels are associated in elderly subjects with a smaller hippocampus and an increased rate of atrophy of the medial temporal lobe and white matter [40,41,42].

A recent prospective University of Oxford study carried out with 107 persons aged between 61 and 87 years shows a significant association between Vitamin B12 status and the size of the brain. In addition to cognitive tests and measurement of the brain with MRI, the plasma Vitamin B12 and holotranscobalamin (holoTC) levels were recorded in the healthy subjects at study onset and once a year. After 5 years, age-related brain atrophy was most advanced in the subjects with the lowest Vitamin B12 status. Compared to study participants with the highest Vitamin B12 baseline levels, the lowest tertile (plasma Vitamin B12 levels <308 pmol/L, plasma holotranscobalamin levels <54 pmol/L) was associated with a risk level raised by more than 6-fold for loss of brain volume (odds ratio 6.17; 95% confidence interval, 1.25–30.47). Of particular note is the fact that atrophy of the brain already occurred with a marginal Vitamin B12 deficiency and not at the Vitamin B12 deficiency level currently defined as manifest disease [34].

In a recent randomized and double-blind interventional study (VITACOG study) involving 168 elderly persons with mild cognitive impairment (age: >70 years), supplementation of Vitamin B12 (500 p, g/day, p.o.), folic acid (0.8 mg/day, p.o.) and vitamin B6 (20 mg/day, p.o.) over a period of 24 months significantly slowed the progression of brain atrophy and the reduction of cognitive performance (by 53.3%, $p = 0.001$) compared with the placebo group in the subjects with homocysteine levels >13 μmol/L. The accelerated rate of brain atrophy in elderly with mild cognitive impairment can be slowed by treatment with homocysteine-lowering B vitamins [43]. Vitamin B12, folic acid and vitamin B6 lower homocysteine, which directly leads to a decrease in gray matter atrophy, thereby slowing cognitive decline. B-vitamin supplementation can slow the atrophy of specific brain regions that are a key component of the Alzheimer's disease process and that are associated with cognitive decline [44].

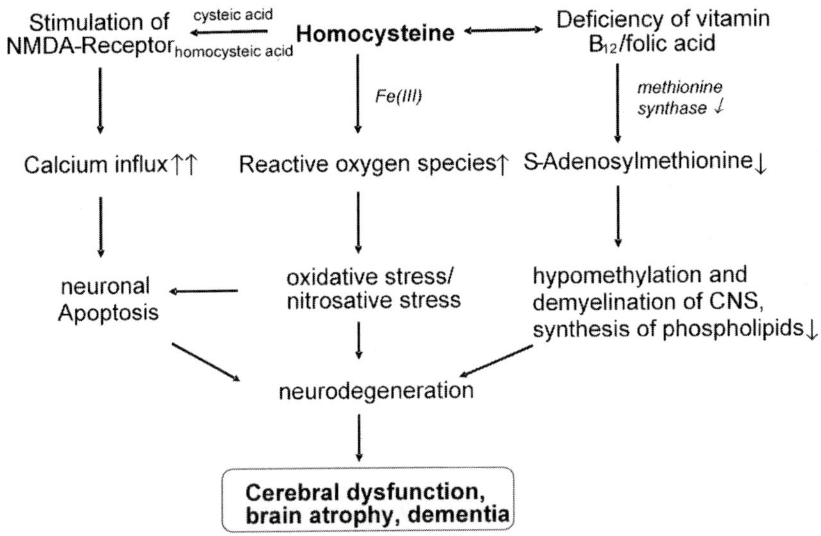

Figure 3. Possible neurotoxic effects of homocysteine (model).

9. CONCLUSIONS

The incidence of a Vitamin B12 deficiency has to date received too little attention in orthodox medical practice. In particular, older persons, patients on long-term medication and those with neurological disorders can all benefit from adjuvant Vitamin B12 administration. Cognitive performance can be improved, and the risk of brain atrophy reduced, by Vitamin B12.

CONFLICTS OF INTEREST

The authors declare no conflict of interest.

REFERENCES

1. Gröber, U. Micronutrients: Metabolic Tuning—Prevention—Therapy. Drug Metab. Drug Interact. 2009,24, 331.
2. Gröber, U. Interactions between drugs and micronutrients. Med. Monatsschr. Pharm. 2006, 29, 26–35.
3. Allen, L.H. How common is vitamin B-12 deficiency? Am. J. Clin. Nutr. 2009, 89, 693S–696S.

4. Standing Committee on the Scientific Evaluation of Dietary Reference Intakes and Its Panel on Folate, Other B Vitamins, and Choline and Subcommittee on Upper Reference Levels of Nutrients; Food and Nutrition Board; Institute of Medicine. In Dietary Reference Intakes for Thiamin, Riboflavin, Niacin, Vitamin B6, Folate, Vitamin B12, Pantothenic Acid, Biotin, and Choline; National Academy Press: Washington, DC, USA, 1998.

5. Healton, E.B.; Savage, D.G. Neurologic aspects of cobalamin deficiency. Medicine (Baltimore) 1991, 70, 229–245.

6. Neurologic Manifestations Related to Deficiency of Vitamin B12. Available online: http://yassermetwally.com/blog/?p=194 (accessed on 10 December 2013).

7. Sethi, N.K.; Robilotti, E.; Sadan, Y. Neurological manifestations of Vitamin B12 deficiency. Internet J. Nutr. Wellness 2005, 2, 1–12.

8. Lorenzl, S.; Vogeser, M.; Müller-Schunk, S. Clinically and MRI documented funicular myelosis in a patient with metabolical Vitamin B12 deficiency but normal Vitamin B12 serum level. J. Neurol. 2003,250, 1010–1011.

9. Dror, D.K.; Allen, L.H. Effect of Vitamin B12 deficiency on neurodevelopment in infants: Current knowledge and possible mechanisms. Nutr. Rev. 2008, 66, 250–255.

10. Dali-Youcef, N.; Andres, E. An update on cobalamin deficiency in adults. QJM 2009, 102, 17–28.

11. Herrmann, W.; Obeid, R. Functional Vitamin B12 deficiency and determination of holotranscobalamin in populations at risk. Clin. Chem. Lab. Med. 2003, 41, 1478–1488.

12. Nexo, E.; Hoffmann-Lucke, E. Holotranscobalamin, a marker of vitamin B-12 status: Analytical aspects and clinical utility. Am. J. Clin. Nutr. 2011, 94, 359S–365S.

13. Schrempf, W.; Eulitz, M.; Neumeister, V.; Siegert, G.; Koch, R.; Reichmann, H.; Storch, A. Utility of measuring Vitamin B12 and its active fraction, holotranscobalamin, in neurological Vitamin B12 deficiency syndromes. J. Neurol. 2011, 258, 393–401.

14. Lindenbaum, J.; Savage, D.G. Diagnosis of cobalamin deficiency: II. Relative sensitivities of serum cobalamin, methylmalonic acid, and total homocysteine concentrations. Am. J. Hematol. 1990, 34, 99–107.

15. Troxler, H.; Hersberger, M.; Baumgartner, M. Methylmalonsäure zur bestimmung des Vitamin B12-mangels. Schweiz. Med. Forum 2008, 8, 823–825. (in German).

16. Savage, D.G.; Lindenbaum, J. Sensitivity of serum methylmalonic acid and total homocysteine determinations for diagnosing cobalamin and folate deficiencies. Am. J. Med. 1994, 96, 239–246.

17. Nationale Verzehrsstudie II (National Nutrtion Survey II), 2008. Available online: http://www.nationale-verzehrsstudie.de (accessed on 4 December 2013).

18. Herrmann, W.; Obeid, R. The usefulness of holotranscobalamin in predicting Vitamin B12 status in different clinical settings. Curr. Drug Metab. 2005, 6, 47–53.

19. Coelhoo, D.; Suormala, T.; Stucki, M.; Lerner-Ellis, J.P.; Rosenblatt, D.S.; Newbold, R.F.; Baumgartner, M.R.; Fowler, B. Gene identification for the cblD defect of Vitamin B12 metabolism. N. Engl. J. Med.2008, 358, 1454–1464.

20. Obeid, R.; Schorr, H.; Eckert, R.; Herrmann, W. Vitamin B12 status in the elderly as judged by available biochemical markers. Clin. Chem. 2004, 50, 238–241.

21. Kaptan, K.; Beyan, C.; Ural, A.U. Helicobacter pylori—Is it a novel causative agent in Vitamin B12 deficiency? Arch. Intern. Med. 2000, 160, 1349–1353.

22. Seal, E.C.; Metz, J.; Flicker, L. A randomized, double-blind, placebo-controlled study of oralVitamin B12 supplementation in older patients with subnormal or borderline serum Vitamin B12 concentrations. J. Am. Geriatr. Soc. 2002, 50, 146–151.

23. De Jager, J.; Kooy, A.; Lehert, P. Long term treatment with metformin in patients with type 2 diabetes and risk of vitamin B-12 deficiency: Randomised placebo controlled trial. BMJ 2010, 340.

24. Sahin, M. Effects of metformin or rosiglitazone on serum concentrations of homocysteine, folate, and Vitamin B12 in patients with type 2 diabetes mellitus. J. Diabetes Complicat. 2007, 21, 118–123.

25. Ting, R.Z. Risk factors of vitamin B(12) deficiency in patients receiving metformin. Arch. Intern. Med.2006, 166, 1975–1979.

26. Moore, E.M.; Mander, A.G.; Ames, D.; Kotowicz, M.A.; Carne, R.P.; Brodaty, H.; Woodward, M.; Boundy, K.; Ellis, K.A.; Bush, A.I.; et al. Increased risk of cognitive impairment in patients with diabetes is associated with metformin. Diabetes Care 2013, 36, 2981–2987.

27. Henoun Loukili, N.; Noel, E. Cobalamin deficiency due to non-immune atrophic gastritis in elderly patients. A report of 25 cases. J. Nutr. Health Aging 2005, 9, 462.

28. Herrmann, W.; Schorr, H. Vitamin B-12 status, particularly holotranscobalamin II and methylmalonic acid concentrations, and hyperhomocysteinemia in vegetarians. Am. J. Clin. Nutr. 2003, 78, 131–136.

29. Lücke, T.; Korenke, G.C. Mütterlicher Vitamin-B12-Mangel: Ursache neurologischer Symptomatik im Säuglingsalter. Z. Geburtshilfe Neonatol. 2007, 211, 157–161. (in German).

30. Solomon, L.R. Disorders of cobalamin (Vitamin B12) metabolism: Emerging concepts in pathophysiology, diagnosis and treatment. Blood Rev. 2007, 21, 113–130.

31. Pietrzik, K.; Golly, I.; Loew, D. Handbuch der Vitamine. Für Prophylaxe, Beratung und Therapie(in German), 1st ed.; Urban & Fischer: Munich, Germany, 2008; pp. 126–146.

32. Eussen, S.J.; de Groot, L.C. Oral cyanocobalamin supplementation in older people with Vitamin B12 deficiency: A dose-finding trial. Arch. Intern. Med. 2005, 165, 1167–1172.

33. Vidal-Alaball, J.; Butler, P.P. Oral Vitamin B12 versus intramuscular Vitamin B12 for Vitamin B12 deficiency. Cochrane Database Syst. Rev. 2009, 3, CD004655.

34. Vogiatzoglou, A. Vitamin B12 status and rate of brain volume loss in community-dwelling elderly.Neurology 2008, 71, 826–832.

35. Riggs, K.M. Relations of Vitamin B-12, Vitamin B-6, folate, and homocysteine to cognitive performance in the Normative Aging Study. Am. J. Clin. Nutr. 1996, 63, 306–314.

36. Blasko, I.; Hinterberger, M.; Kemmler, G. Conversion from mild cognitive impairment to dementia: Influence of folic acid and Vitamin B12 use in the VITA cohort. J. Nutr. Health Aging 2012, 16, 687–694.

37. Vogiatzoglou, A.; Smith, A.D.; Nurk, E. Cognitive function in an elderly population: Interaction between Vitamin B12 status, depression, and apolipoprotein E s4: The Hordaland Homocysteine Study. Psychosom. Med. 2013, 75, 20–29.

38. Budge, M. Plasma total homocysteine and cognitive performance in a volunteer elderly population.Ann. N. Y. Acad. Sci. 2000, 903, 407–410.

39. Seshadri, S. Plasma homocysteine as a risk factor for dementia and Alzheimer's disease. N. Engl. J. Med. 2002, 346, 476–483.

40. Den Heijer, H. Homocysteine and brain atrophy on MRI of non-demented elderly. Brain 2003, 126, 170–175.

41. Williams, J.H. Minimal hippocampal width relates to plasma homocysteine in community-dwelling older people. Age Ageing 2002, 31, 440–444.

42. Scott, T.M. Homocysteine and B vitamins relate to brain volume and white-matter changes in geriatric patients with psychiatric disorders. Am. J. Geriatr. Psychiatry 2004, 12, 631–638.

43. Smith, A.D. Homocysteine-lowering by B vitamins slows the rate of accelerated brain atrophy in mild cognitive impairment: A randomized controlled trial. PLoS One 2010, 5, e12244.

44. Douaud, G.; Refsum, H.; de Jager, C.A.; Jacoby, R.; Nichols, T.E.; Smith, S.M.; Smith, A.D. Preventing Alzheimer's disease-related gray matter atrophy by B-vitamin treatment. Proc. Natl. Acad. Sci. USA2013, 110, 9523–9528.

8

DOCOSAHEXAENOIC ACID AND COGNITION THROUGHOUT THE LIFESPAN

Michael J. Weiser [1,*], Christopher M. Butt [1] and M. Hasan Mohajeri [2]

[1]DSM Nutritional Products, R&D Human Nutrition and Health, Boulder, CO, USA

[2]DSM Nutritional Products, R&D Human Nutrition and Health, Basel, Switzerland

ABSTRACT

Docosahexaenoic acid (DHA) is the predominant omega-3 (*n*-3) polyunsaturated fatty acid (PUFA) found in the brain and can affect neurological function by modulating signal transduction pathways, neurotransmission, neurogenesis, myelination, membrane receptor function, synaptic plasticity, neuroinflammation, membrane integrity and membrane organization. DHA is rapidly accumulated in the brain during gestation and early infancy, and the availability of DHA via transfer from maternal stores impacts the degree of DHA incorporation into neural tissues. The consumption of DHA leads to many positive physiological and behavioral effects, including those on cognition. Advanced cognitive function is uniquely human, and the optimal development and aging of cognitive abilities has profound impacts on quality of life, productivity, and advancement of society in general. However, the modern diet typically lacks appreciable amounts of DHA. Therefore, in modern populations, maintaining optimal levels of DHA in the brain throughout the lifespan likely requires obtaining preformed DHA via dietary or supplemental sources. In this review, we examine the role of DHA in optimal cognition during development, adulthood, and aging with a focus on human evidence and putative mechanisms of action.

Keywords: brain lipids; omega-3 polyunsaturated fatty acids; nutrition; learning; memory; comprehension; development; aging; neurodegeneration

1. INTRODUCTION

The cognitive ability of humans is arguably the most advanced in the entire animal kingdom. Such advancement is believed to be conferred by an expanded cerebral cortex and a highly developed prefrontal cortex, both of which are brain regions important for cognition. There are many different domains to the clinical construct of cognition. These include attention, memory (working and long-term), perception, language, problem solving, comprehension, reasoning, computation, reading and speech. Cognition changes throughout the lifespan matching the development, maturation and aging of the brain. The brain is a lipid-rich organ that consumes 20% of the body's energy, but it only comprises 2% of the body's mass. Over half of the brain's dry weight is comprised of lipids, and it is especially enriched in long-chain omega-3 (*n*-3) polyunsaturated fatty acids (PUFAs), suggesting a key role for these molecules in the optimal development, maturation and aging of neural structures and networks. A substantial amount of literature exists that highlights the crucial role of nutrition in brain development, and thus on brain function and mental performance in humans. Ultimately, the proper functioning of the brain has significant dependence upon maintaining its optimal lipid composition [1].

Quantitatively, docosahexaenoic acid (DHA; 22:6*n*-3) is the most significant *n*-3 PUFA in the brain as both eicosapentaenoic acid (EPA; 20:5*n*-3) and α-linolenic acid (ALA; 18:3*n*-3) are present in only very small quantities. DHA makes up over 90% of the *n*-3 PUFAs in the brain and 10%–20% of its total lipids. DHA is especially concentrated in the gray matter [2]. It is stored primarily in phosphatidylethanolamine (PE) and phosphatidylserine (PS) membrane phospholipids, with smaller amounts also found in phosphatidylcholine (PC; [3]), where it plays an important role in the biosynthesis of PS (DHA-PS) in the brain [4]. DHA is enriched in membranes structures found at synaptic terminals, mitochondria and endoplasmic reticulum [5], and it can ultimately affect cellular characteristics and physiological processes including membrane fluidity, lipid raft function, neurotransmitter release, transmembrane receptor function, gene expression, signal transduction, myelination, neuroinflammation, and neuronal differentiation and growth [6,7,8].

The brain's frontal lobes are particularly responsive to the supply of DHA during development [9]. Decades of work have clearly established the responsibilities of the frontal lobes for executive and higher-order cognitive activities including sustained attention, planning and problem solving [10], and the prefrontal lobe in particular for social, emotional and behavioral development [11]. Therefore, maintaining optimal lipid composition in these brain regions, and specifically DHA levels, is not only important during the development and maturation of the brain from gestation through childhood and adolescence [12,13], but such maintenance is also critical for successful aging of the adult brain [1,14,15,16]. In this review, we will discuss the importance of DHA for optimal neurological health throughout the lifespan, with a particular emphasis on cognitive function.

2. DHA DELIVERY TO THE BRAIN

DHA synthesized *de novo* originates from ALA via a series of desaturations and elongations primarily within the endoplasmic reticulum (ER), with the exception of the last step, a β-oxidation from tetracosahexaenoic acid (24:6*n*-3) that occurs in peroxisomes [17]. ALA is considered an essential nutrient because humans lack the *n*-3 desaturase enzyme required for its production. However, it could be argued that DHA is also an essential nutrient due to inefficiencies of the 5-desaturase and 6-desaturase enzymes (FADS1/2) needed for its biosynthesis, and the competition for these enzymes by the omega-6 (*n*-6) PUFA linoleic acid (LA; 18:2*n*-6). LA is typically consumed in high amounts in modern diets, which exacerbates the increase of *n*-6 PUFAs, as well as the decrease of *n*-3 PUFAs, that are incorporated in peripheral and neural tissues [18]. Therefore, many researchers conclude that preformed DHA consumption is required for reaching and maintaining ideal brain DHA concentrations and related neurological functions [19,20,21].

There is a general consensus on the recommended daily average intake requirements for infants. This consensus is driven by data from randomized control trials (RCTs) that suggest a minimum level of DHA at 0.32% of total fatty acids in formula [22,23,24], which is towards the lower half of the global concentration range found in human milk (0.06% to 1.4% of total fatty acids) [25,26]. The evidence for daily intake requirements for children and adults is still emerging. Nonetheless, current guidelines are in the range of 250 to 500 mg EPA + DHA per day [27,28,29]. However, most people do not consume enough *n*-3 PUFAs, as indicated by average modern daily dietary DHA intakes that are closer to 100 mg per day [21,30,31].

The shift in modern diets towards reduced *n*-3 PUFA intake, increased *n*-6 PUFA consumption, combined with less physical activity has had a detrimental impact on development and aging, especially with regard to cognitive function. The optimal dietary *n*-6 to *n*-3 PUFA ratio is 2:1 and below, while the current Western diet is typically in the range of 10:1 to 25:1 [32]. DHA likely played an important role in the evolution of the human brain, as significant structural changes in the encephalon and gains in cognitive abilities coincided with the use of aquaculture as a substantial portion of the human diet [33,34]. Habitats near aquatic food sources, such as lakes, rivers, and the sea provided rich dietary sources of DHA. It therefore follows that sources of preformed DHA include oils from microalgae, fatty fish (especially salmon, mackerel, sardines, and herring), and fish oil.

The human brain metabolizes approximately 4 mg of DHA per day, resulting in an estimated half-life of brain DHA of 2.5 years [35], much longer than that of DHA in peripheral tissues (e.g., two minutes in plasma; [35]). Importantly, although EPA does have significant acute anti-inflammatory actions in neural tissue [36,37] and the absorbance of EPA and DHA to the brain are similar [38,39], EPA levels are extremely low in brain tissue. This circumstance occurs because EPA is rapidly oxidized and removed from the brain, or it is elongated to the *n*-3 PUFA docosapentaenoic acid (DPAn3; 22:5*n*-3), which acts as a precursor for DHA [40]. However, EPA conversion is not a significant source of DHA [41,42,43]. The rapid β-oxidation of EPA may be the result of increased uptake of EPA by peroxisomes, where long chain fatty acids are initially catabolized before undergoing mitochondrial oxidation [44,45]. Such energetically costly mechanisms suggest a unique requirement for DHA. Therefore, for the purposes of this review we will consider the effects of combined *n*-3 PUFA (DHA + EPA) consumption or supplementation on brain function and cognition as being significantly dependent on DHA.

DHA is acquired during development through gestational placental transfer and from mother's milk during infancy. The levels of DHA in mother's milk can be as high as 1.4% of total fatty acids with a worldwide average of 0.32% of total fatty acids, depending on the mother's diet and number of pregnancies [25,46]. *De novo* DHA synthesis can occur in the liver, however the DHA precursor ALA does not increase plasma DHA in humans (<0.1% conversion efficiency in humans; [47,48,49]). In fact, neither ALA nor EPA is an effective dietary source of DHA due to the minimal *in vivo* production of DHA from these precursors in humans, indicating that preformed DHA is most effective in maintaining sufficient tissue stores [20]. On the other hand, DHA supplementation adds to the peripheral tissue pool of EPA, likely due to retroconversion [50].

DHA accretion in the brain accelerates during the middle of gestation, slows down in infancy, and reaches a plateau in early adulthood [51,52]. Half of the brain's DHA is accumulated during gestation, and the infant brain acquires five-times the level of lipids on a daily basis as the adult brain [53,54]. In adults, accretion is slower, and individuals with red blood cell DHA concentrations on par with the average European or American (~4% of total fatty acids) require 4–6 months of oral DHA supplementation to reach a steady state concentration that is dependent on DHA dose (8%–9% for 1000 mg per day; 5%–6% for 200 mg per day) [41]. Local *de novo* DHA synthesis in the brain is very low, thus DHA levels are maintained via delivery from the blood [55,56]. Circulating levels of DHA in the blood can reach as high as ~5% of that which is ingested orally [57] and ~0.5% of the circulating level is delivered to the central nervous system (CNS; [35]). After oral ingestion of DHA, lipases in the gut deliver unesterified free fatty acid (DHA-FFA) to the small intestine, and processing by the small intestine and liver results in circulating versions of DHA as DHA-triacylglycerides (DHA-TAGs), DHA-PC and DHA-FFA bound to low density lipoprotein (LDL) and albumin. These various forms are dissociated at the blood-brain barrier (BBB) through both active and passive processes that are mediated by endothelial lipases, fatty acid binding proteins (FABPs), and apolipoprotein E (ApoE) [58,59,60,61]. Unesterified DHA freely passes the BBB [39,61], and it appears that the brain derives most of its DHA from the unesterified FFA pool in blood [55]. Within the central nervous system, DHA is transported primarily via FABPs [59,60] and ApoE produced by astrocytes [61]. Membrane-incorporated DHA cycles in and out of the membrane from the phospholipids to the intracellular FFA pool via actions of DHA-coenzyme A (DHA-CoA) [62,63], providing a mechanism to respond to dynamic cellular events and to challenges during development and aging.

3. DHA AND COGNITION IN DEVELOPMENT

3.1. DHA during Gestation and Infancy

DHA is necessary for the growth and maturation of an infant's brain and retina [26,64,65], and as outlined earlier, DHA is conditionally essential for humans since it cannot be synthesized efficiently. This particular point needs to be stressed in that when DHA is needed in large amounts, such as during rapid phases of brain growth, it needs to be consumed via external sources. Furthermore, synthesis of DHA from its precursors in the fetus and placenta is insufficient to meet the demand of rapidly developing neural tissues [26,66,67,68], requiring the delivery of maternal DHA stores via placental transfer and mother's milk during pregnancy and after birth, respectively [66,67]. Therefore, the adequate supply of DHA to the developing brain is largely dependent on the dietary intake of the mother [69], and this supply is very important to the cognitive development of the progeny [70,71].

The observations above have led to the recommendation that pregnant and nursing women should consume at least 200 mg of DHA daily [26,72]. Supplementation with DHA during pregnancy may also benefit both the mother and baby by extending the length of gestation. A Cochrane review in 2006 concluded that maternal *n*-3 PUFA supplementation during pregnancy increased the gestational age by 2.5 days resulting in an average increase of birth weight by 50 g and birth length by 0.5 cm [73]. In alignment with this finding, a meta-analysis revealed that maternal *n*-3 PUFA supplementation was associated with pregnancies averaging 1.6 to 2.6 days closer to term [74]. A further meta-analysis of six RCTs found a 31% and 61% overall reduction of the risk of preterm births (defined as delivery before the 34th week of gestation) in all pregnancies and high-risk pregnancies, respectively, when mothers supplemented with *n*-3 PUFAs [75]. Lastly, maternal supplementation of 400 mg DHA per day in the second half of gestation was beneficial to children's growth tested at 18 months of age. This effect was only seen in children born to first time mothers [76]. These effects on growth and development of the fetus likely have a profound impact on cognitive abilities later in life.

Epidemiological studies have shown the importance of DHA during pregnancy for neuronal development. A large study ($N = 11,875$) showed that a lower intake of seafood, a rich source of DHA, during pregnancy was associated with risk of suboptimal development. In contrast, children born to mothers with a high intake of seafood during pregnancy exhibited greater pro-social behavior, better fine motor and social development scores, and higher verbal intelligence at eight years of age [77]. RCT trials have also provided evidence for a positive effect of DHA supplementation during pregnancy. For instance, taking 200 mg DHA orally per day for four months during pregnancy improved cognitive abilities of children tested at five years of age [78]. In a more recent multicenter RCT, 2399 pregnant women, who were <21 weeks into gestation with singleton pregnancies, were supplemented with 800 mg DHA and 100 mg EPA until delivery. While there was no effect of supplementation on overall mean cognitive scores in the offspring when measured at 18 months, the resulting data indicated that *n*-3 PUFA supplementation lowered the number of preterm births and low birth weights, resulted in fewer admissions to neonatal intensive care units, and reduced the number of children with cognitive scores indicative of delayed cognitive development [79].

Postnatally, the importance of DHA obtained from mother's milk for neuronal development has been reported repeatedly [80,81,82,83,84]. High DHA concentrations in breast milk have been associated with several brain-related positive health benefits in infants. These associations include a better ability to adjust to changes in surrounding [85], better mental development [86,87], improved hand-eye coordination [82], better attention scores [78] and memory performance later in life [88]. Nevertheless, some studies report neutral effects of DHA supplementation of lactating mothers on neurodevelopmental outcomes, results which likely depend on length and timing of DHA supplementation as well as the developmental time points assessed [82].

The DHA content of mother's milk is directly dependent on the mother's diet [26,89,90]. Dietary intake of DHA leads to a dose-dependent increase in DHA levels of breast milk [64,91]. Typically, DHA levels in tissues are higher in breastfed infants when compared to formula-fed infants [92]. Thus, it is not surprising that breastfeeding in the first six months of life is recommended for the optimal development of the baby, provided that the mother's nutritional status is favorable [81,93]. This is an important recommendation, and is corroborated by many studies that attest to superior neuronal development of breastfed infants in comparison to infants fed with formula (for a review, see [90]). For example, the Western Australian Pregnancy Cohort (Raine) Study recruited 2900 pregnant women and followed the live births for 14 years. This long-term study provided evidence that a shorter duration of breastfeeding may be a predictor of adverse developmental mental health outcomes throughout childhood and early adolescence [94].

PUFA supplementation of infant formula is an effective means to realize the benefits of breast milk in infants who are not able to be breastfed. For example, a 2007 meta-study summarized several RCTs that had examined PUFA supplementation of term infants and found a consistent association of DHA (and ARA) supplementation with beneficial effects on visual development in the first year of life [95]. For example, in a prospective, double-blind RCT (the DIAMOND study), 181 infants were enrolled at 1–9 days of age and assigned randomly to receive one of four term infant formulas with one of four levels of DHA: Control (0% DHA, 0% ARA), 0.32% DHA, 0.64% DHA, or 0.96% DHA for 12 months. All DHA-supplemented formulas contained 0.64% ARA. At 18 months of age, children who received DHA-supplemented formulas showed a significantly higher MDI score [96]. In another trial, 420 healthy term infants were randomized to receive a DHA-enriched fish oil supplement (containing at least 250 mg DHA and 60 mg EPA per day) or a placebo from birth to six months. Developmental assessment occurred at 18 months via the Bayley Scales of Infant and Toddler Development and the Child Behavior Checklist. Language assessment occurred at 12 and 18 months via the Macarthur–Bates Communicative Development Inventory. When compared to placebo, the fish oil group had significantly higher erythrocyte and plasma phospholipid DHA levels at six months of age. In a small subset analysis (about 40% of the total population), children in the DHA-enriched fish oil group had significantly higher percentile ranks of both later developing gestures at 12 and 18 months and the total number of gestures [97]. Supplementation of infant formula with DHA and ARA (0.32% and 0.72% of total fatty acids, respectively) provided for the first 17 weeks of life resulted in visual acuity scores similar to breastfed infants who performed better than control formula fed children at four years of age [98]. However, linolenic acid (precursor to DHA) provided via formula was ineffective in a separate trial in a similar measure of visual acuity at 16 and 34 weeks of age [99]. These results suggest that dietary supplementation of DHA during the first year of life likely leads to enhanced cognitive performance.

The studies cited above highlight the importance of maintaining sufficient intake of DHA during both pregnancy and nursing. For example, Helland *et al.* found that the supplementation of pregnant women with *n*-3 PUFAs from Week 18 of pregnancy through lactation conferred a cognitive benefit to their children on an intelligence test at four years of age, and the cognitive benefits in the children correlated significantly with maternal intake of DHA [87]. Furthermore, mathematical modelling revealed that a 100 mg/day increase in maternal DHA intake resulted in a small but significant increase in the intelligence quotient (IQ) of infants between 10 and 39 months of age [100]. Systematic review of maternal DHA supplementation indicated that DHA supported mental development and longer-term cognitive functioning of the child. When DHA was given to pregnant and lactating women, an increase in maternal intake of 1 g per day of DHA increased the child's IQ by 0.8 to 1.8 points [100]. However, when pregnant mothers were supplemented with 4 g of fish oil daily for a shorter period (only during late gestation) no effects were observed on overall mental or psychomotor development when measured at 10 months of age, but did positively affect hand and eye coordination at 2.5 years [101,102]. Similarly, supplementation of infant formula with DHA (0.2% of total fatty acids) for the first four months of infancy had no effect on Bayley Mental Index scores at 12 months of age [103]. These data suggest that the best neurodevelopmental effects of DHA, including those on cognitive performance, might be best achieved by exposure during both pregnancy and lactation [87]. Such findings emphasize the urgency of adequate intake of dietary DHA by pregnant and lactating women to ensure the healthy development of the brain and visual system in their offspring.

3.2. Preterm Infants and DHA

As alluded to above, the rapid growth of the brain during the last trimester of pregnancy requires a relatively high amount of DHA for proper brain development. Due to this circumstance, infants born prematurely are uniquely susceptible to the effects of DHA deficiency. Moreover, the ability of preterm infants to synthesize DHA *de novo* via elongation and desaturation of precursor fatty acids is insufficient to make up for their substantial deficits in DHA tissue levels [104]. Beneficial effects of PUFA supplementation in preterm babies for cognitive [105] and psychomotor development are well-established [106]. For example, Henriksen and colleagues showed that supplementation of expressed mother milk with 32 mg DHA and 31 mg of the *n*-6 PUFA arachidonic acid (ARA; 20:4*n*-6) per day resulted in better cognitive performance of premature infants weighing 1500 g or less at birth when examined six months later [107].

PUFA supplementation of infant formula for preterm infants has consistently demonstrated positive effects on aspects of neurobehavioral development [108]. High-DHA (approximately 1% total fatty acids) formula was compared to standard DHA (approximately 0.3% total fatty acids) from Day 2 to 4 of life until term-corrected age in an RCT of preterm infants. At 18 months of age, infants born at less than 1250g in the high DHA group had a 50% lower likelihood of showing mental delay, as measured by the Bayley Mental Delay Index (MDI), when compared to infants in the standard DHA group [109]. In addition, an overall positive effect of DHA on MDI measured at 18 months of age was only observed among girls, who had less than half the risk of mild mental delay and less than a fifth of the risk of severe mental delay after high DHA supplementation compared to the standard DHA group. This effect was strongest among infants of mothers with low level of education, while no significant effect on mental development was seen among infants of mothers with higher education. However, when the same group was studied seven years later no differences between high-DHA and standard-DHA groups were evident in measures of weight, height, head circumference, or visual function [110,111]. Overall, current data suggest that infants fed formula benefit greatly from standard DHA supplementation and that this is especially important for preterm infants.

3.3. DHA during Infancy and the Relationship to Socioeconomic Conditions

The effect sizes associated with *n*-3 PUFA supplementation on postnatal growth and neurodevelopment in infants are relatively small. This circumstance is likely due to study populations that have been limited to formula-fed infants in high-income countries. Well-designed studies examining the effect of postnatal PUFA supplementation on infant growth in low-income countries are scarce, but some emerging data exist. For instance, a trial in rural Gambia that randomized 183 infants to receive fish oil or olive oil placebo from 3 to 9 months of age resulted in significant increases of growth parameters, such as mid-upper arm circumference and skin thickness in the fish oil group [112].

Interestingly, a recent Cochrane review came to the conclusion that, in high-income countries, "there is inconclusive evidence to support or refute the practice of giving PUFA supplementation to breastfeeding mothers in order to improve neurodevelopment," a similar conclusion to that of Scholtz *et al.* in a review highlighting the need for better study designs in future studies [113]. Nonetheless, the Cochrane review reported moderate positive evidence for measures obtained beyond 24 months related to language development and body weight in children. However, the authors also reported neutral findings for outcomes related to child length, intelligence and problem-solving, psychomotor development, motor development, and visual acuity among other measures. Most positive effects were dependent upon a specific developmental stage (with some realized short-term at birth to 12 months, and/or medium term at 12 to 24 months, and/or long term beyond 24 months). Lastly, no side effects were reportedly associated with *n*-3 PUFA supplementation [114].

The financial resources of the family can impact maternal consumption of DHA via dietary or supplementary means, ultimately affecting the DHA content of breast milk [115]. Studies of children aged 3–5 years reveal a low intake of ALA, DHA, as well as total fat in low-income countries [116,117]. The intake of fatty acids, especially DHA, of pregnant or lactating women and their young children do not meet the recommendations in many low-income countries [117,118]. Conversely, intake of the *n*-3 PUFA precursor ALA during pregnancy generally meets the recommendation in Mexico and the USA, but does not in Chile, Bangladesh, and India [76,117,119,120]. In studies examining the intake of DHA during pregnancy, none of the women in Mexico, Bangladesh or India met recommendations [76,117,120]. Postnatally, a decline in fat intake from 50% to 25% of energy between birth and 18 months is also observed in children in developing countries, which is primarily driven by a decline in intake of mother's milk. This decline intensifies further to about 15% of energy from fat by 2–3 years of age [112]. Importantly, the decline in breast milk intake leads to rapid reduction of dietary *n*-3 PUFA intake whereas the *n*-6 PUFA intake remains relatively stable. Since*n*-6 and *n*-3 PUFA precursors compete for the same elongation and desaturation enzymes, this increase in the *n*-6:*n*-3 ratio results in further decreases in *n*-3 PUFAs beyond what is attributable to total ingested levels. These data highlight the need for DHA supplementation in poorer communities in order to promote optimal cognitive development and possibly socioeconomic development.

3.4. DHA and Cognition in Children

The effect of DHA supplementation on cognition during childhood is controversial, with studies reporting no effects [121] or suggesting an improvement in verbal learning and memory [122]. Lower total plasma *n*-3 PUFA concentrations in boys of 6–12 years are correlated with a greater number of learning and behavior problems [123]. Children, 7–9 years of age, who were given an *n*-3 PUFA enriched diet for six months exhibited significantly higher plasma and red blood cell DHA and EPA levels, and these PUFA levels correlated with superior verbal learning, spelling and reading abilities [122]. Moreover, activation of the prefrontal cortex, a brain region critical for executive function, during a cognitive challenge was enhanced in 8–10-year-old normal boys who were supplemented daily for eight weeks with 1.2 g DHA compared to placebo [124].

In the DHA Oxford Learning and Behavior (DOLAB) study, Richardson *et al.* performed an RCT in 362 healthy school children aged 7–9 years from mainstream primary schools in The United Kingdom that initially underperformed in reading ability (bottom tertile). The children received either 600 mg of algal DHA each day for 16 weeks or a corn or soybean oil placebo designed to match both taste and color. Reading ability, working memory and behavior were measured using British Ability Scales (BAS II) assessments and parent/teacher observations using the Conners' Rating Scales. The results showed subtle reading improvements across the sample, but significant results were observed in the poorest-reading subgroup (lowest quintile). This subgroup experienced an eight-month improvement in reading age after DHA supplementation. Moreover, parent-rated behavior problems (ADHD-type symptoms) were significantly reduced by DHA supplementation. These data are valuable because, in contrast to the well-known critical period for the beneficial actions of DHA on brain function during prenatal development and early life, the DOLAB study showed that DHA can affect brain function well beyond early development in healthy children [125].

In a recent Australian RCT, the question was asked whether supplementation of *n*-3 PUFAs or *n*-6 PUFAs could improve cognitive performance in children [126]. A total of 616 term infants were randomized to receive tuna fish oil (high in *n*-3 PUFAs) or sunflower oil (high in *n*-6 PUFAs) from the time breastfeeding ceased or at the age of six months until the age of five years. Academic performance was measured by the National Assessment Program Literacy and Numeracy (NAPLAN) in school years 3, 5, 7 and 9. Plasma *n*-3 PUFA levels were measured at regular intervals until eight years of age. No significant differences in NAPLAN scores were observed between active and control groups. However, at eight years, *n*-3 PUFA levels in plasma were positively associated with the NAPLAN score measured at five years of age. These data are encouraging, because the observed correlation between the academic performance and *n*-3 PUFA levels was evident even though the supplementation had stopped three years earlier. Furthermore, there are a few issues that may explain the lack of overall effect in NAPLAN scale. First, the attrition of the test cohort was extremely high, and less than half of the original study population completed the final assessment. This situation rendered the final test size of the cohort too small to reveal statistical significance. Secondly, a relatively low amount of *n*-3 PUFAs were supplemented (135 mg of DHA and 32 mg of EPA), making it less likely to find significant effects when assessing cognitive performance. Lastly, the authors acknowledged that the parents of the children in the NAPLAN cohort were slightly older and more likely to be tertiary educated. Moreover, the mothers were more likely to have fully breastfed than the mothers of those who did not consent to or provide NAPLAN data, making it very likely that background intake of *n*-3 and *n*-6 PUFAs would also confound the results in this relatively prosperous group [126].

The positive effects of a diet rich in DHA on cognitive abilities may also have select sex differences. A large cohort of USA children aged 6–16 years of age, who were part of the third National Health and Nutrition Examination Survey (NHANES III), revealed cognitive benefits of higher *n*-3 PUFA consumption in both boys and girls, but the effects were twice as prominent in girls [127]. The authors speculated that the sex differences were due to a greater need for DHA in girls, since they need it not only for their own growth and development, but also in order to accrue peripheral stores destined to provide DHA for the neurodevelopment of their offspring. Indeed, recent animal work indicates greater peripheral synthesis of DHA from its precursor ALA in females as compared to males, an effect that emerges at the time of sexual maturity [128]. Interestingly, while the peripheral levels of DHA (plasma and liver phospholipids) were higher, the brain levels were similar, leading the authors to conclude similarly to Lassek *et al.* [127] that "the additional production of DHA in the liver of sexually mature females was not particularly intended for the brain, but probably for storage and mobilization upon future pregnancy."

There is a fair amount of inconsistency between studies examining the effects *n*-3 PUFA supplementation on cognitive performance in children or infants. Thus, there are several potential reasons for varying results including genetic factors, study design and methodological flaws, *etc.* It is also conceivable that genetic variation in fatty acid metabolism affects PUFA status. Potential genetic and environmental factors are not typically considered when recruiting for a study, and they likely contribute to some of the discrepancies between studies. Several study design considerations may help to reduce data variability. For example, stratification of the study population based on baseline DHA levels may help to avoid ceiling effects. It is also worthwhile to specify whether supplements should be taken on an empty stomach or with food, since certain food matrices can affect the bioavailability of *n*-3 PUFAs [129]. Using sensitive, standardized and validated analytical methods and cognitive tests as well as sufficient statistical power, dose and supplementation duration might also help resolve some supplementation effects. Importantly, the source of DHA (the particular type of fish flesh or oil, organ meat, oils derived from algae or crustaceans such as krill,*etc.*) as well as the ratio of EPA to DHA may affect the overall bioavailability of the *n*-3 PUFAs [130,131,132]. Assessment of tissue levels of DHA is also critically dependent upon the sample assayed, where erythrocytes are preferred over plasma fractions since their DHA content correlates better with peripheral and central concentrations of DHA rather than being heavily influenced by recent dietary intake as is the case for plasma [133]. Furthermore, the consumption and tissue levels of other nutrients should be considered. Very recent research published by Jernerén *et al.*, [134] suggests that homocysteine (Hcy) status, which is affected by B-vitamin status, may also determine the effects of *n*-3 PUFAs on cognitive decline and dementia (discussed in a subsequent section). Thus, B-vitamin status may explain some more of the inconsistent data across *n*-3 PUFAs supplementation trials. Last, but not least: it is important that the supplemented materials are of high quality, the source is completely defined, and the dosages are clearly defined and validated. In summary, assessments of dietary intakes of DHA suggest that pregnant and lactating women and young children often do not meet recommendations. Improving the dietary habits of mothers-to-be, pregnant and lactating women, and infants and children, with a particular focus on consuming high-quality sources of DHA, may be feasible strategies to enhance tissue levels and reap the corresponding developmental and cognitive benefits.

3.5. Mechanisms of DHA Actions during Development

As already outlined, DHA is accumulated in the brain tissue mainly during the second half of pregnancy and during the first two years of life [135]. Currently the consumption of one to two portions of fish per week, including oily fish, which is a rich source of DHA, is recommended. The challenging question is: What are DHA's actions in the developing brain that lead to gains in cognitive function? This question is far from being answered, but several pathways can be identified, many of which overlap with those occurring during adulthood and aging (described in a subsequent section).

DHA, as the most abundant *n*-3 PUFA in the brain and retina, contributes to the structure of brain cell membranes. DHA is also implicated in neurogenesis, neurotransmission, and cell survival within the CNS [136,137]. DHA contributes to cell membrane fluidity [138] and to signal transduction within the CNS by activating cell membrane receptors [135]. DHA also alters gene expression in mammalian brain tissue [139,140] that influences neurite outgrowth and learning and memory

[141,142]. The process of neurite outgrowth in hippocampal neurons is enhanced by DHA, which may in turn promote learning [141]. Growth of neurites requires the accumulation of lipids in new membranes, and DHA helps to organize lipid raft domains in the membrane by pushing cholesterol into these structures important for neurite extension, myelination, and membrane-mediated signaling [143,144,145,146]. Also important for neurite outgrowth, DHA enhances protein kinase B (PKB; also known as Akt) signaling and in turn the mTOR (mechanistic target of rapamycin) complex, which promotes neuronal growth [147,148]. DHA improves learning and memory by facilitating the formation of pre- and postsynaptic proteins that enable synaptic transmission and long-term potentiation (LTP) [149]. Under oxidative stress, DHA can promote repair and growth of neurons by activating peroxisome proliferator-activated receptor gamma (PPAR) and through its activating effect on syntaxin-3 (STX-3) [150].

DHA is deposited within the cerebral cortex at an accelerated rate during the last trimester of gestation and during the first two years after birth, rendering this phase of neuronal development particularly vulnerable to nutritional insufficiencies [151]. This early accelerated rate of DHA deposition coincides with the onset of myelination, a process that is sensitive to DHA accumulation and stores [54,152]. Notably, animal data show that it is physiologically difficult to reverse the effects of early brain DHA depletion [153] and that reduction of *n*-3 PUFAs in the diet negatively affects DHA concentrations within the brain [154,155,156]. Studies in non-human primates show that low DHA levels are associated with deficits in the visual system, in brain functions [157], and in motor capabilities [158]. Furthermore, animal models provide solid evidence that the consequences of dietary DHA deficiency are a high *n*-6 to *n*-3 PUFA ratio in brain fatty acid composition and deficiencies in learning and memory behaviors [142,159], possibly due, in part, to negative impacts on neurite outgrowth and myelination [160].

4. DHA AND COGNITION IN ADULTHOOD AND AGING

4.1. DHA during Adulthood

Cognitive function reaches its peak during middle adulthood. Thus, the detection of measurable effects of DHA consumption requires lengthy observational or treatment periods, and measurable effects are most effectively observed in cases of DHA-insufficiency or cognitive impairment. Indeed, a recent meta-analysis by Abubakari *et al.* [161] of 12 RCTs in adults found no effect of *n*-3 PUFAs on cognitive measures. However, the authors selected trials with subjects exhibiting a multitude of baseline cognitive conditions and/or subjects diagnosed with psychiatric conditions such as depression and schizophrenia, in addition to cognitive test parameters that likely clouded any measurable effect of *n*-3 PUFAs. Another recent meta-analysis of 34 RCTs found no effect of *n*-3 PUFA supplementation on cognitive performance, but the investigators combined trials performed with children, adults and elderly without performing any subgroup analyses that considered age or dose [162]. This limited their conclusions greatly since assessment of cognitive function is highly dependent on developmental timeline and age. Conversely, a well-designed meta-analysis by Yurko-Mauro and colleagues revealed that DHA + EPA supplementation improves episodic memory outcomes in adults with mild memory complaints, an effect primarily attributable to daily DHA doses above 580 mg [163].

Observational studies suggest a correlation between blood levels of DHA and cognition in healthy adults. For example, higher serum DHA levels were associated with better non-verbal reasoning, mental flexibility, working memory and vocabulary in 35 to 54 year-olds with no neuropsychiatric disorders and no supplemental fish oil use [164]. In an extension of this study, Leckie *et al.* [165] found that a higher ratio of DHA to ARA in serum counteracted the negative effect of low physical activity on both working memory and the trail marking task. This is important since physical activity can increase gray matter volumes and reduce the overall risk for Alzheimer's disease [166,167].

RCTs performed with adults report mixed results for DHA and *n*-3 PUFA supplementation on cognition. Supplementation of healthy adults from 18 to 70 years of age with 850 mg DHA and 630 mg EPA daily for 12 weeks had no appreciable effect on cognition [168]. Likewise, supplementation of young adults (18 to 35 year-olds) with 1000 mg DHA and 200 mg EPA for 12 weeks had no significant cognitive effects [169]. Furthermore, daily supplementation with 250 mg DHA and 1740 mg EPA in a college-age cohort for four weeks had no effect on attention, memory, or inhibitory responses [170]. In contrast, another study performed with college-age students and a higher daily DHA dose (480 mg DHA and 720 mg EPA) for four weeks found improvements in verbal learning and memory in the supplemented group despite the very short treatment interval [171]. Similarly, Stonehouse *et al.* (2013) reported improvement of episodic and working memory in 18 to 35 year-olds provided a high-DHA supplement containing 1160 mg DHA and 170 mg EPA daily for six months compared to placebo controls [172]. Interestingly, the effect on episodic memory was driven primarily by the women in the study, whereas the men were largely responsible for the effect on working memory. This is not completely surprising given the existence of sex differences in particular cognitive domains, including spatial working memory [173,174]. Recently, it was also shown that deficits in episodic memory observed in lonely individuals was prevented by *n*-3 PUFA supplementation for four months [175]. These data suggest that detecting cognitive effects in adults likely requires substantial levels of DHA provided for extended treatment periods in an experimental cohort with a relatively narrow age range.

4.2. DHA during Normal Aging

The age demographics of the global population is shifting as lifespans increase over time due in part to advancements in medicine and positive economic development. The segment of people 65 and older is projected to triple to 1.5 billion globally by 2050 (WHO, [176]). Unfortunately, cognitive ability declines naturally with age even in the healthiest of individuals. This decline is typically subtle, but it is nonetheless undesirable and ultimately affects the quality of life. Environmental factors such as diet, exercise, and DHA consumption can positively affect the normal aging process and overall mental health and performance.

Total gray matter volume declines with age [177], matching a parallel decrease in DHA composition [178]. This drop in DHA may be partially due to changes in activities for the enzymes responsible for DHA accretion into phospholipids [179,180] or to shifts in plasma pharmacokinetics [181]. Importantly, *n*-3 PUFA intake is positively correlated with gray matter volume in adults [182] and in brain regions responsible for cognition in normal, elderly adults [183]. During normal aging there is a gradual 10%–15% loss of total neuronal synapses resulting in a cognitive decline that is typically noticed around age 65 [184], and an increasing risk of dementia that is largely negligible before age 60 [185]. However, processes leading to neuronal loss and the impairment of brain functions may be active at much younger ages [186]. For example, in susceptible young adults, an age-related reduction of neurons in the brain is reported [187]. Of note, declines in episodic memory can begin as early as 20 years of age [188].

Support of the clinical findings above can be found in preclinical animal experimentation, in which age-related neuronal loss was shown to begin at the end of adolescence in rat brains [189]. Total synaptic loss is the best correlate of cognitive decline during aging, as overall synaptic density affects cognitive ability [190]. Increases in oxidative stress and inflammation in both nervous and immune systems also occur with aging [191], leading to DNA damage and telomere shortening. DHA has proven synaptic effects that improve synapse strength and numbers, and DHA can help prevent or mitigate oxidative stress and neuroinflammation. These actions will be discussed in more detail in the section on potential DHA mechanisms during aging.

Many observational studies have linked dietary consumption of *n*-3 PUFAs, and DHA in particular, with improvements in cognitive function and/or reductions in cognitive decline in healthy, aging populations. For example, DHA intake was correlated with performance and speed in a verbal learning test performed in a cohort of 45 to 70 year-old healthy individuals [192]. Van Gelder *et al.* [193] performed a prospective study using longitudinal data that calculated EPA + DHA consumption at baseline and at five years later in 70 to 89 year-olds and found that the decline in baseline mini-mental state exam (MMSE) scores was negatively associated with dietary EPA + DHA levels. In a group of 65 to 80 year-old healthy individuals, consumption of greater than 2.1 g of *n*-3 PUFAs per day was associated with better memory and executive function [194]. Velho *et al.* [195] reported that, in a cohort of healthy elderly (over 65 years-old), those that improved their MMSE scores at 8.5 months from baseline had higher consumption of *n*-3 PUFAs than those who did not show improvements. Cross-sectional analysis of elderly Spanish residents ($N = 304$), with an average age of 75 years-old, found a positive correlation between dietary DHA consumption and MMSE scores and that lower DHA intake was a predictor of cognitive impairment [196]. In a large cohort of Chinese adults (average age of 65; part of the Singapore Longitudinal Aging Studies) the daily consumption of fish oil supplements was associated with higher baseline MMSE scores and a lower risk of decline in cognition over a 1.5-year span [197]. Titiova *et al.* [198] more recently reported a prospective observational study in a healthy elderly population where baseline dietary DHA intake levels at age 70 were positively correlated with larger gray matter volume and declarative memory test performance (seven minute screen) at age 75. In a larger prospective study using data from the China Health and Nutrition Survey, Qin *et al.* [199] examined the decline in global cognitive scores over an average of 5.3 years in a group of Chinese adults ($N = 1566$, mean age 63 years old). This study detected a positive correlation between at least one serving of fish per week and slower declines in global cognitive function, composite and verbal memory scores. Finally, most recently del Brutto *et al.* [200] reported on a cohort from a rural middle-to-low income area of Ecuador that cognitive function as measured by Montreal Cognitive Assessment (MoCA) scores was positively related to the number of fish servings per week, and the data suggested an intake of at least four fish servings per week was best.

While dietary DHA intake is a convenient measure, there are inherent flaws in the reliability and accuracy of data acquired from food frequency questionnaires [201]. Therefore, studies that measure tissue levels of DHA could provide a more appropriate measure of current and past DHA status in relation to cognitive performance. Indeed, some but not all observational studies have linked DHA concentrations in blood with overall cognition and select cognitive domains. For example, a recent study measured the cognitive function of 2157 postmenopausal women at baseline and annually for six years and found no correlation between erythrocyte DHA + EPA and cognitive performance in seven cognitive domains at baseline or over time [202]. On the other hand, the Etude du

Vieillissement Arteriel (EVA) cohort of French 63 to 74 year-olds exhibited a reduced risk of overall cognitive decline over four years in those with higher erythrocyte *n*-3 PUFA content [203]. Dullemeijer *et al.* [204] performed a cross-sectional longitudinal study in a Dutch cohort (50–70 years old) and found that higher plasma concentrations of total *n*-3 PUFAs at baseline was associated with less decline in sensorimotor and complex speed-related cognitive domains three years later. However, they did not observe associations with changes in memory, information-processing speed, or word fluency. Whalley *et al.* [205] performed a battery of cognitive tests in a Scottish cohort at ages 64, 66, and 68 years old and found a positive correlation between overall cognitive performance over time and erythrocyte DHA content at baseline (64 years old). Interestingly, this effect was only apparent in the absence of the ApoE4 allele, indicating a genetic influence on the link between peripheral DHA levels and cognition. The Framingham Offspring Study performed in a large group of elderly women (mean age 67 years old) reported that the women with erythrocyte DHA levels (but not EPA) in the lowest quartile had lower brain volume and poorer scores for visual memory, executive function, and abstract thinking than those in the top three quartiles [206]. Lower serum DHA (but not EPA) was also found in a small selected case study group of elderly with reduced performance in the MMSE compared to a cognitively healthy control group [207]. In this study, DHA serum concentrations were correlated with performance on a majority of the tests (memory, attention, and mental flexibility). Recently, Otsuka *et al.* [208] reported results of a cross-sectional longitudinal study in Japanese elderly adults that revealed a link between low serum DHA levels and a greater risk for cognitive decline over a decade as measured by the MMSE. This is especially profound given that elderly Japanese have twice the circulating levels of DHA as compared to similar cohorts in England [208,209]. Overall, these reports provide a suggestive link between the tissue levels of DHA and cognition during aging.

Data from RCTs in healthy aging adults has been historically sparse, but recently more trials have been reported, with some mixed but generally positive effects of DHA and *n*-3 PUFA supplementation on cognition. Initial reports indicated no differences in overall cognition in a group of elderly individuals over 65 years of age receiving daily high or low dose fish oil for 26 weeks, however subgroup analysis did reveal improvements in attention in those with the ApoE4 genotype [210]. Similarly, Dangour *et al.* [209] found no effect of daily *n*-3 PUFA supplementation for two years on cognitive measures in a group of 70 to 79 year old healthy individuals. However, controls in this study did not have a decline in cognitive performance over the two-year span, possibly masking any potential effect of DHA. Additionally, Stough *et al.* [211] reported no effect of daily DHA (252 mg) for 90 days on cognition as measured by Cognitive Drug Research (CDR) scores, however this cohort included a wide range of ages (45 to 80) and a small sample size, making interpretations difficult. In contrast, a combined supplementation of 800 mg DHA and 12 mg lutein for four months in healthy 60 to 80 year-old women, improvements in verbal fluency, memory scores, and rate of learning were seen [212].

More recent RCTs have yielded more compelling results than the earlier work. The Memory Improvement With Docosahexaenoic Acid Study (MIDAS) study, performed by Yurko-Mauro and colleagues [213], provided 900 mg of DHA or a placebo daily for 24 weeks in 485 healthy elderly individuals (mean age of 70 years-old) who had a self-reported mild memory complaint and a MMSE score over 26 (cognitively normal). Those supplemented with DHA exhibited improved episodic and visual recognition memory, but not in executive or working memory, and plasma DHA levels were directly correlated to scores in episodic memory. Furthermore, Vakhapova *et al.* [214,215] reported benefits in an elderly population with memory complaints (non-dementia) of improved immediate recall memory and sustained attention when provided a daily 300 mg DHA-phosphatidylserine supplement for 15 weeks. In a cross-over RCT, Nilsson *et al.* [216] reported a benefit of 3 g daily *n*-3 PUFA supplementation (1050 mg DHA) for five weeks in measures of working memory and selective attention in a group of middle age to elderly subjects. It is intriguing to think that the metric used by Yurko-Mauro *et al.* and Vakhapova *et al.* of a mild subjective memory complaint might be an early correlate of age-related cognitive decline, and their findings could indicate the potential power of DHA supplementation during the long, early and undetectable phases of cognitive impairment and dementia.

Recent RCTs have also associated the cognitive benefits of DHA with neurophysiology or anatomical changes in the brain. Witte *et al.* provided *n*-3 PUFA supplementation (880 mg DHA and 1320 mg EPA) for 26 weeks in healthy subjects (age 50 to 75) and found improvements in executive function, white matter integrity, gray matter volume, and parameters of neurovascular function compared to individuals provided a placebo. Tokuda *et al.* [217] provided 55 to 64 year old Japanese men who already consumed an average of 543 mg DHA per day and had substantial plasma DHA levels (7.0% of fatty acids) a supplement containing 300 mg DHA, 100 mg EPA, and 120 mg of ARA daily for four weeks. They report that treatment prevented a decline in auditory event-related potential (ERP) latencies (a measure of cognitive processing speed) that was observed in the placebo group. Deficits in auditory ERPs are typically observed in Alzheimer's disease [218]. Finally, a very recent study reported by Strike *et al.* [219] suggests that a combined supplemental approach may be beneficial. They gave a group of postmenopausal women between the ages of 60 and 84 years-old a daily supplement containing 1000 mg DHA, 160 mg EPA, 240 mg ginkgo biloba, 60 mg phosphatidylserine, 20 mg tocopherol, 1 mg folic acid, and 20 g vitamin B12 for six months and found a shorter mean psychomotor response latency (a measure of information processing speed) and verbal recognition memory in the treatment group relative to the placebo group. They did not detect any changes in executive function or paired associate learning. In totality, data from studies reporting the effect of dietary DHA consumption, blood DHA concentrations, and supplementation with DHA on parameters of cognition in normal aging individuals provide a substantial argument for obtaining sufficient amounts of DHA via dietary or supplemental means during aging.

4.3. DHA and Cognition in Mild Cognitive Impairment and Dementia

During aging, an increasing share of the resources available for normal cellular maintenance are spent on repair mechanisms needed by the cell to cope with the cumulative effects of oxidative, inflammatory, and other environmental insults. With this shift in demand, any diminished availability of energy, as a result of poor mitochondrial function, may result in neurodegenerative processes that can lead to neuronal loss and eventually to cognitive impairment, dementia, or other neuropsychiatric maladies. Unfortunately, neurodegenerative processes and neuronal cell death occur well before clinical signs of cognitive deficits are confirmed [186]. Dementia is not a disease itself, but rather a group of chronic symptoms that are common to several neuropsychiatric disorders (Alzheimer's disease (AD), Lewy body dementia, Parkinson's disease,*etc.*). Dementia symptoms, such as deficits in memory, language and executive function, lead to poor cognitive function in these individuals [186]. Dementia often results in poor self-care that can lead to inadequate nutrition, which could potentially exacerbate the cognitive deficits. In the most severe cases of dementia there is a loss of functional independence that results in institutionalization. Worst-case scenarios occur in low- and middle-income, developing countries where access to nutrient-dense foods and the ability to afford medical and functional care are leading to exponential growth in dementia prevalence rates [220].

There is a normal degradation in cognitive ability and brain atrophy with age [221,222]. However, the rate of atrophy is markedly higher in mild cognitive impairment (MCI) and dementia. The atrophy rate is especially high in the subgroup of MCI subjects that eventually develop clinically diagnosed AD [223], the most prevalent neurodegenerative disease. Approximately half of all individuals with MCI progress into AD within five years. Furthermore, brain glucose metabolism decreases 10%–15% during normal aging, and the extent of cognitive decline in MCI and AD is associated with the degree of glucose metabolism loss (nearly 35% in some brain regions; [224,225,226]). Preclinical animal studies have convincingly shown that DHA provided over a substantial amount of time can reduce neuronal loss and improve learning and memory as the animals age (For meta-analysis, see [227]). However, studying the effects of DHA on the risk of dementia is very challenging given the low incidence rates in cognitively healthy individuals. Some estimates have called for nearly 50,000 participants at baseline for proper statistical power [228], whereas the rate of cognitive decline during dementia is a more accessible measure. In addition, changes in cognitive domains are dependent upon the specific type of dementia. For example, early changes in AD are seen in episodic memory, whereas in vascular dementia early deficits in executive function are seen. Later stages of dementia and AD involve a multitude of cognitive domains.

Observational studies have linked the consumption of *n*-3 PUFAs with a lower prevalence of dementia [229] and lower overall risk of developing dementia [230]. Albanese *et al.* found a significant dose-dependent decrease in dementia relative to fish intake in a large group of subjects (14,960) residing in middle-to-low income areas [229]. The Three-City cohort study of 8085 French residents over the age of 65 found an inverse relationship between fish consumption and overall risk of dementia over a four-year timeframe in ApoE4 non-carriers (80% of subjects) [230]. Blood levels of DHA have also been tied inversely to mild cognitive impairment (MCI) and dementia. For instance, in the Framingham study, plasma phosphatidylcholine fatty acid content was measured in 899 subjects with an average age of 76 and no dementia at baseline. When re-assessed for cognitive ability nine years later, individuals in the top quartile for baseline plasma PC-DHA levels had a 47% lower risk of all-dementia (grouped with AD) *versus* the other three quartiles combined. No other fatty acid was significantly correlated (including EPA), and food intake surveys revealed that this quartile had an average DHA intake of 180 mg per day [231,232]. Cherubini *et al.* [233] also reported higher levels of plasma DHA in cognitively normal subjects as compared to those with dementia in an aging Italian cohort. With regards to MCI, Milte *et al.* [234] detected higher levels of the *n*-6 PUFA docosapentaenoic acid (DPA*n*-6; 22:5(*n*-6)) in the erythrocytes of MCI patients relative to healthy controls. More recently, Yin *et al.*[235] reported lower blood levels of DHA in amnestic and multi-domain MCI patients as compared to normal control subjects. These are intriguing findings since the DPA*n*-6 replaces DHA in the brain during DHA deficiency in an inefficient attempt to retain function.

Combinations with other nutrients are likely also important. Interesting results from the homocysteine and B vitamins in cognitive impairment (VITACOG) trial were very recently published in which 168 patients with MCI (≥70 years of age) were given placebo or high dose Hcy-lowering B vitamins (folic acid, B6, B12) and assessed for brain atrophy via MRI at baseline and at a two-year follow-up [134]. B vitamin treatment reduced the brain atrophy rates by 40%, an effect only observed in the subgroup with the highest tertile of baseline plasma *n*-3 PUFA levels. Baseline plasma DHA, but not EPA, was a significant predictor of reduced yearly brain atrophy rate in those who took B vitamins, but not in placebo controls. Thus, the DHA status of the MCI patient affected the correlation between vitamin B supplementation and brain atrophy, and possibly cognitive decline by extension. It would be interesting to determine whether the Hcy and DHA status of the MCI subject affects or predicts the rate of conversion from MCI to dementia (about 5%–10% per year). If so, perhaps supplementation with B vitamins in combination with DHA in MCI patients would be an effective prophylactic treatment aimed at reducing the risk of further cognitive decline and the development of dementia.

RCTs have provided evidence indicating largely positive effects of *n*-3 PUFA supplementation on cognitive measures in subjects with MCI or dementia. Early evidence from Terano *et al.* [236] indicated improved dementia scores in patients with moderately severe dementia caused by thrombotic cerebrovascular disorder who received 720 mg of DHA daily for 12 months. Combined supplementation with DHA (240 mg) and ARA (240 mg) daily for 90 days was shown to improve attention and immediate memory in patients with mild cognitive dysfunction [237]. Furthermore, supplementation with 720 mg DHA and 1080 mg EPA daily for 24 weeks in MCI patients improved their scores in the Clinician's Global Impression of Change (CIBC)-plus and Alzheimer's Disease Assessment Scale (ADAS)-cog [238]. Sinn *et al.*[239] administered several doses of *n*-3 PUFAs, including high EPA (1670 mg), high DHA (1550 mg), or high LA (2200 mg) daily for six months in MCI patients and detected improvements only in the high DHA group particularly for Initial Letter Fluency, a measure of fluid thinking ability. The other cognitive measures did not show any differences, but the baseline erythrocyte DHA levels in this study were higher than those of Chiu *et al.* (5% *vs.* 4.2%). In a small, preliminary trial of 25 MCI patients, administration of a high-DHA (1440 mg) supplement also containing small amounts of EPA, tryptophan, phospholipids, and melatonin daily for three months improved MMSE scores, semantic verbal fluency and olfactory sensitivity [240]. The benefits of DHA on cognition in MCI may be dose-dependent since a recent report found no effect of daily supplementation with 180 mg DHA plus 120 mg EPA for 180 days in mild to moderate MCI patients on scores in the MMSE and Abbreviated Mental Test (AMT) [241]. Overall, optimal tissue levels of DHA are important in reducing the likelihood of developing, and improving the symptoms of, MCI and dementia.

4.4. DHA and Cognition in Alzheimer's Disease

AD is a uniquely human, progressive neurological disease resulting in hallmark neuropathology consisting of senile plaques, neurofibrillary tangles, neuronal atrophy, and abnormal brain glucose metabolism. AD accounts for more than 70% of dementia cases and has an estimated worldwide prevalence of about 4.4% of the population over 65 years old [242]. This prevalence is expected to grow from 5.2 million Americans in 2014 to 13.8 million Americans and to 115 million people globally by 2050 (Alzheimer's Association 2014 report [243], WHO 2012 Dementia Report [244]). Deaths from AD rose by 66% in a recent eight-year span, highlighting the lack of efficacious therapeutic options currently available, aging societal demographics, and shifting environmental impacts such as nutrition and physical activity [186]. A recent study in Medicare fee-for-service beneficiaries in the US found that the average total cost for a patient with dementia during the final five years of life was nearly $300,000. The out-of-pocket expenses during the final five years of life were approximately 80 times more for dementia patients when compared to those with heart disease or cancer [245]. Caretaking costs associated with dementia and Alzheimer's in the US is not generally reimbursed by Medicare, and these costs often completely deplete the household wealth of the patient and/or family member caretakers. Unfortunately, no therapeutic cure is currently available, and few investigational new drugs are currently being tested as pharmaceutical companies have traditionally experienced frequent failures in attempts to find efficacy [186,246].

Late onset sporadic AD (LOAD) is the most prevalent form of AD and has the lowest identifiable link to genetics. Therefore, LOAD may be most sensitive to environmental factors, such as diet and DHA intake. This is especially true given that neuropathological changes occur decades before clinically identifiable cognitive deficits, providing a long window of time for the cumulative effects of environmental factors to affect the manifestation of the disease. Traditionally, clinical studies have focused on later stages of the disease, when the disease neuropathology appears to be intractable and resistant to therapeutic approaches. AD affects 32% of people over 85 years old [247], and age is the single greatest risk factor, suggesting that even AD can be considered normal physiological aging.

Brain DHA composition likely plays a role in AD. The brains of non-DHA supplemented Alzheimer's patients have 65–95 nmol/g of unesterified DHA, much less than normal controls (110 nmol/g; [248,249]). Furthermore, deficient liver biosynthesis of DHA has been observed in AD patients, where it appears that DHA biosynthesis halts at the last β-oxidation step from tetracosahexaenoic acid (24:6*n*-3) to DHA [248]. However, this is likely a minimal source of DHA (*vs.* preformed via diet). Recent data also suggests that AD patients have problems processing

DHA [61], therefore the magnitude of DHA's effects may be less in later stages of the disease [250]. Phospholipids PC and PE from various brain regions (particularly hippocampus) in the AD brain have reduced DHA content compared to control brains, further implicating DHA in the etiology of the disease [251,252].

Observational studies are generally supportive of a preventative role of dietary *n*-3 PUFAs and DHA with regards to risk and incidence of AD. Morris *et al.* [253] found that one or more servings of fish per week (or about 200 mg of DHA) was associated with a 60% lower risk of developing AD and that total intake of DHA (but not EPA) was also a determinant of lower AD risk. Furthermore, patients with early stage AD reportedly have lower dietary intakes of *n*-3 PUFAs as compared to healthy individuals [254]. Very recently, the AD Neuroimaging Initiative trial reported results in 229 normal, 397 MCI, and 193 AD patients assessed frequently over a two-year period. They found significant correlations between fish oil supplement use and lower brain atrophy in the hippocampus and cortical gray matter areas across all subjects [255]. There is some inconsistency in the reports of blood DHA levels in AD patients likely due to altered DHA pharmacology and bioavailability and the particular tissue or lipid fraction analyzed. One of the earlier reports indicated that total PL, PC and PE isolated from the plasma of subjects with AD, dementia, or cognitive impairment no dementia (CIND) contained less DHA than found in healthy elderly controls [256]. Furthermore, Tully *et al.* [257] reported that community-living elderly with AD had approximately half the serum concentrations of cholesteryl ester-DHA in comparison to non-dementia controls. These data also indicated that DHA and total saturated fatty acid levels were determinants of the clinical dementia rating. Wang *et al.* [258] found that lower scores on the MMSE in mild and moderate AD patients were associated with lower erythrocyte DHA content. More recently, Lopez *et al.* [259] evaluated an elderly cohort with an average age of 80 years-old for dementia and blood DHA levels. They reported that plasma DHA levels in the highest tertile had 65% reduced odds of all-cause dementia and a 60% reduced chance of AD, an effect that was recapitulated with dietary intake questionnaire data (highest tertile of DHA intake had 72% reduced odds of developing AD). Phillips *et al.* [260] also found positive correlations between composite memory scores (verbal reasoning, contextual, visual, and verbal memory) and overall cognitive status with plasma PC-DHA content across a cohort of normal elderly, cognitive impairment no dementia, and AD subjects.

RCTs investigating the therapeutic potential of DHA in improving the symptoms of AD are scarce and have mixed results at best. One of the first studies was performed by Kotani and colleagues [237], where they supplemented AD patients for 90 days with a daily dose of 240 mg DHA plus 240 mg ARA. This study found no significant changes in the repeatable battery for assessment of neuropsychological status test (RBANS; test five main cognitive domains). It would have been intriguing if Kotani *et al.* had a DHA-alone group to assess any potential counteractive effects of ARA in the combination administration. Subsequently, initial results from the OmegAD study in 174 mild-to-moderate AD patients (mean age 74) provided 1720 mg DHA plus 600 mg EPA daily for six months found no changes in ADAS-cog, MMSE, or Alzheimer's Prevention Initiative (API) tests [261,262]. However, in this early report of the OmegAD study, they did discover that a small subset of AD patients with milder cognitive dysfunction had lower declines in MMSE scores after supplementation. The next reported study was a small preliminary RCT that dosed AD patients for 24 weeks with a daily supplement containing 720 mg of DHA and 1080 mg of EPA. It reported improvements in the Clinician's Interview-Based Impression of Change Scale (CIBIC-plus), but not the ADAS-cog score [238]. Subsequently a large ($N = 295$), multi-center (51) study was reported by Quinn and colleagues [250], where they provided approximately 1000 mg of algal DHA daily for 18 months to patients with mild to moderate AD. They did not detect any differences in ADAS-cog or clinical dementia rating, suggesting that DHA is likely more effective as a prophylactic rather than a therapeutic treatment for AD. Interestingly, DHA-treated subjects in the ApoE4 negative subgroup had less decline in ADAS-cog and MMSE over time *versus* placebo, indicating a potential genotype dependence for DHA's effects in AD. Most recently, updated results have been described by the investigators of the OmegAD study where they have analyzed the levels of *n*-3 PUFAs present in the plasma acquired at baseline and after six months of the high-DHA supplement [263]. They report that increasing plasma levels of DHA in these AD patients was tied to preservation of cognition as measured by ADAS-cog scores. Higher concentrations of plasma DHA resulted in a lower rate of cognitive decline, an effect that was similar across genders. Cerebrospinal fluid (CSF) measures were also obtained in a small group of patients ($N = 33$) from the OmegAD study [264]. The patients in the treatment group had significant increases in the concentration of DHA in the CSF, and these concentrations were inversely correlated to CSF levels of tau (total and phosphorylated) and directly proportional to CSF levels of interleukin (IL)-1 receptor type II (anti-inflammatory effect). Tau levels are elevated during the prodromal phase (total tau) and clinical phase (phosphorylated tau) of AD, and these intriguing results suggest that DHA may be able to mitigate this increase to some extent.

The role of DHA in aging and dementia continues to be an emerging area of research, and more clinical work is certainly warranted given the limited, yet promising results thus far. The later stages of AD are largely intractable, so emphasis should be placed on prophylactic and early-stage therapeutic uses of DHA. Recent negative RCTs in AD patients have been focused on symptomatic effects in already diseased individuals, where significant and irreversible neuronal loss has occurred. DHA effects may be best determined with the disease-modifying effects of DHA focused upon in a large secondary population with mild cognitive complaints (much like the MIDAS study), along with several risk factors for dementia and AD (e.g., ApoE4, cardiovascular disease, low *n*-3 PUFA levels, early plasma AD biomarkers, etc). Pre-planned analysis of DHA effects within specific subgroups (dementia, genotype, baseline indicators, *etc.*) on repeated measures of cognitive function over a substantial amount of time is preferable, but this approach would certainly be costly and time-consuming. Nonetheless, these types of trials are needed to determine the prophylactic effects of DHA and the population subgroups that may benefit the most. In this regard, the ability of DHA to save neurons, and ultimately cognitive capacity, from a seemingly predestined fate could be determined. Considering that the aging of the brain occurs over decades, a general recommendation for maintaining DHA consumption throughout adulthood seems appropriate given the substantial data in aging populations, regardless of cognitive ability or disability.

4.5. Mechanisms of DHA Actions during Aging

Many of the effects of DHA during aging likely occur via several of the same pathways utilized during development. These include those important for neuron growth and survival, maintenance of myelination, reduction or resolution of inflammation, synaptic plasticity, membrane receptor function and lipid raft organization. There is certainly overlap with mechanisms of DHA action between normal and pathological aging. For example, the human brain contains extensive myelination that requires resource-demanding maintenance and repair, and oligodendrocytes, the primary myelinating cell of the CNS, require 2–3 times more energy other brain cell types [265]. It is this homeostatic maintenance and repair of myelin that some researchers consider the weakest link in maintaining brain health in the face of environmental insults and undesirable genetic factors. Myelin sheath abnormalities and defects are observed in normal aging, and white matter is affected in the initial stages of neurodegeneration [266]. Additionally, brain regions myelinated last in development are first to be affected in AD [267], suggesting a unique susceptibility of these late-myelinated areas to white matter defects and degeneration [268]. These observations have led to the hypothesis that amyloidogenesis and hyperphosphorylation of tau might be byproducts of pathological strain on myelin homeostatic processes, rather than causes [269]. The brain sacrifices axonal transport mechanisms for the sake of saving myelin and/or as a result of poor glucose uptake and overwhelmed anti-oxidative capacity of the neuron. These circumstances often result in synaptic loss and axonal degeneration [269,270]. Interestingly, Virtanen *et al.* [271] found that, in a large cohort of elderly subjects (N = 3660; ≥65 years old), higher plasma DHA was associated with a better white matter grade as measured by MRI as well as a lower risk of subclinical brain infarcts. White matter intensities are a predictor of AD conversion in MCI subjects, and silent brain infarcts are associated with more precipitous declines in cognitive functions over time [272].

DHA also likely has cardiovascular benefits throughout adulthood leading to better perfusion of the brain. These benefits include lower blood pressure, improved vasoreactivity, dampened hepatic triglyceride synthesis, and reduced platelet aggregation. The brain is decidedly the most perfused organ of the human body [273], and cardiovascular diseases (common with aging) increase the risk of developing dementia [274]. Vascular dementia is the second most common form of dementia behind AD, and cerebral blood flow reductions and vascular pathologies are often reported in AD [275,276]. Jackson and colleagues [277] provided young adults (mean age of 22 years old) DHA-rich oil, EPA-rich oil, or placebo daily for 12 weeks and reported increased cerebral blood flow as measured by concentration of oxygenated hemoglobin in the DHA group as compared to placebo controls. There was no effect of high-EPA oil treatment on blood flow. This is in line with results from Beydoun *et al.* [278], who observed a positive correlation between *n*-3 PUFAs levels in plasma and cognitive performance in hypertensive and dyslipidemic middle-aged people. Moreover, this correlation was stronger than what was observed in healthy individuals. In addition, Baierle*et al.* [207] found that both increased blood Hcy levels (a risk factor for cardiovascular disease) and reduced serum DHA levels were associated with decreased cognitive performance in healthy elderly. DHA can alter the expression of genes encoding enzymes important for Hcy metabolism, methionine adenosyltransferase (MAT) and methylenetetrahydrofolate reductase (MTHFR) [279], and plasma Hcy and MTHFR polymorphisms are risk factors for dementia [280].

DHA can also affect processes involved in neural plasticity and LTP, which are essential for proper learning and memory function. DHA increases brain-derived neurotrophic factor (BDNF) and both DHA and BDNF affect AkT and ERK/MAPK/CREB signaling pathways to ultimately promote neural plasticity and LTP, which require synaptic modifications that are critical to learning and memory [281,282,283]. Calcium/calmodulin-dependent kinase II (CaMKII) and *N*-methyl-D-aspartate (NMDA) receptor function are essential for the maintenance of LTP and are positively modulated by DHA [284,285]. DHA also mitigates loss of the intracellular scaffold in neurodegeneration models, helping to maintain healthy axons and synaptic structures integral to cognitive function [284,286]. Furthermore, DHA supplementation may improve cognition by enhancing neurogenesis via the retinoid X receptor (RXR) and retinoic acid receptor (RAR), which decrease in expression with age in animal models [287].

The aging of the brain, and especially pathological aging, is also considered to be a consequence of sustained chronic inflammation. Alzheimer patients replete with amyloid plaques who do not exhibit signs of dementia (high pathology controls) have very few signs of neuroinflammation and neurodegeneration [288,289,290], suggesting that an inflammatory component to AD is at minimum partially responsible for the neurodegeneration and dementia. Brains from AD patients exhibit high levels of activated microglia [291], and the degree of cognitive impairment is inversely correlated with the extent of microglial activation [292,293]. Cytokines from these activated microglia break down neural membranes and release pro-inflammatory ARA metabolites, and anti-inflammatory DHA metabolites. Elevated pro-inflammatory cytokines have been identified in the CSF of AD patients [294,295,296], and amyloid beta (Aβ) itself has been shown to induce pro-inflammatory cytokines via activation of inflammasomes [297,298]. DHA can potentially counteract the effects of Aβ since recent data indicates that DHA enhances phagocytosis of Aβ42 by human microglia [299]. Further implicating inflammation in neurodegeneration, gestational immune insult via a viral mimetic, followed by a second similar insult in adulthood, leads to an AD-like phenotype in mice that includes CNS protein aggregates and deficits in cognition [300]. Interestingly, our recent work in this maternal immune activation model has shown that DHA provided throughout development and adulthood significantly dampens subsequent immune stimulation by a second insult with a viral mimetic in adulthood [301].

Important to the resolving process of inflammation, specialized pro-resolving mediators (SPMs) are derivatives of *n*-3 PUFAs and are reduced in AD where brain inflammation is increased [302]. SPMs have anti-inflammatory and pro-resolving characteristics and include protectins, D-series resolvins, and maresins derived from DHA via cyclooxygenase (COX) and lipoxygenase (LOX) pathways [303]. Recent results from the OmegAD trial found that cultured peripheral blood mononuclear cells (PBMCs) from AD patients who received the *n*-3 PUFA supplementation maintained levels of SPMs lipoxin A4 and resolvin D1 in culture over time, even when insulted with Aβ40, whereas PBMCs from control AD patients did not [302]. Furthermore, the effect on SPMs was positively correlated with cognitive changes and changes in transthyretin (prealbumin), which has been shown to inhibit the toxic effects of A. DHA and its metabolites can affect inflammation via several pathways including activation of PPARs, inhibition of nuclear factor kappa-B (NFκB), and activation of the transmembrane receptor GPR120. This is confirmed by observational studies showing that blood levels of *n*-3 PUFAs are associated with lower cytokine levels [304,305], and supplementation with fish oil for 26 weeks changes over 1000 genes in the PBMCs of an elderly cohort, resulting in a more anti-inflammatory gene expression profile [306]. Furthermore, daily supplementation with high-DHA (1700 mg DHA and 600 mg EPA) for six months in AD patients (OmegAD study) reduced the levels of lipopolysaccharide (LPS)-induced cytokines (IL-1B, IL-6) released from isolated PBMCs relative to within-patient baseline levels [307].

During pathological aging of the brain (and even in some cognitively normal elderly), inflammation results in neurofibrillary tangles made of abnormal forms of tau protein that cause alterations in cytoskeletal stability, axonal transport, and loss of synaptic contacts [269,300]. This can affect protein extrusion mechanisms and axonal energy metabolism, resulting in even more phosphorylated tau (p-tau), axonal blockage and leakage, and ultimately cell death. DHA can reduce p-tau levels likely by inhibition of the phosphorylation of tau via c-Jun *N*-terminal kinase 1 (JNK1) [308,309] or Akt through glycogen synthase kinase-3β (GSK3β) [308,310], and help to stabilize white matter [311]. Activation of the Akt pathway promotes cell survival via inhibition of caspase-3, potentially saving neurons during metabolic stress in neurodegeneration. These effects are similar to those seen in developmental reelin-signaling pathways that affect neuronal migration and cortical structure [312,313]. Reelin expression decreases with age and increases with sufficient DHA, and reelin is thought to be involved in the pathogenesis of AD [314]. Both reelin and DHA increase phosphoinositide 3-kinase (PI3K) activity, which activates AkT, which in turn inhibits GSK3β. GSK3β inhibits glycogen production and phosphorylates tau, and ultimately it is this poor glucose uptake and storage in combination with tau hyperphosphorylation that likely precipitates AD pathology [315]. This is in line with emerging data that suggest AD pathophysiology is at least partially mediated by impairments in brain insulin sensitivity, and in glucose metabolism and utilization, that lead to oxidative stress and inflammation [316]. Accordingly, diabetes patients are more likely to develop dementia [317]. Interestingly, animal and *in vitro* studies identify positive effects of DHA on endothelial and glial GLUT1 levels and brain glucose uptake [318,319,320,321].

In AD, extraneuronal Aβ aggregates to form senile plaques that limit plasticity and spine formation, promote loss of memory, increase inflammation via activated microglia, and increase pro-inflammatory cytokines. DHA can inhibit the activity of β-secretase, thereby inhibiting the formation of aggregates [322], and stimulate microglia to phagocytose Aβ peptides *in vitro* [299]. The production of Aβ from amyloid precursor protein (APP) via β-secretase is impacted by the integrity of lipid rafts, which can act to separate APP from the enzyme. Interestingly, there is a decrease in the DHA content of lipid rafts in cortex of AD patients [323], and this decrease in DHA enhances interactions between APP and β-secretase, thereby promoting amyloidogenesis in these subjects [324]. Furthermore, Aβ42 interacts with caveolin-1 containing membranes of erythrocytes in a stronger fashion in DHA-enriched erythrocytes, suggesting that DHA might help with the clearance of Aβ via lipid-raft mediated degradation [325]. These data are corroborated by a cross-sectional study that measured plasma Aβ40 and Aβ42 in 1,219 normal elderly over 65 years of age and found a strong inverse correlation between *n*-3 PUFA intake and plasma Aβ40 and Aβ42 levels [326].

5. GENERAL CONSIDERATIONS AND CONCLUSIONS

There is substantial evidence regarding DHA's importance during pregnancy and infancy on the development of the brain and resulting cognitive function of the child. These effects on learning ability are dependent upon accumulation of DHA during gestation and nursing, and they highlight the need for maternal consumption of dietary or supplemental sources of DHA. Beyond development, a greater understanding of the impact of DHA across decades of life may require piecing together several well-designed longitudinal epidemiological studies. However, epidemiological studies may underestimate the effect of DHA on cognition due to other long-term factors that impact mortality and therefore those potentially helped most by DHA do not reach the age at which cognitive decline or dementia occurs. Nutrition and diet can also be affected by education and socioeconomic factors that can influence cognitive abilities, not to mention a propensity for a healthier lifestyle in general. Such long-term effects on overall health, and in particular on the cardiovascular and neurological systems, will clearly play into an individual's cognitive abilities.

Regarding intervention trials, inter-individual and day-to-day intra-individual variability in DHA consumption likely contributes to an under-estimation of the effect of DHA supplementation on cognition in RCTs. The interventions are comparably much shorter than long-term dietary habits, and RCTs are often not focused on a subject pool that is likely to benefit the most from interventions, such as those with insufficient DHA levels. Accordingly, basal DHA levels must be taken into consideration, as an effect of DHA is likely to be inconsequential in those with adequate tissue levels of DHA. Linking familial risks with early plasma indicators of AD and dementia (C-reactive protein, interleukin-6 and α1-antichymotrypsin, for example) with plasma DHA status may also help to identify those with the most potential for benefit from supplementation. In observational trials, caution must be exercised when examining dietary DHA intake levels and blood concentrations of DHA. Dietary intake only accounts for 66% of the variance in erythrocyte*n*-3 PUFAs concentrations, indicating that age, sex, BMI, physical activity, micronutrient status and other factors are also involved [327]. Similarly, due to the long half-life of brain DHA (2–3 years), fatty acid content of blood compartments may not accurately depict brain lipid composition, especially in the cases of subjects with cognitive impairment or AD and their likely associations with altered DHA processing [61,328]. Even if peripheral levels are comparable between subjects, the availability of DHA for various tissues and cell types may very well be different. This is highlighted by some recent studies utilizing isotopically labeled DHA indicating changes in DHA dynamics within the body as we age. DHA dynamics may also be dependent on genetic factors such as ApoE4 allele carrier status. For example, a single oral 50 mg dose of ^{13}C-DHA took significantly longer to clear from the blood of elderly subjects than young subjects [181]. Furthermore, the half-life of ^{13}C-DHA has been measured as 32 days in ApoE4 carriers, and 140 days in non-carriers [329].

Another consideration is the use of other supplements in boosting the effects of DHA, or acting in synergy, especially with regards to aging. For example, curcumin can boost DHA levels in the brain of animals [330], and it is intriguing to speculate that curcumin could help mitigate the losses in DHA that are associated with the heightened β-oxidation that is observed in AD. As previously mentioned, B vitamins and DHA work in tandem to reduce brain atrophy in MCI patients [134]. B vitamins lower Hcy, an intermediate in some oxidative stress-related pathways that are a risk factor for vascular disease as well as dementia. Hcy can also serve as notification that pathological neurodegenerative processes are occurring. Additionally, B vitamins may promote DHA incorporation into phospholipids, and likely have synergistic anti-inflammatory effects. In support of a combinatorial approach, Scheltens *et al.* [331] observed improvements in memory after daily supplementation with DHA, EPA, phospholipids, choline, uridine monophosphate, vitamin E, vitamin C, selenium, folic acid, vitamin B6, and vitamin B12 for 24 weeks in subjects with mild AD.

The non-steroidal anti-inflammatory drug (NSAID) aspirin might be another potential synergistic agent. Acetylation by aspirin enables COX-2 to initiate the biosynthetic pathway that produces resolvins (D1–D6) from DHA [332,333]. Aspirin-dependent resolvins are potent DHA-derived pro-resolving immune mediators [334]. Resolvin D3 has been shown to be especially effective in the late-resolving phase of inflammation (catabasis), the completion of which is critical to prevent an acute inflammatory response from becoming chronic activation [335]. The use of NSAIDs has been associated with a reduced risk for AD, especially in long-time users [336,337]. However, these effects are not realized in currently diagnosed AD, highlighting the likely importance of preventative supplementation [338].

Overall, DHA appears to have the ability to influence many different signaling pathways, receptor systems, enzyme activities, membrane structures and dynamics that ultimately lead to overall better development, maintenance and aging of the CNS, resulting in optimal cognition throughout the lifespan. These benefits likely require a sustained supply of DHA across development, adolescence and adulthood to build and maintain sufficient pools and/or to replenish depleted neural stores. For those unable to obtain sufficient amounts of DHA via dietary means, supplemental DHA from fish oil or vegetarian (algal oil) sources is ideal. DHA-containing supplements are taken daily by millions of people worldwide and have been shown to be safe and well tolerated even at high doses [339].

ACKNOWLEDGMENTS

Literature search and review, and manuscript composition was made possible via internal support by DSM Nutritional Products Ltd.

CONFLICTS OF INTEREST

Michael J. Weiser, Christopher M. Butt and M. Hasan Mohajeri are employees of DSM Nutritional Products Ltd.

REFERENCES

1. Bryan, J.; Osendarp, S.; Hughes, D.; Calvaresi, E.; Baghurst, K.; van Klinken, J.W. Nutrients for cognitive development in school-aged children. Nutr. Rev. 2004, 62, 295–306.
2. Brenna, J.T.; Diau, G.Y. The influence of dietary docosahexaenoic acid and arachidonic acid on central nervous system polyunsaturated fatty acid composition. Prostaglandins Leukot. Essent. Fat. Acids 2007,77, 247–250.
3. Rapoport, S.I. In vivo fatty acid incorporation into brain phospholipids in relation to signal transduction and membrane remodeling. Neurochem. Res. 1999, 24, 1403–1415.
4. Garcia, M.C.; Ward, G.; Ma, Y.C.; Salem, N., Jr.; Kim, H.Y. Effect of docosahexaenoic acid on the synthesis of phosphatidylserine in rat brain in microsomes and C6 glioma cells. J. Neurochem. 1998, 70, 24–30.
5. Suzuki, H.; Manabe, S.; Wada, O.; Crawford, M.A. Rapid incorporation of docosahexaenoic acid from dietary sources into brain microsomal, synaptosomal and mitochondrial membranes in adult mice.Int. J. Vitam. Nutr. Res. 1997, 67, 272–278.

6. Uauy, R.; Dangour, A.D. Nutrition in brain development and aging: Role of essential fatty acids. Nutr. Rev. 2006, 64, S24–S33.

7. Haubner, L.; Sullivan, J.; Ashmeade, T.; Saste, M.; Wiener, D.; Carver, J. The effects of maternal dietary docosahexaenoic acid intake on rat pup myelin and the auditory startle response. Dev. Neurosci. 2007,29, 460–467.

8. Orr, S.K.; Bazinet, R.P. The emerging role of docosahexaenoic acid in neuroinflammation. Curr. Opin. Investig. Drugs 2008, 9, 735–743.

9. Goustard-Langelier, B.; Guesnet, P.; Durand, G.; Antoine, J.M.; Alessandri, J.M. n-3 and n-6 fatty acid enrichment by dietary fish oil and phospholipid sources in brain cortical areas and nonneural tissues of formula-fed piglets. Lipids 1999, 34, 5–16.

10. Anderson, V.; Fenwick, T.; Manly, T.; Robertson, I. Attentional skills following traumatic brain injury in childhood: A componential analysis. Brain Inj. 1998, 12, 937–949.

11. Barkley, R.A. The executive functions and self-regulation: An evolutionary neuropsychological perspective. Neuropsychol. Rev. 2001, 11, 1–29.

12. Kuratko, C.N.; Barrett, E.C.; Nelson, E.B.; Salem, N. The relationship of docosahexaenoic acid (DHA) with learning and behavior in healthy children: A review. Nutrients 2013, 5, 2777–2810.

13. Stonehouse, W. Does consumption of LC omega-3 PUFA enhance cognitive performance in healthy school-aged children and throughout adulthood? Evidence from clinical trials. Nutrients 2014, 6, 2730–2758.

14. Cederholm, T.; Salem, N., Jr.; Palmblad, J. Omega-3 fatty acids in the prevention of cognitive decline in humans. Adv. Nutr. 2013, 4, 672–676.

15. Joffre, C.; Nadjar, A.; Lebbadi, M.; Calon, F.; Laye, S. n-3 LCPUFA improves cognition: The young, the old and the sick. Prostaglandins Leukot. Essent. Fat. Acids 2014, 91, 1–20.

16. Salem, N.; Vandal, M.; Calon, F. The benefit of docosahexaenoic acid for the adult brain in aging and dementia. Prostaglandins Leukot. Essent. Fat. Acids 2015, 92, 15–22.

17. Sprecher, H.; Luthria, D.L.; Mohammed, B.S.; Baykousheva, S.P. Reevaluation of the pathways for the biosynthesis of polyunsaturated fatty acids. J. Lipid Res. 1995, 36, 2471–2477.

18. Emken, E.A.; Adlof, R.O.; Gulley, R.M. Dietary linoleic acid influences desaturation and acylation of deuterium-labeled linoleic and

linolenic acids in young adult males. Biochim. Biophys. Acta 1994, 1213, 277–288. [CrossRef]

19. Barcelo-Coblijn, G.; Murphy, E.J. Alpha-linolenic acid and its conversion to longer chain n-3 fatty acids: Benefits for human health and a role in maintaining tissue n-3 fatty acid levels. Prog. Lipid Res.2009, 48, 355–374.

20. Brenna, J.T.; Salem, N., Jr.; Sinclair, A.J.; Cunnane, S.C. Alpha-linolenic acid supplementation and conversion to n-3 long-chain polyunsaturated fatty acids in humans. Prostaglandins Leukot. Essent. Fat. Acids 2009, 80, 85–91.

21. Salem, N.; Eggersdorfer, M. Is the world supply of omega-3 fatty acids adequate for optimal human nutrition? Curr. Opin. Clin. Nutr. Metab. Care 2015, 18, 147–154.

22. Birch, E.E.; Carlson, S.E.; Hoffman, D.R.; Fitzgerald-Gustafson, K.M.; Fu, V.L.; Drover, J.R.; Castaneda, Y.S.; Minns, L.; Wheaton, D.K.; Mundy, D.; et al. The diamond (DHA intake and measurement of neural development) study: A double-masked, randomized controlled clinical trial of the maturation of infant visual acuity as a function of the dietary level of docosahexaenoic acid. Am. J. Clin. Nutr.2010, 91, 848–859.

23. Hoffman, D.R.; Boettcher, J.A.; Diersen-Schade, D.A. Toward optimizing vision and cognition in term infants by dietary docosahexaenoic and arachidonic acid supplementation: A review of randomized controlled trials. Prostaglandins Leukot. Essent. Fat. Acids 2009, 81, 151–158.

24. Makrides, M. Is there a dietary requirement for DHA in pregnancy? Prostaglandins Leukot. Essent. Fat. Acids 2009, 81, 171–174.

25. Brenna, J.T.; Varamini, B.; Jensen, R.G.; Diersen-Schade, D.A.; Boettcher, J.A.; Arterburn, L.M. Docosahexaenoic and arachidonic acid concentrations in human breast milk worldwide. Am. J. Clin. Nutr. 2007, 85, 1457–1464.

26. Koletzko, B.; Lien, E.; Agostoni, C.; Bohles, H.; Campoy, C.; Cetin, I.; Decsi, T.; Dudenhausen, J.W.; Dupont, C.; Forsyth, S.; et al. The roles of long-chain polyunsaturated fatty acids in pregnancy, lactation and infancy: Review of current knowledge and consensus recommendations. J. Perinat. Med.2008, 36, 5–14.

27. Aranceta, J.; Pérez-Rodrigo, C. Recommended dietary reference intakes, nutritional goals and dietary guidelines for fat and fatty acids: A systematic review. Br. J. Nutr. 2012, 107, S8–S22.

28. Mosca, L.; Benjamin, E.J.; Berra, K.; Bezanson, J.L.; Dolor, R.J.; Lloyd-Jones, D.M.; Newby, L.K.; Pina, I.L.; Roger, V.L.; Shaw, L.J.; et al.

Effectiveness-based guidelines for the prevention of cardiovascular disease in women—2011 update: A guideline from the American Heart Association. Circulation 2011,123, 1243–1262.

29. World Health Organization; Food and Agriculture Organization of the United Nations. Report of the Joint FAO/WHO Expert Consultation on the Risks and Benefits of Fish Consumption; FAO Fisheries and Aquaculture Report No. 978; WHO: Geneva, Switzerland; FAO: Rome, Italy, 2010; Volume 978, pp. 25–29.

30. Blasbalg, T.L.; Hibbeln, J.R.; Ramsden, C.E.; Majchrzak, S.F.; Rawlings, R.R. Changes in consumption of omega-3 and omega-6 fatty acids in the United States during the 20th century. Am. J. Clin. Nutr.2011, 93, 950–962.

31. Meyer, B.J. Are we consuming enough long chain omega-3 polyunsaturated fatty acids for optimal health? Prostaglandins Leukot. Essent. Fat. Acids 2011, 85, 275–280.

32. Simopoulos, A.P. Evolutionary aspects of diet: The omega-6/omega-3 ratio and the brain. Mol. Neurobiol. 2011, 44, 203–215.

33. Broadhurst, C.L.; Cunnane, S.C.; Crawford, M.A. Rift valley lake fish and shellfish provided brain-specific nutrition for early homo. Br. J. Nutr. 1998, 79, 3–21.

34. Crawford, M.A.; Bloom, M.; Broadhurst, C.L.; Schmidt, W.F.; Cunnane, S.C.; Galli, C.; Gehbremeskel, K.; Linseisen, F.; Lloyd-Smith, J.; Parkington, J. Evidence for the unique function of docosahexaenoic acid during the evolution of the modern hominid brain. Lipids 1999, 34, S39–S47.

35. Umhau, J.C.; Zhou, W.; Carson, R.E.; Rapoport, S.I.; Polozova, A.; Demar, J.; Hussein, N.; Bhattacharjee, A.K.; Ma, K.; Esposito, G.; et al. Imaging incorporation of circulating docosahexaenoic acid into the human brain using positron emission tomography. J. Lipid Res. 2009, 50, 1259–1268.

36. Kawashima, A.; Harada, T.; Imada, K.; Yano, T.; Mizuguchi, K. Eicosapentaenoic acid inhibits interleukin-6 production in interleukin-1beta-stimulated C6 glioma cells through peroxisome proliferator-activated receptor-gamma. Prostaglandins Leukot. Essent. Fat. Acids 2008, 79, 59–65.

37. Orr, S.K.; Trepanier, M.O.; Bazinet, R.P. n-3 polyunsaturated fatty acids in animal models with neuroinflammation. Prostaglandins Leukot. Essent. Fat. Acids 2013, 88, 97–103.

38. Chen, C.T.; Liu, Z.; Ouellet, M.; Calon, F.; Bazinet, R.P. Rapid beta-oxidation of eicosapentaenoic acid in mouse brain: An in situ study. Prostaglandins Leukot. Essent. Fat. Acids 2009, 80, 157–163.

39. Ouellet, M.; Emond, V.; Chen, C.T.; Julien, C.; Bourasset, F.; Oddo, S.; LaFerla, F.; Bazinet, R.P.; Calon, F. Diffusion of docosahexaenoic and eicosapentaenoic acids through the blood-brain barrier: An in situ cerebral perfusion study. Neurochem. Int. 2009, 55, 476–482.

40. Chen, C.T.; Bazinet, R.P. Beta-oxidation and rapid metabolism, but not uptake regulate brain eicosapentaenoic acid levels. Prostaglandins Leukot. Essent. Fat. Acids 2015, 92, 33–40.

41. Arterburn, L.M.; Hall, E.B.; Oken, H. Distribution, interconversion, and dose response of n-3 fatty acids in humans. Am. J. Clin. Nutr. 2006, 83, 1467S–1476S.

42. Chen, C.T.; Domenichiello, A.F.; Trépanier, M.O.; Liu, Z.; Masoodi, M.; Bazinet, R.P. The low levels of eicosapentaenoic acid in rat brain phospholipids are maintained via multiple redundant mechanisms.J. Lipid Res. 2013, 54, 2410–2422.

43. Kaur, G.; Molero, J.C.; Weisinger, H.S.; Sinclair, A.J. Orally administered [14C] DPA and [14C] DHA are metabolised differently to [14C] EPA in rats. Br. J. Nutr. 2013, 109, 441–448.

44. Eaton, S.; Bartlett, K.; Pourfarzam, M. Mammalian mitochondrial beta-oxidation. Biochem. J. 1996, 320, 345–357.

45. Reddy, J.K.; Hashimoto, T. Peroxisomal beta-oxidation and peroxisome proliferator-activated receptor alpha: An adaptive metabolic system. Annu. Rev. Nutr. 2001, 21, 193–230.

46. Morse, N.L. Benefits of docosahexaenoic acid, folic acid, vitamin D and iodine on foetal and infant brain development and function following maternal supplementation during pregnancy and lactation. Nutrients 2012, 4, 799–840.

47. Lin, Y.H.; Llanos, A.; Mena, P.; Uauy, R.; Salem, N., Jr.; Pawlosky, R.J. Compartmental analyses of 2H5-alpha-linolenic acid and CU-eicosapentaenoic acid toward synthesis of plasma labeled 22:6n-3 in newborn term infants. Am. J. Clin. Nutr. 2010, 92, 284–293.

48. Pawlosky, R.J.; Salem, N., Jr. Perspectives on alcohol consumption: Liver polyunsaturated fatty acids and essential fatty acid metabolism. Alcohol 2004, 34, 27–33.

49. Plourde, M.; Cunnane, S.C. Extremely limited synthesis of long chain polyunsaturates in adults: Implications for their dietary essentiality and use as supplements. Appl. Physiol. Nutr. Metab. 2007, 32, 619–634.

50. Egert, S.; Lindenmeier, M.; Harnack, K.; Krome, K.; Erbersdobler, H.F.; Wahrburg, U.; Somoza, V. Margarines fortified with alpha-linolenic acid, eicosapentaenoic acid, or docosahexaenoic acid alter

the fatty acid composition of erythrocytes but do not affect the antioxidant status of healthy adults. J. Nutr. 2012, 142, 1638–1644.

51. Carver, J.; Benford, V.; Han, B.; Cantor, A. The relationship between age and the fatty acid composition of cerebral cortex and erythrocytes in human subjects. Brain Res. Bull. 2001, 56, 79–85. [CrossRef]

52. Martinez, M. Tissue levels of polyunsaturated fatty acids during early human development. J. Pediatr.1992, 120, S129–S138. [CrossRef]

53. Bourre, J.M. Effects of nutrients (in food) on the structure and function of the nervous system: Update on dietary requirements for brain. Part 1: Micronutrients. J. Nutr. Health Aging 2006, 10, 377–385.

54. McNamara, R.K.; Carlson, S.E. Role of omega-3 fatty acids in brain development and function: Potential implications for the pathogenesis and prevention of psychopathology. Prostaglandins Leukot. Essent. Fat. Acids 2006, 75, 329–349.

55. Chen, C.T.; Kitson, A.P.; Hopperton, K.E.; Domenichiello, A.F.; Trepanier, M.O.; Lin, L.E.; Ermini, L.; Post, M.; Thies, F.; Bazinet, R.P. Plasma non-esterified docosahexaenoic acid is the major pool supplying the brain. Sci. Rep. 2015, 5, 15791.

56. DeMar, J.C., Jr.; Ma, K.; Chang, L.; Bell, J.M.; Rapoport, S.I. Alpha-linolenic acid does not contribute appreciably to docosahexaenoic acid within brain phospholipids of adult rats fed a diet enriched in docosahexaenoic acid. J. Neurochem. 2005, 94, 1063–1076.

57. Lemaitre-Delaunay, D.; Pachiaudi, C.; Laville, M.; Pousin, J.; Armstrong, M.; Lagarde, M. Blood compartmental metabolism of docosahexaenoic acid (DHA) in humans after ingestion of a single dose of [13C] DHA in phosphatidylcholine. J. Lipid Res. 1999, 40, 1867–1874.

58. Chen, S.; Subbaiah, P.V. Regioisomers of phosphatidylcholine containing DHA and their potential to deliver DHA to the brain: Role of phospholipase specificities. Lipids 2013, 48, 675–686.

59. Liu, R.Z.; Mita, R.; Beaulieu, M.; Gao, Z.; Godbout, R. Fatty acid binding proteins in brain development and disease. Int. J. Dev. Biol. 2010, 54, 1229–1239.

60. Pan, Y.; Scanlon, M.J.; Owada, Y.; Yamamoto, Y.; Porter, C.J.H.; Nicolazzo, J.A. Fatty acid-binding protein 5 facilitates the blood-brain barrier transport of docosahexaenoic acid. Mol. Pharm. 2015, 12, 4375–4385.

61. Vandal, M.; Alata, W.; Tremblay, C.; Rioux-Perreault, C.; Salem, N.; Calon, F.; Plourde, M. Reduction in DHA transport to the brain of mice expressing human APOE4 compared to APOE 2. J. Neurochem.2014, 129, 516–526.

62. Chen, C.T.; Green, J.T.; Orr, S.K.; Bazinet, R.P. Regulation of brain polyunsaturated fatty acid uptake and turnover. Prostaglandins Leukot. Essent. Fat. Acids 2008, 79, 85–91.

63. Kennedy, E.P.; Weiss, S.B. The function of cytidine coenzymes in the biosynthesis of phospholipides. J. Biol. Chem. 1956, 222, 193–214.

64. Jensen, C.L.; Lapillonne, A. Docosahexaenoic acid and lactation. Prostaglandins Leukot. Essent. Fat. Acids 2009, 81, 175–178.

65. Simmer, K.; Patole, S.K.; Rao, S.C. Long-chain polyunsaturated fatty acid supplementation in infants born at term. Cochrane Database Syst. Rev. 2011. [CrossRef]

66. Chambaz, J.; Ravel, D.; Manier, M.C.; Pepin, D.; Mulliez, N.; Bereziat, G. Essential fatty acids interconversion in the human fetal liver. Biol. Neonate 1985, 47, 136–140.

67. Innis, S.M. Essential fatty acid transfer and fetal development. Placenta 2005, 26, S70–S75.

68. Uauy, R.; Mena, P.; Rojas, C. Essential fatty acids in early life: Structural and functional role. Proc. Nutr. Soc. 2000, 59, 3–15.

69. Grantham-McGregor, S.; Cheung, Y.B.; Cueto, S.; Glewwe, P.; Richter, L.; Strupp, B. Developmental potential in the first 5 years for children in developing countries. Lancet 2007, 369, 60–70. [CrossRef]

70. Farquharson, J.; Cockburn, F.; Patrick, W.A.; Jamieson, E.C.; Logan, R.W. Infant cerebral cortex phospholipid fatty-acid composition and diet. Lancet 1992, 340, 810–813. [CrossRef]

71. Makrides, M.; Neumann, M.; Simmer, K.; Pater, J.; Gibson, R. Are long-chain polyunsaturated fatty acids essential nutrients in infancy? Lancet 1995, 345, 1463–1468. [CrossRef]

72. European Food Safety Authority. Scientific opinion on dietary reference values for fats, including saturated fatty acids, polyunsaturated fatty acids, monounsaturated fatty acids, trans fatty acids, and cholesterol. EFSA J. 2010, 8, 1461.

73. Makrides, M.; Duley, L.; Olsen, S.F. Marine oil, and other prostaglandin precursor, supplementation for pregnancy uncomplicated by pre-eclampsia or intrauterine growth restriction. Cochrane Database Syst Rev. 2006, 3. [CrossRef]

74. Szajewska, H.; Horvath, A.; Koletzko, B. Effect of n-3 long-chain polyunsaturated fatty acid supplementation of women with low-risk pregnancies on pregnancy outcomes and growth measures at birth: A meta-analysis of randomized controlled trials. Am. J. Clin. Nutr. 2006, 83, 1337–1344.

75. Horvath, A.; Koletzko, B.; Szajewska, H. Effect of supplementation of women in high-risk pregnancies with long-chain polyunsaturated fatty acids on pregnancy outcomes and growth measures at birth: A meta-analysis of randomized controlled trials. Br. J. Nutr. 2007, 98, 253–259.

76. Ramakrishnan, U.; Stein, A.D.; Parra-Cabrera, S.; Wang, M.; Imhoff-Kunsch, B.; Juarez-Marquez, S.; Rivera, J.; Martorell, R. Effects of docosahexaenoic acid supplementation during pregnancy on gestational age and size at birth: Randomized, double-blind, placebo-controlled trial in Mexico. Food Nutr. Bull. 2010, 31, S108–S116.

77. Hibbeln, J.R.; Davis, J.M.; Steer, C.; Emmett, P.; Rogers, I.; Williams, C.; Golding, J. Maternal seafood consumption in pregnancy and neurodevelopmental outcomes in childhood (ALSPAC study): An observational cohort study. Lancet 2007, 369, 578–585. [CrossRef]

78. Jensen, C.L.; Voigt, R.G.; Llorente, A.M.; Peters, S.U.; Prager, T.C.; Zou, Y.L.; Rozelle, J.C.; Turcich, M.R.; Fraley, J.K.; Anderson, R.E.; et al. Effects of early maternal docosahexaenoic acid intake on neuropsychological status and visual acuity at five years of age of breast-fed term infants. J. Pediatr.2010, 157, 900–905.

79. Makrides, M.; Gibson, R.A.; McPhee, A.J.; Yelland, L.; Quinlivan, J.; Ryan, P. Effect of DHA supplementation during pregnancy on maternal depression and neurodevelopment of young children: A randomized controlled trial. JAMA 2010, 304, 1675–1683.

80. Belfort, M.B.; Rifas-Shiman, S.L.; Kleinman, K.P.; Guthrie, L.B.; Bellinger, D.C.; Taveras, E.M.; Gillman, M.W.; Oken, E. Infant feeding and childhood cognition at ages 3 and 7 years: Effects of breastfeeding duration and exclusivity. JAMA Pediatr. 2013, 167, 836–844.

81. Hoddinott, P.; Tappin, D.; Wright, C. Breast feeding. BMJ 2008, 336, 881–887.

82. Jensen, C.L.; Voigt, R.G.; Prager, T.C.; Zou, Y.L.; Fraley, J.K.; Rozelle, J.C.; Turcich, M.R.; Llorente, A.M.; Anderson, R.E.; Heird, W.C. Effects of maternal docosahexaenoic acid intake on visual function and neurodevelopment in breastfed term infants. Am. J. Clin. Nutr. 2005, 82, 125–132.

83. Uauy, R.; Hoffman, D.R.; Mena, P.; Llanos, A.; Birch, E.E. Term infant studies of DHA and ara supplementation on neurodevelopment: Results of randomized controlled trials. J. Pediatr. 2003, 143, S17–S25. [CrossRef]

84. Walker, S.P.; Wachs, T.D.; Gardner, J.M.; Lozoff, B.; Wasserman, G.A.; Pollitt, E.; Carter, J.A.; the International Child Development Steering Group. Child development: Risk factors for adverse outcomes in developing countries. Lancet 2007, 369, 145–157. [CrossRef]

85. Hart, S.L.; Boylan, L.M.; Carroll, S.R.; Musick, Y.A.; Kuratko, C.; Border, B.G.; Lampe, R.M. Brief report: Newborn behavior differs with decosahexaenoic acid levels in breast milk. J. Pediatr. Psychol.2006, 31, 221–226.

86. Decsi, T.; Campoy, C.; Koletzko, B. Effect of n-3 polyunsaturated fatty acid supplementation in pregnancy: The nuheal trial. Adv. Exp. Med. Biol. 2005, 569, 109–113.

87. Helland, I.B.; Smith, L.; Saarem, K.; Saugstad, O.D.; Drevon, C.A. Maternal supplementation with very-long-chain n-3 fatty acids during pregnancy and lactation augments children's IQ at 4 years of age. Pediatrics 2003, 111, e39–e44.

88. Boucher, O.; Burden, M.J.; Muckle, G.; Saint-Amour, D.; Ayotte, P.; Dewailly, E.; Nelson, C.A.; Jacobson, S.W.; Jacobson, J.L. Neurophysiologic and neurobehavioral evidence of beneficial effects of prenatal omega-3 fatty acid intake on memory function at school age. Am. J. Clin. Nutr. 2011, 93, 1025–1037.

89. Innis, S.M. Impact of maternal diet on human milk composition and neurological development of infants. Am. J. Clin. Nutr. 2014, 99, 734S–741S.

90. Agostoni, C.; Brunetti, I.; Marco, A.D. Polyunsaturated fatty acids in development in breastfed infants human and neurological. Curr. Pediatr. Rev. 2005, 25–30. [CrossRef]

91. Gibson, R.A.; Neumann, M.A.; Makrides, M. Effect of increasing breast milk docosahexaenoic acid on plasma and erythrocyte phospholipid fatty acids and neural indices of exclusively breast fed infants.Eur. J. Clin. Nutr. 1997, 51, 578–584.

92. Innis, S.M. Dietary (n-3) fatty acids and brain development. J. Nutr. 2007, 137, 855–859.

93. Horta, B.; Bahl, R.; Martines, J.; Victora, C. Evidence on the Long-Term Effects of Breastfeeding; WHO: Geneva, Switzerland, 2007; pp. 1–52.

94. Oddy, W.H.; Kendall, G.E.; Li, J.; Jacoby, P.; Robinson, M.; de Klerk, N.H.; Silburn, S.R.; Zubrick, S.R.; Landau, L.I.; Stanley, F.J. The long-

term effects of breastfeeding on child and adolescent mental health: A pregnancy cohort study followed for 14 years. J. Pediatr. 2010, 156, 568–574.

95. Eilander, A.; Hundscheid, D.C.; Osendarp, S.J.; Transler, C.; Zock, P.L. Effects of n-3 long chain polyunsaturated fatty acid supplementation on visual and cognitive development throughout childhood: A review of human studies. Prostaglandins Leukot. Essent. Fat. Acids 2007, 76, 189–203.

96. Drover, J.R.; Hoffman, D.R.; Castañeda, Y.S.; Morale, S.E.; Garfield, S.; Wheaton, D.H.; Birch, E.E. Cognitive function in 18-month-old term infants of the diamond study: A randomized, controlled clinical trial with multiple dietary levels of docosahexaenoic acid. Early Hum. Dev. 2011, 87, 223–230.

97. Meldrum, S.J.; D'Vaz, N.; Simmer, K.; Dunstan, J.A.; Hird, K.; Prescott, S.L. Effects of high-dose fish oil supplementation during early infancy on neurodevelopment and language: A randomised controlled trial. Br. J. Nutr. 2012, 108, 1443–1454.

98. Birch, E.E.; Garfield, S.; Castañeda, Y.; Hughbanks-Wheaton, D.; Uauy, R.; Hoffman, D. Visual acuity and cognitive outcomes at 4 years of age in a double-blind, randomized trial of long-chain polyunsaturated fatty acid-supplemented infant formula. Early Hum. Dev. 2007, 83, 279–284.

99. Makrides, M.; Neumann, M.A.; Jeffrey, B.; Lien, E.L.; Gibson, R.A. A randomized trial of different ratios of linoleic to alpha-linolenic acid in the diet of term infants: Effects on visual function and growth. Am. J. Clin. Nutr. 2000, 71, 120–129.

100. Cohen, J.T.; Bellinger, D.C.; Connor, W.E.; Shaywitz, B.A. A quantitative analysis of prenatal intake ofn-3 polyunsaturated fatty acids and cognitive development. Am. J. Prev. Med. 2005, 29, 366–374.

101. Tofail, F.; Kabir, I.; Hamadani, J.D.; Chowdhury, F.; Yesmin, S.; Mehreen, F.; Huda, S.N. Supplementation of fish-oil and soy-oil during pregnancy and psychomotor development of infants. J. Health Popul. Nutr. 2006, 24, 48–56.

102. Dunstan, J.A.; Simmer, K.; Dixon, G.; Prescott, S.L. Cognitive assessment of children at age 2 (1/2) years after maternal fish oil supplementation in pregnancy: A randomised controlled trial. Arch. Dis. Child. Fetal Neonatal Ed. 2008, 93, F45–F50.

103. Scott, D.T.; Janowsky, J.S.; Carroll, R.E.; Taylor, J.A.; Auestad, N.; Montalto, M.B. Formula supplementation with long-chain polyunsaturated fatty acids: Are there developmental benefits?Pediatrics 1998, 102, E59.

104. Hoffman, D.R.; Uauy, R. Essentiality of dietary omega 3 fatty acids for premature infants: Plasma and red blood cell fatty acid composition. Lipids 1992, 27, 886–895.

105. Fewtrell, M.S.; Abbott, R.A.; Kennedy, K.; Singhal, A.; Morley, R.; Caine, E.; Jamieson, C.; Cockburn, F.; Lucas, A. Randomized, double-blind trial of long-chain polyunsaturated fatty acid supplementation with fish oil and borage oil in preterm infants. J. Pediatr. 2004, 144, 471–479.

106. Clandinin, M.T.; van Aerde, J.E.; Merkel, K.L.; Harris, C.L.; Springer, M.A.; Hansen, J.W.; Diersen-Schade, D.A. Growth and development of preterm infants fed infant formulas containing docosahexaenoic acid and arachidonic acid. J. Pediatr. 2005, 146, 461–468.

107. Henriksen, C.; Haugholt, K.; Lindgren, M.; Aurvåg, A.K.; Rønnestad, A.; Grønn, M.; Solberg, R.; Moen, A.; Nakstad, B.; Berge, R.K.; et al. Improved cognitive development among preterm infants attributable to early supplementation of human milk with docosahexaenoic acid and arachidonic acid.Pediatrics 2008, 121, 1137–1145.

108. Makrides, M.; Collins, C.T.; Gibson, R.A. Impact of fatty acid status on growth and neurobehavioural development in humans. Mater. Child Nutr. 2011, 7 (Suppl. 2), 80–88.

109. Makrides, M.; Gibson, R.A.; McPhee, A.J.; Collins, C.T.; Davis, P.G.; Doyle, L.W.; Simmer, K.; Colditz, P.B.; Morris, S.; Smithers, L.G.; et al. Neurodevelopmental outcomes of preterm infants fed high-dose docosahexaenoic acid: A randomized controlled trial. JAMA 2009, 301, 175–182.

110. Collins, C.T.; Gibson, R.A.; Anderson, P.J.; McPhee, A.J.; Sullivan, T.R.; Gould, J.F.; Ryan, P.; Doyle, L.W.; Davis, P.G.; McMichael, J.E.; et al. Neurodevelopmental outcomes at 7 years' corrected age in preterm infants who were fed high-dose docosahexaenoic acid to term equivalent: A follow-up of a randomised controlled trial. BMJ Open 2015, 5.

111. Molloy, C.S.; Stokes, S.; Makrides, M.; Collins, C.T.; Anderson, P.J.; Doyle, L.W. Long-term effect of high-dose supplementation with DHA on visual function at school age in children born at <33 wk gestational age: Results from a follow-up of a randomized controlled trial. Am. J. Clin. Nutr. 2016, 103, 268–275.

112. Prentice, A.M.; van der Merwe, L. Impact of fatty acid status on immune function of children in low-income countries. Mater. Child Nutr. 2011, 7 (Suppl. S2), 89–98.

113. Scholtz, S.A.; Colombo, J.; Carlson, S.E. Clinical overview of effects of dietary long-chain polyunsaturated fatty acids during the perinatal period. Nestle Nutr. Inst. Workshop Ser. 2013, 77, 145–154.

114. Delgado-Noguera, M.F.; Calvache, J.A.; Bonfill Cosp, X.; Kotanidou, E.P.; Galli-Tsinopoulou, A. Supplementation with long chain polyunsaturated fatty acids (LCPUFA) to breastfeeding mothers for improving child growth and development. Cochrane Database Syst. Rev. 2015, 7. [CrossRef]

115. Michaelsen, K.F.; Dewey, K.G.; Perez-Exposito, A.B.; Nurhasan, M.; Lauritzen, L.; Roos, N. Food sources and intake of n-6 and n-3 fatty acids in low-income countries with emphasis on infants, young children (6–24 months), and pregnant and lactating women. Mater. Child Nutr. 2011, 7 (Suppl. 2), 124–140.

116. Barbarich, B.N.; Willows, N.D.; Wang, L.; Clandinin, M.T. Polyunsaturated fatty acids and anthropometric indices of children in rural China. Eur. J. Clin. Nutr. 2006, 60, 1100–1107.

117. Yakes, E.A.; Arsenault, J.E.; Islam, M.M.; Hossain, M.B.; Ahmed, T.; German, J.B.; Gillies, L.A.; Rahman, A.S.; Drake, C.; Jamil, K.M.; et al. Intakes and breast-milk concentrations of essential fatty acids are low among Bangladeshi women with 24–48-month-old children. Br. J. Nutr. 2011, 105, 1660–1670.

118. Huybregts, L.F.; Roberfroid, D.A.; Kolsteren, P.W.; van Camp, J.H. Dietary behaviour, food and nutrient intake of pregnant women in a rural community in Burkina Faso. Mater. Child Nutr. 2009, 5, 211–222.

119. Mardones, F.; Urrutia, M.T.; Villarroel, L.; Rioseco, A.; Castillo, O.; Rozowski, J.; Tapia, J.L.; Bastias, G.; Bacallao, J.; Rojas, I. Effects of a dairy product fortified with multiple micronutrients and omega-3 fatty acids on birth weight and gestation duration in pregnant Chilean women. Public Health Nutr.2008, 11, 30–40.

120. Muthayya, S.; Eilander, A.; Transler, C.; Thomas, T.; van der Knaap, H.C.; Srinivasan, K.; van Klinken, B.J.; Osendarp, S.J.; Kurpad, A.V. Effect of fortification with multiple micronutrients and n-3 fatty acids on growth and cognitive performance in Indian schoolchildren: The champion (children's health and mental performance influenced by optimal nutrition) study. Am. J. Clin. Nutr. 2009, 89, 1766–1775.

121. Osendarp, S.J.; Baghurst, K.I.; Bryan, J.; Calvaresi, E.; Hughes, D.; Hussaini, M.; Karyadi, S.J.; van Klinken, B.J.; van der Knaap, H.C.; Lukito, W.; et al. Effect of a 12-mo micronutrient intervention on learning and memory in well-nourished and marginally nourished school-aged children: 2 parallel, randomized, placebo-controlled

studies in Australia and Indonesia. Am. J. Clin. Nutr. 2007, 86, 1082–1093.

122. Dalton, A.; Wolmarans, P.; Witthuhn, R.C.; van Stuijvenberg, M.E.; Swanevelder, S.A.; Smuts, C.M. A randomised control trial in schoolchildren showed improvement in cognitive function after consuming a bread spread, containing fish flour from a marine source. Prostaglandins Leukot. Essent. Fat. Acids 2009, 80, 143–149.

123. Stevens, L.J.; Zentall, S.S.; Abate, M.L.; Kuczek, T.; Burgess, J.R. Omega-3 fatty acids in boys with behavior, learning, and health problems. Physiol. Behav. 1996, 59, 915–920. [CrossRef]

124. McNamara, R.K.; Able, J.; Jandacek, R.; Rider, T.; Tso, P.; Eliassen, J.C.; Alfieri, D.; Weber, W.; Jarvis, K.; DelBello, M.P.; et al. Docosahexaenoic acid supplementation increases prefrontal cortex activation during sustained attention in healthy boys: A placebo-controlled, dose-ranging, functional magnetic resonance imaging study. Am. J. Clin. Nutr. 2010, 91, 1060–1067.

125. Richardson, A.J.; Burton, J.R.; Sewell, R.P.; Spreckelsen, T.F.; Montgomery, P. Docosahexaenoic acid for reading, cognition and behavior in children aged 7–9 years: A randomized, controlled trial (the DOLAB study). PLoS ONE 2012, 7, e43909.

126. Brew, B.K.; Toelle, B.G.; Webb, K.L.; Almqvist, C.; Marks, G.B.; Investigators, C. Omega-3 supplementation during the first 5 years of life and later academic performance: A randomised controlled trial. Eur. J. Clin. Nutr. 2015, 69, 419–424.

127. Lassek, W.D.; Gaulin, S.J.C. Sex differences in the relationship of dietary fatty acids to cognitive measures in American children. Front. Evol. Neurosci. 2011, 3, 5.

128. Extier, A.; Langelier, B.; Perruchot, M.H.; Guesnet, P.; van Veldhoven, P.P.; Lavialle, M.; Alessandri, J.M. Gender affects liver desaturase expression in a rat model of n-3 fatty acid repletion. J. Nutr. Biochem. 2010, 21, 180–187.

129. Schram, L.B.; Nielsen, C.J.; Porsgaard, T.; Nielsen, N.S.; Holm, R.; Mu, H. Food matrices affect the bioavailability of (n-3) polyunsaturated fatty acids in a single meal study in humans. Food Res. Int.2007, 40, 1062–1068. [CrossRef]

130. Ulven, S.M.; Kirkhus, B.; Lamglait, A.; Basu, S.; Elind, E.; Haider, T.; Berge, K.; Vik, H.; Pedersen, J.I. Metabolic effects of krill oil are essentially similar to those of fish oil but at lower dose of EPA and DHA, in healthy volunteers. Lipids 2011, 46, 37–46.

131. Elvevoll, E.O.; Barstad, H.; Breimo, E.S.; Brox, J.; Eilertsen, K.E.; Lund, T.; Olsen, J.O.; Osterud, B. Enhanced incorporation of n-3 fatty acids from fish compared with fish oils. Lipids 2006, 41, 1109–1114.

132. Arterburn, L.M.; Oken, H.A.; Hoffman, J.P.; Bailey-Hall, E.; Chung, G.; Rom, D.; Hamersley, J.; McCarthy, D. Bioequivalence of docosahexaenoic acid from different algal oils in capsules and in a DHA -fortified food. Lipids 2007, 42, 1011–1024.

133. Kuratko, C.N.; Salem, N., Jr. Biomarkers of DHA status. Prostaglandins Leukot. Essent. Fat. Acids 2009,81, 111–118.

134. Jernerén, F.; Elshorbagy, A.K.; Oulhaj, A.; Smith, S.M.; Refsum, H.; Smith, A.D. Brain atrophy in cognitively impaired elderly: The importance of long-chain ω-3 fatty acids and B vitamin status in a randomized controlled trial. Am. J. Clin. Nutr. 2015, 102, 215–221.

135. McCann, J.C.; Ames, B.N. Is docosahexaenoic acid, an n-3 long-chain polyunsaturated fatty acid, required for development of normal brain function? An overview of evidence from cognitive and behavioral tests in humans and animals. Am. J. Clin. Nutr. 2005, 82, 281–295.

136. Fleith, M.; Clandinin, M.T. Dietary pufa for preterm and term infants: Review of clinical studies. Crit. Rev. Food Sci Nutr. 2005, 45, 205–229.

137. Innis, S.M. Fatty acids and early human development. Early Hum. Dev. 2007, 83, 761–766.

138. Youdim, K.A.; Martin, A.; Joseph, J.A. Essential fatty acids and the brain: Possible health implications.Int. J. Dev. Neurosci. 2000, 18, 383–399. [CrossRef]

139. De Urquiza, A.M.; Liu, S.; Sjoberg, M.; Zetterstrom, R.H.; Griffiths, W.; Sjovall, J.; Perlmann, T. Docosahexaenoic acid, a ligand for the retinoid X receptor in mouse brain. Science 2000, 290, 2140–2144.

140. Kitajka, K.; Sinclair, A.J.; Weisinger, R.S.; Weisinger, H.S.; Mathai, M.; Jayasooriya, A.P.; Halver, J.E.; Puskas, L.G. Effects of dietary omega-3 polyunsaturated fatty acids on brain gene expression. Proc. Natl. Acad. Sci. USA 2004, 101, 10931–10936.

141. Calderon, F.; Kim, H.Y. Docosahexaenoic acid promotes neurite growth in hippocampal neurons. J. Neurochem. 2004, 90, 979–988.

142. Moriguchi, T.; Greiner, R.S.; Salem, N., Jr. Behavioral deficits associated with dietary induction of decreased brain docosahexaenoic acid concentration. J. Neurochem. 2000, 75, 2563–2573.

143. Lingwood, B.E.; Healy, G.N.; Sullivan, S.M.; Pow, D.V.; Colditz, P.B. MAP2 provides reliable early assessment of neural injury in the newborn piglet model of birth asphyxia. J. Neurosci. Methods 2008,171, 140–146.

144. Shaikh, S.R.; Dumaual, A.C.; Castillo, A.; LoCascio, D.; Siddiqui, R.A.; Stillwell, W.; Wassall, S.R. Oleic and docosahexaenoic acid differentially phase separate from lipid raft molecules: A comparative NMR, DSC, AFM, and detergent extraction study. Biophys. J. 2004, 87, 1752–1766.

145. Stillwell, W.; Shaikh, S.R.; Zerouga, M.; Siddiqui, R.; Wassall, S.R. Docosahexaenoic acid affects cell signaling by altering lipid rafts. Reprod. Nutr. Dev. 2005, 45, 559–579.

146. Vinson, M.; Rausch, O.; Maycox, P.R.; Prinjha, R.K.; Chapman, D.; Morrow, R.; Harper, A.J.; Dingwall, C.; Walsh, F.S.; Burbidge, S.A.; et al. Lipid rafts mediate the interaction between myelin-associated glycoprotein (MAG) on myelin and MAG-receptors on neurons. Mol. Cell. Neurosci. 2003, 22, 344–352. [CrossRef]

147. Jaworski, J.; Sheng, M. The growing role of mtor in neuronal development and plasticity. Mol. Neurobiol. 2006, 34, 205–219. [CrossRef]

148. Jin, Y.; Sui, H.J.; Dong, Y.; Ding, Q.; Qu, W.H.; Yu, S.X.; Jin, Y.X. Atorvastatin enhances neurite outgrowth in cortical neurons in vitro via up-regulating the Akt/mTOR and Akt/GSK-3beta signaling pathways. Acta Pharmacol. Sin. 2012, 33, 861–872.

149. Cao, D.; Kevala, K.; Kim, J.; Moon, H.S.; Jun, S.B.; Lovinger, D.; Kim, H.Y. Docosahexaenoic acid promotes hippocampal neuronal development and synaptic function. J. Neurochem. 2009, 111, 510–521.

150. Darios, F.; Davletov, B. Omega-3 and omega-6 fatty acids stimulate cell membrane expansion by acting on syntaxin 3. Nature 2006, 440, 813–817.

151. Nyaradi, A.; Li, J.; Hickling, S.; Foster, J.; Oddy, W.H. The role of nutrition in children's neurocognitive development, from pregnancy through childhood. Front. Hum. Neurosci. 2013, 7, 97.

152. Bourre, J.M.; Pascal, G.; Durand, G.; Masson, M.; Dumont, O.; Piciotti, M. Alterations in the fatty acid composition of rat brain cells (neurons, astrocytes, and oligodendrocytes) and of subcellular fractions (myelin and synaptosomes) induced by a diet devoid of n-3 fatty acids. J. Neurochem. 1984, 43, 342–348.

153. Ikemoto, A.; Ohishi, M.; Sato, Y.; Hata, N.; Misawa, Y.; Fujii, Y.; Okuyama, H. Reversibility of n-3 fatty acid deficiency-induced

alterations of learning behavior in the rat: Level of n-6 fatty acids as another critical factor. J. Lipid Res. 2001, 42, 1655–1663.

154. Brenna, J.T. Animal studies of the functional consequences of suboptimal polyunsaturated fatty acid status during pregnancy, lactation and early post-natal life. Mater. Child Nutr. 2011, 7 (Suppl. 2), 59–79.

155. Diau, G.Y.; Hsieh, A.T.; Sarkadi-Nagy, E.A.; Wijendran, V.; Nathanielsz, P.W.; Brenna, J.T. The influence of long chain polyunsaturate supplementation on docosahexaenoic acid and arachidonic acid in baboon neonate central nervous system. BMC Med. 2005, 3, 11.

156. Luchtman, D.W.; Song, C. Cognitive enhancement by omega-3 fatty acids from child-hood to old age: Findings from animal and clinical studies. Neuropharmacology 2013, 64, 550–565.

157. Neuringer, M.; Connor, W.E.; Lin, D.S.; Barstad, L.; Luck, S. Biochemical and functional effects of prenatal and postnatal omega 3 fatty acid deficiency on retina and brain in rhesus monkeys. Proc. Natl. Acad. Sci. USA 1986, 83, 4021–4025.

158. Champoux, M.; Hibbeln, J.R.; Shannon, C.; Majchrzak, S.; Suomi, S.J.; Salem, N., Jr.; Higley, J.D. Fatty acid formula supplementation and neuromotor development in rhesus monkey neonates. Pediatr. Res.2002, 51, 273–281.

159. Greiner, R.S.; Moriguchi, T.; Hutton, A.; Slotnick, B.M.; Salem, N., Jr. Rats with low levels of brain docosahexaenoic acid show impaired performance in olfactory-based and spatial learning tasks. Lipids1999, 34 (Suppl. 1), S239–S243.

160. Novak, E.M.; Dyer, R.A.; Innis, S.M. High dietary omega-6 fatty acids contribute to reduced docosahexaenoic acid in the developing brain and inhibit secondary neurite growth. Brain Res. 2008,1237, 136–145.

161. Abubakari, A.R.; Naderali, M.M.; Naderali, E.K. Omega-3 fatty acid supplementation and cognitive function: Are smaller dosages more beneficial? Int. J. Gen. Med. 2014, 7, 463–473.

162. Jiao, J.; Li, Q.; Chu, J.; Zeng, W.; Yang, M.; Zhu, S. Effect of n-3 PUFA supplementation on cognitive function throughout the life span from infancy to old age: A systematic review and meta-analysis of randomized controlled trials. Am. J. Clin. Nutr. 2014, 100, 1422–1436.

163. Yurko-Mauro, K.; Alexander, D.D.; van Elswyk, M.E. Docosahexaenoic acid and adult memory: A systematic review and meta-analysis. PLoS ONE 2015, 10, e0120391.

164. Muldoon, M.F.; Ryan, C.M.; Sheu, L.; Yao, J.K.; Conklin, S.M.; Manuck, S.B. Serum phospholipid docosahexaenonic acid is associated with cognitive functioning during middle adulthood. J. Nutr.2010, 140, 848–853.

165. Leckie, R.L.; Manuck, S.B.; Bhattacharjee, N.; Muldoon, M.F.; Flory, J.M.; Erickson, K.I. Omega-3 fatty acids moderate effects of physical activity on cognitive function. Neuropsychologia 2014, 59, 103–111.

166. Erickson, K.I.; Voss, M.W.; Prakash, R.S.; Basak, C.; Szabo, A.; Chaddock, L.; Kim, J.S.; Heo, S.; Alves, H.; White, S.M.; et al. Exercise training increases size of hippocampus and improves memory. Proc. Natl. Acad. Sci. USA 2011, 108, 3017–3022.

167. Erickson, K.I.; Weinstein, A.M.; Lopez, O.L. Physical activity, brain plasticity, and Alzheimer's disease. Arch. Med. Res. 2012, 43, 615–621.

168. Rogers, P.J.; Appleton, K.M.; Kessler, D.; Peters, T.J.; Gunnell, D.; Hayward, R.C.; Heatherley, S.V.; Christian, L.M.; McNaughton, S.A.; Ness, A.R. No effect of n-3 long-chain polyunsaturated fatty acid (EPA and DHA) supplementation on depressed mood and cognitive function: A randomised controlled trial. Br. J. Nutr. 2008, 99, 421–431.

169. Jackson, P.A.; Deary, M.E.; Reay, J.L.; Scholey, A.B.; Kennedy, D.O. No effect of 12 weeks' supplementation with 1 g DHA -rich or EPA-rich fish oil on cognitive function or mood in healthy young adults aged 18–35 years. Br. J. Nutr. 2012, 107, 1232–1243.

170. Antypa, N.; van der Does, A.J.W.; Smelt, A.H.M.; Rogers, R.D. Omega-3 fatty acids (fish-oil) and depression-related cognition in healthy volunteers. J. Psychopharmacol. 2009, 23, 831–840.

171. Karr, J.E.; Grindstaff, T.R.; Alexander, J.E. Omega-3 polyunsaturated fatty acids and cognition in a college-aged population. Exp. Clin. Psychopharmacol. 2012, 20, 236–242.

172. Stonehouse, W.; Conlon, C.A.; Podd, J.; Hill, S.R.; Minihane, A.M.; Haskell, C.; Kennedy, D. DHA supplementation improved both memory and reaction time in healthy young adults: A randomized controlled trial. Am. J. Clin. Nutr. 2013, 94, 1134–1143.

173. Duff, S.J.; Hampson, E. A sex difference on a novel spatial working memory task in humans. Brain Cogn. 2001, 47, 470–493.

174. Halpern, D.F. Mapping cognitive processes onto the brain: Mind the gap. Brain Cogn. 2000, 42, 128–130.

175. Jaremka, L.M.; Derry, H.M.; Bornstein, R.; Prakash, R.S.; Peng, J.; Belury, M.A.; Andridge, R.R.; Malarkey, W.B.; Kiecolt-Glaser, J.K.

Omega-3 supplementation and loneliness-related memory problems. Psychosom. Med. 2014, 76, 650–658.

176. Services, H. Global Health and Aging; National Institutes of Health: Bethesda, MD, USA, 2011; Volume 1, pp. 273–277.

177. Resnick, S.M.; Pham, D.L.; Kraut, M.A.; Zonderman, A.B.; Davatzikos, C. Longitudinal magnetic resonance imaging studies of older adults: A shrinking brain. J. Neurosci. 2003, 23, 3295–3301.

178. McNamara, R.K.; Liu, Y.; Jandacek, R.; Rider, T.; Tso, P. The aging human orbitofrontal cortex: Decreasing polyunsaturated fatty acid composition and associated increases in lipogenic gene expression and stearoyl-CoA desaturase activity. Prostaglandins Leukot. Essent. Fat. Acids 2008, 78, 293–304.

179. Andre, A.; Juaneda, P.; Sebedio, J.L.; Chardigny, J.M. Plasmalogen metabolism-related enzymes in rat brain during aging: Influence of n-3 fatty acid intake. Biochimie 2006, 88, 103–111.

180. Giusto, N.M.; Salvador, G.A.; Castagnet, P.I.; Pasquare, S.J.; de Boschero, M.G.I. Age-associated changes in central nervous system glycerolipid composition and metabolism. Neurochem. Res. 2002, 27, 1513–1523.

181. Plourde, M.; Chouinard-Watkins, R.; Vandal, M.; Zhang, Y.; Lawrence, P.; Brenna, J.T.; Cunnane, S.C. Plasma incorporation, apparent retroconversion and beta-oxidation of 13c-docosahexaenoic acid in the elderly. Nutr. Metab. 2011, 8, 5.

182. Conklin, S.M.; Gianaros, P.J.; Brown, S.M.; Yao, J.K.; Hariri, A.R.; Manuck, S.B.; Muldoon, M.F. Long-chain omega-3 fatty acid intake is associated positively with corticolimbic gray matter volume in healthy adults. Neurosci. Lett. 2007, 421, 209–212.

183. Raji, C.A.; Erickson, K.I.; Lopez, O.L.; Kuller, L.H.; Gach, H.M.; Thompson, P.M.; Riverol, M.; Becker, J.T. Regular fish consumption and age-related brain gray matter loss. Am. J. Prev. Med. 2014, 47, 444–451.

184. Masliah, E.; Crews, L.; Hansen, L. Synaptic remodeling during aging and in Alzheimer's disease. J. Alzheimer's Dis. 2006, 9, 91–99.

185. Jorm, A.F.; Jolley, D. The incidence of dementia: A meta-analysis. Neurology 1998, 51, 728–733.

186. Mohajeri, M.H.; Troesch, B.; Weber, P. Inadequate supply of vitamins and DHA in the elderly: Implications for brain aging and Alzheimer-type dementia. Nutrition 2015, 31, 261–275.

187. Reiman, E.M.; Quiroz, Y.T.; Fleisher, A.S.; Chen, K.; Velez-Pardo, C.; Jimenez-Del-Rio, M.; Fagan, A.M.; Shah, A.R.; Alvarez, S.; Arbelaez,

A.; et al. Brain imaging and fluid biomarker analysis in young adults at genetic risk for autosomal dominant Alzheimer's disease in the presenilin 1 E280A kindred: A case-control study. Lancet Neurol. 2012, 11, 1048–1056. [CrossRef]

188. Ellinson, M.; Thomas, J.; Patterson, A. A critical evaluation of the relationship between serum vitamin B, folate and total homocysteine with cognitive impairment in the elderly. J. Hum. Nutr. Diet. 2004, 17, 371–383.

189. Adunsky, A.; Arinzon, Z.; Fidelman, Z.; Krasniansky, I.; Arad, M.; Gepstein, R. Plasma homocysteine levels and cognitive status in long-term stay geriatric patients: A cross-sectional study. Arch. Gerontol. Geriatr. 2005, 40, 129–138.

190. Terry, R.D. Cell death or synaptic loss in Alzheimer disease. J. Neuropathol. Exp. Neurol. 2000, 59, 1118–1119.

191. Vida, C.; Gonzalez, E.M.; de la Fuente, M. Increase of oxidation and inflammation in nervous and immune systems with aging and anxiety. Curr. Pharm. Des. 2014, 20, 4656–4678.

192. Kalmijn, S.; van Boxtel, M.P.J.; Ocke, M.; Verschuren, W.M.M.; Kromhout, D.; Launer, L.J. Dietary intake of fatty acids and fish in relation to cognitive performance at middle age. Neurology 2004, 62, 275–280.

193. Van Gelder, B.M.; Tijhuis, M.; Kalmijn, S.; Kromhout, D. Fish consumption, n-3 fatty acids, and subsequent 5-y cognitive decline in elderly men: The Zutphen elderly study. Am. J. Clin. Nutr. 2007,85, 1142–1147.

194. Eskelinen, M.H.; Ngandu, T.; Helkala, E.L.; Tuomilehto, J.; Nissinen, A.; Soininen, H.; Kivipelto, M. Fat intake at midlife and cognitive impairment later in life: A population-based caide study. Int. J. Geriatr. Psychiatry 2008, 23, 741–747.

195. Velho, S.; Marques-Vidal, P.; Baptista, F.; Camilo, M.E. Dietary intake adequacy and cognitive function in free-living active elderly: A cross-sectional and short-term prospective study. Clin. Nutr. 2008, 27, 77–86.

196. González, S.; Huerta, J.M.; Fernández, S.; Patterson, A.M.; Lasheras, C. The relationship between dietary lipids and cognitive performance in an elderly population. Int. J. Food Sci. Nutr. 2010, 61, 217–225.

197. Gao, Q.; Niti, M.; Feng, L.; Yap, K.B.; Ng, T.P. Omega-3 polyunsaturated fatty acid supplements and cognitive decline: Singapore longitudinal aging studies. J. Nutr. Health Aging 2011, 15, 32–35.

198. Titova, O.E.; Sjögren, P.; Brooks, S.J.; Kullberg, J.; Ax, E.; Kilander, L.; Riserus, U.; Cederholm, T.; Larsson, E.M.; Johansson, L.; et al. Dietary intake of eicosapentaenoic and docosahexaenoic acids is linked to gray matter volume and cognitive function in elderly. Age 2013, 35, 1495–1505.

199. Qin, B.; Plassman, B.L.; Edwards, L.J.; Popkin, B.M.; Adair, L.S.; Mendez, M.A. Fish intake is associated with slower cognitive decline in Chinese older adults. J. Nutr. 2014, 144, 1579–1585.

200. Del Brutto, O.H.; Mera, R.M.; Gillman, J.; Zambrano, M.; Ha, J.E. Oily fish intake and cognitive performance in community-dwelling older adults: The Atahualpa project. J. Community Health 2016,41, 82–86.

201. Archer, E.; Pavela, G.; Lavie, C.J. The inadmissibility of what we eat in America and nhanes dietary data in nutrition and obesity research and the scientific formulation of national dietary guidelines.Mayo Clin. Proc. 2015, 90, 911–926.

202. Ammann, E.M.; Pottala, J.V.; Harris, W.S.; Espeland, M.A.; Wallace, R.; Denburg, N.L.; Carnahan, R.M.; Robinson, J.G. Omega-3 fatty acids and domain-specific cognitive aging: Secondary analyses of data from WHISCA. Neurology 2013, 81, 1484–1491.

203. Heude, B.; Ducimetiere, P.; Berr, C. Cognitive decline and fatty acid composition of erythrocyte membranes—The EVA study. Am. J. Clin. Nutr. 2003, 77, 803–808.

204. Dullemeijer, C.; Durga, J.; Brouwer, I.A.; van de Rest, O.; Kok, F.J.; Brummer, R.J.M.; van Boxtel, M.P.; Verhoef, P. n-3 fatty acid proportions in plasma and cognitive performance in older adults. Am. J. Clin. Nutr. 2007, 86, 1479–1485.

205. Whalley, L.J.; Deary, I.J.; Starr, J.M.; Wahle, K.W.; Rance, K.A.; Bourne, V.J.; Fox, H.C. n-3 fatty acid erythrocyte membrane content, APOEe4, and cognitive variation: An observational follow-up study in late adulthood. Am. J. Clin. Nutr. 2008, 87, 449–454.

206. Tan, Z.S.; Harris, W.S.; Beiser, a.S.; Au, R.; Himali, J.J.; Debette, S.; Pikula, A.; Decarli, C.; Wolf, P.A.; Vasan, R.S.; et al. Red blood cell ω-3 fatty acid levels and markers of accelerated brain aging. Neurology2012, 78, 658–664.

207. Baierle, M.; Vencato, P.; Oldenburg, L.; Bordignon, S.; Zibetti, M.; Trentini, C.; Duarte, M.; Veit, J.; Somacal, S.; Emanuelli, T.; et al. Fatty acid status and its relationship to cognitive decline and homocysteine levels in the elderly. Nutrients 2014, 6, 3624–3640.

208. Otsuka, R.; Tange, C.; Nishita, Y.; Kato, Y.; Imai, T.; Ando, F.; Shimokata, H. Serum docosahexaenoic and eicosapentaenoic acid

and risk of cognitive decline over 10 years among elderly Japanese. Eur. J. Clin. Nutr. 2014, 68, 503–509.

209. Dangour, A.D.; Allen, E.; Elbourne, D.; Fasey, N.; Fletcher, A.E.; Hardy, P.; Holder, G.E.; Knight, R.; Letley, L.; Richards, M.; et al. Effect of 2-y n-3 long-chain polyunsaturated fatty acid supplementation on cognitive function in older people: A randomized, double-blind, controlled trial. Am. J. Clin. Nutr.2010, 91, 1725–1732.

210. Van de Rest, O.; Geleijnse, J.M.; Kok, F.J.; van Staveren, W.A.; Dullemeijer, C.; OldeRikkert, M.G.M.; Beekman, A.T.F.; de Groot, C.P.G.M. Effect of fish oil on cognitive performance in older subjects: A randomized, controlled trial. Neurology 2008, 71, 430–438.

211. Stough, C.; Downey, L.; Silber, B.; Lloyd, J.; Kure, C.; Wesnes, K.; Camfield, D. The effects of 90-day supplementation with the omega-3 essential fatty acid docosahexaenoic acid (DHA) on cognitive function and visual acuity in a healthy aging population. Neurobiol. Aging 2012, 33, e821–e823.

212. Johnson, E.J.; Mcdonald, K.; Caldarella, S.M.; Chung, H.Y.; Troen, A.M.; Snodderly, D.M. Cognitive findings of an exploratory trial of docosahexaenoic acid and lutein supplementation in older women.Nutr. Neurosci. 2008, 11, 75–83.

213. Yurko-Mauro, K.; McCarthy, D.; Rom, D.; Nelson, E.B.; Ryan, A.S.; Blackwell, A.; Salem, N.; Stedman, M. Beneficial effects of docosahexaenoic acid on cognition in age-related cognitive decline. Alzheimer's Dement. 2010, 6, 456–464.

214. Vakhapova, V.; Cohen, T.; Richter, Y.; Herzog, Y.; Kam, Y.; Korczyn, A.D. Phosphatidylserine containing omega-3 fatty acids may improve memory abilities in nondemented elderly individuals with memory complaints: Results from an open-label extension study. Dement. Geriatr. Cogn. Disord.2014, 38, 39–45.

215. Vakhapova, V.; Cohen, T.; Richter, Y.; Herzog, Y.; Korczyn, A.D. Phosphatidylserine containing ω–3 fatty acids may improve memory abilities in non-demented elderly with memory complaints: A double-blind placebo-controlled trial. Dement. Geriatr. Cogn. Disord. 2010, 29, 467–474.

216. Nilsson, A.; Radeborg, K.; Salo, I.; Bjorck, I. Effects of supplementation with n-3 polyunsaturated fatty acids on cognitive performance and cardiometabolic risk markers in healthy 51 to 72 years old subjects: A randomized controlled cross-over study. Nutr. J. 2012, 11, 99.

217. Tokuda, H.; Sueyasu, T.; Kontani, M.; Kawashima, H.; Shibata, H.; Koga, Y. Low doses of long-chain polyunsaturated fatty acids affect cognitive function in elderly Japanese men: A randomized controlled trial. J. Oleo Sci. 2015, 2015, 1–12.

218. Ito, J. Auditory event-related potentials (ERPs) in healthy subjects and patients with dementia. Jpn. J. Clin. Path. (Rinsho Byori). 1991, 39, 859–864.

219. Strike, S.C.; Carlisle, A.; Gibson, E.L.; Dyall, S.C. A high omega-3 fatty acid multinutrient supplement benefits cognition and mobility in older women: A randomized, double-blind, placebo-controlled pilot study. J. Gerontol. Ser. A Biol. Sci. Med. Sci. 2016, 71, 236–242.

220. Maestre, G.E. Assessing dementia in resource-poor regions. Curr. Neurol. Neurosci. Rep. 2012, 12, 511–519.

221. Deary, I.J.; Corley, J.; Gow, A.J.; Harris, S.E.; Houlihan, L.M.; Marioni, R.E.; Penke, L.; Rafnsson, S.B.; Starr, J.M. Age-associated cognitive decline. Br. Med. Bull. 2009, 92, 135–152.

222. Scahill, R.I.; Frost, C.; Jenkins, R.; Whitwell, J.L.; Rossor, M.N.; Fox, N.C. A longitudinal study of brain volume changes in normal aging using serial registered magnetic resonance imaging. Arch. Neurol.2003, 60, 989–994.

223. Risacher, S.L.; Saykin, A.J.; West, J.D.; Shen, L.; Firpi, H.A.; McDonald, B.C.; Alzheimer's Disease Neuroimaging Initiative. Baseline MRI predictors of conversion from MCI to probable AD in the ADNI cohort. Curr. Alzheimer Res. 2009, 6, 347–361.

224. Kalpouzos, G.; Eustache, F.; de la Sayette, V.; Viader, F.; Chetelat, G.; Desgranges, B. Working memory and FDG-PET dissociate early and late onset Alzheimer disease patients. J. Neurol. 2005, 252, 548–558.

225. Landau, S.M.; Harvey, D.; Madison, C.M.; Koeppe, R.A.; Reiman, E.M.; Foster, N.L.; Weiner, M.W.; Jagust, W.J.; the Alzheimer's Disease Neuroimaging Initiative. Associations between cognitive, functional, and FDG-PET measures of decline in AD and MCI. Neurobiol. Aging 2011, 32, 1207–1218.

226. Nugent, S.; Tremblay, S.; Chen, K.W.; Ayutyanont, N.; Roontiva, A.; Castellano, C.A.; Fortier, M.; Roy, M.; Courchesne-Loyer, A.; Bocti, C.; et al. Brain glucose and acetoacetate metabolism: A comparison of young and older adults. Neurobiol. Aging 2014, 35, 1386–1395.

227. Hooijmans, C.R.; Pasker-de Jong, P.C.M.; de Vries, R.B.M.; Ritskes-Hoitinga, M. The effects of long-term omega-3 fatty acid supplementation on cognition and Alzheimer's pathology in animal

models of Alzheimer's disease: A systematic review and meta-analysis. J. Alzheimer's Dis. 2012, 28, 191–209.

228. Richard, E.; Andrieu, S.; Solomon, A.; Mangialasche, F.; Ahtiluoto, S.; van Charante, E.P.M.; Coley, N.; Fratiglioni, L.; Neely, A.S.; Vellas, B.; et al. Methodological challenges in designing dementia prevention trials—The European dementia prevention initiative (EDPI). J. Neurol. Sci. 2012, 322, 64–70.

229. Albanese, E.; Dangour, A.D.; Uauy, R.; Acosta, D.; Guerra, M.; Guerra, S.S.G.; Huang, Y.; Jacob, K.S.; Rodriguez, J.L.D.; Noriega, L.H.; et al. Dietary fish and meat intake and dementia in latin America, China, and India: A 10/66 dementia research group population-based study. Am. J. Clin. Nutr. 2009,90, 392–400.

230. Barberger-Gateau, P.; Raffaitin, C.; Letenneur, L.; Berr, C.; Tzourio, C.; Dartigues, J.F.; Alpe, A. Dietary patterns and risk of dementia. Neurology 2007, 69, 1921–1930.

231. Schaefer, E.J.; Bongard, V.; Beiser, A.S.; Lamon-Fava, S.; Robins, S.J.; Au, R.; Tucker, K.L.; Kyle, D.J.; Wilson, P.W.F.; Wolf, P.A. Plasma phosphatidylcholine docosahexaenoic acid content and risk of dementia and Alzheimer disease: The Framingham heart study. Arch. Neurol. 2006, 63, 1545–1550.

232. Wu, S.; Ding, Y.; Wu, F.; Li, R.; Hou, J.; Mao, P. Omega-3 fatty acids intake and risks of dementia and Alzheimer's disease: A meta-analysis. Neurosci. Biobehav. Rev. 2015, 48, 1–9.

233. Cherubini, A.; Andres-Lacueva, C.; Martin, A.; Lauretani, F.; Iorio, A.D.; Bartali, B.; Corsi, A.; Bandinelli, S.; Mattson, M.P.; Ferrucci, L. Low plasma n-3 fatty acids and dementia in older persons: The InCHIANTI study. J. Gerontol. Ser. A Biol. Sci. Med. Sci. 2007, 62, 1120–1126. [CrossRef]

234. Milte, C.M.; Sinn, N.; Street, S.J.; Buckley, J.D.; Coates, A.M.; Howe, P.R.C. Erythrocyte polyunsaturated fatty acid status, memory, cognition and mood in older adults with mild cognitive impairment and healthy controls. Prostaglandins Leukot. Essent. Fat. Acids 2011, 84, 153–161.

235. Yin, Y.; Fan, Y.; Lin, F.; Xu, Y.; Zhang, J. Nutrient biomarkers and vascular risk factors in subtypes of mild cognitive impairment: A cross-sectional study. J. Nutr. Health Aging 2015, 19, 39–47.

236. Terano, T.; Fujishiro, S.; Ban, T.; Yamamoto, K.; Tanaka, T.; Noguchi, Y.; Tamura, Y.; Yazawa, K.; Hirayama, T. Docosahexaenoic acid supplementation improves the moderately severe dementia from thrombotic cerebrovascular diseases. Lipids 1999, 34, S345–S346.

237. Kotani, S.; Sakaguchi, E.; Warashina, S.; Matsukawa, N.; Ishikura, Y.; Kiso, Y.; Sakakibara, M.; Yoshimoto, T.; Guo, J.; Yamashima, T. Dietary supplementation of arachidonic and docosahexaenoic acids improves cognitive dysfunction. Neurosci. Res. 2006, 56, 159–164.

238. Chiu, C.C.; Su, K.P.; Cheng, T.C.; Liu, H.C.; Chang, C.J.; Dewey, M.E.; Stewart, R.; Huang, S.Y. The effects of omega-3 fatty acids monotherapy in Alzheimer's disease and mild cognitive impairment: A preliminary randomized double-blind placebo-controlled study. Prog. Neuro Psychopharmacol. Biol. Psychiatry 2008, 32, 1538–1544.

239. Sinn, N.; Milte, C.M.; Street, S.J.; Buckley, J.D.; Coates, A.M.; Petkov, J.; Howe, P.R.C. Effects of n-3 fatty acids, EPA v. DHA, on depressive symptoms, quality of life, memory and executive function in older adults with mild cognitive impairment: A 6-month randomised controlled trial. Br. J. Nutr. 2012,107, 1682–1693.

240. Cazzola, R.; Rondanelli, M.; Faliva, M.; Cestaro, B. Effects of DHA - phospholipids, melatonin and tryptophan supplementation on erythrocyte membrane physico-chemical properties in elderly patients suffering from mild cognitive impairment. Exp. Gerontol. 2012, 47, 974–978.

241. Mahmoudi, M.; Hedayat, M.; Sharifi, F.; Mirarefin, M.; Nazari, N.; Mehrdad, N.; Ghaderpanahi, M.; Tajalizadekhoob, Y.; Badamchizade, Z.; Larijani, B.; et al. Effect of low dose ω-3 poly unsaturated fatty acids on cognitive status among older people: A double-blind randomized placebo-controlled study.J. Diabetes Metab. Disord. 2014, 13, 34.

242. Ward, A.; Crean, S.; Mercaldi, C.J.; Collins, J.M.; Boyd, D.; Cook, M.N.; Arrighi, H.M. Prevalence of apolipoprotein E4 genotype and homozygotes (APOE e4/4) among patients diagnosed with Alzheimer's disease: A systematic review and meta-analysis. Neuroepidemiology 2012, 38, 1–17.

243. Alzheimer's Association. 2014 Alzheimer's disease facts and figures. Alzheimer's Dement. 2014, 10, e47–e92.

244. World Health Organization. Dementia: A public health priority. In Dementia; WHO: Geneva, Switzerland, 2012; p. 112.

245. Kelley, A.S.; McGarry, K.; Gorges, R.; Skinner, J.S. The burden of health care costs for patients with dementia in the last 5 years of life. Ann. Intern. Med. 2015, 163, 729–736.

246. Cummings, J.L.; Zhong, K. Repackaging FDA-approved drugs for degenerative diseases: Promises and challenges. Expert Rev. Clin. Pharmacol. 2014, 7, 161–165.

247. Hebert, L.E.; Weuve, J.; Scherr, P.A.; Evans, D.A. Alzheimer disease in the United States (2010–2050) estimated using the 2010 census. Neurology 2013, 80, 1778–1783.

248. Astarita, G.; Jung, K.M.; Berchtold, N.C.; Nguyen, V.Q.; Gillen, D.L.; Head, E.; Cotman, C.W.; Piomelli, D. Deficient liver biosynthesis of docosahexaenoic acid correlates with cognitive impairment in Alzheimer's disease. PLoS ONE 2010, 5, e12538.

249. Igarashi, M.; Ma, K.; Gao, F.; Kim, H.W.; Rapoport, S.I.; Rao, J.S. Disturbed choline plasmalogen and phospholipid fatty acid concentrations in Alzheimer's disease prefrontal cortex. J. Alzheimer's Dis.2011, 24, 507–517.

250. Quinn, J.F.; Raman, R.; Thomas, R.G.; Yurko-Mauro, K.; Nelson, E.B.; van Dyck, C.; Galvin, J.E.; Emond, J.; Jack, C.R., Jr.; Weiner, M.; et al. Docosahexaenoic acid supplementation and cognitive decline in Alzheimer disease: A randomized trial. JAMA 2010, 304, 1903–1911.

251. Prasad, M.R.; Lovell, M.A.; Yatin, M.; Dhillon, H.; Markesbery, W.R. Regional membrane phospholipid alterations in Alzheimer's disease. Neurochem. Res. 1998, 23, 81–88.

252. Soderberg, M.; Edlund, C.; Kristensson, K.; Dallner, G. Fatty acid composition of brain phospholipids in aging and in Alzheimer's disease. Lipids 1991, 26, 421–425.

253. Morris, M.C.; Evans, D.a.; Bienias, J.L.; Tangney, C.C.; Bennett, D.a.; Wilson, R.S.; Aggarwal, N.; Schneider, J. Consumption of fish and n-3 fatty acids and risk of incident Alzheimer disease. Arch. Neurol. 2003, 60, 940–946.

254. Shatenstein, B.; Kergoat, M.J.; Reid, I. Poor nutrient intakes during 1-year follow-up with community-dwelling older adults with early-stage Alzheimer dementia compared to cognitively intact matched controls. J. Am. Diet. Assoc. 2007, 107, 2091–2099.

255. Daiello, L.A.; Gongvatana, A.; Dunsiger, S.; Cohen, R.A.; Ott, B.R. Association of fish oil supplement use with preservation of brain volume and cognitive function. Alzheimer's Dement. 2015, 11, 226–235.

256. Conquer, J.; Tierney, M.; Zecevic, J.; Bettger, W.; Fisher, R. Fatty acid analysis of blood plasma of patients with Alzheimer's disease, other types of dementia, and cognitive impairment. Lipids 2000, 35, 1305–1312.

257. Tully, A.M.; Roche, H.M.; Doyle, R.; Fallon, C.; Bruce, I.; Lawlor, B.; Coakley, D.; Gibney, M.J. Low serum cholesteryl ester-

docosahexaenoic acid levels in Alzheimer's disease: A case–control study. Br. J. Nutr. 2003, 89, 483.

258. Wang, W.; Shinto, L.; Connor, W.E.; Quinn, J.F. Nutritional biomarkers in Alzheimer's disease: The association between carotenoids, n-3 fatty acids, and dementia severity. J. Alzheimer's Dis. 2008, 13, 31–38.

259. Lopez, L.B.; Kritz-Silverstein, D.; Barrett-Connor, E. High dietary and plasma levels of the omega-3 fatty acid docosahexaenoic acid are associated with decreased dementia risk: The rancho bernardo study. J. Nutr. Health Aging 2011, 15, 25–31.

260. Phillips, M.A.; Childs, C.E.; Calder, P.C.; Rogers, P.J. Lower omega-3 fatty acid intake and status are associated with poorer cognitive function in older age: A comparison of individuals with and without cognitive impairment and Alzheimer's disease. Nutr. Neurosci. 2012, 15, 271–278.

261. Freund-Levi, Y.; Basun, H.; Cederholm, T.; Faxen-Irving, G.; Garlind, A.; Grut, M.; Vedin, I.; Palmblad, J.; Wahlund, L.O.; Eriksdotter-Jonhagen, M. Omega-3 supplementation in mild to moderate Alzheimer's disease: Effects on neuropsychiatric symptoms. Int. J. Geriatr. Psychiatry 2008, 23, 161–169.

262. Freund-Levi, Y.; Eriksdotter-Jonhagen, M.; Cederholm, T.; Basun, H.; Faxen-Irving, G.; Garlind, A.; Vedin, I.; Vessby, B.; Wahlund, L.O.; Palmblad, J. Omega-3 fatty acid treatment in 174 patients with mild to moderate Alzheimer disease: Omegad study: A randomized double-blind trial. Arch. Neurol.2006, 63, 1402–1408.

263. Eriksdotter, M.; Vedin, I.; Falahati, F.; Freund-Levi, Y.; Hjorth, E.; Faxen-Irving, G.; Wahlund, L.O.; Schultzberg, M.; Basun, H.; Cederholm, T.; et al. Plasma fatty acid profiles in relation to cognition and gender in Alzheimer's disease patients during oral omega-3 fatty acid supplementation: The omegad study. J. Alzheimer's Dis. 2015, 48, 805–812.

264. Levi, Y.F.; Vedin, I.; Cederholm, T.; Basun, H.; Irving, G.F.; Eriksdotter, M.; Hjorth, E.; Schultzberg, M.; Vessby, B.; Wahlund, L.O.; et al. Transfer of omega-3 fatty acids across the blood-brain barrier after dietary supplementation with a docosahexaenoic acid-rich omega-3 fatty acid preparation in patients with Alzheimer's disease: The omegad study. J. Intern. Med. 2014, 275, 428–436.

265. Connor, J.R.; Menzies, S.L. Relationship of iron to oligodendrocytes and myelination. Glia 1996, 17, 83–93. [CrossRef]

266. Peters, A. Age-related changes in oligodendrocytes in monkey cerebral cortex. J. Comp. Neurol. 1996,371, 153–163. [CrossRef]

267. Thompson, P.M.; Hayashi, K.M.; de Zubicaray, G.; Janke, A.L.; Rose, S.E.; Semple, J.; Herman, D.; Hong, M.S.; Dittmer, S.S.; Doddrell, D.M.; et al. Dynamics of gray matter loss in Alzheimer's disease. J. Neurosci. 2003, 23, 994–1005.

268. Braak, H.; Braak, E. Development of Alzheimer-related neurofibrillary changes in the neocortex inversely recapitulates cortical myelogenesis. Acta Neuropathol. 1996, 92, 197–201.

269. Bartzokis, G. Alzheimer's disease as homeostatic responses to age-related myelin breakdown.Neurobiol. Aging 2011, 32, 1341–1371.

270. Beal, M.F. Oxidative damage as an early marker of Alzheimer's disease and mild cognitive impairment. Neurobiol. Aging 2005, 26, 585–586.

271. Virtanen, J.K.; Siscovick, D.S.; Lemaitre, R.N.; Longstreth, W.T.; Spiegelman, D.; Rimm, E.B.; King, I.B.; Mozaffarian, D. Circulating omega-3 polyunsaturated fatty acids and subclinical brain abnormalities on MRI in older adults: The Cardiovascular Health Study. J. Am. Heart Assoc. 2013, 2.

272. Vermeer, S.E.; Prins, N.D.; den Heijer, T.; Hofman, A.; Koudstaal, P.J.; Breteler, M.M. Silent brain infarcts and the risk of dementia and cognitive decline. N. Engl. J. Med. 2003, 348, 1215–1222.

273. Peters, A.; Schweiger, U.; Pellerin, L.; Hubold, C.; Oltmanns, K.M.; Conrad, M.; Schultes, B.; Born, J.; Fehm, H.L. The selfish brain: Competition for energy resources. Neurosci. Biobehav. Rev. 2004, 28, 143–180.

274. Whitmer, R.A.; Sidney, S.; Selby, J.; Johnston, S.C.; Yaffe, K. Midlife cardiovascular risk factors and risk of dementia in late life. Neurology 2005, 64, 277–281.

275. Kalaria, R.N.; Ballard, C. Overlap between pathology of Alzheimer disease and vascular dementia.Alzheimer Dis. Assoc. Disord. 1999, 13 (Suppl. 3), S115–S123.

276. Roher, A.E.; Debbins, J.P.; Malek-Ahmadi, M.; Chen, K.; Pipe, J.G.; Maze, S.; Belden, C.; Maarouf, C.L.; Thiyyagura, P.; Mo, H.; et al. Cerebral blood flow in Alzheimer's disease. Vasc. Health Risk Manag.2012, 8, 599–611.

277. Jackson, P.A.; Reay, J.L.; Scholey, A.B.; Kennedy, D.O. DHA -rich oil modulates the cerebral haemodynamic response to cognitive tasks in healthy young adults: A near ir spectroscopy pilot study. Br. J. Nutr. 2012, 107, 1093–1098.

278. Beydoun, M.A.; Kaufman, J.S.; Satia, J.A.; Rosamond, W.; Folsom, A.R. Plasma n-3 fatty acids and the risk of cognitive decline in older

adults: The atherosclerosis risk in communities study. Am. J. Clin. Nutr. 2007, 85, 1103–1111.

279. Huang, T.; Hu, X.; Khan, N.; Yang, J.; Li, D. Effect of polyunsaturated fatty acids on homocysteine metabolism through regulating the gene expressions involved in methionine metabolism. Sci. World J.2013, 2013.

280. Ford, A.H.; Flicker, L.; Alfonso, H.; Hankey, G.J.; Norman, P.E.; van Bockxmeer, F.M.; Almeida, O.P. Plasma homocysteine and MTHFRC677T polymorphism as risk factors for incident dementia. J. Neurol. Neurosurg. Psychiatry 2012, 83, 70–75.

281. Boneva, N.B.; Yamashima, T. New insights into "GPR40-CREB interaction in adult neurogenesis" specific for primates. Hippocampus 2012, 22, 896–905.

282. Brand, A.; Schonfeld, E.; Isharel, I.; Yavin, E. Docosahexaenoic acid-dependent iron accumulation in oligodendroglia cells protects from hydrogen peroxide-induced damage. J. Neurochem. 2008, 105, 1325–1335.

283. Gururajan, A.; van den Buuse, M. Is the mtor-signalling cascade disrupted in schizophrenia? J. Neurochem. 2014, 129, 377–387.

284. Calon, F.; Lim, G.P.; Morihara, T.; Yang, F.; Ubeda, O.; Salem, N.; Frautschy, S.A.; Cole, G.M. Dietaryn-3 polyunsaturated fatty acid depletion activates caspases and decreases NMDA receptors in the brain of a transgenic mouse model of Alzheimer's disease. Eur. J. Neurosci. 2005, 22, 617–626.

285. Nishikawa, M.; Kimura, S.; Akaike, N. Facilitatory effect of docosahexaenoic acid on N-methyl-D-aspartate response in pyramidal neurones of rat cerebral cortex. J. Physiol. 1994, 475, 83–93.

286. Calon, F.; Lim, G.P.; Yang, F.; Morihara, T.; Teter, B.; Ubeda, O.; Rostaing, P.; Triller, A.; Salem, N., Jr.; Ashe, K.H.; et al. Docosahexaenoic acid protects from dendritic pathology in an Alzheimer's disease mouse model. Neuron 2004, 43, 633–645.

287. Dyall, S.C.; Michael, G.J.; Michael-Titus, A.T. Omega-3 fatty acids reverse age-related decreases in nuclear receptors and increase neurogenesis in old rats. J. Neurosci. Res. 2010, 88, 2091–2102.

288. Maarouf, C.L.; Daugs, I.D.; Kokjohn, T.A.; Walker, D.G.; Hunter, J.M.; Kruchowsky, J.C.; Woltjer, R.; Kaye, J.; Castano, E.M.; Sabbagh, M.N.; et al. Alzheimer's disease and non-demented high pathology control nonagenarians: Comparing and contrasting the biochemistry of cognitively successful aging.PLoS ONE 2011, 6, e27291.

289. Morimoto, K.; Horio, J.; Satoh, H.; Sue, L.; Beach, T.; Arita, S.; Tooyama, I.; Konishi, Y. Expression profiles of cytokines in the brains of Alzheimer's disease (ad) patients compared to the brains of non-demented patients with and without increasing ad pathology. J. Alzheimer's Dis. 2011, 25, 59–76.

290. Parachikova, A.; Agadjanyan, M.G.; Cribbs, D.H.; Blurton-Jones, M.; Perreau, V.; Rogers, J.; Beach, T.G.; Cotman, C.W. Inflammatory changes parallel the early stages of Alzheimer disease. Neurobiol. Aging 2007, 28, 1821–1833.

291. Cagnin, A.; Brooks, D.J.; Kennedy, A.M.; Gunn, R.N.; Myers, R.; Turkheimer, F.E.; Jones, T.; Banati, R.B. In-vivo measurement of activated microglia in dementia. Lancet 2001, 358, 461–467. [CrossRef]

292. Edison, P.; Archer, H.A.; Gerhard, A.; Hinz, R.; Pavese, N.; Turkheimer, F.E.; Hammers, A.; Tai, Y.F.; Fox, N.; Kennedy, A.; et al. Microglia, amyloid, and cognition in Alzheimer's disease: An [11C](R)PK11195-PET and [11C]PIB-PET study. Neurobiol. Dis. 2008, 32, 412–419.

293. Yokokura, M.; Mori, N.; Yagi, S.; Yoshikawa, E.; Kikuchi, M.; Yoshihara, Y.; Wakuda, T.; Sugihara, G.; Takebayashi, K.; Suda, S.; et al. In vivo changes in microglial activation and amyloid deposits in brain regions with hypometabolism in Alzheimer's disease. Eur. J. Nucl. Med. Mol. Imaging 2011, 38, 343–351.

294. Blum-Degen, D.; Muller, T.; Kuhn, W.; Gerlach, M.; Przuntek, H.; Riederer, P. Interleukin-1 beta and interleukin-6 are elevated in the cerebrospinal fluid of Alzheimer's and de novo parkinson's disease patients. Neurosci. Lett. 1995, 202, 17–20. [CrossRef]

295. Heneka, M.T.; Kummer, M.P.; Latz, E. Innate immune activation in neurodegenerative disease. Nat. Rev. Immunol. 2014, 14, 463–477.

296. Jia, J.P.; Meng, R.; Sun, Y.X.; Sun, W.J.; Ji, X.M.; Jia, L.F. Cerebrospinal fluid tau, abeta1–42 and inflammatory cytokines in patients with Alzheimer's disease and vascular dementia. Neurosci. Lett.2005, 383, 12–16.

297. Halle, A.; Hornung, V.; Petzold, G.C.; Stewart, C.R.; Monks, B.G.; Reinheckel, T.; Fitzgerald, K.A.; Latz, E.; Moore, K.J.; Golenbock, D.T. The NALP3 inflammasome is involved in the innate immune response to amyloid-beta. Nat. Immunol. 2008, 9, 857–865.

298. Salminen, A.; Ojala, J.; Suuronen, T.; Kaarniranta, K.; Kauppinen, A. Amyloid-beta oligomers set fire to inflammasomes and induce Alzheimer's pathology. J. Cell. Mol. Med. 2008, 12, 2255–2262.

299. Hjorth, E.; Zhu, M.; Toro, V.C.; Vedin, I.; Palmblad, J.; Cederholm, T.; Freund-Levi, Y.; Faxen-Irving, G.; Wahlund, L.O.; Basun, H.; et al. Omega-3 fatty acids enhance phagocytosis of Alzheimer's disease-related amyloid-β42 by human microglia and decrease inflammatory markers. J. Alzheimer's Dis. 2013,35, 697–713.

300. Krstic, D.; Knuesel, I. Deciphering the mechanism underlying late-onset Alzheimer disease. Nat. Rev. Neurol. 2012, 9, 25–34.

301. Weiser, M.J.; Mucha, B.; Denheyer, H.; Atkinson, D.; Schanz, N.; Vassiliou, E.; Benno, R.H. Dietary docosahexaenoic acid alleviates autistic-like behaviors resulting from maternal immune activation in mice. Prostaglandins Leukot. Essent. Fat. Acids 2015.

302. Wang, X.; Hjorth, E.; Vedin, I.; Eriksdotter, M.; Freund-Levi, Y.; Wahlund, L.O.; Cederholm, T.; Palmblad, J.; Schultzberg, M. Effects of n-3 FA supplementation on the release of proresolving lipid mediators by blood mononuclear cells: The OmegAD study. J. Lipid Res. 2015, 56, 674–681.

303. Serhan, C.N.; Dalli, J.; Colas, R.A.; Winkler, J.W.; Chiang, N. Protectins and maresins: New pro-resolving families of mediators in acute inflammation and resolution bioactive metabolome. Biochim. Biophys. Acta 2015, 1851, 397–413.

304. Alfano, C.M.; Imayama, I.; Neuhouser, M.L.; Kiecolt-Glaser, J.K.; Smith, A.W.; Meeske, K.; McTiernan, A.; Bernstein, L.; Baumgartner, K.B.; Ulrich, C.M.; et al. Fatigue, inflammation, and omega-3 and omega-6 fatty acid intake among breast cancer survivors. J. Clin. Oncol. 2012, 30, 1280–1287.

305. Kiecolt-Glaser, J.K.; Belury, M.A.; Andridge, R.; Malarkey, W.B.; Glaser, R. Omega-3 supplementation lowers inflammation and anxiety in medical students: A randomized controlled trial. Brain Behav. Immun. 2011, 25, 1725–1734.

306. Bouwens, M.; Bromhaar, M.G.; Jansen, J.; Muller, M.; Afman, L.A. Postprandial dietary lipid-specific effects on human peripheral blood mononuclear cell gene expression profiles. Am. J. Clin. Nutr. 2010,91, 208–217.

307. Vedin, I.; Cederholm, T.; Levi, Y.F.; Basun, H.; Garlind, A.; Irving, G.F. Effects of docosahexaenoic acid-rich n-3 fatty acid supplementation on cytokine release from blood mononuclear leukocytes: The OmegAD study. Am. Soc. Nutr. 2008, 87, 1616–1622.

308. Green, K.N.; Martinez-Coria, H.; Khashwji, H.; Hall, E.B.; Yurko-Mauro, K.A.; Ellis, L.; LaFerla, F.M. Dietary docosahexaenoic acid and docosapentaenoic acid ameliorate amyloid-beta and tau pathology via a mechanism involving presenilin 1 levels. J.Neurosci. 2007, 27, 4385–4395.

309. Ma, Q.L.; Yang, F.; Rosario, E.R.; Ubeda, O.J.; Beech, W.; Gant, D.J.; Chen, P.P.; Hudspeth, B.; Chen, C.; Zhao, Y.; et al. Beta-amyloid oligomers induce phosphorylation of tau and inactivation of insulin receptor substrate via c-Jun N-terminal kinase signaling: Suppression by omega-3 fatty acids and curcumin. J. Neurosci. 2009, 29, 9078–9089.

310. Calon, F. Omega-3 polyunsaturated fatty acids in Alzheimer's disease: Key questions and partial answers. Curr. Alzheimer Res. 2011, 8, 470–478.

311. Janssen, C.I.F.; Kiliaan, A.J. Long-chain polyunsaturated fatty acids (LCPUFA) from genesis to senescence: The influence of LCPUFA on neural development, aging, and neurodegeneration. Prog. Lipid Res. 2014, 53, 1–17.

312. Krstic, D.; Pfister, S.; Notter, T.; Knuesel, I. Decisive role of reelin signaling during early stages of Alzheimer's disease. Neuroscience 2013, 246, 108–116.

313. Yavin, E.; Himovichi, E.; Eilam, R. Delayed cell migration in the developing rat brain following maternal omega 3 alpha linolenic acid dietary deficiency. Neuroscience 2009, 162, 1011–1022.

314. Botella-Lopez, A.; Burgaya, F.; Gavin, R.; Garcia-Ayllon, M.S.; Gomez-Tortosa, E.; Pena-Casanova, J.; Urena, J.M.; del Rio, J.A.; Blesa, R.; Soriano, E.; et al. Reelin expression and glycosylation patterns are altered in Alzheimer's disease. Proc. Natl. Acad. Sci. USA 2006, 103, 5573–5578.

315. Medhi, B.; Chakrabarty, M. Insulin resistance: An emerging link in Alzheimer's disease. Neurol. Sci.2013, 34, 1719–1725.

316. De la Monte, S.M.; Tong, M. Brain metabolic dysfunction at the core of Alzheimer's disease. Biochem. Pharmacol. 2014, 88, 548–559.

317. Velayudhan, L.; Poppe, M.; Archer, N.; Proitsi, P.; Brown, R.G.; Lovestone, S. Risk of developing dementia in people with diabetes and mild cognitive impairment. Br. J. Psychiatry 2010, 196, 36–40.

318. Hennebelle, M.; Harbeby, E.; Tremblay, S.; Chouinard-Watkins, R.; Pifferi, F.; Plourde, M.; Guesnet, P.; Cunnane, S.C. Challenges to determining whether DHA can protect against age-related cognitive decline. Clin. Lipidol. 2015, 10, 91–102. [CrossRef]

319. Pifferi, F.; Jouin, M.; Alessandri, J.M.; Haedke, U.; Roux, F.; Perriere, N.; Denis, I.; Lavialle, M.; Guesnet, P. n-3 fatty acids modulate brain glucose transport in endothelial cells of the blood-brain barrier. Prostaglandins Leukot. Essent. Fat. Acids 2007, 77, 279–286.

320. Pifferi, F.; Jouin, M.; Alessandri, J.M.; Roux, F.; Perriere, N.; Langelier, B.; Lavialle, M.; Cunnane, S.; Guesnet, P. n-3 long-chain

fatty acids and regulation of glucose transport in two models of rat brain endothelial cells. Neurochem. Int. 2010, 56, 703–710.

321. Pifferi, F.; Roux, F.; Langelier, B.; Alessandri, J.M.; Vancassel, S.; Jouin, M.; Lavialle, M.; Guesnet, P. (n-3) polyunsaturated fatty acid deficiency reduces the expression of both isoforms of the brain glucose transporter glut1 in rats. J. Nutr. 2005, 135, 2241–2246.

322. Grimm, M.O.; Kuchenbecker, J.; Grosgen, S.; Burg, V.K.; Hundsdorfer, B.; Rothhaar, T.L.; Friess, P.; de Wilde, M.C.; Broersen, L.M.; Penke, B.; et al. Docosahexaenoic acid reduces amyloid beta production via multiple pleiotropic mechanisms. J. Biol. Chem. 2011, 286, 14028–14039.

323. Martín, V.; Fabelo, N.; Santpere, G.; Puig, B.; Marín, R.; Ferrer, I.; Díaz, M. Lipid alterations in lipid rafts from Alzheimer's disease human brain cortex. J. Alzheimer's Dis. 2010, 19, 489–502.

324. Fabelo, N.; Martin, V.; Marin, R.; Moreno, D.; Ferrer, I.; Diaz, M. Altered lipid composition in cortical lipid rafts occurs at early stages of sporadic Alzheimer's disease and facilitates APP/BACE1 interactions. Neurobiol. Aging 2014, 35, 1801–1812.

325. Hashimoto, M.; Hossain, S.; Katakura, M.; al Mamun, A.; Shido, O. The binding of aβ1–42 to lipid rafts of rbc is enhanced by dietary docosahexaenoic acid in rats: Implicates to Alzheimer's disease. Biochim. Biophys. Acta 2015, 1848, 1402–1409.

326. Gu, Y.; Schupf, N.; Cosentino, S.A.; Luchsinger, J.A.; Scarmeas, N. Nutrient intake and plasma beta-amyloid. Neurology 2012, 78, 1832–1840.

327. Flock, M.R.; Skulas-Ray, A.C.; Harris, W.S.; Etherton, T.D.; Fleming, J.A.; Kris-Etherton, P.M. Determinants of erythrocyte omega-3 fatty acid content in response to fish oil supplementation: A dose-response randomized controlled trial. J. Am. Heart Assoc. 2013, 2.

328. Cunnane, S.C.; Schneider, J.A.; Tangney, C.; Tremblay-Mercier, J.; Fortier, M.; Bennett, D.A.; Morris, M.C. Plasma and brain fatty acid profiles in mild cognitive impairment and Alzheimer's disease. J. Alzheimer's Dis. 2012, 29, 691–697.

329. Chouinard-Watkins, R.; Rioux-Perreault, C.; Fortier, M.; Tremblay-Mercier, J.; Zhang, Y.; Lawrence, P.; Vohl, M.C.; Perron, P.; Lorrain, D.; Brenna, J.T.; et al. Disturbance in uniformly 13c-labelled DHA metabolism in elderly human subjects carrying the apoE ε4 allele. Br. J. Nutr. 2013, 110, 1751–1759.

330. Wu, A.; Noble, E.E.; Tyagi, E.; Ying, Z.; Zhuang, Y.; Gomez-Pinilla, F. Curcumin boosts DHA in the brain: Implications for the prevention of anxiety disorders. Biochim. Biophys. Acta 2015, 1852, 951–961.

331. Scheltens, P.; Twisk, J.W.R.; Blesa, R.; Scarpini, E.; von Arnim, C.A.F.; Bongers, A.; Harrison, J.; Swinkels, S.H.N.; Stam, C.J.; de Waal, H.; et al. Efficacy of souvenaid in mild Alzheimer's disease: Results from a randomized, controlled trial. J. Alzheimer's Dis. 2012, 31, 225–236.

332. Serhan, C.N. Lipoxins and aspirin-triggered 15-epi-lipoxin biosynthesis: An update and role in anti-inflammation and pro-resolution. Prostaglandins Other Lipid Mediat. 2002, 68–69, 433–455. [CrossRef]

333. Serhan, C.N.; Gotlinger, K.; Hong, S.; Arita, M. Resolvins, docosatrienes, and neuroprotectins, novel omega-3-derived mediators, and their aspirin-triggered endogenous epimers: An overview of their protective roles in catabasis. Prostaglandins Other Lipid Mediat. 2004, 73, 155–172.

334. Ariel, A.; Serhan, C.N. Resolvins and protectins in the termination program of acute inflammation.Trends Immunol. 2007, 28, 176–183.

335. Dalli, J.; Winkler, J.W.; Colas, R.A.; Arnardottir, H.; Cheng, C.Y.; Chiang, N.; Petasis, N.A.; Serhan, C.N. Resolvin D3 and aspirin-triggered resolvin D3 are potent immunoresolvents. Chem. Biol. 2013, 20, 188–201.

336. Esiri, M.M. The interplay between inflammation and neurodegeneration in CNS disease. J. Neuroimmunol. 2007, 184, 4–16.

337. Etminan, M.; Gill, S.; Samii, A. Effect of non-steroidal anti-inflammatory drugs on risk of Alzheimer's disease: Systematic review and meta-analysis of observational studies. BMJ 2003, 327, 128.

338. Lehrer, S. Nasal nsaids for Alzheimer's disease. Am. J. Alzheimers Dis. Other Dement. 2014, 29, 401–403.

339. Lewis, M.D.; Bailes, J. Neuroprotection for the warrior: Dietary supplementation with omega-3 fatty acids. Mil. Med. 2011, 176, 1120–1127.

INDEX

O

P

S

V